Computational Methods in Engineering & the Sciences

This Series publishes books on all aspects of computational methods used in engineering and the sciences. With emphasis on simulation through mathematical modelling, the Series accepts high quality content books across different domains of engineering, materials, and other applied sciences. The Series publishes monographs, contributed volumes, professional books, and handbooks, spanning across cutting edge research as well as basics of professional practice. The topics of interest include the development and applications of computational simulations in the broad fields of Solid & Structural Mechanics, Fluid Dynamics, Heat Transfer, Electromagnetics, Multiphysics, Optimization, Stochastics with simulations in and for Structural Health Monitoring, Energy Systems, Aerospace Systems, Machines and Turbines. Climate Prediction, Effects of Earthquakes, Geotechnical Systems, Chemical and Biomolecular Systems, Molecular Biology, Nano and Microfluidics, Materials Science, Nanotechnology, Manufacturing and 3D printing, Artificial Intelligence, Internet-of-Things.

Timon Rabczuk · Huilong Ren · Xiaoying Zhuang

Computational Methods Based on Peridynamics and Nonlocal Operators

Theory and Applications

Springer

Timon Rabczuk
Weimar, Germany

Xiaoying Zhuang
Leibniz University Hannover
Hannover, Germany

Huilong Ren
Department of Mathematics and Physics
Leibniz University
Hannover, Germany

ISSN 2662-4869 ISSN 2662-4877 (electronic)
Computational Methods in Engineering & the Sciences
ISBN 978-3-031-20905-5 ISBN 978-3-031-20906-2 (eBook)
https://doi.org/10.1007/978-3-031-20906-2

This Springer imprint is published by the registered company Springer Nature Switzerland AG
The registered company address is: Gewerbestrasse 11, 6330 Cham, Switzerland

Preface

Feynman once said, calculus is the language of God. Calculus uses the partial differential derivatives and integrals to account for various physical phenomena. On one hand, it is well known that many physical problems or physical theories are formulated concisely by partial differential equations. No matter how complicated the physical phenomena appear, the PDEs to describe the mechanism are just one/several lines of equations. Mathematically, PDEs are a combinational result of partial differential derivatives of different orders. Partial differential derivatives are defined at a point without size, in this sense, the PDEs model can be viewed as a local model. On the other hand, integral expression is defined in a finite domain, which consists of infinite points. In this sense, the integral can be viewed as a nonlocal model. Or we can say the local model corresponds to differential equations while the nonlocal model is related to integral equations. Physically, our world contains not just local models but also a lot of nonlocal models, for example, the universal gravitation, Quantum entanglement and social network. Nonlocal viewpoints offer us a new perspective to understand the world. The role of integral equations in solving physical problems is underestimated for a long time. It is time to view the world non-locally.

Among the nonlocal models, Peridynamics (PD) attracts much attention in the field of fracture mechanics. One key feature of PD is the nonlocality, which is quite different from the ideas in conventional methods such as FEM and meshless methods. However, conventional PD suffers from problems such as constant horizon, explicit algorithm, hourglass mode. In this book, by examining the nonlocality with scrutiny, we proposed several new concepts such as Dual-Horizon (DH) in PD, Dual-Support (DS) in smoothed particle hydrodynamics (SPH), nonlocal operators and operator energy functional. The conventional PD (SPH) is incorporated in the DH-PD (DS-SPH), which can adopt an inhomogeneous discretization and inhomogeneous support domains. The DH-PD (DS-SPH) can be viewed as some fundamental improvement on the conventional PD (SPH). Dual formulation of PD and SPH allows h-adaptivity while satisfying the conservation of linear momentum, angular momentum and energy.

By developing the concept of nonlocality further, we introduced the nonlocal operator method as a generalization of DH-PD (DS-SPH). Combined with energy functional of various physical models, the nonlocal forms based on dual-support concepts are derived. In addition, the variation of the energy functional allows implicit formulation of the nonlocal theory. The introduction of dual-support is a mathematical innovation. Traditional integral equation is formulated in an integral domain. Dual-support is actually the dual-integral domain in mathematics. It is as well a natural consequence of the nonlocal model under the variationally derivation. Meanwhile, we developed the higher order nonlocal operator method which is capable of solving higher order partial differential equations on arbitrary domains in higher dimensional space.

One prominent feature of nonlocal operator methods is the compatibility of variational principle and weighted residual method. NOM caters for a direct numerical implementation based on weighted residual method for many physical problems. In continuous form, NOM can be used to derive many nonlocal models in strong form (or integral form). The nonlocal operator can be regarded as the integral form, “equivalent” to the differential form in the sense of a nonlocal interaction model. The nonlocal operator plays an equivalent role as the derivatives of the shape functions in the meshless methods or those of the finite element method. Based on the variational principle, the residual and the tangent stiffness matrix can be obtained with ease by a series of matrix multiplications.

The physical models solved by NOM include linear elasticity with fractures (e.g. Kalthoff–Winkler experiment and plate with branching cracks in 2D/3D), Kirchhoff–Love plate, hyperelasticity, gradient elasticity, higher order gradient elasticity with large deformation, Von Karman thin plate equations, N-d Poisson equation (N=2-5), 1D Schrodinger equation, 2D electrostatic problem, Maxwell equations, Cahn–Hilliard equation and others.

We also provide some numerical examples written in Mathematica language in Github, including the static linear elasticity in 2D/3D, explicit nonlocal elasticity, nonlocal thin plate with explicit time integration and NOM with explicit phase field scheme (https://github.com/hl-ren/NOM_explicit_Phase_Field, https://github.com/hl-ren/Nonlocal_thin_plate, https://github.com/hl-ren/Nonlocal_elasticity).

Weimar, Germany — Timon Rabczuk
Hannover, Germany — Huilong Ren
Hannover, Germany — Xiaoying Zhuang

Contents

About the Authors

Timon Rabczuk is chaired professor of Computational Mechanics at the Bauhaus Universität Weimar, Germany. He has published more than 600 SCI papers, mainly about computational methods for the solution of partial differential equations. He is the editor-in-chief of CMC-Computers, Materials and Continua, an associated editor of International Journal of Impact Engineering and Applied Physics A, and an executive editor of FSCE-Frontiers of Structural and Civil Engineering. He was listed as one of ISI Highly Cited Researchers in Computer Science and Engineering from 2014 up to now.

Huilong Ren is currently a research fellow at the Institut für Photonik of Leibniz Universität Hannover, Germany. He obtained his Ph.D. degree with the thesis "Dual-horizon peridynamics and nonlocal operator method" at Institut für Strukturmechanik (ISM) of Bauhaus-Universität Weimar. He is interested in nonlocality-based methods (peridynamics, nonlocal operator method), numerical methods (FEM, IGA, meshless methods), and automatic code generation for numerical solutions of partial differential equations.

Xiaoying Zhuang obtained her Ph.D. degree in Durham University in computational mechanics and received her bachelor and master degrees in civil engineering in Tongji University Shanghai. Her research interests include computational methods for multiscale and multiphysics modelling and integrated materials engineering for advanced materials and functional materials such as metamaterials and polymeric nanocomposite.

Chapter 1
Introduction

As the range of physical phenomena to be simulated in engineering practice broadens, numerical methods play an important role in various fields. Numerical methods are the main tool to find the approximate solutions of physical problems, where the exact solutions are unavailable in most cases. Since various boundary conditions can be taken into account, the numerical method has advantages over the experimental method due to its low cost and more detailed information of the model. In terms of whether the mesh is used or not, numerical methods can be generally divided into two categories: mesh-based method and meshless method. Mesh-based methods include finite element method (Feng 1965; Zienkiewicz et al. 1977), finite volume method (Versteeg and Malalasekera 2007) and so on. In the field of meshless methods, there are Smoothed Particle Hydrodynamics (SPH) (Lucy 1977) finite difference method (Gürlebeck and Hommel 2003; Taflove and Hagness 2005) Moving Least Square (MLS) approximations (Lancaster and Salkauskas 1981) Element Free Galerkin method (EFG) (Belytschko et al. 1994), Reproducing Kernel Particle Method (RKPM) (Liu et al. 1995), Partition of Unity (PU) (Babuška and Melenk 1997) Hp-cloud method (Duarte and Oden 1996) Optimal Transportation Method (OTM) (Li et al. 2010) and generalized finite difference method (Liszka and Orkisz 1980; Perrone and Kao 1975), to name a few.

With the increasing complexity of the engineering problems such as crack propagation, high velocity impact and higher order continuity in higher order PDEs, greater demands are being placed on the numerical methods. What characteristics will a good numerical method possess? There is no exact answer to this question but some good characteristics include: (1) the low cost of discretization of the computational domain is preferred; (2) higher order continuity if possible; (3) numerical result should be insensitive to the mesh distortion; (4) both implicit and explicit numerical implementation are allowed; (5) the method should be compatible with the variational principle or the weighted residual method.

Continuity of a function The function f is said to be of (differentiability) class C^k if the derivatives f', f'', ..., $f^{(k)}$ exist and are continuous (the continuity is implied by differentiability for all the derivatives except for $f^{(k)}$). The function f is said to

T. Rabczuk et al., *Computational Methods Based on Peridynamics and Nonlocal Operators*, Computational Methods in Engineering & the Sciences,
https://doi.org/10.1007/978-3-031-20906-2_1

be of class C^∞, or smooth, if it has derivatives of all orders (Warner 2013). Higher order continuity has significant meanings for various numerical methods. Higher continuity indicates that higher order partial differential equations can be solved. The conventional finite element shape function is of C^0 continuity at element edges, which has difficulty in solving higher order partial differential equations where $C^k (k \geq 1)$ continuity is required. Mixed finite element scheme can be used to overcome the requirement of higher order continuity at the cost of extra degrees of freedom, which however makes the implementation (including numerical scheme, solving techniques) tedious and may suffer from the stability issues. Different from finite element method, many meshless methods are of C^∞ continuity and are suitable for higher order partial differential equations.

Cost of discretization Mesh is the basis for the finite element method. There are two types of mesh for the domain description, Lagrangian mesh and Eulerian mesh. Lagrangian mesh deforming with the displacement field has good computational efficiency for moderate deformation problems. When encountered with large distortion, the numerical scheme of Lagrangian mesh suffers from slow convergence, inaccurate result or suspension of computation due to zero element size that leads to zero time increment. In this case, Lagrangian mesh requires expensive remeshing. Eulerian mesh is fixed in the space, being independent of the geometrical description of the problems, thus is free of mesh distortion. However Eulerian mesh fails to capture the position of free surface, deformed boundary and moving material interfaces accurately.

The generation of a mesh of tetrahedra, pyramids, prisms or hexahedral requires a special geometrical algorithm. The mesh on a complex geometrical domain requires much human intervention, which becomes the bottleneck of computer aided engineering. Furthermore, for complex problems involving moving mesh or remeshing around certain domain (i.e. problems with crack propagation, metal forming process, high/hyper-velocity impact with extreme large deformation), the geometrical algorithm and numerical algorithm are coupled, which makes the problem extremely difficult to solve. In addition, some element types (i.e. tetrahedron, triangle element) are not suitable for dynamic analysis due to over-stiffness of these elements. Quadrilateral element suffers from hourglass mode when single Gauss point integration is employed.

Different from the mesh-based method, the computational domain in the meshless method is discretized into particles or nodes. The cost of the discretization in meshless is much less than that in mesh-based methods. The set of nodes in the meshless method is orderless, thus the connectivity between nodes and the node consistency on adjoint edges of two elements are not required. The adaptive refinement on nodes set can be implemented at a lower cost than that in FEM. The meshless method is less sensitive to the distortion of the discretization compared with mesh-based methods. Therefore, the meshless method shows great advantages in modeling complex problems involving discontinuity, crack propagation.

Implicit or explicit method In numerical analysis, there are generally two approaches, implicit method and explicit method, to obtain the numerical approximations to the solutions of a time-dependent ordinary or partial differential equations.

Explicit method calculates the state of a system at current time from the history state of the system, while implicit method calculates the solution by solving equations with both current state and history state of a system (Pareschi and Russo 2000). Mathematically, let $Y(t - \Delta t)$ be the history state, current system state $Y(t)$ is solved by explicit method as

$$Y(t) = F(Y(t - \Delta t)), \tag{1.1}$$

and by implicit method as

$$G(Y(t), Y(t - \Delta t)) = 0. \tag{1.2}$$

Explicit method is simple but is limited by the time increment for numerical instability issues. Compared with explicit methods, implicit methods are relatively harder in numerical implementation. Implicit method can adopt a very large time increment step and is suitable for stiff problems. For static, quasi-static or dynamic problems with long duration, implicit methods may take much less computational time.

Compatibility with variational principle or weighted residual method (WRM) Variational principle and weighted residual method are the basis of various numerical methods, such as finite element methods and meshless methods. Some physical problems can be described by the Lagrangian, which can be applied with variational principle to obtain the governing equations and the associated boundary conditions. While other physical problems are expressed by a set of partial differential equations, in this case, the Lagrangian or the energy functional can not be established. In order to solve these problems, the weighted residual method can be applied.

Since variational principle or WRM is closely related to various physical problems, a method based on the variational principle or weighted residual method can be regarded as a general numerical method. General numerical method is applicable for different physical problems, while non-general methods have a very narrow field of application.

Different from FEM and meshless methods with many salient features described above, peridynamics (PD) as a special method can only be solved with explicit methods. For different physical models, the PD formulation should be derived from scratch. PD has little connection with the variational principle and weighted residual method, although the cost of discretization is low compared with other methods.

FEMs and meshless methods are usually based on shape functions to approximate the solutions, while PD is based on the nonlocal interaction between particles in the solid domain. In physics, the principle of locality can be understood as an object being directly influenced only by its immediate surroundings. A theory based on the principle of locality is said to be a "local theory". The examples include the directly internal force between two points with infinitesimal distance in a solid and contact force due to mechanical contact. In contrast with the principle of locality, action at a distance is the concept that an object can be affected without being physically touched by another object. That is, it is the nonlocal interaction of objects that are separated in space (Hesse 1955). For example, Coulomb's law and Newton's law of universal gravitation are such early theories.

The concept of action at a distance has led to significant developments in physics, from the concept of a field, to descriptions of quantum entanglement and the mediator particles of the Standard Model (Hesse 1955). In the continuum mechanic field, Eringen in his book (Eringen 2002) stated that "the nonlocal continuum field theories are concerned with the physics of material bodies whose behavior at a material point is influenced by the state of all points of the body". In the spirit of nonlocal interaction in nonlocal continuum field theories, various of nonlocal mechanical models have been proposed, for examples, nonlocal linear elasticity (Rogula 1982), nonlocal fluid dynamics, nonlocal electromagnetic theory (Eringen 2002), peridynamics (Silling 2000), nonlocal damage model (Bažant and Jirásek 2002).

In mathematics, the nonlocal theories are formulated in integral equations, while the ingredient in local theories is various differential operators defined as a point. Nonlocality takes into account the internal length scale of the computational model. When the length scale reduces to zero, the nonlocality theory recovers the local theory. Nonlocality is usually used to consider the physical phenomenon such as nonlocal interaction, memory effect, history effect. The application of nonlocality in the design of numerical methods remains unclear.

The purpose of this book is to derive some numerical methods based on the nonlocality. We found that nonlocality can be used to devise some general numerical methods with power similar to that in FEM and meshless methods. The methods derived by nonlocality are variationally consistent and have most of the salient features described above. More specifically, we derived the dual-horizon peridynamics (DH-PD), dual-support smoothed particle hydrodynamics (DS-SPH), low order nonlocal operator method (NOM) and higher order NOM.

1.1 Overview of Meshless Method

In this section, we review briefly some common meshless methods such as SPH, MLS, EFG, RKPM, LME and etc. SPH as one of the most popular mesh-free methods, was invented in 1977 for modeling astrophysical problems (Gingold and Monaghan 1977; Lucy 1977). MLS is firstly proposed by Lancaster and Salkauskas (1981) for surface reconstruction. Latter Nayroles et al. (1992) applied the MLS to solve solid mechanics and called their method diffuse element method (DEM). Belytschko et al. (1994) improved the DEM by adding some higher order terms in the derivative calculation of the shape functions. They applied this method in the solid mechanics and heat conduction problems and named the implementation as EFG method. Liu et al. proposed the RKPM as an improvement of the continuous SPH approximation. A more general concept for RKPM and EFG is the partition of unity developed by Duarte and Oden (Babuška and Melenk 1997). Based on the idea of partition of unity, Duarte and Oden developed the Hp-cloud method (Duarte and Oden 1996). In order to maintain the Kronecker-delta property of shape function on the boundary in meshless method, Arroyo et al. developed the Local Maximum-Entropy schemes (LME) (Arroyo and Ortiz 2006) based on the unbiased statistical inference. For more complete review of meshless methods, we refer to (Belytschko et al. 1996; Huerta et al. 2018; Nguyen et al. 2008).

1.1.1 Smoothed Particle Hydrodynamics

SPH method has been successfully applied to solve problems such as impact in solids (Allahdadi et al. 1993; Rabczuk and Eibl 2003; Randles and Libersky 1996), dynamic fracture and fragmentation (Monaghan 2000; Rabczuk et al. 2004), multiple-phase flows (Cleary 1998; Hu and Adams 2006; Monaghan 1997), free-surface flows in fluid dynamics (Fang et al. 2009; Monaghan 1994), Magnetohydrodynamics (Monaghan 1988, 2000; Price and Monaghan 2004), solidification and phase-transition (Blom 2014; Monaghan et al. 2005).

The basic idea of SPH is to approximate a function $\mathbf{u}(\mathbf{x})$ on domain by a convolution

$$\mathbf{u}_h(\mathbf{x}) = \int_\Omega C_\rho \phi\left(\frac{\mathbf{x}-\mathbf{x}'}{\rho}\right)\mathbf{u}(\mathbf{x}')\mathrm{d}V_{\mathbf{x}'}, \tag{1.3}$$

where $\mathbf{u}_h$ is the approximation of $\mathbf{u}$, ϕ is the kernel function, ρ is the dilation parameter, C_ρ is the normalization constant such that

$$\int_\Omega C_\rho \phi\left(\frac{\mathbf{x}'}{\rho}\right)\mathrm{d}V_{\mathbf{x}'} = 1. \tag{1.4}$$

The discrete form of SPH can be written as

$$\mathbf{u}_h(\mathbf{x}) = \sum_a C_\rho \phi\left(\frac{\mathbf{x}-\mathbf{x}_a}{\rho}\right)\mathbf{u}_a V_a, \tag{1.5}$$

where $\mathbf{x}_a$ and V_a are the particle and associated particle volume in SPH.

According to Ref. (Monaghan 1994), the kernel function should satisfy the conditions below:

(i) $w(\mathbf{x}-\mathbf{x}_a, \rho) > 0$, on a subdomain Ω_a of Ω.
(ii) $w(\mathbf{x}-\mathbf{x}_a, \rho) = 0$, in domain Ω/Ω_a of Ω.
(iii) $w(\mathbf{x}-\mathbf{x}_a, \rho) \to \delta(\mathbf{x}-\mathbf{x}_a)$ the Dirac delta function, as $\rho \to 0$.
(iv) $w(\mathbf{x}-\mathbf{x}_a, \rho)$ is a monotonically decreasing function w.r.t. $\|\mathbf{x}-\mathbf{x}_a\|$.

In order to reflect the isotropy, the kernel function is written as a function of the normalized radius r_a, where $r_a = \frac{\|\mathbf{x}-\mathbf{x}_a\|}{\max_b(\|\mathbf{x}-\mathbf{x}_b\|)}$. Some common kernel functions include cubic spline, Gaussian and quartic spline, quintic spline, Wendland kernel (Wendland 1995), to name only a few.

Cubic spline:

$$w(r) = \begin{cases} \frac{2}{3} - 4r^2 + 4r^3 & \text{for } r \leq \frac{1}{2} \\ \frac{4}{3} - 4r + 4r^2 - \frac{4}{3}r^3 & \text{for } \frac{1}{2} < r \leq 1 \\ 0 & \text{for } r > 1 \end{cases}$$

Gaussian:

$$w(r) = \begin{cases} \frac{\exp(-(\alpha r)^2) - \exp(-\alpha^2)}{1 - \exp(-\alpha^2)} & \text{for } r \leq 1 \\ 0 & \text{for } r > 1 \end{cases}$$

Quartic spline: $w(r) = \begin{cases} 1 - 6r^2 + 8r^3 - 3r^4 & \text{for } r \leq 1 \\ 0 & \text{for } r > 1 \end{cases}$

For more discussions on these kernel functions, the interested reader is referred to Dehnen and Aly (2003), Liu and Liu (2012).

SPH in its continuous form fulfills zero- and first-order completeness and hence the 'standard' kernel function as given in Eq. (1.3) is sufficient. The zeroth-order and first-order completeness are given as

$$\sum_a C_\rho \phi \left(\frac{\mathbf{x} - \mathbf{x}_a}{\rho} \right) V_a = 1, \tag{1.6}$$

$$\sum_a C_\rho \phi \left(\frac{\mathbf{x} - \mathbf{x}_a}{\rho} \right) \mathbf{x}_a V_a = \mathbf{x}. \tag{1.7}$$

However, not even zero-order completeness is guaranteed for the discrete SPH form as shown by Nguyen et al. (2008). Therefore, many correction methods have been developed such as the symmetrization proposed by Monaghan (1988), Randles & Libersky correction (Randles and Libersky 1996), Johnson & Beissel correction (Johnson and Beissel 1996) and Krongauz-Belytschko corrected derivatives (Krongauz and Belytschko 1997).

1.1.2 Moving Least Square (MLS) Method

The basic idea of MLS scheme is to approximate $u(\mathbf{x})$ at $\mathbf{x}$ based on a polynomial least-square fit of u in a neighborhood of $\mathbf{x}$ (Huerta et al. 2018). Namely, the MLS approximation $u^h(\mathbf{x})$ of a function $u(\mathbf{x})$ at a point $\mathbf{x}$ is defined by

$$u^h(\mathbf{x}) = \sum_b^m p_b(\mathbf{x}) a_b(\mathbf{x}) = \boldsymbol{p}^T(\mathbf{x}) \boldsymbol{a}(\mathbf{x}), \tag{1.8}$$

where $\boldsymbol{p}(\mathbf{x})$ is a vector of basis functions defined in space coordinates $\mathbf{x}^T = \{x, y\}$. m is the number of basis functions.

Consider the basic constraint functional

$$J = \sum_{a}^{n} w(\mathbf{x} - \mathbf{x}_a)\left(\boldsymbol{p}^T(\mathbf{x}_a)\boldsymbol{a}(\mathbf{x}) - u_a\right)^2, \tag{1.9}$$

where $w(\mathbf{x} - \mathbf{x}_a)$ is the compact supported weighting function.

The minimization of J, $\delta J = 0$, leads to

$$\boldsymbol{a}(\mathbf{x}) = \boldsymbol{A}(\mathbf{x})^{-1}\boldsymbol{B}(\mathbf{x})\boldsymbol{u}_e, \tag{1.10}$$

where $\boldsymbol{u}_e$ is the vector of unknowns, $\boldsymbol{A}(\mathbf{x})$ is called the moment matrix, and matrices $\boldsymbol{A}$ and $\boldsymbol{B}$ are defined by

$$\boldsymbol{A}(\mathbf{x}) = \sum_{a}^{n} w(\mathbf{x} - \mathbf{x}_a)\boldsymbol{p}(\mathbf{x}_a)\boldsymbol{p}^T(\mathbf{x}_a),$$

$$\boldsymbol{B}(\mathbf{x}) = [w(\mathbf{x} - \mathbf{x}_1)\boldsymbol{p}(\mathbf{x}_1), w(\mathbf{x} - \mathbf{x}_1)\boldsymbol{p}(\mathbf{x}_1), \ldots, w(\mathbf{x} - \mathbf{x}_n)\boldsymbol{p}(\mathbf{x}_n)].$$

Then MLS approximation can be written as

$$u^h(\mathbf{x}) = \boldsymbol{p}^T(\mathbf{x})\boldsymbol{A}(\mathbf{x})^{-1}\boldsymbol{B}(\mathbf{x})\boldsymbol{u}_e = \phi(\mathbf{x})\boldsymbol{u}_e, \tag{1.11}$$

where ϕ is the set of shape functions,

$$\phi = \boldsymbol{p}^T\boldsymbol{A}^{-1}\boldsymbol{B}. \tag{1.12}$$

The first and second derivatives of the shape function can be calculated by

$$\phi_{,i} = \boldsymbol{p}_{,i}\boldsymbol{A}^{-1}\boldsymbol{B} + \boldsymbol{p}(\boldsymbol{A}^{-1}\boldsymbol{B}_{,i} - \boldsymbol{G}_i\boldsymbol{B}) \tag{1.13}$$

$$\begin{aligned}\phi_{,ij} &= (\boldsymbol{p}\boldsymbol{G}_i - \boldsymbol{p}_{,i}\boldsymbol{A}^{-1})(\boldsymbol{A}_{,j}\boldsymbol{A}^{-1}\boldsymbol{B} - \boldsymbol{B}_{,j}) - \boldsymbol{p}(\boldsymbol{G}_{ij}\boldsymbol{B} - \boldsymbol{A}^{-1}\boldsymbol{B}_{,ij}) \\ &\quad + (\boldsymbol{p}\boldsymbol{G}_j - \boldsymbol{p}_{,j}\boldsymbol{A}^{-1})(\boldsymbol{A}_{,i}\boldsymbol{A}^{-1}\boldsymbol{B} - \boldsymbol{B}_{,i}) + \boldsymbol{p}_{,ij}\boldsymbol{A}^{-1}\boldsymbol{B},\end{aligned} \tag{1.14}$$

where

$$\boldsymbol{G}_i = \boldsymbol{A}^{-1}\boldsymbol{A}_{,i}\boldsymbol{A}^{-1}, \; \boldsymbol{G}_{ij} = \boldsymbol{A}^{-1}\boldsymbol{A}_{,ij}\boldsymbol{A}^{-1}. \tag{1.15}$$

In the calculation of derivatives of shape function, the inverse of A is involved. We can see that the number of terms in the higher order derivative of shape function increases exponentially. For third-order or fourth-order derivatives, the calculation is very expensive. In addition, MLS often requires a higher order numerical integration, which is more expensive than that in FEM.

1.1.3 Reproducing Kernel Particle Method (RKPM)

The continuous RK approximation of a function $u(\mathbf{x})$ on a domain Ω is constructed by the product of a kernel function Φ_a with a compact support, and a correction function composed of a linear combination of the basis function (Liu et al. 1995). In the case of 1-D, RK approximation can be written as

$$u^h(x) = \int_\Omega \Psi(x; x-y)u(y)\mathrm{d}V_y, \tag{1.16}$$

where

$$\Psi(x; x-y) = \Phi_a(x-y)C(x; x-y), \tag{1.17}$$

where $\Phi_a(x-y)$ is the kernel function with compact support; $C(x; x-y)$ is a correction function for the enforcement of the reproducing conditions, which is defined by

$$C(x; x-y) = b_0(x) + b_1(x)(x-y) + \cdots + b_N(x)(x-y)^N = \mathbf{b}^\mathrm{T}(x)\mathbf{H}(x-y), \tag{1.18}$$

where

$$\mathbf{b}^\mathrm{T}(x) = [b_0(x), b_1(x), \ldots, b_N(x)], \; \mathbf{H}^\mathrm{T}(x-y) = \left[1, x-y, \ldots, (x-y)^N\right]. \tag{1.19}$$

$b_i(x)$ is calculated by satisfying the reproducing conditions,

$$\int_\Omega C(x; x-y)\Phi_a(x-y)\mathbf{H}(x-y)\mathrm{d}y = \mathbf{H}(0), \tag{1.20}$$

Inserting Eq. 1.18 into Eq. 1.20 leads to

$$\left[\int_\Omega \mathbf{H}(x-y)\Phi_a(x-y)\mathbf{H}^\mathrm{T}(x-y)\mathrm{d}V_y\right]\mathbf{b}(x) = \mathbf{H}(0). \tag{1.21}$$

Then $\mathbf{b}(x)$ can be solved as

$$\mathbf{b}(x) = \mathbf{M}(x)^{-1}\mathbf{H}(0) \tag{1.22}$$

$$\mathbf{M}(x) = \int_\Omega \mathbf{H}(x-y)\mathbf{H}^\mathrm{T}(x-y)\Phi_a(x-y)\mathrm{d}V_y. \tag{1.23}$$

Then Eq. 1.16 can be written as (Chen et al. 1998)

$$u^R(x) = \int_{\Omega} C(x; x-y)\Phi_a(x-y)f(y)\mathrm{d}y \tag{1.24}$$

$$= \mathbf{H}^{\mathrm{T}}(0)\mathbf{M}^{-1}(x)\int_{\Omega}\mathbf{H}(x-y)\Phi_a(x-y)u(y)\mathrm{d}V_y. \tag{1.25}$$

1.1.4 Essential Boundary Conditions

In order to enforce essential boundary conditions in meshfree method, various techniques have been proposed, among which include, Lagrange multipliers (Belytschko et al. 1994), modified variational principles (Belytschko et al. 1994), penalty method (Bonet and Kulasegaram 2000; Zhu and Atluri 1998), perturbed Lagrangian (Chu and Moran 1995), coupling to finite element (Huerta and Fernández-Méndez 2000), modified shape functions (Wagner and Liu 2000). The method based on shape function modification can be difficult for complex domains and for the integration of the weak form. Mixed interpolations on FEM and meshfree preserve consistency and continuity (Huerta and Fernández-Méndez 2000) and can be implemented with ease.

The Lagrange multiplier method is general and applicable to all kinds of problems but the number of dimensions of the algebraic equations is enlarged. Lagrange multiplier converts the original problem into a saddle point problem. Stiffness matrix becomes neither positive definite nor negative definite, which cannot be solved by standard linear solvers. Meanwhile, the resolution of the multiplier λ should be adequate, too coarse would lead to undesired solution, too finer would result in singular system equations. Lagrange multiplier and the unknown field must satisfy the Babuska-Brezzi stability conditions for the purpose of convergence of the approximation (Babuška 1973; Brezzi 1974).

Penalty method enforces the constraint by converting the constraint into a quadratic functional with a large enough penalty parameter and attaching this functional to the original functional. No additional degrees of freedom are introduced and the positive definite property of the tangent stiffness matrix is preserved. The modified functional is

$$\Pi^* = \Pi + \alpha\int_{\partial\Omega}(u-u_d)^2\mathrm{d}S, \tag{1.26}$$

where Π is the original functional, α is a large number.

Two drawbacks of the penalty method are, the Dirichlet boundary condition is weakly imposed and the tangent stiffness matrix is poorly conditioned. In order to achieve the convergence of the penalty method, the order of magnitude of the penalty must be selected with caution. The choice of the penalty parameter α can be found in Babuska (1973).

Penalty method is sensitive to the penalty parameter, which can be overcome by Nitsche's method (Nitsche 1975). Nitsche's method has been used in a range of fields, i.e. finite element method (Juntunen and Stenberg 2009), incompressible elasticity problems (Hansbo and Larson 2002), Navier-Stokes problem (Becker 2002). Nitsche's method takes into account the terms in natural boundary condition as well as the terms of the essential boundary condition. In other words, the natural boundary terms in conjugate with the essential terms are added in the weak form. By doing so, the penalty parameter can be small, thus making the tangent stiffness matrix well-conditioned. Nitsche's method can be viewed as a consistent improvement of the penalty method. In addition, the LBB conditions in the Lagrange multiplier are avoided. One drawback of Nitsche's method is that the deduction of the weak form is not as straightforward as the Lagrange method and penalty method.

1.1.5 Maximum-Entropy Meshfree Approximations

Arroyo and Ortiz (2006) proposed maximum-entropy meshfree approximations, whose basis function is constructed by solving a statistical inference problem based on the nodes distribution. From the information-theoretical point of view (Jaynes 1957), the basis function can be viewed as the optimal probability distribution with additional constraints of 0th and 1st order consistency conditions being satisfied. Optimal probability distribution is given by the maximum-entropy principle

$$\text{Maximize}\quad -\sum_I N_I(\mathbf{x})\log N_I(\mathbf{x}),\ \text{subject to}\sum_I N_I(\mathbf{x}) = 1,\quad \sum_I N_I(\mathbf{x})\mathbf{x}_I = \mathbf{x}. \tag{1.27}$$

One implicit condition for the above equation is the non-negativity of the basis function. For polygonal finite element approximation (Sukumar 2004), the above principle leads to the least biased choice of the basis functions consistent with the constraints. In the context of meshless method, Arroyo and Ortiz (2006) added a controllable degree of locality (the interpolated value $u(\mathbf{x})$ depends more on the coefficients u_I whose node coordinate $\mathbf{x}_I$ closer to $\mathbf{x}$) and proposed the variational characterization of the basis function

$$\begin{aligned}\text{Minimize}\quad &\sum_I N_I(\mathbf{x})\log N_I(\mathbf{x}) + \sum_I \beta_I N_I(\mathbf{x})\,|\mathbf{x}-\mathbf{x}_I|^2, \\ &\text{subject to}\sum_I N_I(\mathbf{x}) = 1,\quad \sum_I N_I(\mathbf{x})\mathbf{x}_I = \mathbf{x}.\end{aligned} \tag{1.28}$$

The parameter β_I in the second term controls the width of the basis function. When $\beta_I \to +\infty$, the basis function is equivalent to the linear shape function of the Delaunay triangulation on the nodes (Arroyo and Ortiz, 2006).

Along with the variational definition and duality method, the basis function can be written as

$$N_I(\mathbf{x}) = \frac{1}{Z} \exp\left[-\beta_I \left|\mathbf{x} - \mathbf{x}_I\right|^2 + \lambda \cdot (\mathbf{x} - \mathbf{x}_I)\right] \tag{1.29}$$

$$\text{where } Z = \sum_J \exp\left[-\beta_J \left|\boldsymbol{x} - \boldsymbol{x}_J\right|^2 + \lambda \cdot (\boldsymbol{x} - \boldsymbol{x}_J)\right], \tag{1.30}$$

and λ is the Lagrange multiplier enforcing linear consistency. The above explicit basis function belongs to the local maximum entropy (LME) approximations. One advantage of LME approximants over MLS approximants is the weak version of Kronecker delta property at the boundary (Arroyo and Ortiz 2006).

1.1.6 Peridynamics

Peridynamics (PD) proposed by Silling (2000) can be viewed as a nonlocal theory. It was alleged that this theory can model the continuous and discontinuous problems in a unified formulation. For mechanical problems with crack propagation, PD can model the crack by simply introducing the breakage of bonds in the horizon when a certain crack criterion is satisfied. Although the formulation is restricted by explicit time integration, PD can model the crack very intuitively, in contrast with other meshless methods and finite element methods where the description of the fracture is very cumbersome. There are general two types of peridynamics, bond-based peridynamics (BB-PD) and state-based peridynamics (SB-PD), where the latter can be further divided into ordinary state-based peridynamics (OSB-PD) and non-ordinary state-based peridynamics (NOSB-PD). BB-PD can model very limited elastic material because of Poisson ratio restriction, while OSB-PD can model isotropic linear elastic material. NOSB-PD is based on the continuum mechanics; various materials can be modeled by NOSB-PD. PD also shows some similarity with smoothed particle hydrodynamic (Ganzenmüller et al. 2015).

PD can be viewed as a generalization of molecule dynamics. The molecule dynamics are suitable for the modeling of interaction between molecules in the micro-world, while PD discretize the computational domain into particles and take into account the nonlocal interaction between particles in a way similar to the Coulomb force between molecules with electric charge. By summing the whole interaction from its neighbours, the particle moves according to Newton's laws. The summation of nonlocal interactions is the discrete form of the integral equation. PD is based on integral equations, rather than the differential equations. When the discontinuity is encountered, the differential operator cannot be well defined, which leads to the very cumbersome implementation of numerical algorithms. On the other hand, the integral form is defined on an explicit domain with internal length scale. Local form is defined at a point; the description of discontinuity requires the Heaviside function. One reason is that the point has no volume or shape, which is abstract and difficult to handle. For integral equations, the integral domain at a point can be processed with a geometrical algorithm. One can modify the integral domain based on the physical problem. The volume and shape can be viewed as the internal parameters to express the discontinuity. In PD, the integral domain is denoted by horizon. By removing the neighbors in the horizon based on the certain crack criterion, the crack forms naturally.

Several key concepts in PD including bond and horizon. Bond is the medium to transmit interaction between two particles, horizon is the range where the nonlocal interaction is taken into account. For each particle, the interaction of two points beyond the horizon is assumed to be zero. The horizon is a region defined by

$$\mathcal{H}_{\mathbf{x}} = \{\mathbf{x}' : \|\mathbf{x} - \mathbf{x}'\| \leq h\}. \tag{1.31}$$

For elastic mechanics, the set of governing equations in local differential form is

$$\nabla \cdot \boldsymbol{\sigma} + \mathbf{b} = \rho \ddot{\mathbf{u}}, \tag{1.32}$$

where $\boldsymbol{\sigma}$ is the Cauchy stress tensor. While in PD, the integral equation based on the bond forces can be written as

$$\int_{H_{\mathbf{x}}} \mathbf{f}(\mathbf{x}, \mathbf{x}') \mathrm{d}V_{\mathbf{x}'} + \mathbf{b} = \rho \ddot{\mathbf{u}}, \tag{1.33}$$

where $\mathbf{f}(\mathbf{x}, \mathbf{x}')$ is the bond force vector between point $\mathbf{x}$ and $\mathbf{x}'$.

One method to improve the numerical efficiency of PD is the FEM and PD coupling scheme to take advantage of both methods, where regions of interest are modeled using peridynamics while remaining regions are modeled by the finite element method. Macek and Silling (2007) implemented the peridynamic model in a conventional finite element analysis (FEA) code using truss elements. Kilic and Madenci (2010) introduced an overlap region to couple the bond-based peridynamics and FEM. Ni et al. (2019) developed a method to couple FEM meshes with ordinary state-based peridynamics, which combined the advantages of peridynamics for crack modeling and robust finite element methods for continuum problems.

In order to remove the zero-energy mode in the non-ordinary state-based PD, several techniques have been proposed, for example, sub-horizon scheme (Chowdhury et al. 2019), stabilized peridynamics formulation (Li et al. 2018), peridynamics with stress point (Luo and Sundararaghavan 2018), higher-order approximation scheme (Yaghoobi and Chorzepa 2017).

Pasetto et al. (2018) proposed a reproducing kernel enhanced approach for peridynamics. The basis functions reproducing conditions from the reproducing kernel particle method (Liu et al. 1995) is applied to the displacement field. This approach requires the numerical integration on the horizon in the polar coordinate system. Convergence rate is improved at the cost of more computational cost. Peridynamics has been applied to solve other problems such as phase field fracture (Roy et al. 2017), flexoelectricity problems (Roy and Roy 2019).

1.1.7 Peridynamic Differential Operator Method

A further development of PD is the Peridynamic differential operator method (PDDO) proposed by Madenci et al. (2019). Consider the Taylor series expansion of a scalar field $f(\mathbf{x}') = f(\mathbf{x} + \boldsymbol{\xi})$ with multivariables, which can be expressed as

$$f(\mathbf{x}+\boldsymbol{\xi}) = \sum_{n_1=0}^{N} \sum_{n_2=0}^{N-n_1} \cdots \sum_{n_N=0}^{N-n_1-\cdots-n_{N-1}} \frac{1}{n_1! n_2! \cdots n_N!} \xi_1^{n_1} \xi_2^{n_2} \cdots \xi_M^{n_N} \frac{\partial^{n_1+n_2+\cdots+n_N} f(\mathbf{x})}{\partial x_1^{n_1} \partial x_2^{n_2} \cdots \partial x_M^{n_N}} + R(N, \mathbf{x}), \tag{1.34}$$

where $\boldsymbol{\xi} = \mathbf{x}' - \mathbf{x}$ and $R(N, \mathbf{x})$ is the remainder.

Based on the property of the orthogonal function, $g_N^{p_1 p_2 \cdots p_N}(\boldsymbol{\xi})$, the PD nonlocal expression for the partial derivatives of any order can be derived as

$$\frac{\partial^{p_1+p_2+\cdots+p_N} f(\mathbf{x})}{\partial x_1^{p_1} \partial x_2^{p_2} \ldots \partial x_M^{p_N}} = \int_{H_\mathbf{x}} f(\mathbf{x}+\boldsymbol{\xi}) g_N^{p_1 p_2 \cdots p_N}(\boldsymbol{\xi}) dx_1 dx_2 \cdots dx_M. \tag{1.35}$$

The orthogonality property of the orthogonal function can be stated as

$$\frac{1}{n_1! n_2! \cdots n_N!} \int_{H_\mathbf{x}} \xi_1^{n_1} \xi_2^{n_2} \cdots \xi_M^{n_M} g_N^{p_1 p_2 \cdots p_N}(\boldsymbol{\xi}) dx_1 dx_2 \cdots dx_M = \delta_{n_1 p_1} \delta_{n_2 p_2} \cdots \delta_{n_{N-1} p_{N-1}} \delta_{n_N p_N}, \tag{1.36}$$

in which p_i denotes the order of differentiation with respect to the variable x_i with $i = 1 \ldots M$.

The orthogonal function can be explicitly written as

$$g_N^{p_1 p_1 \cdots p_N}(\boldsymbol{\xi}) = \sum_{q_1=0}^{N} \sum_{q_2=0}^{N-q_1} \cdots \sum_{q_N=0}^{N-q_1-\cdots-q_{N-1}} a_{q_1 q_2 \cdots q_N}^{p_1 p_2 \cdots p_N} w_{q_1 q_2 \cdots q_N}(|\boldsymbol{\xi}|) \xi_1^{q_1} \xi_2^{q_2} \cdots \xi_M^{q_N}, \tag{1.37}$$

where $w_{q_1 q_2 \cdots q_N}(|\boldsymbol{\xi}|)$ is the weight function associated with each term $\xi_1^{q_1} \xi_2^{q_2} \cdots \xi_M^{q_N}$ in the polynomial expansion.

The unknown coefficients, $a_{q_1 q_2 \cdots q_M}^{p_1 p_2 \cdots p_M}$, can be determined from the solution of

$$\sum_{q_1=0}^{N} \sum_{q_2=0}^{N-q_1} \cdots \sum_{q_N=0}^{N-q_1-\cdots-q_{N-1}} A_{(n_1 n_2 \cdots n_N)(q_1 q_2 \cdots q_N)} a_{q_1 q_2 \cdots q_N}^{p_1 p_1 \cdots p_N} = b_{n_1 n_2 \cdots n_N}^{p_1 p_2 \cdots p_N}, \tag{1.38}$$

where

$$A_{(n_1n_2\cdots n_N)(q_1q_2\cdots q_N)} = \int_{H_{\mathbf{x}}} w_{q_1q_2\cdots q_N}(|\xi|)\xi_1^{n_1+q_1}\xi_2^{n_2+q_2}\cdots\xi_M^{n_N+q_N}dx_1dx_2\cdots dx_M \tag{1.39}$$

$$b_{n_1n_2\cdots n_N}^{p_1p_2\cdots p_N} = n_1!n_2!\cdots n_M!\delta_{n_1p_1}\delta_{n_2p_2}\cdots\delta_{n_Np_N}. \tag{1.40}$$

1.2 Brief Review of Nonlocal Theories

Nonlocal models exist in many fields, ranging from nanoscale to macroscale, from continuum mechanics to fracture mechanics. The research on nonlocal theories is vast, for example, nonlocal elastic continuum (Eringen 1983, 1972; Eringen and Edelen 1972), peridynamics considering discontinuities and long-range forces (Silling 2000), wave propagation analysis in temperature-dependent inhomogeneous nonlocal strain gradient nanoplates (Ebrahimi et al. 2016), the bending, buckling and vibration of nonlocal beam theory (Aydogdu 2009; Li and Hu 2015; Reddy 2007; Sahmani et al. 2018), nonlocal strain gradient nanoplate model in thermal environment (Farajpour et al. 2016), nonlocal elastic nano-beams (Barretta and de Sciarra 2018), the vibrations of nonlocal rods (Xu et al. 2017) and double-walled carbon nanotubes (Zhang et al. 2005), nonlocal friction laws in elasticity (Oden and Pires 1983), nonlocal third-order shear deformation plate (Aghababaei and Reddy 2009), nonlocal model for fracture modeling (Bažant and Planas 2019; Bažant and Jirásek 2002), to name a few. In nonlocal elasticity, the stress tensor is based on the integral of the "local" stress field in a domain, in contrast with the local elasticity defining the stress based on the strain field at a point. The nonlocal elasticity proposed by Eringen and Edelen (1972) has the form

$$U_0\left(\varepsilon_{ij}, \varepsilon'_{ij}, \alpha_0\right) = \frac{1}{2}\varepsilon_{ij}C_{ijkl}\int_V \alpha_0\left(\left|\mathbf{x}-\mathbf{x}'\right|, e_0a\right)\varepsilon'_{kl}dV', \tag{1.41}$$

where ε_{ij} and ε'_{ij} are the strain tensors at point $\mathbf{x}$ and point $\mathbf{x}'$, respectively, C_{ijkl} is the elastic modulus tensor of classical elasticity, e_0 is the nonlocal material constant, a is an internal characteristic length and α_0 is the principal attenuation kernel function related to the nonlocality effect between the point $\mathbf{x}$ and neighboring points $\mathbf{x}'$ within a domain V.

One generalization of nonlocal elasticity considers the gradient of the strain tensor (Lim et al. 2015)

$$\begin{aligned} U_0\left(\varepsilon_{ij}, \varepsilon'_{ij}, \alpha_0; \varepsilon_{ij,m}, \varepsilon'_{ij,m}, \alpha_1\right) &= \frac{1}{2}\varepsilon_{ij}C_{ijkl}\int_V \alpha_0\left(|\mathbf{x}-\mathbf{x}'|, e_0a\right)\varepsilon'_{kl}dV' \\ &+ \frac{l^2}{2}\varepsilon_{ij,m}C_{ijkl}\int_V \alpha_1\left(|\mathbf{x}-\mathbf{x}'|, e_1a\right)\varepsilon'_{kl,m}dV', \end{aligned} \tag{1.42}$$

where l is a material length scale introduced to determine the significance of higher-order strain gradient field, $\alpha_1\left(\left|\mathbf{x}-\mathbf{x}'\right|, e_1 a\right)$ is an additional attenuation kernel function to describe the nonlocal effect of the first-order strain gradient field and e_1 is the related material constant. Based on the variational derivation, the governing equations and their boundary conditions can be derived accordingly.

1.3 Energy Form and Variational Principle

Many physical problems allow for an energy form or a functional description of the system. The governing equations are the Euler–Lagrange equations derived from the functional. The functional has some advantages in terms of numerical implementation. Consider a functional dependent on functions $\mathbf{u}$ and their derivatives $\nabla\mathbf{u}$ over the domain Ω, which can be conceptually written as

$$\Pi(\mathbf{u}, \nabla\mathbf{u}) = \int_{\Omega} \phi(\mathbf{u}_{\mathbf{x}}, \nabla\mathbf{u}_{\mathbf{x}}) dV_{\mathbf{x}}, \tag{1.43}$$

where $\phi(\mathbf{u}_{\mathbf{x}}, \nabla\mathbf{u}_{\mathbf{x}})$ is the functional density defined at point $\mathbf{x}$. The variation of Π is

$$\begin{aligned} \delta\Pi &= \int_{\Omega} \frac{\partial\phi}{\partial\mathbf{u}} \cdot \delta\mathbf{u} + \frac{\partial\phi}{\partial\nabla\mathbf{u}} : \nabla\delta\mathbf{u} dV_{\mathbf{x}} \\ &= \int_{\Omega} \frac{\partial\phi}{\partial\mathbf{u}} \cdot \delta\mathbf{u} - \nabla \cdot \frac{\partial\phi}{\partial\nabla\mathbf{u}} \cdot \delta\mathbf{u} dV_{\mathbf{x}} + \int_{\partial\Omega} \mathbf{n} \cdot \frac{\partial\phi}{\partial\nabla\mathbf{u}} \cdot \delta\mathbf{u} dS_{\mathbf{x}}. \end{aligned} \tag{1.44}$$

The terms on the boundary should be addressed together with the boundary conditions and we omit these terms for simplicity. For any $\delta\mathbf{u}$, $\delta\Pi = 0$ leads to the Euler-Lagrange equation

$$\frac{\partial\phi}{\partial\mathbf{u}} - \nabla \cdot \frac{\partial\phi}{\partial\nabla\mathbf{u}} = 0. \tag{1.45}$$

Usually, the energy functional is suitable for direct numerical implementation. Based on certain numerical scheme (e.g. finite element method, meshless method), the function $\mathbf{u}$ after discretization can be written as

$$\mathbf{u}_{\mathbf{x}} = \boldsymbol{N}_{\mathbf{x}}\boldsymbol{U}_{\mathbf{x}}, \nabla\mathbf{u}_{\mathbf{x}} = \boldsymbol{M}_{\mathbf{x}}\boldsymbol{U}_{\mathbf{x}}, \tag{1.46}$$

where $\boldsymbol{N}_{\mathbf{x}}$ is a set of shape functions in the support of $\mathbf{x}$ and $\boldsymbol{M}_{\mathbf{x}}$ denotes the derivative of the $\boldsymbol{N}_{\mathbf{x}}$.

Consider the first variation and second variation of Π, we have

$$\delta\Pi = \sum_{\Omega}(\frac{\partial\phi}{\partial\mathbf{u_x}}\delta\mathbf{u_x} + \frac{\partial\phi}{\partial\mathbf{u_x}}\nabla\delta\mathbf{u_x})\Delta V_{\mathbf{x}} \tag{1.47}$$

$$= \sum_{\Omega}(\frac{\partial\phi}{\partial\mathbf{u_x}}N_{\mathbf{x}}\delta U_{\mathbf{x}} + \frac{\partial\phi}{\partial\mathbf{u_x}}M_{\mathbf{x}}\delta U_{\mathbf{x}})\Delta V_{\mathbf{x}} \tag{1.48}$$

$$\delta^2\Pi = \sum_{\Omega}(\delta U_{\mathbf{x}}^T N_{\mathbf{x}}^T \frac{\partial^2\phi}{\partial\mathbf{u_x}\partial\mathbf{u_x}} N_{\mathbf{x}}\delta U_{\mathbf{x}} + (\delta U_{\mathbf{x}})^T M_{\mathbf{x}}^T \frac{\partial^2\phi}{\partial\nabla\mathbf{u_x}\partial\nabla\mathbf{u_x}} M_{\mathbf{x}}\delta U_{\mathbf{x}})\Delta V_{\mathbf{x}}. \tag{1.49}$$

Here $\Delta V_{\mathbf{x}}$ can be understood as the weight of the Gauss point or the volume associated with that point. Therefore, the residual vector and tangent stiffness matrix can be extracted as

$$R_g = \sum_{\Omega}\left(\frac{\partial\phi}{\partial\mathbf{u_x}}N_{\mathbf{x}} + \frac{\partial\phi}{\partial\mathbf{u_x}}M_{\mathbf{x}}\right)\Delta V_{\mathbf{x}} \tag{1.50}$$

$$K_g = \sum_{\Omega}\left(N_{\mathbf{x}}^T \frac{\partial^2\phi}{\partial\mathbf{u_x}\partial\mathbf{u_x}} N_{\mathbf{x}} + M_{\mathbf{x}}^T \frac{\partial^2\phi}{\partial\nabla\mathbf{u_x}\partial\nabla\mathbf{u_x}} M_{\mathbf{x}}\right)\Delta V_{\mathbf{x}}. \tag{1.51}$$

The essential boundary conditions and natural boundary conditions can be enforced by modifying the entries in the residual vector and tangent stiffness matrix.

1.4 Weak Form and Weighted Residual Method

Many problems in engineering and physics are proposed in the form of the partial differential equations and the boundary conditions which should be satisfied by the unknown field functions. Generally, as shown in Fig. 1.1a, the unknown field $\mathbf{u}(\mathbf{x})$ should satisfy

$$\mathbf{A}(\mathbf{u}) = \begin{bmatrix} \Lambda_1(\mathbf{u}) \\ \Lambda_2(\mathbf{u}) \\ \vdots \end{bmatrix} = \mathbf{0}, \mathbf{x} \in \Omega, \tag{1.52}$$

where Ω can be volume/area domains. Meanwhile, the unknown function $\mathbf{u}$ should satisfy the boundary conditions

$$\mathbf{B}(\mathbf{u}) = \begin{bmatrix} B_1(\mathbf{u}) \\ B_2(\mathbf{u}) \\ \vdots \end{bmatrix} = \mathbf{0}, \mathbf{x} \in \partial\Omega. \tag{1.53}$$

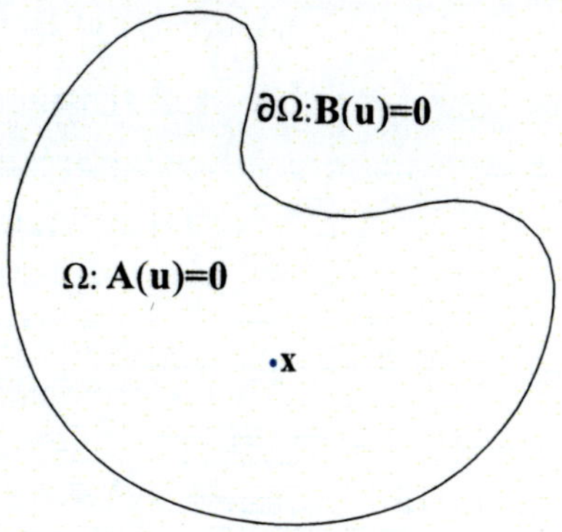

Fig. 1.1 Computational domain Ω

The equivalent integral form of Eqs. 1.52, 1.53 is

$$\int_{\Omega} \mathbf{v}^T \mathbf{A}(\mathbf{u}) d\Omega + \int_{\partial\Omega} \bar{\mathbf{v}}^T \mathbf{B}(\mathbf{u}) d\Gamma = 0, \forall \mathbf{v}, \bar{\mathbf{v}} \in H_0^1. \tag{1.54}$$

The "weak" form of integral form can be obtained by integration by part

$$\int_{\Omega} \mathbf{C}^T(\mathbf{v}) \mathbf{D}(\mathbf{u}) d\Omega + \int_{\partial\Omega} \mathbf{E}^T(\bar{\mathbf{v}}) \mathbf{F}(\mathbf{u}) d\Gamma = 0, \tag{1.55}$$

where $\mathbf{C}, \mathbf{D}, \mathbf{E}, \mathbf{F}$ are the differential operators. The weak form in Eq. 1.55 decreases the continuity of $\mathbf{u}$ by improving the continuity of $\mathbf{v}$.

For convenience, we consider second-order PDEs and assume $\mathbf{C}(\mathbf{v}, \nabla\mathbf{v}), \mathbf{D}(\mathbf{u}, \nabla\mathbf{u})$, $\mathbf{E}(\mathbf{v}, \nabla\mathbf{v}), \mathbf{D}(\mathbf{u}, \nabla\mathbf{u})$ are functional dependent on the unknowns $\mathbf{v}, \nabla\mathbf{v}, \mathbf{u}, \nabla\mathbf{u}$.

Based on certain numerical scheme (e.g. finite element method, meshless method), the field value and their derivative in discrete form can be expressed with matrix

$$\mathbf{v}_{\mathbf{x}} = \boldsymbol{N}_{\mathbf{x}} \boldsymbol{V}_{\mathbf{x}}, \nabla\mathbf{v}_{\mathbf{x}} = \boldsymbol{M}_{\mathbf{x}} \boldsymbol{V}_{\mathbf{x}} \tag{1.56}$$

$$\mathbf{u}_{\mathbf{x}} = \boldsymbol{N}_{\mathbf{x}} \boldsymbol{U}_{\mathbf{x}}, \nabla\mathbf{u}_{\mathbf{x}} = \boldsymbol{M}_{\mathbf{x}} \boldsymbol{U}_{\mathbf{x}}, \tag{1.57}$$

where $\boldsymbol{N}_{\mathbf{x}}$ is a set of shape functions in the support of $\mathbf{x}$ and $\boldsymbol{M}_{\mathbf{x}}$ denotes the derivative of the $\boldsymbol{N}_{\mathbf{x}}$; $\boldsymbol{V}_{\mathbf{x}}$ is a set of unknowns in the support of $\mathbf{x}$. Here we interpolate the trial function and test function with the same shape functions.

The functional for Eq. 1.55 can be written as

$$\Pi = \int_{\Omega} \mathbf{C}^T(\mathbf{v}, \nabla\mathbf{v}) \mathbf{D}(\mathbf{u}, \nabla\mathbf{u}) d\Omega + \int_{\partial\Omega} \mathbf{E}^T(\bar{\mathbf{v}}, \nabla\bar{\mathbf{v}}) \mathbf{F}(\mathbf{u}, \nabla\mathbf{u}) d\Gamma \tag{1.58}$$

$$= \sum_{\mathbf{x}\in\Omega} \mathbf{C}^T(\mathbf{v}_{\mathbf{x}}, \nabla\mathbf{v}_{\mathbf{x}}) \mathbf{D}(\mathbf{u}_{\mathbf{x}}, \nabla\mathbf{u}_{\mathbf{x}}) \Delta V_{\mathbf{x}} + \sum_{\mathbf{x}\in\partial\Omega} \mathbf{E}^T(\bar{\mathbf{v}}_{\mathbf{x}}, \nabla\bar{\mathbf{v}}_{\mathbf{x}}) \mathbf{F}(\mathbf{u}_{\mathbf{x}}, \nabla\mathbf{u}_{\mathbf{x}}) \Delta S_{\mathbf{x}}. \tag{1.59}$$

Consider the variation of Π with respect to $\mathbf{v}$ field

$$\delta\Pi = \int_{\Omega} \delta\mathbf{C}^T(\mathbf{v}, \nabla\mathbf{v})\mathbf{D}(\mathbf{u}, \nabla\mathbf{u}) + \int_{\partial\Omega} \delta\mathbf{E}^T(\bar{\mathbf{v}}, \nabla\bar{\mathbf{v}})\mathbf{F}(\mathbf{u}, \nabla\mathbf{u})d\Gamma, \tag{1.60}$$

$$= \sum_{\mathbf{x}\in\Omega} \delta\mathbf{C}^T(\mathbf{v}, \nabla\mathbf{v})\mathbf{D}(\mathbf{u}, \nabla\mathbf{u})\Delta V_{\mathbf{x}} + \sum_{\mathbf{x}\in\partial\Omega} \delta\mathbf{E}^T(\bar{\mathbf{v}}, \nabla\bar{\mathbf{v}})\mathbf{F}(\mathbf{u}, \nabla\mathbf{u})\Delta S_{\mathbf{x}}, \tag{1.61}$$

where

$$\delta\mathbf{C}(\mathbf{v}_{\mathbf{x}}, \nabla\mathbf{v}_{\mathbf{x}}) = \frac{\partial\mathbf{C}}{\partial\mathbf{v}_{\mathbf{x}}}\delta\mathbf{v}_{\mathbf{x}} + \frac{\partial\mathbf{C}}{\partial\nabla\mathbf{v}_{\mathbf{x}}}\nabla\delta\mathbf{v}_{\mathbf{x}} = \frac{\partial\mathbf{C}}{\partial\mathbf{v}_{\mathbf{x}}}\boldsymbol{N}_{\mathbf{x}}\delta\boldsymbol{V}_{\mathbf{x}} + \frac{\partial\mathbf{C}}{\partial\nabla\mathbf{v}_{\mathbf{x}}}\boldsymbol{M}_{\mathbf{x}}\delta\boldsymbol{V}_{\mathbf{x}} \tag{1.62}$$

$$\delta\mathbf{E}(\mathbf{v}_{\mathbf{x}}, \nabla\mathbf{v}_{\mathbf{x}}) = \frac{\partial\mathbf{E}}{\partial\mathbf{v}_{\mathbf{x}}}\delta\mathbf{v}_{\mathbf{x}} + \frac{\partial\mathbf{E}}{\partial\nabla\mathbf{v}_{\mathbf{x}}}\nabla\delta\mathbf{v}_{\mathbf{x}} = \frac{\partial\mathbf{E}}{\partial\mathbf{v}_{\mathbf{x}}}\boldsymbol{N}_{\mathbf{x}}\delta\boldsymbol{V}_{\mathbf{x}} + \frac{\partial\mathbf{E}}{\partial\nabla\mathbf{v}_{\mathbf{x}}}\boldsymbol{M}_{\mathbf{x}}\delta\boldsymbol{V}_{\mathbf{x}}. \tag{1.63}$$

The residual vector can be extracted as

$$\boldsymbol{R}_g = \sum_{\mathbf{x}\in\Omega}\left(\frac{\partial\mathbf{C}}{\partial\mathbf{v}_{\mathbf{x}}}\boldsymbol{N}_{\mathbf{x}} + \frac{\partial\mathbf{C}}{\partial\nabla\mathbf{v}_{\mathbf{x}}}\boldsymbol{M}_{\mathbf{x}}\right)^T\mathbf{D}(\mathbf{u}_{\mathbf{x}}, \nabla\mathbf{u}_{\mathbf{x}})\Delta V_{\mathbf{x}} + \sum_{\mathbf{x}\in\partial\Omega}\left(\frac{\partial\mathbf{E}}{\partial\mathbf{v}_{\mathbf{x}}}\boldsymbol{N}_{\mathbf{x}} + \frac{\partial\mathbf{E}}{\partial\nabla\mathbf{v}_{\mathbf{x}}}\boldsymbol{M}_{\mathbf{x}}\right)^T\mathbf{F}(\mathbf{u}_{\mathbf{x}}, \nabla\mathbf{u}_{\mathbf{x}})\Delta S_{\mathbf{x}}. \tag{1.64}$$

Consider the variation of $\delta\Pi$ with respect to $\mathbf{u}$ field

$$\delta^2\Pi = \int_{\Omega} \delta\mathbf{C}^T(\mathbf{v}, \nabla\mathbf{v})\delta\mathbf{D}(\mathbf{u}, \nabla\mathbf{u})d\Omega + \int_{\partial\Omega} \delta\mathbf{E}^T(\bar{\mathbf{v}}, \nabla\bar{\mathbf{v}})\delta\mathbf{F}(\mathbf{u}, \nabla\mathbf{u})d\Gamma, \tag{1.65}$$

$$= \sum_{\mathbf{x}\in\Omega} \delta\mathbf{C}^T(\mathbf{v}_{\mathbf{x}}, \nabla\mathbf{v}_{\mathbf{x}})\delta\mathbf{D}(\mathbf{u}_{\mathbf{x}}, \nabla\mathbf{u}_{\mathbf{x}})\Delta V_{\mathbf{x}} + \sum_{\mathbf{x}\in\partial\Omega} \delta\mathbf{E}^T(\bar{\mathbf{v}}_{\mathbf{x}}, \nabla\bar{\mathbf{v}}_{\mathbf{x}})\delta\mathbf{F}(\mathbf{u}_{\mathbf{x}}, \nabla\mathbf{u}_{\mathbf{x}})\Delta S_{\mathbf{x}}, , \tag{1.66}$$

where

$$\delta\mathbf{D}(\mathbf{u}, \nabla\mathbf{u}) = \frac{\partial\mathbf{D}}{\partial\mathbf{u}}\delta\mathbf{u} + \frac{\partial\mathbf{D}}{\partial\nabla\mathbf{u}}\nabla\delta\mathbf{u} \tag{1.67}$$

$$\delta\mathbf{F}(\mathbf{u}, \nabla\mathbf{u}) = \frac{\partial\mathbf{F}}{\partial\mathbf{u}}\delta\mathbf{u} + \frac{\partial\mathbf{F}}{\partial\nabla\mathbf{u}}\nabla\delta\mathbf{u}. \tag{1.68}$$

$$\delta\mathbf{D}(\mathbf{u}_{\mathbf{x}}, \nabla\mathbf{u}_{\mathbf{x}}) = \frac{\partial\mathbf{D}}{\partial\mathbf{u}_{\mathbf{x}}}\delta\mathbf{u}_{\mathbf{x}} + \frac{\partial\mathbf{D}}{\partial\nabla\mathbf{u}_{\mathbf{x}}}\nabla\delta\mathbf{u}_{\mathbf{x}} = \frac{\partial\mathbf{D}}{\partial\mathbf{u}_{\mathbf{x}}}\boldsymbol{N}_{\mathbf{x}}\delta\boldsymbol{U}_{\mathbf{x}} + \frac{\partial\mathbf{D}}{\partial\nabla\mathbf{u}_{\mathbf{x}}}\boldsymbol{M}_{\mathbf{x}}\delta\boldsymbol{U}_{\mathbf{x}} \tag{1.69}$$

$$\delta\mathbf{F}(\mathbf{u}_{\mathbf{x}}, \nabla\mathbf{u}_{\mathbf{x}}) = \frac{\partial\mathbf{F}}{\partial\mathbf{u}_{\mathbf{x}}}\delta\mathbf{u}_{\mathbf{x}} + \frac{\partial\mathbf{F}}{\partial\nabla\mathbf{u}_{\mathbf{x}}}\nabla\delta\mathbf{u}_{\mathbf{x}} = \frac{\partial\mathbf{F}}{\partial\mathbf{u}_{\mathbf{x}}}\boldsymbol{N}_{\mathbf{x}}\delta\boldsymbol{U}_{\mathbf{x}} + \frac{\partial\mathbf{F}}{\partial\nabla\mathbf{u}_{\mathbf{x}}}\boldsymbol{M}_{\mathbf{x}}\delta\boldsymbol{U}_{\mathbf{x}}. \tag{1.70}$$

The tangent stiffness matrix can be extracted from $\delta^2 \Pi$ as

$$\begin{aligned} \boldsymbol{K}_g = & \sum_{\mathbf{x}\in\Omega} \Big(\boldsymbol{N}_{\mathbf{x}}^T (\frac{\partial \mathbf{C}}{\partial \mathbf{v}_{\mathbf{x}}})^T + \boldsymbol{M}_{\mathbf{x}}^T (\frac{\partial \mathbf{C}}{\partial \nabla \mathbf{v}_{\mathbf{x}}})^T \Big) \Big(\frac{\partial \mathbf{D}}{\partial \mathbf{u}_{\mathbf{x}}} \boldsymbol{N}_{\mathbf{x}} + \frac{\partial \mathbf{D}}{\partial \nabla \mathbf{u}_{\mathbf{x}}} \boldsymbol{M}_{\mathbf{x}} \Big) \Delta V_{\mathbf{x}} \\ & + \sum_{\mathbf{x}\in\partial\Omega} \Big(\boldsymbol{N}_{\mathbf{x}}^T \Big(\frac{\partial \mathbf{E}}{\partial \mathbf{v}_{\mathbf{x}}}\Big)^T + \boldsymbol{M}_{\mathbf{x}}^T \Big(\frac{\partial \mathbf{E}}{\partial \nabla \mathbf{v}_{\mathbf{x}}}\Big)^T \Big) \Big(\frac{\partial \mathbf{F}}{\partial \mathbf{u}_{\mathbf{x}}} \boldsymbol{N}_{\mathbf{x}} + \frac{\partial \mathbf{F}}{\partial \nabla \mathbf{u}_{\mathbf{x}}} \boldsymbol{M}_{\mathbf{x}} \Big) \Delta S_{\mathbf{x}}. \end{aligned} \tag{1.71}$$

It can be seen that the weighted residual method obtains the residual and global tangent stiffness matrix by the matrix multiplication. When the residual and tangent stiffness matrix of the linear/nonlinear problem are obtained, various numerical techniques in linear algebra can be used to find the solution.

1.5 Outline of the Book

The content of the book is based on several papers published. We have developed several numerical methods, such as dual-horizon peridynamics, dual-support smoothed particle hydrodynamics and nonlocal operator methods. All these methods are based on the dual concept of the support. Each of these methods represents a fundamental improvement over the conventional version. The rest of the chapters is listed as follows.

In Chap. 2, we derive the dual-horizon peridynamics. The new formulation derived from Newton's third law takes into account the direct force from the horizon domain and reaction force from the dual-horizon domain. We prove that the DH-PD satisfies the conservation of linear momentum and angular momentum. The dual-horizon peridynamics allows for an arbitrary horizon for each particle and the discretization can be nonuniform. Some numerical examples at the end of this chapter are presented to demonstrate the performance of the dual-horizon formulation of peridynamics.

In Chap. 3, we formally introduce the concept of support and dual-support and derive the nonlocal operator method, which is featured with three nonlocal operators, e.g. nonlocal gradient, nonlocal curl and nonlocal divergence. These nonlocal operators converge to the corresponding local ones when the support degenerates to one point. The operator matrices for these operators are derived explicitly. The residuals and tangent stiffness matrices of the four typical functionals based on the four common operators are studied. With the aid of variational principle, and dual-support concept, the nonlocal strong forms for diffusion equation, solid mechanics, Maxwell's equations are obtained. Some numerical examples including the 1D Schrodinger equation, Poisson equation, linear elasticity and nonlinear elasticity are validated with the exact solutions or numerical result by other methods.

In Chap. 4, the nonlocal operator method is developed in the electromagnetic field. We briefly reviewed the Maxwell equations and then derived the energy functional for the electric field. Based on the energy functional, we derived the residual and tangent stiffness matrix of Maxwell's equation. And then one waveguide problem is solved.

In Chap. 5, we extend the first order NOM to higher order NOM by using a higher order Taylor series expansion. The first order NOM is a special case of the higher order NOM. The higher order NOM can be derived by two methods, one is the weighted Taylor series expansion, the other is the minimization of operator energy functional for Taylor series expansion. For a special higher order quadratic functional, we derived the residual, tangent stiffness matrix and the nonlocal strong form. In the section of numerical examples, we applied the current method to problems formulated by strong form and weak form. These examples include the Poisson equation in 2–5 dimensional space, the Von Karman plate equations and the phase-field fracture model.

In Chap. 6, we develop the nonlocal operator method further by adopting numerical integration. The NOM in Chaps. 4, 5 are particle-based, which has some difficulty in enforcing boundary conditions accurately. The NOM with interpolation property can solve these problems when an accurate numerical integration scheme is adopted. These boundary conditions are formulated with modified variational principles. The gradient solid problem is solved by the current method. The numerical examples are compared with the exact solution, and good agreement is obtained.

In Chap. 7, the dual-support smoothed particle hydrodynamics are presented as a special case of the nonlocal operator method in Chap. 3. Several numerical examples in linear/nonlinear elasticity are presented.

In Chap. 8, we apply the variational principle/weighted residual method based on nonlocal operator method for the derivation of nonlocal forms for elasticity, thin plate, gradient elasticity, electro-magneto-elasticity and phase field fracture method. The nonlocal governing equations are expressed as integral forms of support and dual-support. The first example shows that the nonlocal elasticity has the same form as dual-horizon non-ordinary state-based peridynamics. The derivation is simple and general and it can efficiently convert many local physical models into their corresponding nonlocal forms. In addition, a criterion based on the instability of the nonlocal gradient is proposed for the fracture modelling in linear elasticity. Several numerical examples are presented to validate nonlocal elasticity and the nonlocal thin plate.

In Chap. 9, In this work, we present a nonlocal operator method (NOM) for dynamic fracture exploiting an explicit phase field model. The nonlocal strong forms of the phase field and the associated mechanical model are derived as integral forms by variational principle. The equations are decoupled and solved in time by an explicit scheme employing the Verlet-velocity algorithm for the mechanical field and an adaptive sub-step scheme for the phase field model. The sub-step scheme reduces phase field residual adaptively in a few sub-steps and thus achieves a rate-independent phase field model. The explicit scheme avoids the calculation of the anisotropic stiffness tensor in the implicit phase field model. One advantage of the NOM is its ease in implementation. The method does not require any shape functions and the associated matrices and vectors are obtained automatically after defining the energy of the system. Hence, the approach can be easily extended to more complex coupled problems. Several numerical examples are presented to demonstrate the performance of the current method.

In Chap. 10, we present a general finite deformation higher-order gradient elasticity theory. The governing equations of the higher-order gradient solid along with boundary conditions of various orders are derived from a variational principle using integration by parts on the surface. The objectivity of the energy functional is achieved by carefully selecting the invariants under rigid-body transformation. The third-order gradient solid theory includes more than 10.000 material parameters. However, under certain simplifications, the material parameters can be greatly reduced; down to 3. With this simplified formulation, we develop a nonlocal operator method and apply it to several numerical examples. The numerical analysis shows that the high gradient solid theory exhibits a stiffer response compared to a 'conventional' hyperelastic solid. The numerical tests also demonstrate the capability of the nonlocal operator method in solving higher-order physical problems.

In the appendices, a simple tutorial of Mathematica is provided, in which some frequently used functions in Mathematica are described and several useful programs in NOM are included. Meanwhile, the symmetry of higher order tensors is discussed.

References

Aghababaei R, Reddy J (2009) Nonlocal third-order shear deformation plate theory with application to bending and vibration of plates. J Sound Vib 326(1–2):277–289

Allahdadi F, Carney T, Hipp J, Libersky L, Petschek A (1993) High strain lagrangian hydrodynamics: a three dimensional SPH code for dynamic material response. Technical report, DTIC Document

Arroyo M, Ortiz M (2006) Local maximum-entropy approximation schemes: a seamless bridge between finite elements and meshfree methods. Int J Numer Meth Eng 65(13):2167–2202

Aydogdu M (2009) A general nonlocal beam theory: its application to nanobeam bending, buckling and vibration. Phys E 41(9):1651–1655

Babuška I (1973) The finite element method with Lagrangian multipliers. Numer Math 20(3):179–192

Babuška I (1973) The finite element method with penalty. Math Comput 27(122):221–228

Babuška I, Melenk JM (1997) The partition of unity method. Int J Numer Meth Eng 40(4):727–758

Barretta R, de Sciarra FM (2018) Constitutive boundary conditions for nonlocal strain gradient elastic nano-beams. Int J Eng Sci 130:187–198

Bažant ZP, Planas J (2019) Fracture and size effect in concrete and other quasibrittle materials. Routledge

Bažant Z, Jirásek M (2002) Nonlocal integral formulations of plasticity and damage: survey of progress. J Eng Mech 128(11):1119–1149

Becker R (2002) Mesh adaptation for dirichlet flow control via nitsche's method. Commun Numer Methods Eng 18(9):669–680

Belytschko T, Lu YY, Gu L (1994) Element-free Galerkin methods. Int J Numer Meth Eng 37(2):229–256

Belytschko T, Krongauz Y, Organ D, Fleming M, Krysl P (1996) Meshless methods: an overview and recent developments. Comput Methods Appl Mech Eng 139(1–4):3–47

Blom T (2014) Solidification using smoothed particle hydrodynamics

Bonet J, Kulasegaram S (2000) Correction and stabilization of smooth particle hydrodynamics methods with applications in metal forming simulations. Int J Numer Meth Eng 47(6):1189–1214

Brezzi F (1974) On the existence, uniqueness and approximation of saddle-point problems arising from Lagrangian multipliers. Publications mathématiques et informatique de Rennes S4:1–26

Chen J, Pan C, Roque C, Wang H (1998) A Lagrangian reproducing kernel particle method for metal forming analysis. Comput Mech 22(3):289–307

Chowdhury SR, Roy P, Roy D, Reddy J (2019) A modified peridynamics correspondence principle: Removal of zero-energy deformation and other implications. Comput Methods Appl Mech Eng 346:530–549

Chu Y, Moran B (1995) A computational model for nucleation of solid-solid phase transformations. Modell Simul Mater Sci Eng 3(4):455

Cleary P (1998) Modelling confined multi-material heat and mass flows using SPH. Appl Math Model 22(12):981–993

Dehnen W, Aly H (2012) Improving convergence in smoothed particle hydrodynamics simulations without pairing instability. Mon Not R Astron Soc 425(2):1068–1082

Duarte CA, Oden JT (1996) H-p clouds-an h-p meshless method. Numer Meth Partial Differ Equ Int J 12(6):673–705

Ebrahimi F, Barati MR, Dabbagh A (2016) A nonlocal strain gradient theory for wave propagation analysis in temperature-dependent inhomogeneous nanoplates. Int J Eng Sci 107:169–182

Eringen AC (1983) On differential equations of nonlocal elasticity and solutions of screw dislocation and surface waves. J Appl Phys 54(9):4703–4710. ISSN 0021-8979. https://doi.org/10.1063/1.332803

Eringen AC (2002) Nonlocal continuum field theories. Springer Science & Business Media

Eringen A (1972) Nonlocal polar elastic continua. Int J Eng Sci 10(1):1–16

Eringen AC, Edelen D (1972) On nonlocal elasticity. Int J Eng Sci 10(3):233–248

Fang J, Parriaux A, Rentschler M, Ancey C (2009) Improved SPH methods for simulating free surface flows of viscous fluids. Appl Numer Math 59(2):251–271

Farajpour A, Yazdi MH, Rastgoo A, Mohammadi M (2016) A higher-order nonlocal strain gradient plate model for buckling of orthotropic nanoplates in thermal environment. Acta Mech 227(7):1849–1867

Feng K (1965) Finite difference schemes based on variational principles. Appl. Math. Comput. Math 2:238–262

Ganzenmüller GC, Hiermaier S, May M (2015) On the similarity of meshless discretizations of peridynamics and smooth-particle hydrodynamics. Comput Struct 150:71–78

Gingold R, Monaghan J (1977) Smoothed particle hydrodynamics: theory and application to non-spherical stars. Mon Not R Astron Soc 181(3):375–389

Gürlebeck K, Hommel A (2003) Finite difference Cauchy-Riemann operators and their fundamental solutions in the complex case. In: Singular integral operators, factorization and applications, pp 101–115. Springer

Hansbo P, Larson MG (2002) Discontinuous Galerkin methods for incompressible and nearly incompressible elasticity by Nitsche's method. Comput Methods Appl Mech Eng 191(17–18):1895–1908

Hesse MB (1955) Action at a distance in classical physics. Isis 46(4):337–353

Hu X, Adams N (2006) A multi-phase SPH method for macroscopic and mesoscopic flows. J Comput Phys 213(2):844–861

Huerta A, Fernández-Méndez S (2000) Enrichment and coupling of the finite element and meshless methods. Int J Numer Meth Eng 48(11):1615–1636

Huerta A, Belytschko T, Fernández-Méndez S, Rabczuk T, Zhuang X, Arroyo M (2018) Meshfree methods. Encyclopedia of Computational Mechanics 2nd edn, pp 1–38

Jaynes ET (1957) Information theory and statistical mechanics. Phys Rev 106(4):620

Johnson G, Beissel S (1996) Normalized smoothing functions for SPH impact computations. Int J Numer Meth Eng 39(16):2725–2741

Juntunen M, Stenberg R (2009) Nitsche's method for general boundary conditions. Math Comput 78:1353–1374. ISSN 0025-5718. https://doi.org/10.1090/S0025-5718-08-02183-2

Kilic B, Madenci E (2010) Coupling of peridynamic theory and the finite element method. J Mech Mater Struct 5(5):707–733

Krongauz Y, Belytschko T (1997) Consistent pseudo-derivatives in meshless methods. Comput Methods Appl Mech Eng 146(3):371–386

Lancaster P, Salkauskas K (1981) Surfaces generated by moving least squares methods. Math Comput 37(155):141–158

Li L, Hu Y (2015) Buckling analysis of size-dependent nonlinear beams based on a nonlocal strain gradient theory. Int J Eng Sci 97:84–94

Li B, Habbal F, Ortiz M (2010) Optimal transportation meshfree approximation schemes for fluid and plastic flows. Int J Numer Meth Eng 83(12):1541–1579

Li P, Hao Z, Zhen W (2018) A stabilized non-ordinary state-based peridynamic model. Comput Methods Appl Mech Eng 339:262–280

Lim C, Zhang G, Reddy J (2015) A higher-order nonlocal elasticity and strain gradient theory and its applications in wave propagation. J Mech Phys Solids 78:298–313

Liszka T, Orkisz J (1980) The finite difference method at arbitrary irregular grids and its application in applied mechanics. Comput Struct 11(1):83–95. ISSN 0045-7949. https://doi.org/10.1016/0045-7949(80)90149-2

Liu W, Jun S, Zhang Y (1995) Reproducing kernel particle methods. Int J Numer Meth Fluids 20(8–9):1081–1106

Liu G, Liu M (2003) Smoothed particle hydrodynamics: a meshfree particle method. World Scientific

Lucy L (1977) A numerical approach to the testing of fusion processes. Astron J 82:1013–1024

Luo J, Sundararaghavan V (2018) Stress-point method for stabilizing zero–energy modes in non–ordinary state–based peridynamics. Int J Solids Struct

Macek RW, Silling SA (2007) Peridynamics via finite element analysis. Finite Elem Anal Des 43(15):1169–1178

Madenci E, Barut A, Dorduncu M (2019) Peridynamic differential operator for numerical analysis. Springer

Monaghan J (1988) An introduction to SPH. Comput Phys Commun 48(1):89–96

Monaghan J (1994) Simulating free surface flows with SPH. J Comput Phys 110(2):399–406

Monaghan J (1997) Implicit SPH drag and dusty gas dynamics. J Comput Phys 138(2):801–820

Monaghan J (2000) SPH without a tensile instability. J Comput Phys 159(2):290–311

Monaghan J, Huppert H, Worster M (2005) Solidification using smoothed particle hydrodynamics. J Comput Phys 206(2):684–705

Nayroles B, Touzot G, Villon P (1992) Generalizing the finite element method: diffuse approximation and diffuse elements. Comput Mech 10(5):307–318

Nguyen VP, Rabczuk T, Bordas S, Duflot M (2008) Meshless methods: a review and computer implementation aspects. Math Comput Simul 79(3):763–813

Nitsche J (1975) Allgemeinere randwertprobleme. In: Vorlesungen uber Minimalflachen, pp 431–535. Springer, Berlin, Heidelberg. https://doi.org/10.1007/978-3-642-65619-4_6

Ni T, Zaccariotto M, Zhu Q, Galvanetto U (2019) Coupling of fem and ordinary state-based peridynamics for brittle failure analysis in 3d. In: Mechanics of advanced materials and structures, pp 1–16

Oden JT, Pires E (1983) Nonlocal and nonlinear friction laws and variational principles for contact problems in elasticity

Pareschi L, Russo G (2000) Implicit-explicit Runge-Kutta schemes for stiff systems of differential equations. Recent Trends Numer Anal 3:269–289

Pasetto M, Leng Y, Chen J, Foster JT, Seleson P (2018) A reproducing kernel enhanced approach for peridynamic solutions. In: Computer methods in applied mechanics and engineering

Perrone N, Kao R (1975) A general finite difference method for arbitrary meshes. Comput Struct 5(1):45–5. ISSN 0045-7949. https://doi.org/10.1016/0045-7949(75)90018-8

Price D, Monaghan J (2004) Smoothed particle magnetohydrodynamics–i. algorithm and tests in one dimension. Mon Not R Astron Soc 348(1):123–138

Rabczuk T, Eibl J (2003) Simulation of high velocity concrete fragmentation using SPH/MLSPH. Int J Numer Meth Eng 56(10):1421–1444

Rabczuk T, Eibl J, Stempniewski L (2004) Numerical analysis of high speed concrete fragmentation using a meshfree Lagrangian method. Eng Fract Mech 71(4):547–556

Randles P, Libersky L (1996) Smoothed particle hydrodynamics: some recent improvements and applications. Comput Methods Appl Mech Eng 139(1–4):375–408

Reddy J (2007) Nonlocal theories for bending, buckling and vibration of beams. Int J Eng Sci 45(2–8):288–307

Rogula D (1982) Introduction to nonlocal theory of material media. In Nonlocal theory of material media, pp 123–222. Springer

Roy P, Roy D (2019) Peridynamics model for flexoelectricity and damage. Appl Math Model 68:82–112

Roy P, Deepu S, Pathrikar A, Roy D, Reddy J (2017) Phase field based peridynamics damage model for delamination of composite structures. Compos Struct 180:972–993

Sahmani S, Aghdam MM, Rabczuk T (2018) Nonlocal strain gradient plate model for nonlinear large-amplitude vibrations of functionally graded porous micro/nano-plates reinforced with gpls. Compos Struct 198:51–62

Silling S (2000) Reformulation of elasticity theory for discontinuities and long-range forces. J Mech Phys Solids 48(1):175–209. https://doi.org/10.2172/1895

Sukumar N (2004) Construction of polygonal interpolants: a maximum entropy approach. Int J Numer Meth Eng 61(12):2159–2181

Taflove A, Hagness SC (2005) Computational electrodynamics: the finite-difference time-domain method. Artech House

Versteeg H, Malalasekera W (2007) An introduction to computational fluid dynamics: the finite volume method. Pearson Education

Wagner GJ, Liu WK (2000) Application of essential boundary conditions in mesh-free methods: a corrected collocation method. Int J Numer Meth Eng 47(8):1367–1379

Warner F (2013) Foundations of differentiable manifolds and Lie groups, vol 94. Springer Science & Business Media

Wendland H (1995) Piecewise polynomial, positive definite and compactly supported radial functions of minimal degree. Adv Comput Math 4(1):389–396

Xu X-J, Zheng M-L, Wang X-C (2017) On vibrations of nonlocal rods: Boundary conditions, exact solutions and their asymptotics. Int J Eng Sci 119:217–231

Yaghoobi A, Chorzepa MG (2017) Higher-order approximation to suppress the zero-energy mode in non-ordinary state-based peridynamics. Comput Struct 188:63–79

Zhang Y, Liu G, Xie X (2005) Free transverse vibrations of double-walled carbon nanotubes using a theory of nonlocal elasticity. Phys Rev B 71(19):195404

Zhu T, Atluri S (1998) A modified collocation method and a penalty formulation for enforcing the essential boundary conditions in the element free galerkin method. Comput Mech 21(3):211–222

Zienkiewicz OC, Taylor RL, Nithiarasu P, Zhu J (1997) The finite element method, vol 3. McGraw-hill London

Chapter 2
Dual-Horizon Peridynamics

In PD, the crack is part of the solution and not part of the problem. No representation of the crack topology is needed. Hence, complex fracture patterns including crack branching and coalescence of multiple cracks are captured simply through the breakage of the bonds between particles. What all PD formulations have in common: the horizon sizes are required to be constant to avoid spurious wave reflections and ghost forces between particles. However, in many applications, the spatial distribution of the particles with changing horizon sizes is necessary for computational efficiency, e.g. for adaptive refinement, multiscale modeling and multibody analysis, to name a few. In other words, in order to achieve acceptable accuracy, the entire numerical model has to be discretized with respect to the highest particle resolution locally required, and the horizon size used accordingly. In this chapter, we propose a new PD formulation that removes the issue of varying horizons and ghost force. The new approach is based on the concept of horizon and dual-horizon. We confine the present work to solving solid mechanics problems. The content of the chapter includes the equation of motion of traditional peridynamics, the reason for the phenomenon of the ghost forces, the concept of the horizon and dual-horizon, and the derivation of the equation of motion for varying horizons. The balance of linear momentum and the balance of angular momentum of the present PD formulation are proved. The applications of the new formulation for BB PD, OSB PD and NOSB PD are described and some numerical examples are presented.

2.1 Conventional Peridynamics

Consider a solid in the initial and current configuration as shown in Fig. 2.1. Let $\mathbf{x}$ be material coordinates in the initial configuration Ω_0, and $\mathbf{y} := \mathbf{y}(\mathbf{x}, t)$ and $\mathbf{y}' := \mathbf{y}(\mathbf{x}', t)$ be the spatial coordinates in the current configuration Ω_t, respectively; $\boldsymbol{\xi} := \mathbf{x}' - \mathbf{x}$ is initial bond vector, the relative distance vector between $\mathbf{x}$ and $\mathbf{x}'$; $\mathbf{u} := \mathbf{u}(\mathbf{x}, t)$ and $\mathbf{u}' := \mathbf{u}(\mathbf{x}', t)$ are the displacement vectors for $\mathbf{x}$ and $\mathbf{x}'$, respectively; $\boldsymbol{\eta} := \mathbf{u}' - \mathbf{u}$ is

T. Rabczuk et al., *Computational Methods Based on Peridynamics and Nonlocal Operators*, Computational Methods in Engineering & the Sciences,
https://doi.org/10.1007/978-3-031-20906-2_2

Fig. 2.1 The configuration for deformed body

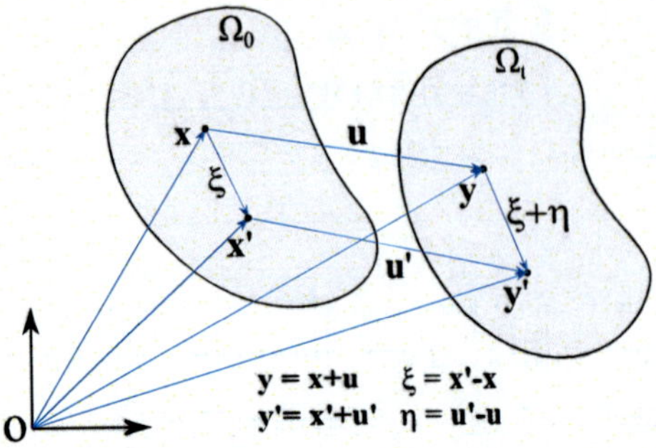

the relative displacement vector for bond $\boldsymbol{\xi}$; $\mathbf{y}\langle\boldsymbol{\xi}\rangle := \mathbf{y}(\mathbf{x}', t) - \mathbf{y}(\mathbf{x}, t) = \boldsymbol{\xi} + \boldsymbol{\eta}$ is the current bond vector for bond $\boldsymbol{\xi}$.

The horizon of point $\mathbf{x}$ in traditional PD represents a domain centered at $\mathbf{x}$, in which the distance from any point $\mathbf{x}'$ to $\mathbf{x}$ is smaller than δ, the radius of all horizons, as is shown in Fig. 2.2. It should be noted that the radius of the horizon for each particle in traditional PD must be the same. Otherwise, spurious wave reflections emerge, as will be shown in the next section.

The equation of motion at any time t for bond-based peridynamics is formulated as (Silling and Askari 2005)

$$\rho\ddot{\mathbf{u}}(\mathbf{x}, t) = \int_{H_{\mathbf{x}}} \mathbf{f}(\boldsymbol{\eta}, \boldsymbol{\xi})\, \mathrm{d}V_{\mathbf{x}'} + \mathbf{b}(\mathbf{x}, t) \tag{2.1}$$

or in discrete form by

$$\rho\ddot{\mathbf{u}}(\mathbf{x}, t) = \sum_{\mathbf{x}' \in H_{\mathbf{x}}} \mathbf{f}(\boldsymbol{\eta}, \boldsymbol{\xi})\, \Delta V_{\mathbf{x}'} + \mathbf{b}(\mathbf{x}, t) \tag{2.2}$$

where $H_{\mathbf{x}}$ denotes the horizon (spherical domain) belonging to $\mathbf{x}$, $\mathbf{b}(\mathbf{x}, t)$ denotes the body force and ρ is mass density in the reference configuration; $\mathbf{f}(\boldsymbol{\eta}, \boldsymbol{\xi})$ is a pairwise

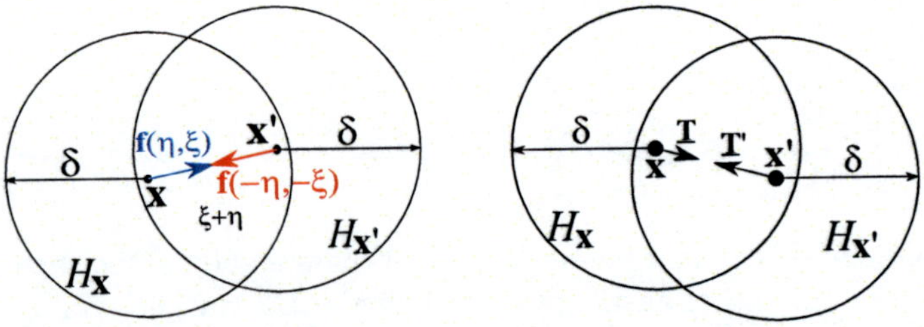

(a) Conventional BB-PD (b) Conventional SB-PD

Fig. 2.2 The bond force in conventional peridynamics

force function that computes the force vector (per unit volume squared), as shown in Fig. 2.2a, and it is assumed $\mathbf{f}(\boldsymbol{\eta}, \boldsymbol{\xi}) = -\mathbf{f}(-\boldsymbol{\eta}, -\boldsymbol{\xi})$. BB-PD actually assumes a spring link between any pair of particles. Equation (2.1) is commonly discretized by structured grids, unstructured grids and Voronoi cells where each particle is assigned a certain volume.

The pairwise force function $\mathbf{f}(\boldsymbol{\eta}, \boldsymbol{\xi})$ in conventional BB-PD is defined as

$$\mathbf{f}(\boldsymbol{\eta}, \boldsymbol{\xi}) = \frac{\partial w(\boldsymbol{\eta}, \boldsymbol{\xi})}{\partial \boldsymbol{\eta}} \qquad \forall \boldsymbol{\eta}, \boldsymbol{\xi} \tag{2.3}$$

where $w(\boldsymbol{\eta}, \boldsymbol{\xi})$ is the micro-potential energy in a single bond. For a microelastic homogeneous and isotropic material, $\mathbf{f}(\boldsymbol{\eta}, \boldsymbol{\xi})$ can be specified as

$$\mathbf{f}(\boldsymbol{\eta}, \boldsymbol{\xi}) = cs \cdot \frac{\boldsymbol{\eta} + \boldsymbol{\xi}}{\|\boldsymbol{\eta} + \boldsymbol{\xi}\|}, \tag{2.4}$$

where c is the micro-modulus, and s is the bond stretch calculated by

$$s = \frac{\|\boldsymbol{\eta} + \boldsymbol{\xi}\| - \|\boldsymbol{\xi}\|}{\|\boldsymbol{\xi}\|}. \tag{2.5}$$

In fact, $\mathbf{f}(\boldsymbol{\eta}, \boldsymbol{\xi})$ in BD-PD can be written as

$$\mathbf{f}(\boldsymbol{\eta}, \boldsymbol{\xi}) = \frac{1}{2}\mathbf{f}(\boldsymbol{\eta}, \boldsymbol{\xi}) - \frac{1}{2}\mathbf{f}(-\boldsymbol{\eta}, -\boldsymbol{\xi}). \tag{2.6}$$

The bond-based peridynamics in Eqs. (2.1) and (2.2) is valid only for a linear elastic micro-mechanics model. For problems with more general and complex material models, the concept of "state-based" peridynamics was introduced where the equation of motion is formulated as (Silling et al. 2007)

$$\rho\ddot{\mathbf{u}}(\mathbf{x}, t) = \int_{H_{\mathbf{x}}} \left(\underline{\mathbf{T}}[\mathbf{x}, t]\langle\mathbf{x}' - \mathbf{x}\rangle - \underline{\mathbf{T}}[\mathbf{x}', t]\langle\mathbf{x} - \mathbf{x}'\rangle\right) \mathrm{d}V_{\mathbf{x}'} + \mathbf{b}(\mathbf{x}, t). \tag{2.7}$$

And in discrete form

$$\rho\ddot{\mathbf{u}}(\mathbf{x}, t) = \sum_{\mathbf{x}' \in H_{\mathbf{x}}} \left(\underline{\mathbf{T}}[\mathbf{x}, t]\langle\mathbf{x}' - \mathbf{x}\rangle - \underline{\mathbf{T}}[\mathbf{x}', t]\langle\mathbf{x} - \mathbf{x}'\rangle\right) \Delta V_{\mathbf{x}'} + \mathbf{b}(\mathbf{x}, t). \tag{2.8}$$

The key difference between SB-PD and BB-PD is the use of $\underline{\mathbf{T}}[\mathbf{x}, t]\langle\mathbf{x}' - \mathbf{x}\rangle$ (abbreviated as $\underline{\mathbf{T}}$) acting on $\mathbf{x}$, which is known as the force state vector for SB-PD (Silling et al. 2007). Similarly, $\underline{\mathbf{T}}[\mathbf{x}', t]\langle\mathbf{x} - \mathbf{x}'\rangle$ (abbreviated as $\underline{\mathbf{T}}'$) is the force vector acting on $\mathbf{x}'$, as shown in Fig. 2.2b. When comparing Eq. (2.7) with Eq. (2.1), it can be seen that the main difference between BB-PD and SB-PD is the way in computing the bond force. The expression $\underline{\mathbf{T}}[\mathbf{x}, t]\langle\mathbf{x}' - \mathbf{x}\rangle - \underline{\mathbf{T}}[\mathbf{x}', t]\langle\mathbf{x} - \mathbf{x}'\rangle$ indicates that the net bond force density on $\mathbf{x}$ comprises two parts, the force state vector $\underline{\mathbf{T}}$ and the negative force state vector $-\underline{\mathbf{T}}'$. Equations (2.1) and (2.7) show the key idea of PD to unify continuous and discontinuous media within a single consistent set of equations.

2.2 Ghost Force and Spurious Wave Reflection in Peridynamics

In the traditional PD theory, the force exerted on a particle is the summation of all the pairwise forces from the particles falling inside its horizon. The horizon sizes must be set constant for all particles so that the bond forces are always pairwise. On the other hand, the sizes of the particles that are assigned with mass quantity can vary (Silling 2000; Silling et al. 2007). When the horizon sizes vary between particles, Eqs. (2.1) and (2.7) are no longer valid. We illustrate the reason for the phenomenon as follows.

For BB-PD, consider a particle $\mathbf{x}'$ falling inside the horizon of particle $\mathbf{x}$, see Fig. 2.3a, and as $\mathbf{x} \notin H_{\mathbf{x}'}$, $\mathbf{f}(-\boldsymbol{\eta}, -\boldsymbol{\xi})$ does not exist; however, as $\mathbf{x}' \in H_{\mathbf{x}}$, $\mathbf{f}(\boldsymbol{\eta}, \boldsymbol{\xi}) \neq 0$ still exists, which is known as the "ghost force" resulting from unbalanced internal force (Silling et al. 2014).

For SB-PD, consider a particle $\mathbf{x}'$ falling inside the horizon of particle $\mathbf{x}$; see Fig. 2.3b. Based on Eq. (2.7), as $\mathbf{x}' \in H_{\mathbf{x}}$, the net bond force density on $\mathbf{x}$ is $\underline{\mathbf{T}}[\mathbf{x}, t]\langle \mathbf{x}' - \mathbf{x}\rangle - \underline{\mathbf{T}}[\mathbf{x}', t]\langle \mathbf{x} - \mathbf{x}'\rangle \neq 0$. Likewise, when computing the net bond force density for particle $\mathbf{x}'$, $\underline{\mathbf{T}}[\mathbf{x}', t]\langle \mathbf{x} - \mathbf{x}'\rangle - \underline{\mathbf{T}}[\mathbf{x}, t]\langle \mathbf{x}' - \mathbf{x}\rangle$ does not exist as $\mathbf{x}$ is not inside the horizon of $\mathbf{x}'$. Consequently, the net bond force density only exists unilaterally. The balance of linear momentum and balance of angular momentum in this case are violated, and hence spurious wave reflections will occur in peridynamics simulations.

Efforts have been made by researchers to explore the possibility of making peridynamics suitable for nonuniform spatial discretization. Bobaru et al. studied the convergence and adaptive refinement in 1D peridynamics (Bobaru et al. 2009) and multiscale modeling in 2D BB-PD (Bobaru and Ha 2011). Dipasquale et al. recently (Dipasquale et al. 2014) introduced a trigger based on the damage state of the material

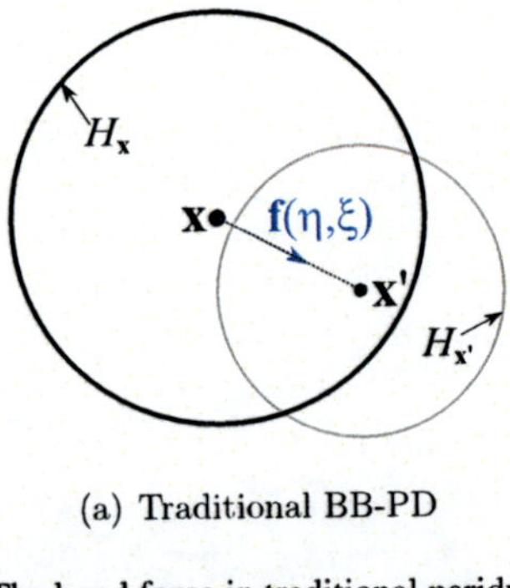

(a) Traditional BB-PD

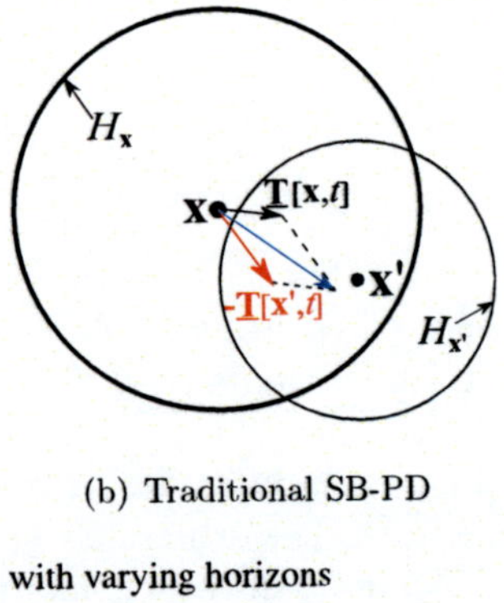

(b) Traditional SB-PD

Fig. 2.3 The bond force in traditional peridynamics with varying horizons for 2D refinement of BB-PD. Refined peridynamics was developed in Askari et al. (2008), Bobaru et al. (2009), Bobaru and Ha (2011), Dipasquale et al. (2014) and Yu et al. (2011), but it is restricted to BB-PD formulation and, to the authors' knowledge, has only been used for one- and 2D problems. In these works, the spurious wave reflection or ghost force problem is not solved, and the refinement is performed by checking the spurious reflection is within an acceptable range compared to the magnitude of the whole wave. Recently, the *partial stress* was proposed in Silling et al. (2014) to remove the ghost force for varying horizons. However, the method imposes certain restrictions on the deformation of the body and requires the computation of partial stress which is complicated. It impairs the simplicity of traditional peridynamics, especially for BB-PD and OSB-PD. Besides, the method does not completely remove the ghost force but with a small residual which is believed to be acceptable. Though the slice method was devised in the same work, it is applicable only to piecewise constant sizes of horizons, and requires additional computation to enforce the consistency between particles close to the interface.

2.3 Governing Equations Based on Horizon and Dual-Horizon

In this section, the concept of *horizon* and *dual-horizon* will be introduced to formulate the balance equations for particles with varying horizons. It is applied to all peridynamics formulations. Let $\mathbf{f}_{\mathbf{xx}'}(\boldsymbol{\eta}, \boldsymbol{\xi})$ denote the force vector density acting on $\mathbf{x}$ due to bond $\mathbf{xx}'$ in the horizon of $\mathbf{x}$. The projection of force state of $\mathbf{x}$ on bond $\boldsymbol{\xi}$ leads to $\mathbf{f}_{\mathbf{xx}'}$. Note that $\mathbf{f}_{\mathbf{xx}'}(\boldsymbol{\eta}, \boldsymbol{\xi})$ aims at replacing the notation $\underline{\mathbf{T}}[\mathbf{x}, t]\langle \mathbf{x}' - \mathbf{x} \rangle$ in conventional SB-PD. For BB-PD, based on Eq. (2.6), $\mathbf{f}_{\mathbf{xx}'}(\boldsymbol{\eta}, \boldsymbol{\xi})$ replaces $\frac{1}{2}\mathbf{f}(\boldsymbol{\eta}, \boldsymbol{\xi})$. More comparisons of the notations will be presented in Sect. 2.3.2. In the next subsections, we will derive the governing equations in a discrete form for our DH-PD formulation and compare it to the traditional PD, which will be then written into an integral form to compare with the conventional peridynamics.

2.3.1 Horizon and Dual-Horizon

We will begin by restating the original concept of horizon in peridynamics. In peridynamics theory, the particles interact with each other within a finite distance. A particle is considered to have an influence over other particles within a small domain centering that particle called "horizon". The radius of the domain is known as the radius of "horizon"; see Fig. 2.2. Note that constant-horizon radii for all particles is a compulsory condition of traditional peridynamics. For discretizations with varying horizons, a single horizon is not sufficient to define the interactions between particles, and the new concept of "horizon" and "dual-horizon" is introduced subsequently.

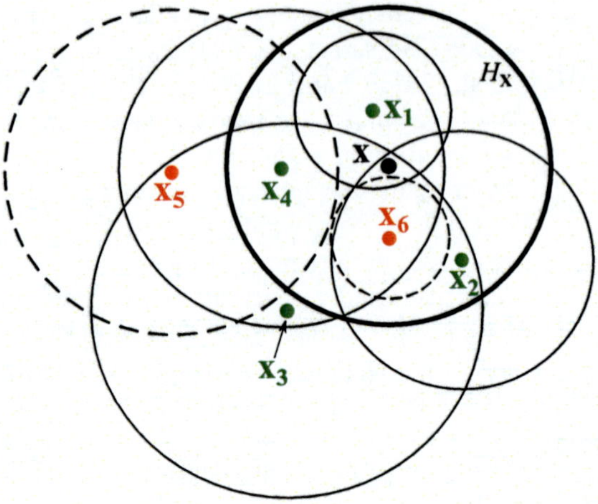

Fig. 2.4 Schematic diagram for horizon and dual-horizon, all circles above are horizons. The green points $\{\mathbf{x}_1, \mathbf{x}_2, \mathbf{x}_3, \mathbf{x}_4\} \in H'_{\mathbf{x}}$, whose horizons are denoted by thin solid line; the red points $\{\mathbf{x}_5, \mathbf{x}_6\} \notin H'_{\mathbf{x}}$, whose horizons are denoted by dashed line

Horizon

The *horizon* $H_{\mathbf{x}}$ is the domain where any particle $\mathbf{x}'$ inside forms bond $\mathbf{xx}'$. The bond $\mathbf{xx}'$ exerts direct force density $\mathbf{f}_{\mathbf{xx}'}$ on $\mathbf{x}$ and reaction force density $-\mathbf{f}_{\mathbf{xx}'}$ on $\mathbf{x}'$. For example, as $\{\mathbf{x}_1, \mathbf{x}_2, \mathbf{x}_4, \mathbf{x}_6\} \in H_{\mathbf{x}}$ as shown in Fig. 2.4, $\mathbf{x}$ will undertake direct forces from bonds $\mathbf{xx}_1$, $\mathbf{xx}_2$, $\mathbf{xx}_4$ and $\mathbf{xx}_6$. The horizon can be viewed as a direct force horizon. The reaction force acting on $\mathbf{x}'$ follows Newton's third law. As shown in Fig. 2.5, the reaction force $-\mathbf{f}_{\mathbf{xx}'}(\boldsymbol{\eta}, \boldsymbol{\xi})$ exerted on $\mathbf{x}'$ by $\mathbf{x}$ is in the opposite direction of the forces applied on $\mathbf{x}$ due to bond $\mathbf{xx}'$.

Dual-horizon

Dual-horizon is defined as a union of the points whose horizons include $\mathbf{x}$, denoted by

$$H'_{\mathbf{x}} = \{\mathbf{x}' | \mathbf{x} \in H_{\mathbf{x}'}\}. \tag{2.9}$$

In the notation of dual-horizon $H'_{\mathbf{x}}$, the superscript prime indicates "dual", and the subscript $\mathbf{x}$ denotes the object particle. It can be understood as the set of horizons that belong to the particles who can "see" $\mathbf{x}$ in their horizons. As shown in Fig. 2.4, the dual-horizon with respect to $\mathbf{x}$ contains particles $\mathbf{x}_1$, $\mathbf{x}_2$, $\mathbf{x}_3$ and $\mathbf{x}_4$, whose horizons are denoted by thin solid circles. Particles $\mathbf{x}_5$ and $\mathbf{x}_6$ are not included in the dual-horizon of $\mathbf{x}$ since their horizons do not include $\mathbf{x}$. In this case, $\mathbf{x}$ becomes the object "observed" by the other particles. If $\mathbf{x}$ is within the horizon of $\mathbf{x}'$, then $\mathbf{x}'$ has an effect on $\mathbf{x}$, corresponding to the passive or reaction effect in the horizon defined previously. Particle $\mathbf{x}$ receives the reaction forces from other particles in the *dual-horizon* of $\mathbf{x}$, and in this sense it is considered as "dual" corresponding to *horizon*. For any point $\mathbf{x}$, the shape of $H'_{\mathbf{x}}$ is arbitrary, and depends on the sizes and shapes of horizons as well as the locations of the particles. Note that the horizon can take other shapes than a circle or sphere.

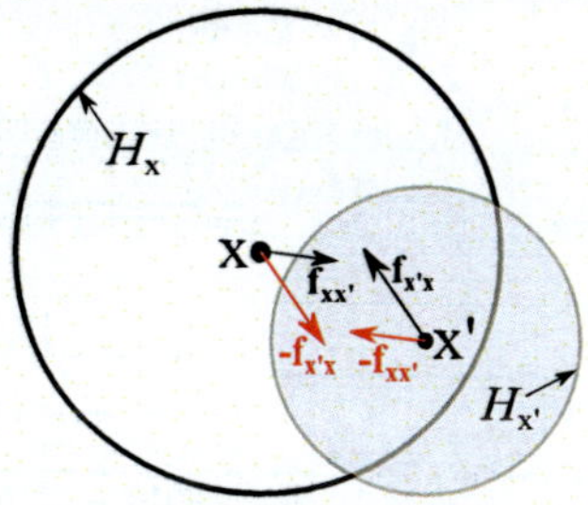

Fig. 2.5 Force vector in peridynamics with varying horizons, where $\mathbf{f}_{\mathbf{x}'\mathbf{x}} = \mathbf{f}_{\mathbf{x}'\mathbf{x}}(-\boldsymbol{\eta}, -\boldsymbol{\xi})$, $\mathbf{f}_{\mathbf{xx}'} = \mathbf{f}_{\mathbf{xx}'}(\boldsymbol{\eta}, \boldsymbol{\xi})$. Reaction force $-\mathbf{f}_{\mathbf{x}'\mathbf{x}}$ on $\mathbf{x}$ due to bond $\mathbf{x}'\mathbf{x}$ exerting direct force $\mathbf{f}_{\mathbf{x}'\mathbf{x}}$ on $\mathbf{x}'$ as $\mathbf{x}' \in H_{\mathbf{x}}$

The bond inside the dual-horizon ($H'_{\mathbf{x}}$) is termed as "dual-bond", and this is corresponding to the bond in the horizon ($H_{\mathbf{x}'}$). It can be seen that the bond of one particle can become the dual-bond for another particle interacting with it.

For models with constant horizons, as $\mathbf{x}' \in H_{\mathbf{x}} \Leftrightarrow \mathbf{x} \in H_{\mathbf{x}'}$, $H'_{\mathbf{x}}$ is equal to $H_{\mathbf{x}}$ and therefore horizon and dual-horizon degenerate to the horizon in traditional peridynamics.

The concept of two horizons solves the horizon variable issue simply and directly by taking into account the interactions between particles of varying horizon sizes. For example, in the framework of dual-horizon peridynamics, the forces between particle $\mathbf{x}$ and $\mathbf{x}'$ with varying horizons are shown in Fig. 2.5. As particle $\mathbf{x}' \in H_{\mathbf{x}}$, there is a force $\mathbf{f}_{\mathbf{xx}'}(\boldsymbol{\eta}, \boldsymbol{\xi}) \neq 0$ on bond $\mathbf{xx}'$ due to the force state on $\mathbf{x}$. We can say $\mathbf{x}$ undertakes a direct force $\mathbf{f}_{\mathbf{xx}'}(\boldsymbol{\eta}, \boldsymbol{\xi})$ from horizon $H_{\mathbf{x}}$ of $\mathbf{x}$, and $\mathbf{x}'$ undertakes a reaction force $-\mathbf{f}_{\mathbf{xx}'}(\boldsymbol{\eta}, \boldsymbol{\xi})$ from its dual-horizon $H'_{\mathbf{x}'}$; likewise, as $\mathbf{x} \notin H_{\mathbf{x}'}$, $\mathbf{f}_{\mathbf{x}'\mathbf{x}}(-\boldsymbol{\eta}, -\boldsymbol{\xi}) = 0$. The net force density on $\mathbf{x}$ is $\mathbf{f}_{\mathbf{xx}'}(\boldsymbol{\eta}, \boldsymbol{\xi}) - \mathbf{f}_{\mathbf{x}'\mathbf{x}}(-\boldsymbol{\eta}, -\boldsymbol{\xi}) = \mathbf{f}_{\mathbf{xx}'}(\boldsymbol{\eta}, \boldsymbol{\xi})$, and the net force density on $\mathbf{x}'$ is $\mathbf{f}_{\mathbf{x}'\mathbf{x}}(-\boldsymbol{\eta}, -\boldsymbol{\xi}) - \mathbf{f}_{\mathbf{xx}'}(\boldsymbol{\eta}, \boldsymbol{\xi}) = -\mathbf{f}_{\mathbf{xx}'}(\boldsymbol{\eta}, \boldsymbol{\xi})$. $\mathbf{f}_{\mathbf{xx}'}(\boldsymbol{\eta}, \boldsymbol{\xi}) - \mathbf{f}_{\mathbf{x}'\mathbf{x}}(-\boldsymbol{\eta}, -\boldsymbol{\xi}) = -(\mathbf{f}_{\mathbf{x}'\mathbf{x}}(-\boldsymbol{\eta}, -\boldsymbol{\xi}) - \mathbf{f}_{\mathbf{xx}'}(\boldsymbol{\eta}, \boldsymbol{\xi}))$ shows the antisymmetry of net force density, therefore no ghost force exists.

Forces by horizon and dual-horizon

In dual-horizon peridynamics formulation, the horizons are differentiated between how a particle receives and exerts forces with other particles. Under this new concept, computing the force density between a pair of particles, denoted as particles $\mathbf{x}$ and $\mathbf{x}'$, is determined by whether $\mathbf{x}'$ is within the horizon of $\mathbf{x}$, and whether $\mathbf{x}'$ is inside the dual-horizon of $\mathbf{x}$. It can be easily seen that the force density vector has the following property:

$$\begin{aligned} \text{if} \quad \mathbf{x}' \in H_{\mathbf{x}} \text{ or } \mathbf{x} \in H'_{\mathbf{x}'}, \ \mathbf{f}_{\mathbf{xx}'}(\boldsymbol{\eta}, \boldsymbol{\xi}) \neq 0\,, \\ \text{else} \quad \mathbf{f}_{\mathbf{xx}'}(\boldsymbol{\eta}, \boldsymbol{\xi}) = 0\,. \end{aligned} \tag{2.10}$$

For any bond between two particles $\mathbf{x}$ and $\mathbf{x}'$ belonging to a domain denoted as Ω in Fig. 2.6, the direct force $\mathbf{f}_{\mathbf{xx}'}(\boldsymbol{\eta}, \boldsymbol{\xi})$ acting on $\mathbf{x}$ due to bond $\mathbf{xx}'$ can be computed by two approaches as follows.

Approach 1 computes the force in terms of $\mathbf{x}$,

Fig. 2.6 Double summation of force, where $\mathbf{f}_{\mathbf{x}'\mathbf{x}} = \mathbf{f}_{\mathbf{x}'\mathbf{x}}(-\boldsymbol{\eta}, -\boldsymbol{\xi})$, $\mathbf{f}_{\mathbf{x}\mathbf{x}'} = \mathbf{f}_{\mathbf{x}\mathbf{x}'}(\boldsymbol{\eta}, \boldsymbol{\xi})$

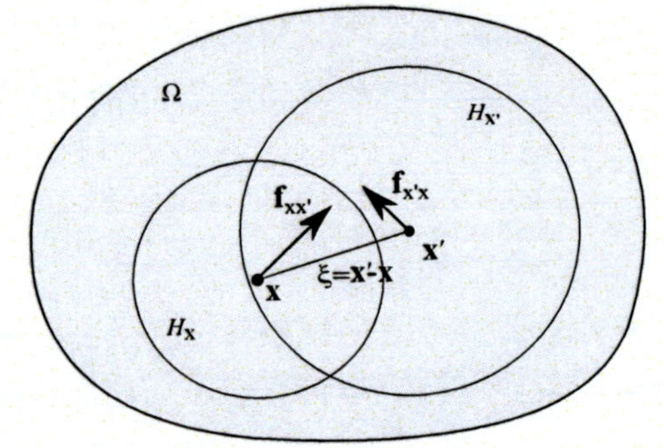

$$\mathbf{f}_{\mathbf{x}\mathbf{x}'}(\boldsymbol{\eta}, \boldsymbol{\xi}) = \begin{cases} \neq 0 & \text{if } \mathbf{x}' \in H_\mathbf{x} \\ 0 & \text{if } \mathbf{x}' \notin H_\mathbf{x} \end{cases},$$

and **Approach 2** is formulated with respect to $\mathbf{x}'$,

$$\mathbf{f}_{\mathbf{x}\mathbf{x}'}(\boldsymbol{\eta}, \boldsymbol{\xi}) = \begin{cases} \neq 0 & \text{if } \mathbf{x} \in H'_{\mathbf{x}'} \\ 0 & \text{if } \mathbf{x} \notin H'_{\mathbf{x}'} \end{cases}.$$

For any domain Ω in Fig. 2.6, the computation of the direct forces that take place between particles (no reaction force considered yet) shall undertake all forces from any particle that belongs to Ω. There are two approaches to achieving that. The first approach is by summing all the forces the particles undertake in the horizon,

$$\sum_{\mathbf{x}\in\Omega} \left(\sum_{\mathbf{x}'\in H_\mathbf{x}} \mathbf{f}_{\mathbf{x}\mathbf{x}'}(\boldsymbol{\eta}, \boldsymbol{\xi})\, \Delta V_{\mathbf{x}'} \right) \Delta V_\mathbf{x}. \tag{2.11}$$

And the second is to add up all the forces that one particle applies to other particles in its dual-horizon,

$$\sum_{\mathbf{x}\in\Omega} \left(\sum_{\mathbf{x}'\in H'_\mathbf{x}} -\mathbf{f}_{\mathbf{x}'\mathbf{x}}(-\boldsymbol{\eta}, -\boldsymbol{\xi})\, \Delta V_{\mathbf{x}'} \right) \Delta V_\mathbf{x}\,. \tag{2.12}$$

Since the total force for any Ω is independent of the approach chosen to compute it, Eqs. (2.11) and (2.12) shall be opposite

$$\sum_{\mathbf{x}\in\Omega} \sum_{\mathbf{x}'\in H_\mathbf{x}} \mathbf{f}_{\mathbf{x}\mathbf{x}'}(\boldsymbol{\eta}, \boldsymbol{\xi})\, \Delta V_{\mathbf{x}'} \Delta V_\mathbf{x} = -\sum_{\mathbf{x}\in\Omega} \sum_{\mathbf{x}'\in H'_\mathbf{x}} -\mathbf{f}_{\mathbf{x}'\mathbf{x}}(-\boldsymbol{\eta}, -\boldsymbol{\xi})\, \Delta V_{\mathbf{x}'} \Delta V_\mathbf{x}\,. \tag{2.13}$$

When the discretization is sufficiently fine, the summation shall approximate the integral and Eq. (2.13) becomes

$$\int_{\mathbf{x}\in\Omega}\int_{\mathbf{x}'\in H_{\mathbf{x}}}\mathbf{f}_{\mathbf{x}\mathbf{x}'}(\boldsymbol{\eta},\boldsymbol{\xi})\,\mathrm{d}V_{\mathbf{x}'}\mathrm{d}V_{\mathbf{x}} = \int_{\mathbf{x}\in\Omega}\int_{\mathbf{x}'\in H'_{\mathbf{x}}}\mathbf{f}_{\mathbf{x}'\mathbf{x}}(-\boldsymbol{\eta},-\boldsymbol{\xi})\,\mathrm{d}V_{\mathbf{x}'}\mathrm{d}V_{\mathbf{x}}\,. \tag{2.14}$$

Equation (2.14) indicates that the global summation of terms in dual-horizon is equal to the global summation of terms in the horizon.

2.3.2 *Equation of Motion for Peridynamics with Horizon Variable*

In the following derivation, $\mathbf{x}'-\mathbf{x}$ will be denoted as $\boldsymbol{\xi}$, and $\mathbf{u}(\mathbf{x}',t)-\mathbf{u}(\mathbf{x},t)$ as $\boldsymbol{\eta}$, hence $\boldsymbol{\xi}+\boldsymbol{\eta}$ represents the current relative position vector between the particles. The internal forces that are exerted on each particle from the other particles should include two parts, namely the forces from the horizon and the forces from the dual-horizon. The other forces applied to a particle include the body force and the inertial force. Let $\Delta V_{\mathbf{x}}$ denote the volume associated with $\mathbf{x}$. The body force for particle $\mathbf{x}$ can be expressed as $\mathbf{b}(\mathbf{x},t)\Delta V_{\mathbf{x}}$, where $\mathbf{b}(\mathbf{x},t)$ is the body force density. The inertia is denoted by $\rho\ddot{\mathbf{u}}(\mathbf{x},t)\Delta V_{\mathbf{x}}$, where ρ is the density associated with $\mathbf{x}$. At any time t, for $\mathbf{x}'$ in the horizon of $\mathbf{x}$, the force vector of $\tilde{\mathbf{f}}_{\mathbf{x}\mathbf{x}'}$ is defined as

$$\tilde{\mathbf{f}}_{\mathbf{x}\mathbf{x}'} := \mathbf{f}_{\mathbf{x}\mathbf{x}'}(\boldsymbol{\eta},\boldsymbol{\xi})\cdot\Delta V_{\mathbf{x}}\cdot\Delta V_{\mathbf{x}'}\,, \tag{2.15}$$

where $\mathbf{f}_{\mathbf{x}\mathbf{x}'}(\boldsymbol{\eta},\boldsymbol{\xi})$ is the force density in the conventional peridynamics with a unit of force per volume squared; $\tilde{\mathbf{f}}_{\mathbf{x}\mathbf{x}'}$ is the force vector acting on particle $\mathbf{x}$ due to bond $\mathbf{x}\mathbf{x}'$. The forces from the horizon are direct forces as they are applied to $\mathbf{x}$. The total force applied to $\mathbf{x}$ from its horizon, $H_{\mathbf{x}}$, can be computed by

$$\sum_{\mathbf{x}'\in H_{\mathbf{x}}}\tilde{\mathbf{f}}_{\mathbf{x}\mathbf{x}'} = \sum_{\mathbf{x}'\in H_{\mathbf{x}}}\mathbf{f}_{\mathbf{x}\mathbf{x}'}(\boldsymbol{\eta},\boldsymbol{\xi})\cdot\Delta V_{\mathbf{x}}\cdot\Delta V_{\mathbf{x}'}\,. \tag{2.16}$$

For any $\mathbf{x}'$ inside the dual-horizon of $\mathbf{x}$, the force $\tilde{\mathbf{f}}_{\mathbf{x}'\mathbf{x}}$ acting on $\mathbf{x}'$ due to bond $\mathbf{x}'\mathbf{x}$ is defined as

$$\tilde{\mathbf{f}}_{\mathbf{x}'\mathbf{x}} := \mathbf{f}_{\mathbf{x}'\mathbf{x}}(-\boldsymbol{\eta},-\boldsymbol{\xi})\cdot\Delta V_{\mathbf{x}'}\cdot\Delta V_{\mathbf{x}} \tag{2.17}$$

where $\mathbf{f}_{\mathbf{x}'\mathbf{x}}(-\boldsymbol{\eta},-\boldsymbol{\xi})$ is the force density per volume squared; $\tilde{\mathbf{f}}_{\mathbf{x}'\mathbf{x}}$ is the force vector acting on particle $\mathbf{x}'$ due to bond $\mathbf{x}'\mathbf{x}$ ($\mathbf{x}'$ inside dual-horizon of $\mathbf{x}$). The total force on $\mathbf{x}$ from dual-horizon $H'_{\mathbf{x}}$ is the summation of reaction force $-\tilde{\mathbf{f}}_{\mathbf{x}'\mathbf{x}}$ (i.e. based on Newton's third law)

$$-\sum_{\mathbf{x}'\in H'_{\mathbf{x}}}\tilde{\mathbf{f}}_{\mathbf{x}'\mathbf{x}} = \sum_{\mathbf{x}'\in H'_{\mathbf{x}}}-\mathbf{f}_{\mathbf{x}'\mathbf{x}}(-\boldsymbol{\eta},-\boldsymbol{\xi})\cdot\Delta V_{\mathbf{x}}\cdot\Delta V_{\mathbf{x}'}. \tag{2.18}$$

By summing over all forces on particle **x**, including inertial force, body force, direct force in Eq. (2.16) and reaction force in Eq. (2.18), we obtain the equation of motion

$$\rho\ddot{\mathbf{u}}(\mathbf{x},t)\Delta V_{\mathbf{x}} = \sum_{\mathbf{x}'\in H_{\mathbf{x}}} \tilde{\mathbf{f}}_{\mathbf{x}\mathbf{x}'} + \left(\sum_{\mathbf{x}'\in H'_{\mathbf{x}}} -\tilde{\mathbf{f}}_{\mathbf{x}'\mathbf{x}}\right) + \mathbf{b}(\mathbf{x},t)\Delta V_{\mathbf{x}} \,. \tag{2.19}$$

Substituting Eqs. (2.16) and (2.18) into Eq. (2.19) leads to

$$\rho\ddot{\mathbf{u}}(\mathbf{x},t)\Delta V_{\mathbf{x}} = \sum_{\mathbf{x}'\in H_{\mathbf{x}}} \mathbf{f}_{\mathbf{x}\mathbf{x}'}(\boldsymbol{\eta},\boldsymbol{\xi})\Delta V_{\mathbf{x}'}\Delta V_{\mathbf{x}} - \sum_{\mathbf{x}'\in H'_{\mathbf{x}}} \mathbf{f}_{\mathbf{x}'\mathbf{x}}(-\boldsymbol{\eta},-\boldsymbol{\xi})\Delta V_{\mathbf{x}'}\Delta V_{\mathbf{x}} + \mathbf{b}(\mathbf{x},t)\Delta V_{\mathbf{x}} \,. \tag{2.20}$$

As the volume $\Delta V_{\mathbf{x}}$ associated with particle **x** is independent of the summation, we can eliminate $\Delta V_{\mathbf{x}}$ in Eq. (2.20), yielding the governing equation based on **x**:

$$\rho\ddot{\mathbf{u}}(\mathbf{x},t) = \sum_{\mathbf{x}'\in H_{\mathbf{x}}} \mathbf{f}_{\mathbf{x}\mathbf{x}'}(\boldsymbol{\eta},\boldsymbol{\xi})\Delta V_{\mathbf{x}'} - \sum_{\mathbf{x}'\in H'_{\mathbf{x}}} \mathbf{f}_{\mathbf{x}'\mathbf{x}}(-\boldsymbol{\eta},-\boldsymbol{\xi})\Delta V_{\mathbf{x}'} + \mathbf{b}(\mathbf{x},t) \,. \tag{2.21}$$

When the discretization is sufficiently fine, the summation is approximating the integration of the force on the dual-horizon and horizon,

$$\lim_{\Delta V_{\mathbf{x}'}\to 0} \sum_{\mathbf{x}'\in H_{\mathbf{x}}} \mathbf{f}_{\mathbf{x}\mathbf{x}'}(\boldsymbol{\eta},\boldsymbol{\xi})\Delta V_{\mathbf{x}'} = \int_{\mathbf{x}'\in H_{\mathbf{x}}} \mathbf{f}_{\mathbf{x}\mathbf{x}'}(\boldsymbol{\eta},\boldsymbol{\xi})\,\mathrm{d}V_{\mathbf{x}'} \tag{2.22}$$

and

$$\lim_{\Delta V_{\mathbf{x}'}\to 0} \sum_{\mathbf{x}'\in H'_{\mathbf{x}}} \mathbf{f}_{\mathbf{x}'\mathbf{x}}(-\boldsymbol{\eta},-\boldsymbol{\xi})\Delta V_{\mathbf{x}'} = \int_{\mathbf{x}'\in H'_{\mathbf{x}}} \mathbf{f}_{\mathbf{x}'\mathbf{x}}(-\boldsymbol{\eta},-\boldsymbol{\xi})\,\mathrm{d}V_{\mathbf{x}'} \,. \tag{2.23}$$

Thus, the integral form of the equation of motion in dual-horizon peridynamics (DH-PD) is given as

$$\rho\ddot{\mathbf{u}}(\mathbf{x},t) = \int_{\mathbf{x}'\in H_{\mathbf{x}}} \mathbf{f}_{\mathbf{x}\mathbf{x}'}(\boldsymbol{\eta},\boldsymbol{\xi})\,\mathrm{d}V_{\mathbf{x}'} - \int_{\mathbf{x}'\in H'_{\mathbf{x}}} \mathbf{f}_{\mathbf{x}'\mathbf{x}}(-\boldsymbol{\eta},-\boldsymbol{\xi})\,\mathrm{d}V_{\mathbf{x}'} + \mathbf{b}(\mathbf{x},t) \,. \tag{2.24}$$

Note that Eq. (2.24) is also valid for particles near the boundary area.

When the horizons are set constant, as $\mathbf{x}' \in H'_{\mathbf{x}} \Leftrightarrow \mathbf{x}' \in H_{\mathbf{x}}$, we have $H'_{\mathbf{x}} = H_{\mathbf{x}}$ and Eq. (2.24) degenerates to

$$\rho\ddot{\mathbf{u}}(\mathbf{x},t) = \int_{\mathbf{x}'\in H_{\mathbf{x}}} \mathbf{f}_{\mathbf{x}\mathbf{x}'}(\boldsymbol{\eta},\boldsymbol{\xi}) - \mathbf{f}_{\mathbf{x}'\mathbf{x}}(-\boldsymbol{\eta},-\boldsymbol{\xi})\,\mathrm{d}V_{\mathbf{x}'} + \mathbf{b}(\mathbf{x},t) \,. \tag{2.25}$$

For bond-based PD, according to Eq. (2.6), the pairwise function $\mathbf{f}(\boldsymbol{\eta}, \boldsymbol{\xi}) = \frac{1}{2}\mathbf{f}(\boldsymbol{\eta}, \boldsymbol{\xi}) - \frac{1}{2}\mathbf{f}(-\boldsymbol{\eta}, -\boldsymbol{\xi})$. When $\mathbf{f}(\boldsymbol{\eta}, \boldsymbol{\xi})$ in Eq. (2.1) is interpreted as the net force density of bond and dual-bond in DH-PD, in other words, $\frac{1}{2}\mathbf{f}(\boldsymbol{\eta}, \boldsymbol{\xi}) \rightarrow \mathbf{f}_{\mathbf{x}\mathbf{x}'}(\boldsymbol{\eta}, \boldsymbol{\xi})$ and $-\frac{1}{2}\mathbf{f}(-\boldsymbol{\eta}, -\boldsymbol{\xi}) \rightarrow -\mathbf{f}_{\mathbf{x}'\mathbf{x}}(-\boldsymbol{\eta}, -\boldsymbol{\xi})$, Eq. (2.1) turns into Eq. (2.25). When $\underline{\mathbf{T}}[\mathbf{x}, t]\langle \mathbf{x}' - \mathbf{x}\rangle$ in Eq. (2.7) is interpreted as the force density vector acting on $\mathbf{x}$ due to bond $\mathbf{x}'$, and $-\underline{\mathbf{T}}[\mathbf{x}', t]\langle \mathbf{x} - \mathbf{x}'\rangle$ as the reaction force acting on $\mathbf{x}$, in other words, $\underline{\mathbf{T}}[\mathbf{x}, t]\langle \mathbf{x}' - \mathbf{x}\rangle \rightarrow \mathbf{f}_{\mathbf{x}\mathbf{x}'}(\boldsymbol{\eta}, \boldsymbol{\xi})$ and $-\underline{\mathbf{T}}[\mathbf{x}', t]\langle \mathbf{x} - \mathbf{x}'\rangle \rightarrow -\mathbf{f}_{\mathbf{x}'\mathbf{x}}(-\boldsymbol{\eta}, -\boldsymbol{\xi})$, Eq. (2.7) turns into Eq. (2.25). Based on the new notation of DH-PD, all three types of traditional peridynamics are expressed by Eq. (2.25). Therefore, the traditional peridynamics is a special case of the present dual-horizon peridynamics.

For the implementation of the present peridynamics formulation, for any particle $\mathbf{x}$, the force density $\mathbf{f}_{\mathbf{x}'\mathbf{x}}(-\boldsymbol{\eta}, -\boldsymbol{\xi})$ in $H'_{\mathbf{x}}$ can be determined when calculating the force in $H_{\mathbf{x}'}$ for $\mathbf{x}'$. Therefore, it is not necessary to know exactly the dual-horizon geometry, and the formulation can be implemented with minor modification of any peridynamics codes.

2.3.3 *Proof of Basic Physical Principles*

Balance of linear momentum

The internal forces shall satisfy the balance of linear momentum for any bounded body Ω given in discrete form by

$$
\begin{aligned}
&\sum_{\Omega}(\rho\ddot{\mathbf{u}}(\mathbf{x}, t) - \mathbf{b}(\mathbf{x}, t))\Delta V_{\mathbf{x}} \\
&= \sum_{\Omega}\sum_{\mathbf{x}'\in H_{\mathbf{x}}} \mathbf{f}_{\mathbf{x}\mathbf{x}'}(\boldsymbol{\eta}, \boldsymbol{\xi})\,\Delta V_{\mathbf{x}'}\Delta V_{\mathbf{x}} - \sum_{\Omega}\sum_{\mathbf{x}'\in H'_{\mathbf{x}}} \mathbf{f}_{\mathbf{x}'\mathbf{x}}(-\boldsymbol{\eta}, -\boldsymbol{\xi})\,\Delta V_{\mathbf{x}'}\Delta V_{\mathbf{x}} \\
&= \mathbf{0}
\end{aligned}
\tag{2.26}
$$

or integral form

$$
\begin{aligned}
&\int_{\Omega}(\rho\ddot{\mathbf{u}}(\mathbf{x}, t) - \mathbf{b}(\mathbf{x}, t))\mathrm{d}V_{\mathbf{x}} \\
&= \int_{\Omega}\int_{\mathbf{x}'\in H_{\mathbf{x}}} \mathbf{f}_{\mathbf{x}\mathbf{x}'}(\boldsymbol{\eta}, \boldsymbol{\xi})\,\mathrm{d}V_{\mathbf{x}'}\mathrm{d}V_{\mathbf{x}} - \int_{\Omega}\int_{\mathbf{x}'\in H'_{\mathbf{x}}} \mathbf{f}_{\mathbf{x}'\mathbf{x}}(-\boldsymbol{\eta}, -\boldsymbol{\xi})\,\mathrm{d}V_{\mathbf{x}'}\mathrm{d}V_{\mathbf{x}} \\
&= \mathbf{0}\,.
\end{aligned}
\tag{2.27}
$$

For simplicities, let $\mathbf{f}_{\mathbf{x}\mathbf{x}'}$ represent $\mathbf{f}_{\mathbf{x}\mathbf{x}'}(\boldsymbol{\eta}, \boldsymbol{\xi})$, and $\mathbf{f}_{\mathbf{x}'\mathbf{x}}$ denote $\mathbf{f}_{\mathbf{x}'\mathbf{x}}(-\boldsymbol{\eta}, -\boldsymbol{\xi})$.

Proof Based on Eq. (2.14), (2.26) is satisfied, and the conservation of linear momentum is proved.

In fact, as the forces are differentiated here between direct and reaction forces, i.e. a direct force from $\mathbf{x}$ corresponds to a reaction force of $\mathbf{x}'$, the forces are always pairwise but with opposite direction and of the same magnitude. Hence, it is natural that the balance of linear momentum is satisfied.

Balance of angular momentum

To satisfy the balance of angular momentum for any bounded body Ω, it is required that

$$\begin{aligned}
&\sum_{\Omega} \mathbf{y} \times (\rho \ddot{\mathbf{u}}(\mathbf{x}, t) - \mathbf{b}(\mathbf{x}, t)) \Delta V_{\mathbf{x}} \\
&\quad = \sum_{\Omega} \mathbf{y} \times \left(\sum_{\mathbf{x}' \in H_{\mathbf{x}}} \mathbf{f}_{\mathbf{x}\mathbf{x}'}(\boldsymbol{\eta}, \boldsymbol{\xi}) \, \Delta V_{\mathbf{x}'} - \sum_{\mathbf{x}' \in H'_{\mathbf{x}}} \mathbf{f}_{\mathbf{x}'\mathbf{x}}(-\boldsymbol{\eta}, -\boldsymbol{\xi}) \, \Delta V_{\mathbf{x}'} \right) \Delta V_{\mathbf{x}} \\
&\quad = \sum_{\Omega} \sum_{\mathbf{x}' \in H_{\mathbf{x}}} \mathbf{y} \times \mathbf{f}_{\mathbf{x}\mathbf{x}'}(\boldsymbol{\eta}, \boldsymbol{\xi}) \, \Delta V_{\mathbf{x}'} \Delta V_{\mathbf{x}} - \sum_{\Omega} \sum_{\mathbf{x}' \in H'_{\mathbf{x}}} \mathbf{y} \times \mathbf{f}_{\mathbf{x}'\mathbf{x}}(-\boldsymbol{\eta}, -\boldsymbol{\xi}) \, \Delta V_{\mathbf{x}'} \Delta V_{\mathbf{x}} \\
&\quad = \mathbf{0}
\end{aligned} \tag{2.28}$$

or

$$\begin{aligned}
&\int_{\Omega} \mathbf{y} \times (\rho \ddot{\mathbf{u}}(\mathbf{x}, t) - \mathbf{b}(\mathbf{x}, t)) \mathrm{d} V_{\mathbf{x}} \\
&\quad = \int_{\Omega} \mathbf{y} \times \left(\int_{\mathbf{x}' \in H_{\mathbf{x}}} \mathbf{f}_{\mathbf{x}\mathbf{x}'}(\boldsymbol{\eta}, \boldsymbol{\xi}) \, \mathrm{d} V_{\mathbf{x}'} - \int_{\mathbf{x}' \in H'_{\mathbf{x}}} \mathbf{f}_{\mathbf{x}'\mathbf{x}}(-\boldsymbol{\eta}, -\boldsymbol{\xi}) \, \mathrm{d} V_{\mathbf{x}'} \right) \mathrm{d} V_{\mathbf{x}} \\
&\quad = \int_{\Omega} \int_{\mathbf{x}' \in H_{\mathbf{x}}} \mathbf{y} \times \mathbf{f}_{\mathbf{x}\mathbf{x}'}(\boldsymbol{\eta}, \boldsymbol{\xi}) \, \mathrm{d} V_{\mathbf{x}'} \mathrm{d} V_{\mathbf{x}} - \int_{\Omega} \int_{\mathbf{x}' \in H'_{\mathbf{x}}} \mathbf{y} \times \mathbf{f}_{\mathbf{x}'\mathbf{x}}(-\boldsymbol{\eta}, -\boldsymbol{\xi}) \, \mathrm{d} V_{\mathbf{x}'} \mathrm{d} V_{\mathbf{x}} \\
&\quad = \mathbf{0}.
\end{aligned} \tag{2.29}$$

For simplicities, let $\mathbf{f}_{\mathbf{x}\mathbf{x}'}$ again represent $\mathbf{f}_{\mathbf{x}\mathbf{x}'}(\boldsymbol{\eta}, \boldsymbol{\xi})$, and $\mathbf{f}_{\mathbf{x}'\mathbf{x}}$ represent $\mathbf{f}_{\mathbf{x}'\mathbf{x}}(-\boldsymbol{\eta}, -\boldsymbol{\xi})$.

Proposition: In the dual-horizon peridynamics, suppose a constitutive model of the form

$$\mathbf{f} = \hat{\mathbf{f}}(\mathbf{y}, \boldsymbol{\Lambda}) \tag{2.30}$$

where $\hat{\mathbf{f}} : \mathcal{V} \to \mathcal{V}$ is bounded and Riemann-integrable on Hilbert function space and $\mathcal{V}$ is the vector state; $\boldsymbol{\Lambda}$ denotes all variables other than the current deformation vector state. If

$$\sum_{\mathbf{x}' \in H_{\mathbf{x}}} \mathbf{y}\langle \mathbf{x}' - \mathbf{x} \rangle \times \mathbf{f}_{\mathbf{x}\mathbf{x}'} \, \Delta V_{\mathbf{x}'} = 0 \qquad \forall \mathbf{y} \in \mathcal{V}\,, \tag{2.31}$$

or

$$\int_{\mathbf{x}' \in H_{\mathbf{x}}} \mathbf{y}\langle \mathbf{x}' - \mathbf{x} \rangle \times \mathbf{f}_{\mathbf{x}\mathbf{x}'} \, dV_{\mathbf{x}'} = 0 \qquad \forall \mathbf{y} \in \mathcal{V}\,, \tag{2.32}$$

where $\mathbf{f}_{\mathbf{x}\mathbf{x}'} = \mathbf{f}_{\mathbf{x}\mathbf{x}'}(\boldsymbol{\eta}, \boldsymbol{\xi})$ is the force vector density acting on $\mathbf{x}$ and $H_{\mathbf{x}}$ is the horizon of $\mathbf{x}$, then the balance of angular momentum, Eq. (2.29) holds for any deformation of Ω for any given constitutive model.

Proof As the summation over the domain Ω is equivalent to the integrand of the same expression over that domain, we use the discrete form as

$$\begin{aligned}
&\sum_{\mathbf{x} \in \Omega} \mathbf{y} \times (\rho \ddot{\mathbf{u}}(\mathbf{x}, t) - \mathbf{b}(\mathbf{x}, t))\Delta V_{\mathbf{x}} \\
&= \sum_{\mathbf{x} \in \Omega} (\mathbf{x} + \mathbf{u}) \times (\rho \ddot{\mathbf{u}}(\mathbf{x}, t) - \mathbf{b}(\mathbf{x}, t))\Delta V_{\mathbf{x}} \\
&= \sum_{\mathbf{x} \in \Omega} \sum_{\mathbf{x}' \in H_{\mathbf{x}}} (\mathbf{x} + \mathbf{u}) \times \mathbf{f}_{\mathbf{x}\mathbf{x}'} \, \Delta V_{\mathbf{x}'} \Delta V_{\mathbf{x}} - \sum_{\mathbf{x} \in \Omega} \sum_{\mathbf{x}' \in H'_{\mathbf{x}}} (\mathbf{x} + \mathbf{u}) \times \mathbf{f}_{\mathbf{x}'\mathbf{x}} \, \Delta V_{\mathbf{x}'} \Delta V_{\mathbf{x}} \\
&= \sum_{\mathbf{x} \in \Omega} \sum_{\mathbf{x} \in H_{\mathbf{x}}} (\mathbf{x} + \mathbf{u}) \times \mathbf{f}_{\mathbf{x}\mathbf{x}'} \, \Delta V_{\mathbf{x}'} \Delta V_{\mathbf{x}} - \sum_{\mathbf{x} \in \Omega} \sum_{\mathbf{x}' \in H_{\mathbf{x}}} (\mathbf{x}' + \mathbf{u}') \times \mathbf{f}_{\mathbf{x}\mathbf{x}'} \, \Delta V_{\mathbf{x}'} \Delta V_{\mathbf{x}}. \\
&= \sum_{\mathbf{x} \in \Omega} \sum_{\mathbf{x}' \in H_{\mathbf{x}}} (\mathbf{x} + \mathbf{u} - \mathbf{x}' - \mathbf{u}') \times \mathbf{f}_{\mathbf{x}\mathbf{x}'} \, \Delta V_{\mathbf{x}'} \Delta V_{\mathbf{x}} \\
&= -\sum_{\mathbf{x} \in \Omega} \Big(\sum_{\mathbf{x}' \in H_{\mathbf{x}}} \mathbf{y}\langle \mathbf{x}' - \mathbf{x} \rangle \times \mathbf{f}_{\mathbf{x}\mathbf{x}'} \Delta V_{\mathbf{x}'} \Big) \Delta V_{\mathbf{x}} = 0.
\end{aligned} \tag{2.33}$$

In the fourth step, the global summation in dual-horizon is converted into the global summation in the horizon based on Eq. (2.14). In the last step, Eq. (2.31) is applied. Therefore, the balance of angular momentum over the entire analysis domain is satisfied. □

It can be seen that the conservation of angular momentum somehow depends **only** on the horizon; the dual-horizon is not involved. The latter is only needed in Eqs. (2.21) and (2.24). The bond-based, ordinary-based and non-ordinary-based peridynamics all satisfy the balance of angular momentum in the traditional horizon concept (see proof in Silling et al. (2007)). This conclusion can be also illustrated for the BB-PD and OSB-PD since the internal forces $\mathbf{f}_{\mathbf{x}\mathbf{x}'}(\boldsymbol{\eta}, \boldsymbol{\xi})$ and $\mathbf{f}_{\mathbf{x}'\mathbf{x}}(-\boldsymbol{\eta}, -\boldsymbol{\xi})$ are parallel to the bond vector in the current configuration; see Fig. 2.7.

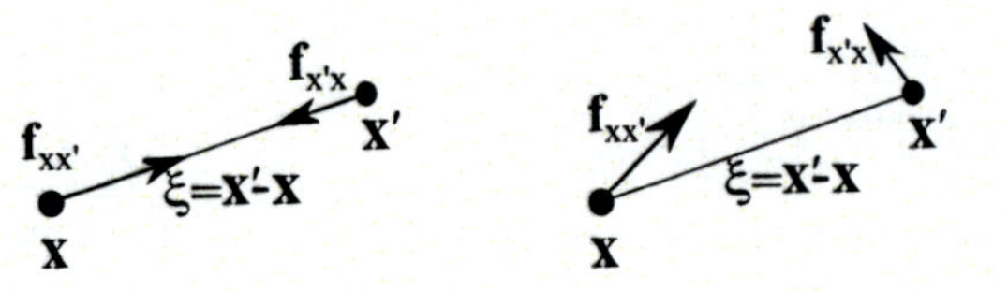

Fig. 2.7 Force vector of BB-PD, OSB-PD and NOSB-PD, where $\mathbf{f}_{\mathbf{x}'\mathbf{x}} = \mathbf{f}_{\mathbf{x}'\mathbf{x}}(-\boldsymbol{\eta}, -\boldsymbol{\xi})$, $\mathbf{f}_{\mathbf{x}\mathbf{x}'} = \mathbf{f}_{\mathbf{x}\mathbf{x}'}(\boldsymbol{\eta}, \boldsymbol{\xi})$

2.4 Dual-Horizon Peridynamics

The dual-horizon formulation will now be applied to all existing peridynamics, namely BB-PD, OSB-PD and NOSB-PD. The calibration of constitutive parameters with respect to the continuum model and some issues concerning the implementations will be discussed for 2D and 3D problems.

2.4.1 Dual-Horizon Bond-Based Peridynamics

In the bond-based peridynamics theory, the energy per unit volume in the body at a given time t is given by Silling and Askari (2005)

$$W = \frac{1}{2} \int_{H_{\mathbf{x}}} w(\boldsymbol{\eta}, \boldsymbol{\xi}) \mathrm{d}V_{\boldsymbol{\xi}} \,. \tag{2.34}$$

For the BB-PD theory, by enforcing the strain energy density being equal to the strain energy density in the classical theory of elasticity (Gerstle et al. 2007; Silling and Askari 2005), we obtain the expression of micro-modulus $C(\delta)$ as

$$\begin{aligned} C(\delta) &= \frac{3E}{\pi\delta^3(1-\nu)} && \text{plane stress} \\ C(\delta) &= \frac{3E}{\pi\delta^3(1+\nu)(1-2\nu)} && \text{plane strain} \\ C(\delta) &= \frac{3E}{\pi\delta^4(1-2\nu)} && \text{3D}\,, \end{aligned} \tag{2.35}$$

where δ is the horizon radius associated with that particle, ν is Poisson's ratio. Note that the value of $C(\delta)$ takes half of the micro-modulus c used in the BB-PD since the bond energy for varying horizons is determined by both horizon and dual-horizon. When the horizon takes the same value as the dual-horizon, the bond energy between two particles is reduced to that of the traditional BB-PD.

Let $w_0(\boldsymbol{\xi}) = C(\delta)s_0^2(\delta)\xi/2$ denote the work required to break a single bond, where $s_0(\delta)$ is the critical bond stretch. By breaking half of all the bonds connected to a given particle along the fracture surface and equalizing the breaking bond energy with the critical energy release rate G_0 (Dipasquale et al. 2014; Gerstle et al. 2007), we can get the expression between the critical energy release rate G_0 and the critical bond stretch $s_0(\delta)$:

$$
\begin{aligned}
s_0(\delta) &= \sqrt{\frac{4\pi G_0}{9E\delta}} && \text{plane stress} \\
s_0(\delta) &= \sqrt{\frac{5\pi G_0}{12E\delta}} && \text{plane strain} \\
s_0(\delta) &= \sqrt{\frac{5G_0}{6E\delta}} && \text{3D}\,.
\end{aligned}
\tag{2.36}
$$

Both the micro-modulus $C(\delta)$ and the critical stretch $s_0(\delta)$ are derived from the local continuum mechanics theory, and they depend on the horizon radii for variable horizons.

In the implementation of the BB-PD, the fracture is introduced by removing particles from the neighbor list once the bond stretch exceeds the critical bond stretch s_0. In order to specify whether a bond is broken or not, a history-dependent scalar-valued function μ is introduced (Silling and Askari 2005),

$$
\mu(t, \boldsymbol{\xi}) = \begin{cases} 1 & \text{if } s(t', \boldsymbol{\xi}) < s_0 \text{ for all } 0 \le t' \le t, \\ 0 & \text{otherwise.} \end{cases}
\tag{2.37}
$$

The local damage at $\mathbf{x}$ is defined as

$$
\phi(\mathbf{x}, t) = 1 - \frac{\int_{H_{\mathbf{x}}} \mu(\mathbf{x}, t, \boldsymbol{\xi}) \mathrm{d}V_\xi}{\int_{H_{\mathbf{x}}} \mathrm{d}V_\xi}\,.
\tag{2.38}
$$

The damage formulation of Eq. (2.38) is also applicable to OSB-PD. For any particle with dual-horizon ($H'_{\mathbf{x}}$) and horizon ($H_{\mathbf{x}}$), the direct force $\mathbf{f}_{\mathbf{x}\mathbf{x}'}(\boldsymbol{\eta}, \boldsymbol{\xi})$ and the reaction force $\mathbf{f}_{\mathbf{x}'\mathbf{x}}(-\boldsymbol{\eta}, -\boldsymbol{\xi})$ in Eq. (2.21) or (2.24) are computed by the following expressions, respectively,

$$
\mathbf{f}_{\mathbf{x}\mathbf{x}'}(\boldsymbol{\eta}, \boldsymbol{\xi}) = C(\delta_{\mathbf{x}}) \cdot s_{\mathbf{x}\mathbf{x}'} \cdot \frac{\boldsymbol{\eta} + \boldsymbol{\xi}}{\|\boldsymbol{\eta} + \boldsymbol{\xi}\|}, \qquad \forall \mathbf{x}' \in H_{\mathbf{x}}
\tag{2.39}
$$

$$
\mathbf{f}_{\mathbf{x}'\mathbf{x}}(-\boldsymbol{\eta}, -\boldsymbol{\xi}) = C(\delta_{\mathbf{x}'}) \cdot s_{\mathbf{x}\mathbf{x}'} \cdot \frac{-(\boldsymbol{\eta} + \boldsymbol{\xi})}{\|\boldsymbol{\eta} + \boldsymbol{\xi}\|}, \qquad \forall \mathbf{x}' \in H'_{\mathbf{x}}\,,
\tag{2.40}
$$

where $C(\delta_{\mathbf{x}'})$ and $C(\delta_{\mathbf{x}})$ are the micro-modulus based on $\delta_{\mathbf{x}'}$ and $\delta_{\mathbf{x}}$ computed from Eq. (2.35) respectively, and $s_{\mathbf{x}\mathbf{x}'}$ is the stretch between particles $\mathbf{x}$ and $\mathbf{x}'$.

2.4.2 *Dual-Horizon Ordinary State-Based Peridynamics*

The concept of "state" for peridynamics was firstly introduced in Silling et al. (2007). A state of order m is a function

$$\underline{\mathbf{A}} : H \rightarrow \mathbf{T}_m, \xi \mapsto \underline{\mathbf{A}}\langle\xi\rangle\ , \tag{2.41}$$

where H is the horizon domain, $\mathbf{T}_m$ denotes the set of all tensors of order m and $\langle\square\rangle$ indicates the vector to which a state operates. In this chapter, the scalar state and vector state are used if not specified otherwise.

The bond force density $\underline{t}_{\mathbf{x}'\mathbf{x}}$ both in 2D and in 3D OSB-PD can be written in a unified expression taking into account the dimension number as

$$\underline{t}_{\mathbf{x}\mathbf{x}'} = \frac{nK\theta_{\mathbf{x}}}{m_{\mathbf{x}}}\omega\langle\boldsymbol{\xi}\rangle \cdot \xi + \frac{n(n+2)G}{m_{\mathbf{x}}}\omega\langle\boldsymbol{\xi}\rangle \underline{e}^d\langle\boldsymbol{\xi}\rangle\ , \tag{2.42}$$

where $n \in \{2, 3\}$ is the dimensional number, K is the bulk modulus, G the shear modulus, $\underline{e}^d\langle\boldsymbol{\xi}\rangle = \underline{e}\langle\boldsymbol{\xi}\rangle - \frac{\theta_{\mathbf{x}}\xi}{n}$, $m_{\mathbf{x}} = \int_{H_{\mathbf{x}}} \omega\langle\boldsymbol{\xi}\rangle\, \xi \cdot \xi\, \mathrm{d}V_{\xi}$, $\theta_{\mathbf{x}} = \frac{n}{m_{\mathbf{x}}} \int_{H_{\mathbf{x}}} \omega\langle\boldsymbol{\xi}\rangle\, \xi \cdot \underline{e}\langle\boldsymbol{\xi}\rangle \mathrm{d}V_{\xi}$, $\underline{e}\langle\boldsymbol{\xi}\rangle = \|\boldsymbol{\xi} + \boldsymbol{\eta}\| - \|\boldsymbol{\xi}\|$ and $\mathrm{d}V_{\xi}$ are the area in 2D, and volume in 3D, respectively. For any particle $\mathbf{x}$, $\mathbf{f}_{\mathbf{x}\mathbf{x}'}(\boldsymbol{\eta}, \boldsymbol{\xi})$ and $\mathbf{f}_{\mathbf{x}'\mathbf{x}}(-\boldsymbol{\eta}, -\boldsymbol{\xi})$ in Eqs. (2.21) and (2.24) can be calculated by

$$\mathbf{f}_{\mathbf{x}\mathbf{x}'}(\boldsymbol{\eta}, \boldsymbol{\xi}) = \underline{t}_{\mathbf{x}\mathbf{x}'} \cdot \frac{\boldsymbol{\eta} + \boldsymbol{\xi}}{\|\boldsymbol{\eta} + \boldsymbol{\xi}\|}, \qquad \forall \mathbf{x}' \in H_{\mathbf{x}}\ , \tag{2.43}$$

and

$$\mathbf{f}_{\mathbf{x}'\mathbf{x}}(-\boldsymbol{\eta}, -\boldsymbol{\xi}) = \underline{t}_{\mathbf{x}'\mathbf{x}} \cdot \frac{-\boldsymbol{\eta} - \boldsymbol{\xi}}{\|\boldsymbol{\eta} + \boldsymbol{\xi}\|}, \qquad \forall \mathbf{x}' \in H'_{\mathbf{x}}\ . \tag{2.44}$$

Here, $\boldsymbol{\xi} = \mathbf{x}' - \mathbf{x}$, $\boldsymbol{\eta} = \mathbf{u}' - \mathbf{u}$, and $\mathbf{x}$ and $\mathbf{x}'$ are the coordinate vectors for $\mathbf{x}$ and $\mathbf{x}'$ in the material configuration, respectively.

2.4.3 *Dual-Horizon Non-ordinary State-Based Peridynamics*

NOSB-PD uses the deformation gradient from classic continuum mechanics. It offers the possibility to consistently include existing constitutive models. With the help of the shape tensor proposed in the NOSB-PD, the fundamental tensors in continuum mechanics can be easily introduced in peridynamics, including the deformation gradient tensor or velocity gradient (Silling and Lehoucq 2010; Warren et al. 2008). The shape tensor for particle $\mathbf{x}$ is defined as

$$\mathbf{K}_{\mathbf{x}} = \int_{\mathbf{x}' \in H_{\mathbf{x}}} \omega\langle \boldsymbol{\xi} \rangle \, \boldsymbol{\xi} \otimes \boldsymbol{\xi} \mathrm{d}V_{\mathbf{x}'} \tag{2.45}$$

where $\boldsymbol{\xi} = \mathbf{x}' - \mathbf{x}$ is the bond in the reference configuration, $H_{\mathbf{x}}$ is the horizon for particle $\mathbf{x}$ and $\omega\langle \boldsymbol{\xi} \rangle$ is the influence function. The deformation tensor for particle $\mathbf{x}$ is given by

$$\mathbf{F}_{\mathbf{x}} = \frac{\partial \mathbf{y}}{\partial \mathbf{x}} = \int_{\mathbf{x}' \in H_{\mathbf{x}}} \omega\langle \boldsymbol{\xi} \rangle \mathbf{y}\langle \boldsymbol{\xi} \rangle \otimes \boldsymbol{\xi} \, \mathrm{d}V_{\mathbf{x}'} \cdot \mathbf{K}_{\mathbf{x}}^{-1} , \tag{2.46}$$

where $\mathbf{y} := \mathbf{y}(\mathbf{x}, t)$ is the spatial coordinate, $\mathbf{x}$ is the material coordinate and $\boldsymbol{\xi} = \mathbf{x}' - \mathbf{x}$. The spatial velocity gradient $\mathbf{L}_{\mathbf{x}}$ for particle $\mathbf{x}$ is defined in the current configuration using the chain rule,

$$\mathbf{L}_{\mathbf{x}} := \frac{\partial \mathbf{v}}{\partial \mathbf{y}} = \frac{\partial \mathbf{v}}{\partial \mathbf{x}} \cdot \frac{\partial \mathbf{x}}{\partial \mathbf{y}} = \dot{\mathbf{F}}_{\mathbf{x}} \mathbf{F}_{\mathbf{x}}^{-1} , \tag{2.47}$$

where $\mathbf{v} := \mathbf{v}(\mathbf{x}, t) = \frac{\partial \mathbf{y}}{\partial t}(\mathbf{x}, t)$ is the velocity vector and $\dot{\mathbf{F}}_{\mathbf{x}}$ is the rate of deformation gradient. The nonlocal form of the velocity gradient can be written as

$$\begin{aligned}
\mathbf{L}_{\mathbf{x}} :&= \int_{\mathbf{x}' \in H_{\mathbf{x}}} \omega\langle \boldsymbol{\xi} \rangle \, \mathbf{v}\langle \boldsymbol{\xi} \rangle \otimes \boldsymbol{\xi} \, \mathrm{d}V_{\mathbf{x}'} \cdot \mathbf{K}_{\mathbf{x}}^{-1} \cdot \left(\int_{\mathbf{x}' \in H_{\mathbf{x}}} \omega\langle \boldsymbol{\xi} \rangle \, \mathbf{y}\langle \boldsymbol{\xi} \rangle \otimes \boldsymbol{\xi} \, \mathrm{d}V_{\mathbf{x}'} \cdot \mathbf{K}_{\mathbf{x}}^{-1} \right)^{-1} \\
&= \int_{\mathbf{x}' \in H_{\mathbf{x}}} \omega\langle \boldsymbol{\xi} \rangle \, \mathbf{v}\langle \boldsymbol{\xi} \rangle \otimes \boldsymbol{\xi} \, \mathrm{d}V_{\mathbf{x}'} \cdot \mathbf{K}_{\mathbf{x}}^{-1} \cdot \mathbf{K}_{\mathbf{x}} \left(\int_{\mathbf{x}' \in H_{\mathbf{x}}} \omega\langle \boldsymbol{\xi} \rangle \, \mathbf{y}\langle \boldsymbol{\xi} \rangle \otimes \boldsymbol{\xi} \, \mathrm{d}V_{\mathbf{x}'} \right)^{-1} \cdot \\
&= \int_{\mathbf{x}' \in H_{\mathbf{x}}} \omega\langle \boldsymbol{\xi} \rangle \, \mathbf{v}\langle \boldsymbol{\xi} \rangle \otimes \boldsymbol{\xi} \, \mathrm{d}V_{\mathbf{x}'} \cdot \left(\int_{\mathbf{x}' \in H_{\mathbf{x}}} \omega\langle \boldsymbol{\xi} \rangle \, \mathbf{y}\langle \boldsymbol{\xi} \rangle \otimes \boldsymbol{\xi} \, \mathrm{d}V_{\mathbf{x}'} \right)^{-1}
\end{aligned} \tag{2.48}$$

where $\boldsymbol{\xi} = \mathbf{x}' - \mathbf{x}$, $\mathbf{v}\langle \boldsymbol{\xi} \rangle = \mathbf{v}(\mathbf{x}', t) - \mathbf{v}(\mathbf{x}, t)$ and $\mathbf{y}\langle \boldsymbol{\xi} \rangle = \mathbf{y}(\mathbf{x}', t) - \mathbf{y}(\mathbf{x}, t)$. It can be seen that the velocity gradient does not require the shape tensor, which means the shape tensor is not necessary to define the velocity gradient. The incremental deformation gradient can also be obtained without the shape tensor $\mathbf{K}$.

Let $\Delta \mathbf{u}$ denote the displacement increment with respect to the previous time step. The incremental spatial deformation gradient $\mathbf{C}_{\mathbf{x}}$ in continuum mechanics is defined as

$$\mathbf{C}_{\mathbf{x}} := \frac{\partial (\Delta \mathbf{u})}{\partial \mathbf{y}} = \frac{\partial (\Delta \mathbf{u})}{\partial \mathbf{x}} \cdot \frac{\partial \mathbf{x}}{\partial \mathbf{y}} = \frac{\partial (\Delta \mathbf{u})}{\partial \mathbf{x}} \cdot \mathbf{F}_{\mathbf{x}}^{-1} . \tag{2.49}$$

The nonlocal counterpart is given by

$$\mathbf{C}_\mathbf{x} := \int_{\mathbf{x}' \in H_\mathbf{x}} \omega\langle\boldsymbol{\xi}\rangle \Delta\mathbf{u}\langle\boldsymbol{\xi}\rangle \otimes \boldsymbol{\xi} \, dV_{\mathbf{x}'} \cdot \left(\int_{\mathbf{x}' \in H_\mathbf{x}} \omega\langle\boldsymbol{\xi}\rangle \mathbf{y}\langle\boldsymbol{\xi}\rangle \otimes \boldsymbol{\xi} \, dV_{\mathbf{x}'}\right)^{-1} \quad .$$

Many material models in non-ordinary state-based peridynamics accounting for the geometrical and material nonlinear problems are solved based on the incremental spatial deformation gradient. The non-ordinary bond force is computed based on Piola–Kirchhoff method as

$$\underline{\mathbf{T}}_{\mathbf{x}\mathbf{x}'} = \omega\langle\boldsymbol{\xi}\rangle \, \mathbf{P}_\mathbf{x} \cdot \mathbf{K}_\mathbf{x}^{-1} \cdot \boldsymbol{\xi} \tag{2.50}$$

$$\mathbf{P}_\mathbf{x} = J_\mathbf{x} \sigma_\mathbf{x} \mathbf{F}_\mathbf{x}^{-T}, \; J_\mathbf{x} = \det \mathbf{F}_\mathbf{x} \tag{2.51}$$

where $\boldsymbol{\xi} = \mathbf{x}' - \mathbf{x}$, $\sigma_\mathbf{x}$ is Cauchy stress tensor. Note that $\underline{\mathbf{T}}_{\mathbf{x}\mathbf{x}'}$ corresponds to $\underline{\mathbf{T}}[\mathbf{x}', t]\langle\mathbf{x} - \mathbf{x}'\rangle$ in Eq. (2.7) in conventional SB-PD, where the subscripts represent the bond force acting on $\mathbf{x}'$ due to $\mathbf{x}$.

Let us define the deformation gradient $\mathbf{F}'_\mathbf{x}$

$$\mathbf{F}'_\mathbf{x} := \mathbf{F}_\mathbf{x} \cdot \mathbf{K}_\mathbf{x} = \int_{\mathbf{x}' \in H_\mathbf{x}} \omega\langle\boldsymbol{\xi}\rangle \mathbf{y}\langle\boldsymbol{\xi}\rangle \otimes \boldsymbol{\xi} \, dV_{\mathbf{x}'} \cdot \mathbf{K}_\mathbf{x}^{-1} \cdot \mathbf{K}_\mathbf{x} = \int_{\mathbf{x}' \in H_\mathbf{x}} \omega\langle\boldsymbol{\xi}\rangle \mathbf{y}\langle\boldsymbol{\xi}\rangle \otimes \boldsymbol{\xi} \, dV_\mathbf{x} \, . \tag{2.52}$$

Therefore

$$J_\mathbf{x} = \det \mathbf{F}_\mathbf{x} = \frac{\det \mathbf{F}'_\mathbf{x}}{\det \mathbf{K}_\mathbf{x}} \, . \tag{2.53}$$

By substituting Eqs. (2.53), (2.52) and (2.51) to (2.50), we obtain

$$\begin{aligned} \underline{\mathbf{T}}_{\mathbf{x}\mathbf{x}'} &= \omega\langle\boldsymbol{\xi}\rangle \; J_\mathbf{x} \sigma_\mathbf{x} \mathbf{F}_\mathbf{x}^{-T} \cdot \mathbf{K}_\mathbf{x}^{-1} \cdot \boldsymbol{\xi} \\ &= \omega\langle\boldsymbol{\xi}\rangle \frac{\det \mathbf{F}'_\mathbf{x}}{\det \mathbf{K}_\mathbf{x}} \cdot \sigma_\mathbf{x} \cdot (\mathbf{F}'_\mathbf{x} \cdot \mathbf{K}_\mathbf{x}^{-1})^{-T} \cdot \mathbf{K}_\mathbf{x}^{-1} \cdot \boldsymbol{\xi} \\ &= \omega\langle\boldsymbol{\xi}\rangle \frac{\det \mathbf{F}'_\mathbf{x}}{\det \mathbf{K}_\mathbf{x}} \cdot \sigma_\mathbf{x} \cdot \mathbf{F}'^{-T}_\mathbf{x} \cdot \mathbf{K}_\mathbf{x} \cdot \mathbf{K}_\mathbf{x}^{-1} \cdot \boldsymbol{\xi} \\ &= \omega\langle\boldsymbol{\xi}\rangle \frac{\det \mathbf{F}'_\mathbf{x}}{\det \mathbf{K}_\mathbf{x}} \cdot \sigma_\mathbf{x} \cdot \mathbf{F}'^{-T}_\mathbf{x} \cdot \boldsymbol{\xi} \, . \end{aligned} \tag{2.54}$$

The shape tensor is only needed for $J_\mathbf{x} = \det \mathbf{F}_\mathbf{x}$, while the computation of the net bond force density does not require the shape tensor. For any particle with horizon $H_\mathbf{x}$ and dual-horizon $H'_\mathbf{x}$, the direct force $\mathbf{f}_{\mathbf{x}\mathbf{x}'}(\boldsymbol{\eta}, \boldsymbol{\xi})$ and the reaction force $\mathbf{f}_{\mathbf{x}'\mathbf{x}}(-\boldsymbol{\eta}, -\boldsymbol{\xi})$ in Eq. (2.21) or (2.24) can be calculated by

Table 2.1 Comparison of the traditional peridynamics and dual-horizon peridynamics, where $\mathbf{f}_{\mathbf{x}'\mathbf{x}} = \mathbf{f}_{\mathbf{x}'\mathbf{x}}(-\boldsymbol{\eta}, -\boldsymbol{\xi})$, $\mathbf{f}_{\mathbf{x}\mathbf{x}'} = \mathbf{f}_{\mathbf{x}\mathbf{x}'}(\boldsymbol{\eta}, \boldsymbol{\xi})$ and $\mathbf{n} = (\boldsymbol{\eta} + \boldsymbol{\xi})/(\|\boldsymbol{\eta} + \boldsymbol{\xi}\|)$ is the unit direction vector for bond $\boldsymbol{\xi}$ in current configuration. Note that the traditional peridynamics is formulated with notation in DH-PD

Model	Traditional peridynamics $\rho\ddot{\mathbf{u}} = \int_{H_{\mathbf{x}}} (\mathbf{f}_{\mathbf{x}\mathbf{x}'} - \mathbf{f}_{\mathbf{x}'\mathbf{x}}) dV_{\mathbf{x}'} + \mathbf{b}$	Dual-horizon peridynamics $\rho\ddot{\mathbf{u}} = \int_{H_{\mathbf{x}}} \mathbf{f}_{\mathbf{x}\mathbf{x}'} dV_{\mathbf{x}'} - \int_{H'_{\mathbf{x}}} \mathbf{f}_{\mathbf{x}'\mathbf{x}} dV_{\mathbf{x}'} + \mathbf{b}$
BB-PD	$\mathbf{f}_{\mathbf{x}\mathbf{x}'} = C(\delta) \cdot s_{\mathbf{x}\mathbf{x}'} \cdot \mathbf{n}$, $\mathbf{f}_{\mathbf{x}'\mathbf{x}} = C(\delta) \cdot s_{\mathbf{x}\mathbf{x}'} \cdot (-\mathbf{n}), \forall \mathbf{x}' \in H_{\mathbf{x}}$	$\mathbf{f}_{\mathbf{x}\mathbf{x}'} = C(\delta_{\mathbf{x}'}) \cdot s_{\mathbf{x}\mathbf{x}'} \cdot \mathbf{n}, \forall \mathbf{x}' \in H_{\mathbf{x}}$; $\mathbf{f}_{\mathbf{x}'\mathbf{x}} = C(\delta_{\mathbf{x}}) \cdot s_{\mathbf{x}\mathbf{x}'} \cdot (-\mathbf{n}), \forall \mathbf{x}' \in H'_{\mathbf{x}}$
OSB-PD	$\mathbf{f}_{\mathbf{x}\mathbf{x}'} = \underline{t}_{\mathbf{x}\mathbf{x}'} \cdot \mathbf{n}$, $\mathbf{f}_{\mathbf{x}'\mathbf{x}} = \underline{t}_{\mathbf{x}'\mathbf{x}} \cdot (-\mathbf{n}), \forall \mathbf{x}' \in H_{\mathbf{x}}$	$\mathbf{f}_{\mathbf{x}\mathbf{x}'} = \underline{t}_{\mathbf{x}\mathbf{x}'} \cdot \mathbf{n}, \forall \mathbf{x}' \in H_{\mathbf{x}}$; $\mathbf{f}_{\mathbf{x}'\mathbf{x}} = \underline{t}_{\mathbf{x}'\mathbf{x}} \cdot (-\mathbf{n}), \forall \mathbf{x}' \in H'_{\mathbf{x}}$
NOSB-PD	$\mathbf{f}_{\mathbf{x}\mathbf{x}'} = \underline{\mathbf{T}}_{\mathbf{x}\mathbf{x}'}$, $\mathbf{f}_{\mathbf{x}'\mathbf{x}} = \underline{\mathbf{T}}_{\mathbf{x}'\mathbf{x}}, \forall \mathbf{x}' \in H_{\mathbf{x}}$	$\mathbf{f}_{\mathbf{x}\mathbf{x}'} = \underline{\mathbf{T}}_{\mathbf{x}\mathbf{x}'}, \forall \mathbf{x}' \in H_{\mathbf{x}}$; $\mathbf{f}_{\mathbf{x}'\mathbf{x}} = \underline{\mathbf{T}}_{\mathbf{x}'\mathbf{x}}, \forall \mathbf{x}' \in H'_{\mathbf{x}}$

$$\mathbf{f}_{\mathbf{x}\mathbf{x}'}(\boldsymbol{\eta}, \boldsymbol{\xi}) = \underline{\mathbf{T}}_{\mathbf{x}\mathbf{x}'} \qquad \forall \mathbf{x}' \in H_{\mathbf{x}} , \tag{2.55}$$

and

$$\mathbf{f}_{\mathbf{x}'\mathbf{x}}(-\boldsymbol{\eta}, -\boldsymbol{\xi}) = \underline{\mathbf{T}}_{\mathbf{x}'\mathbf{x}} \qquad \forall \mathbf{x}' \in H'_{\mathbf{x}} , \tag{2.56}$$

where $\boldsymbol{\xi} = \mathbf{x}' - \mathbf{x}$, and $\mathbf{x}$ and $\mathbf{x}'$ are the coordinate vectors for particle $\mathbf{x}$ and particle $\mathbf{x}'$ in the material configuration, respectively.

A comparison between traditional peridynamics and DH-PD is shown in Table 2.1.

2.5 Numerical Examples

In all numerical examples, the Verlet algorithm is adopted for the time integration; each particle's horizon radius is $\delta = 3\Delta x$, Δx is the particle size associated with that particle and the influence function $\omega\langle\boldsymbol{\xi}\rangle = 1$; the bond micro-modulus $C(\delta_{\mathbf{x}})$ is calculated based on Eq. (2.35), and the force state density $\underline{t}$ in DH-OSBPD based on Eq. (2.42).

2.5.1 *Wave Propagation in 1D Homogeneous Bar*

Consider a 1D bar with length of L =1 m, one-third of the bar in the middle is discretized with $\Delta x_1 = 4.19 \times 10^{-3}$ m and the other parts discretized with $\Delta x_2 = 10\Delta x_1$. The horizon size is selected as $\delta_i = 3.015\Delta x_i$, as shown in Fig. 2.8. Both ends of the bar are free. The initial displacement field is given by

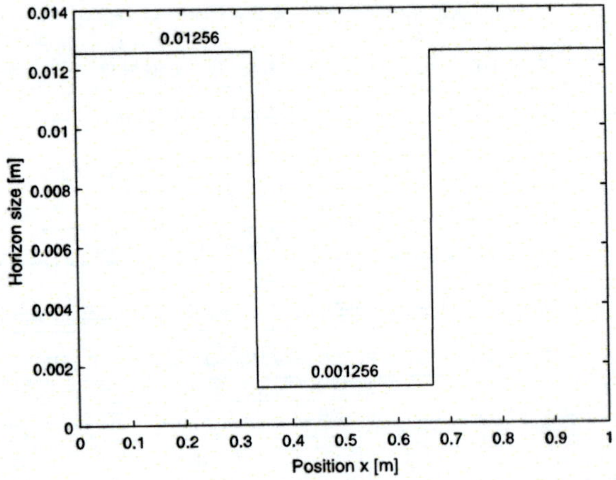

Fig. 2.8 The horizon size distribution

$$u(x, 0) = 0.002 \exp(-\frac{(x - 0.5L)^2}{1000}). \tag{2.57}$$

The bar is solved with DH-PD (Case I) and traditional PD formulation (Case II). The wave profiles at different times are shown in Fig. 2.9 for DH-PD and Fig. 2.10 for traditional PD. It can be seen that the horizon size interfaces have limited influence on the wave profiles in Case I, while the spurious wave emerged for traditional PD formulation. Two points $x = 0.32$ m and $x = 0.34$ m are selected to analyze the wave profile, and the displacements at different times are shown in Fig. 2.11. It shows that the incident wave passed the interface, and the magnitude of the reflected wave is smaller than 3.17% of the magnitude of the incident wave.

On the other hand, for the traditional PD without any treatment on variable horizons, when the incident wave approached the interface, the deformation of the fine material points caused the ghost forces among the coarse material points. The ghost forces give rise to the spurious wave in the domain with coarse material points, as shown in Fig. 2.10 ($T = 0.2$ s). In addition, the wave in the fine material points is hard to pass the interface but is to be reflected due to the fact that its horizon size is much smaller than that of coarse material points.

2.5.2 *2D Wave Reflection in a Rectangular Plate*

Consider a rectangular plate with dimensions of 0.1×0.04 m^2 (see Fig. 2.12). Young's modulus, density and Poisson's ratio used for the plate are $E = 1$, $\rho = 1$ and $\nu = 0$, respectively. Note that this is actually a 1D problem solved in 2D. The initial displacement applied to the plate is described by the following equation:

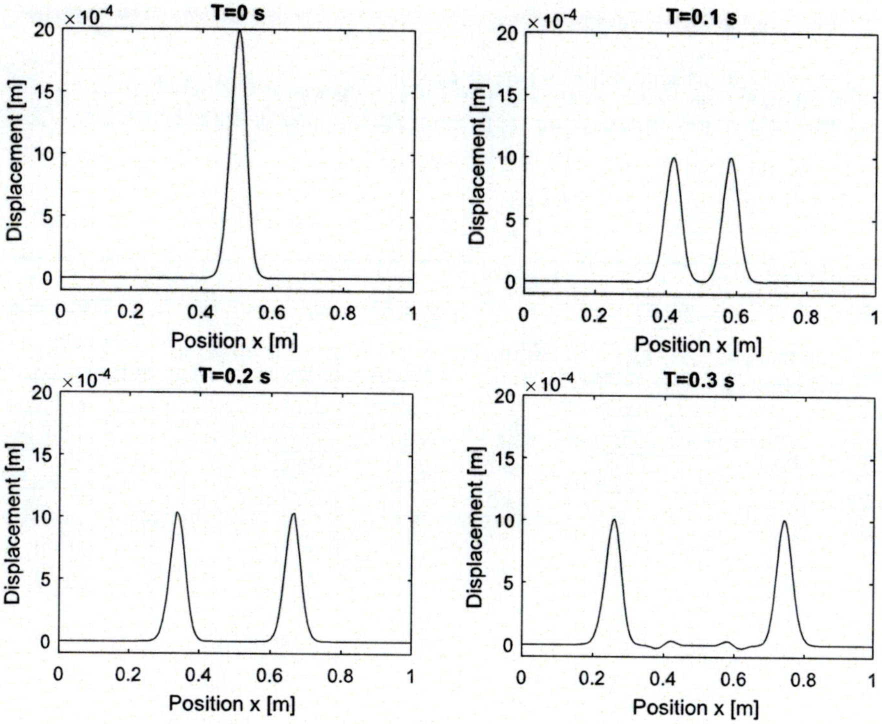

Fig. 2.9 The wave profiles for Case I at different times

$$u_0(x, y) = 0.0002 \exp[-(\frac{x}{0.01})^2], \quad v_0(x, y) = 0, \quad x \in [0, 0.1], y \in [-0.02, 0.02]\,, \tag{2.58}$$

where u_0 and v_0 denote the displacement in the x and y directions, respectively. The wave speed is $v = \sqrt{E/\rho} = 1$ m/s. At any time step, the L_2 error in the displacement is given by

$$\|\text{err}\|_{L_2} = \frac{\|\mathbf{u}^h - \mathbf{u}_{\text{analytic}}\|}{\|\mathbf{u}_{\text{analytic}}\|}\,, \tag{2.59}$$

with

$$\|\mathbf{u}\| = \left(\int_{\Omega_0} \mathbf{u} \cdot \mathbf{u}\, d\Omega_0\right)^{\frac{1}{2}}\,.$$

Three models for solving this problem, namely model I, II and III, are devised; see Fig. 2.13a, b. In model I, the plate is discretized with 4000 particles and all particles have the same horizon sizes of 0.001 m; see Fig. 2.13a. For models II and III, we

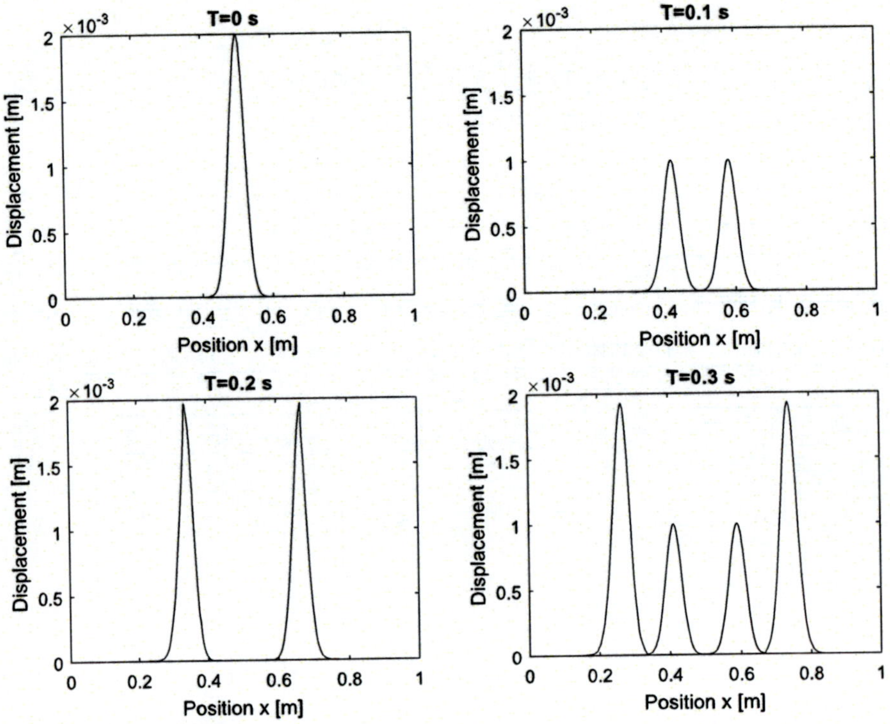

Fig. 2.10 The wave profiles for Case II at different times

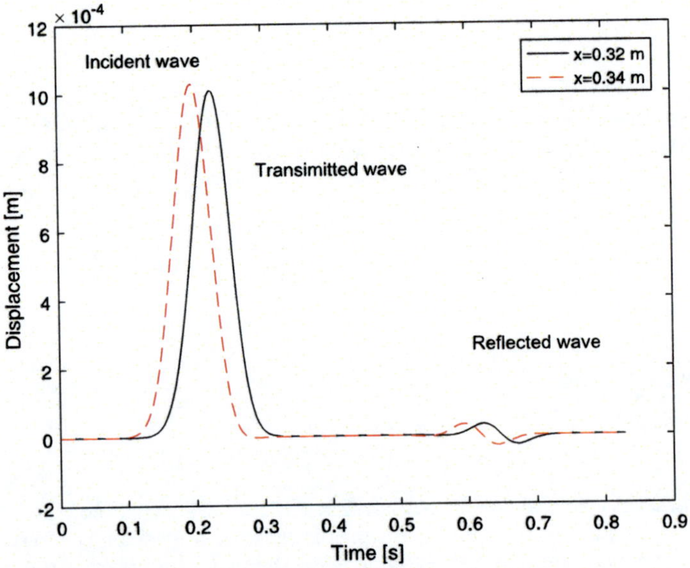

Fig. 2.11 The wave for points $x = 0.32\,\mathrm{m}$ and $x = 0.34\,\mathrm{m}$

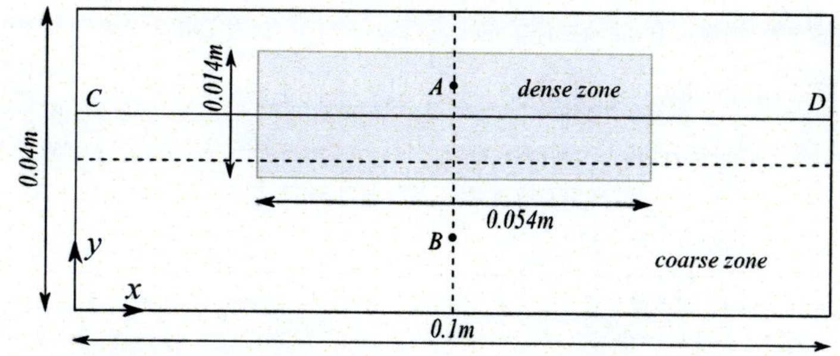

Fig. 2.12 Setup of the plate

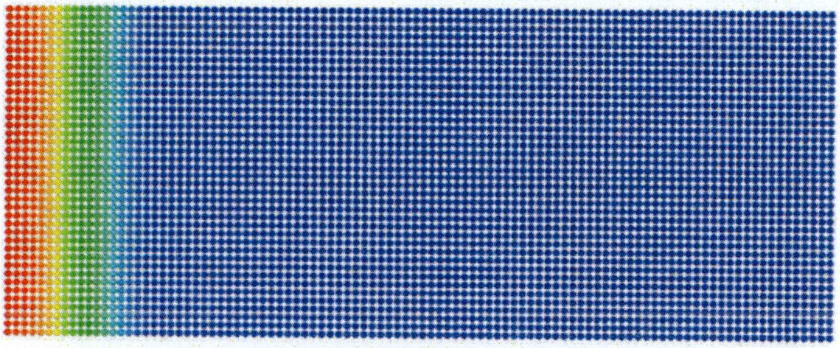

(a) Particles distribution of model I

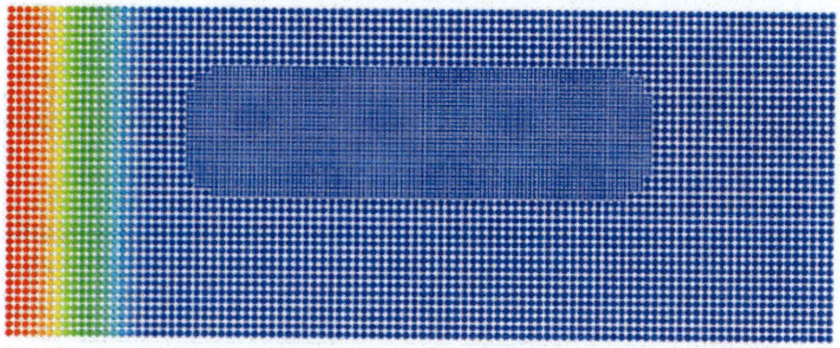

(b) Particles distribution of model II and III

Fig. 2.13 The discretizations of model I, II and III

adopted a discretization of varying horizons as shown in Fig. 2.13b. The horizon radius associated with each particle is set as 3 times the particle size for all models. Therefore, along the interface between the coarse and fine discretization, the horizon sizes vary. The minimal particle size of the 3 models is $\Delta x = 5 \times 10^{-4}$ m, yielding a critical time increment $\Delta t_{max} = \Delta x / v = 5 \times 10^{-4}$ s. The total time is set at 0.5 s with time increment $\Delta t = 5 \times 10^{-4}$ s.

As BB-PD cannot model solids with Poisson's ratio $\nu = 0$, the OSB-PD model is adopted. Model I was solved with the OSB-PD with constant horizons. Model

Table 2.2 Values of the peridynamics parameters for the mixed model

Model	$\Delta x = \Delta y$	$10^3 \cdot \delta$	Particle numbers
I	0.001	3	4000
II	0.001,0.0005	3/1.5	8628
III	0.001,0.0005	3/1.5	8628

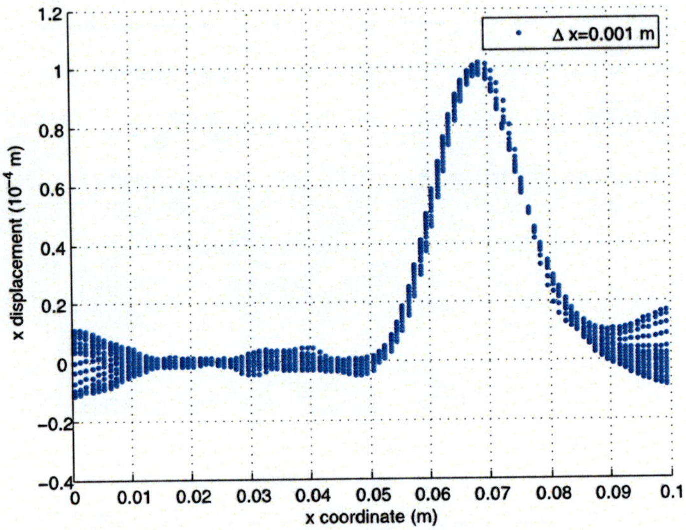

Fig. 2.14 Displacement wave u_x versus x coordinate of model I at step 650

II is the traditional OSB-PD with variable horizons (without additional treatment for ghost force), and model III is our dual-horizon formulation for OSB-PD with variable horizons. The parameters used for the three models are listed in Table 2.2.

Figures 2.14, 2.15 and 2.16 show the displacement in the x-direction along the x-coordinate of all particles. The red star points are particles with $\Delta x = 0.0005$ m, and the blue dot points refer to $\Delta x = 0.001$ m.

The displacement in model III (present dual-horizon PD) is almost identical to that of model I (PD with constant horizon), as shown in Figs. 2.14 and 2.16. Spurious wave reflections are observed for model II, as shown in Fig. 2.15. At step 650, several wave peaks are observed in model II and the maximum displacement is lower than that of models I and III. Therefore, the displacement waves in model II were affected by spurious wave reflections and the results deviate greatly from results of I and III.

We compute the L_2 errors of the wave profile along the line-CD (see Fig. 2.12). The L_2 error (see Table 2.3) of the present peridynamics formulation can achieve almost the same accuracy as the conventional peridynamics with the constant horizon.

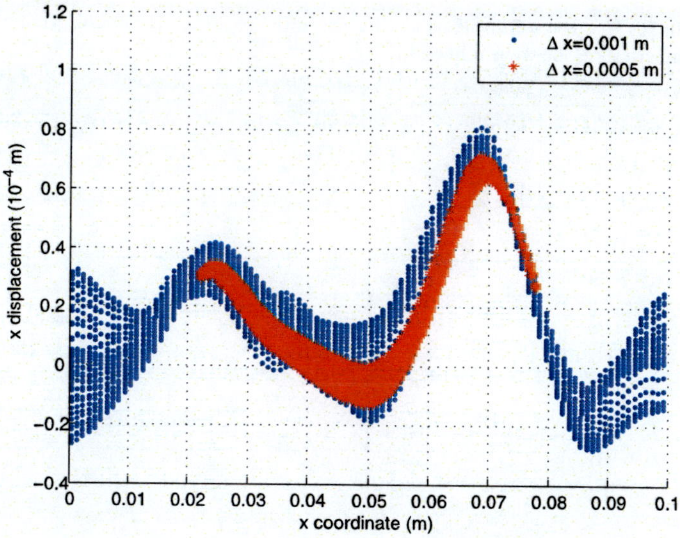

Fig. 2.15 Displacement wave u_x versus x coordinate of model II at step 650

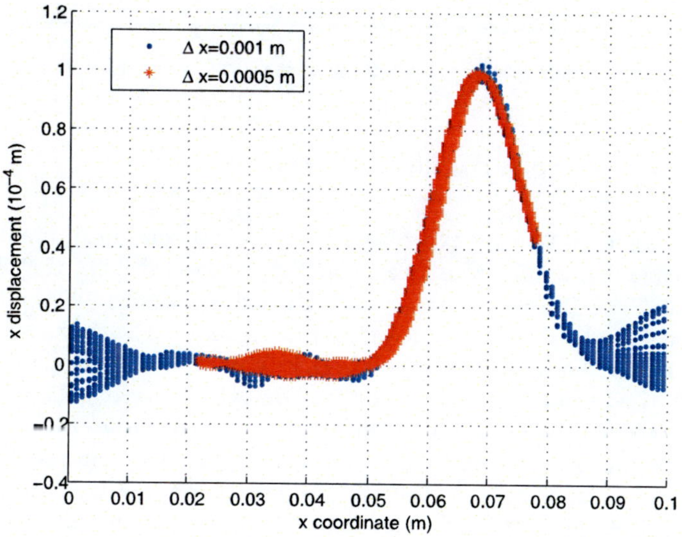

Fig. 2.16 Displacement wave u_x versus x coordinate of model III at step 650

Table 2.3 L_2 error of displacement for the initial condition (Gauss distribution of displacement) before and after the wave reflection

Model	Step 100	Step 200	Step 650	Step 990
I	0.0410	0.0452	0.1277	0.2498
II	0.1149	0.1738	0.7638	1.3233
III	0.0447	0.0500	0.1253	0.2871

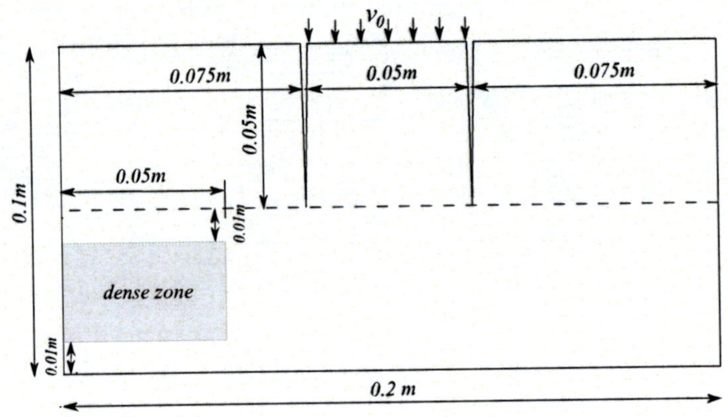

Fig. 2.17 Kalthoff–Winkler's experimental setup

2.5.3 *Kalthoff–Winkler Experiment*

The Kalthoff–Winkler experiment is a classical benchmark problem for dynamic fracture (Batra and Ravinsankar 2000; Belytschko et al. 2003; Li et al. 2002; Rabczuk et al. 2007, 2010; Song et al. 2006; Xu and Needleman 1994). The geometry and model used for the test are depicted in Fig. 2.17. The thickness of the specimen is 0.01 m. For a plate made of steel 18Ni1900 subjected to an impact loading at the speed of $v_0 = 32$ m/s, brittle fracture was observed (Silling 2003). The crack propagates from the end of the initial crack at an angle around 70° versus the initial crack direction. The material parameters used are the same as in Silling (2003), i.e. the elastic modulus $E = 190$ GPa, $\rho = 7800\,\text{kg/m}^3$, $\nu = 0.25$ and the energy release rate $G_0 = 6.9 \times 10^4$ J/m^2. The impact loading was imposed by applying an initial velocity at $v_0 = 22$ m/s to the first three layers of particles in the domain as shown in Fig. 2.17; all other boundaries are free. The plate is discretized with two different particle sizes, namely $\Delta x_{\text{coarse}} = 1.5625 \times 10^{-3}$ m for the coarse subdomain and $\Delta x_{\text{fine}} = 0.5\Delta x_{\text{coarse}} = 7.8125 \times 10^{-4}$ m for the fine subdomain located in the left bottom corner of the model; see Fig. 2.17. Along the thickness of the plate (in a direction perpendicular to the plane surface), four layers of particles in the coarse subdomain and eight layers in the fine subdomain are employed. The total number of particles is 57,968.

The crack propagation speed is computed by

$$V_{l-0.5} = \frac{\|\mathbf{x}_l - \mathbf{x}_{l-1}\|}{t_l - t_{l-1}} , \tag{2.60}$$

where $\mathbf{x}_l$ and $\mathbf{x}_{l-1}$ are the positions of the crack tip at the times t_l and t_{l-1}, respectively. An expression for the Rayleigh wave speed c_R (Graff 1975) is given as

$$\frac{c_R}{c_s} = \frac{0.87 + 1.12\nu}{1 + \nu} , \tag{2.61}$$

where ν is Poisson's ratio, $c_s = \sqrt{\mu/\rho}$ is the shear wave s peed and μ is the shear modulus. The crack starts to propagate at 26.3 μs. The highest crack speed in the simulation is 1530 m/s, about 54.4% of the Rayleigh speed (2799.2 m/s). The crack patterns are nearly symmetrical, as shown in Figs. 2.18, 2.19 and 2.20. Figure 2.22 illustrates the crack speed is very close to that predicted by the peridynamics formulation using constant horizons ($\Delta x_{\text{uniform}} = 1.5625 \times 10^{-3}$ m). The crack propagates initially at an angle of 65.7° in the fine subdomain with respect to the original crack and 65.8° in the coarse subdomain. The crack propagation angle of two other methods (Belytschko et al. 2003; Rabczuk et al. 2010) are 65.1° and 63.8° as shown in Fig. 2.21, respectively.

We also tested the influence of the inhomogeneous discretization on the crack pattern in Kalthoff–Winkler problem. In order to create the irregular material points distribution in the computational domain, we utilize the Abaqus software to mesh the geometrical model with different mesh seed settings, and then convert the finite element mesh into material points. For example, in the case of a triangle element or tetrahedron element, each element's area (or volume) is evenly allocated to its nodes, as shown in Figs. 2.23 and 2.24. Every node (acted as material point) is associated with a unique volume or mass. The material point size Δx is the equivalent diameter,

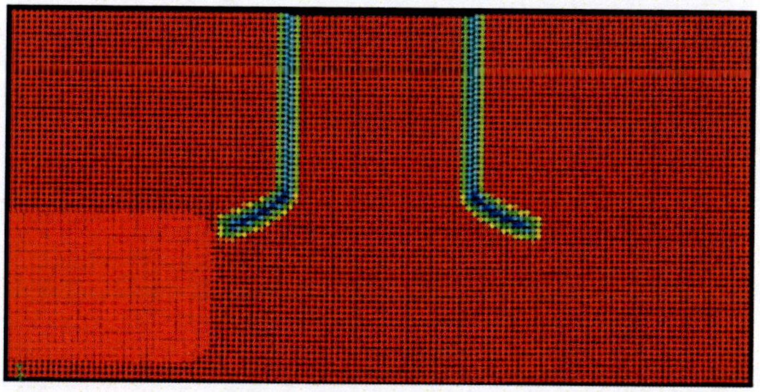

Fig. 2.18 The crack pattern of Kalthoff–Winkler plate by the present dual-horizon peridynamics at step 350

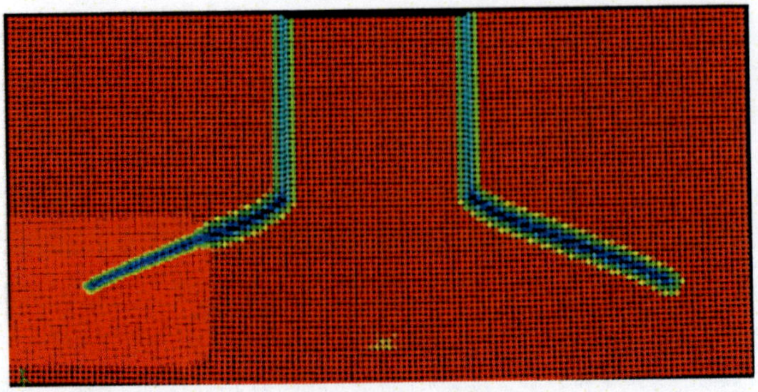

Fig. 2.19 The crack pattern of Kalthoff–Winkler simulation by the present dual-horizon peridynamics at step 650

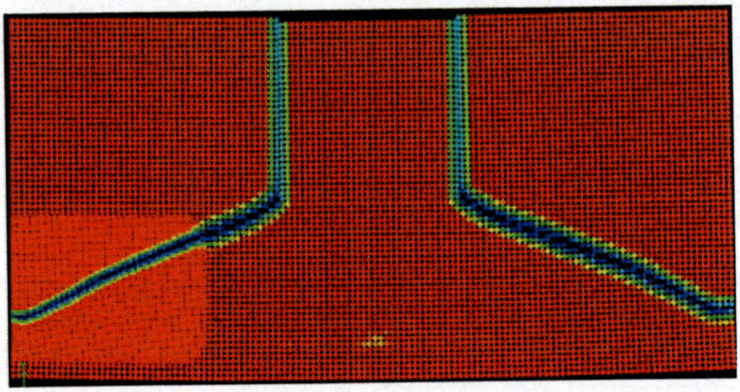

Fig. 2.20 The crack pattern of Kalthoff–Winkler simulation by the present dual-horizon peridynamics at step 875

Fig. 2.21 The crack paths of the Kalthoff–Winkler problem predicted by DH-PD, XFEM and CPM

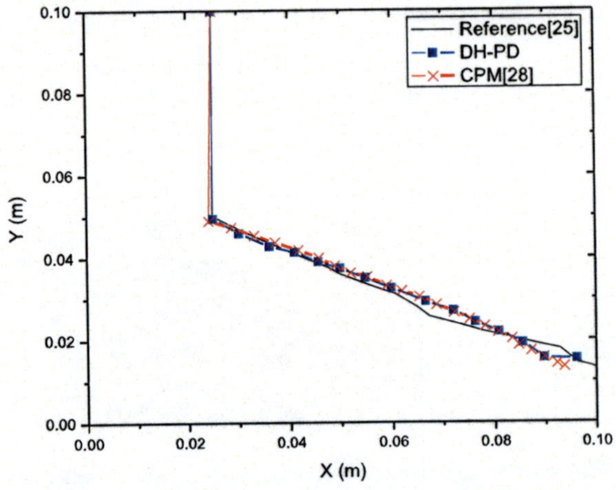

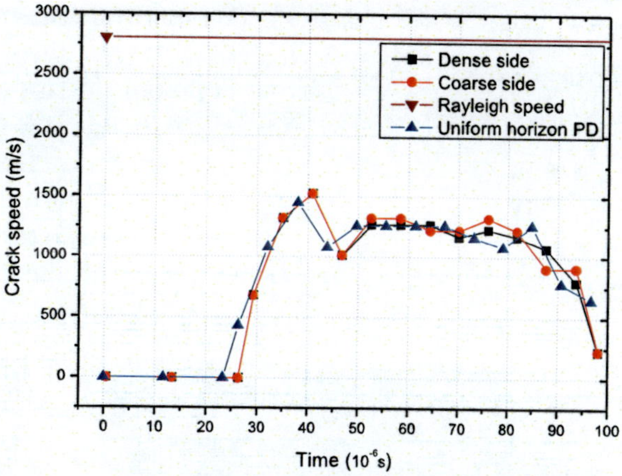

Fig. 2.22 The crack speed of Kalthoff–Winkler simulation by the present dual-horizon PD

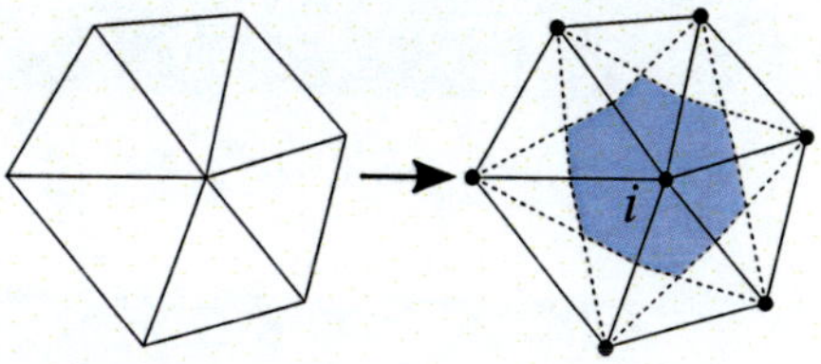

Fig. 2.23 Triangle elements to material points

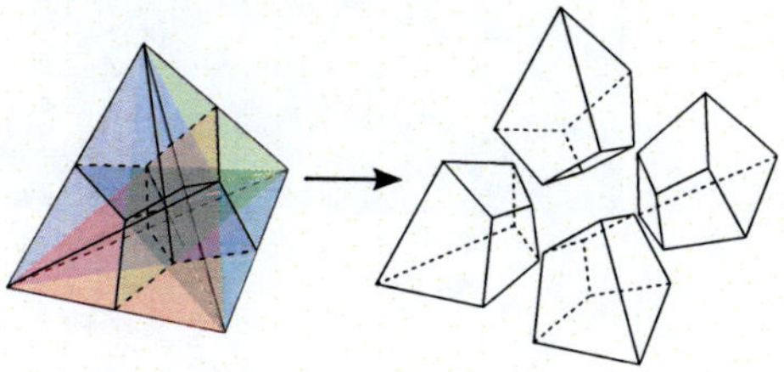

Fig. 2.24 Tetrahedron element to material points

which is calculated based on its volume by assuming the material point is a circle or sphere.

Six cases as shown in Table 2.4 are tested. Cases I, II are in two dimensions with the same material points distribution where the ratio $\Delta x_{max}/\Delta x_{min} = 3.1$, while Cases III–VI in three dimensions with particle size ratio $\Delta x_{max}/\Delta x_{min} = 4.4$. Cases I and III are based on the traditional constant-horizon peridynamics (CH-PD), while Case II and Case IV are carried out with the DH-PD. Case V is conducted to test the influence of spurious wave reflections when variable horizons are used in traditional PD. Case VI is used to test the effect of volume correction given in Appendix A. A coarse particle discretization with a bias toward a "wrong" crack path for Case I–II and a finer particle discretization for Case III–VI are used. In Cases II and IV–VI, the radius of each material point's horizon is selected as three times the material point size, e.g. $\delta_i = 3.015\ \Delta x_i$, while in Cases I and III $\delta = 3.015\ \Delta x_{max}$.

Table 2.4 The parameters in 4 simulations of Kalthoff–Winkler experiment

Case	$\delta_{max}/\delta_{min}$	Type	Volume correction
I	1	CH-PD	Yes
II	3.1	DH-PD	Yes
III	1	CH-PD	Yes
IV	4.4	DH-PD	Yes
V	4.4	CH-PD	Yes
VI	4.4	DH-PD	No

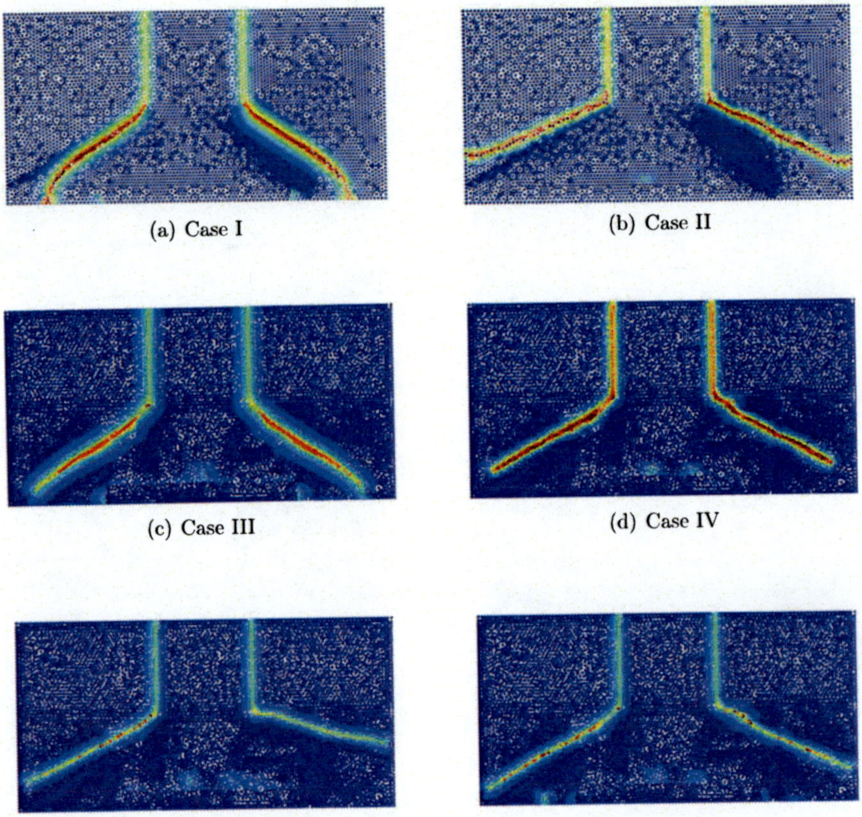

(a) Case I (b) Case II

(c) Case III (d) Case IV

(e) Case V (f) Case VI

Fig. 2.25 The crack patterns in the simulations of Kalthoff–Winkler experiment

Table 2.5 The angle of crack propagation with respect to the original crack

	Case I	Case II	Case III	Case IV	Case V	Case VI
Left	56.1°	67.6°	56.4°	61.4°	62.5°	58.2°
Right	57.6°	64.7°	56.3°	62.7°	76.1°	65.3°

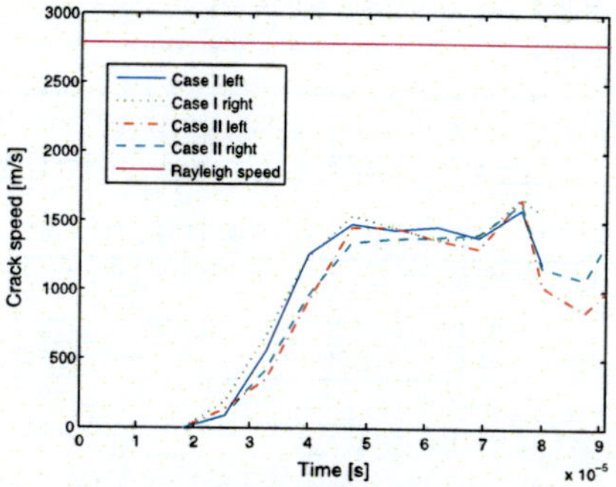

(a) Case I and Case II

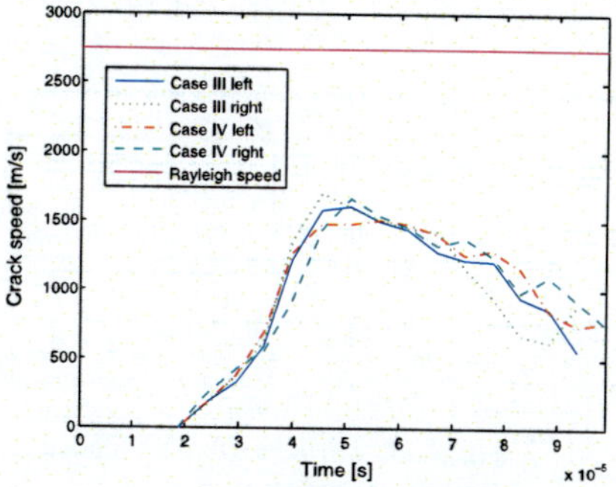

(b) Case III and Case IV

Fig. 2.26 The crack propagation speed in the simulations of Kalthoff–Winkler experiment

The angle of the crack propagation with respect to the original crack direction is shown in Table 2.5. For relatively low speed impact (e.g. $\leq$30 m/s), the crack angle is approximately 70°. Table 2.5 shows that the result by DH-PD is better than that by CH-PD.

The results in Fig. 2.25 and Table 2.5 show that the crack pattern both in 2D and 3D cases by the traditional constant-horizon bond-based peridynamics differs from the experiment, while the crack pattern obtained by the DH-PD agrees well with the experiment besides the introduction of a mesh bias toward an "incorrect" crack path. The comparison of Case IV and Case V shows the crack pattern is affected by spurious wave reflections. The comparison of Case IV and Case VI shows the volume correction has some positive effect on crack propagation, therefore, volume correction is recommended.

The crack starts to propagate at 20 μs. The crack speed in all four case simulations is shown in Fig. 2.26. The maximum crack propagation speeds are 1669 m/s, 1658 m/s, 1699 m/s and 1664 m/s for Cases I, II, III and IV, respectively. All of them are within the limit of 75% of Rayleigh's speed. Therefore, it can be concluded that with the present formulation, the random material points distribution has limited influence on the crack propagation when simulated by DH-PD.

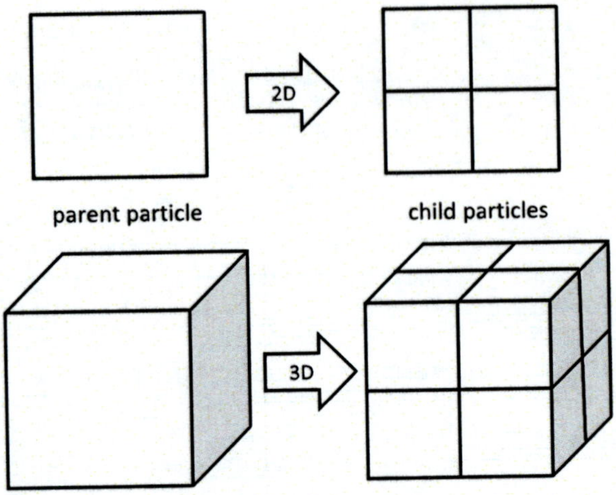

Fig. 2.27 Particle splitting for Adaptive refined peridynamics

2.5.4 Adaptive Refined Peridynamics

The present dual-horizon peridynamics formulation provides the possibility for adaptive refinement within a simple and unified framework. Two examples of adaptively refined peridynamics will be tested in this section: the Kalthoff–Winkler experiment in Sect. 2.5.3 and the plate with pre-crack subjected to traction. The threshold for the adaptive refinement is determined by both the damage-state criteria and energy state, as proposed in Dipasquale et al. (2014). The adaptive refinement procedure consists of two steps:

(1) Search for the particles that exceed the threshold values. Note that the refinement is not only applied to the particles above the threshold but also to the neighboring particles. In this way, it can be ensured that the crack tip always remains inside the refined zone.
(2) Split the particle, named as parent particle, into small particles, named as child particles. The properties that will be mapped from the parent particles to the child particles include mass, volume, coordinate, displacement and velocity.

The method to split particles for structured discretization is shown in Fig. 2.27. Splitting the particle results in halving the maximum time step. In all examples, we restrict each particle from being allowed to split only once in the entire analysis.

3D adaptive refined Kalthoff–Winkler simulation

We now repeat the Kalthoff–Winkler experiment from Sect. 2.5.3. However, the initial discretization is based on uniform horizon size. The initial particle size is $\Delta x_{\text{initial}} = 1.5625 \times 10^{-3}$ m. After refinement, the particle size is reduced to $\Delta x_{\text{refined}} = 7.8125 \times 10^{-4}$ m around the crack. The total particles increased from

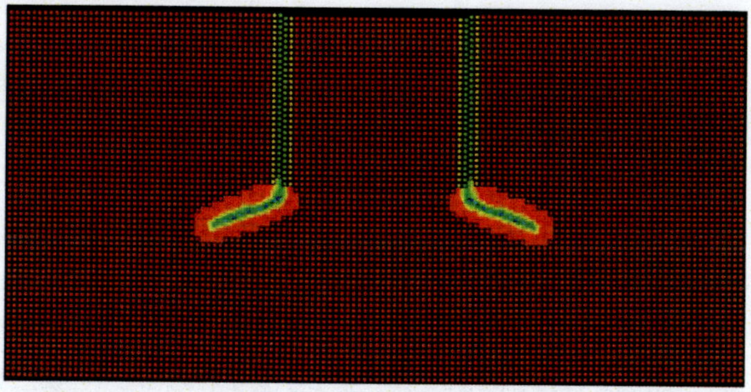

Fig. 2.28 The crack pattern of adaptive refined Kalthoff–Winkler experiment at step 420

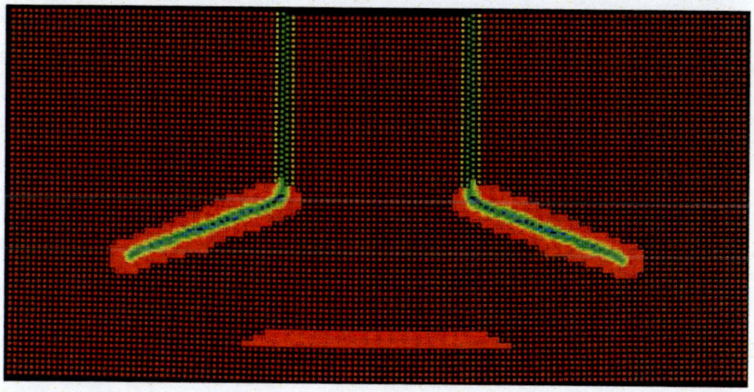

Fig. 2.29 The crack pattern of adaptive refined Kalthoff–Winkler experiment at step 600

32,768 to 54,860 at the end of the simulation. A uniform refinement would result in 32,768×8=262,144 particles.

The crack patterns at certain steps are plotted in Figs. 2.28, 2.29 and 2.30. The crack tip is always contained inside the refined zone. According to Fig. 2.31, the maximum crack propagation speed reaches 1268 m/s, and the average speed is 1047.6 m/s. The crack propagation takes place at an angle of approximately 66.5° with respect to the initial crack direction.

Plate with pre-crack subjected to traction

In the last example, we will test the capability of the new formulation in modeling crack branching. Hence, a plate with a pre-crack subjected to traction is considered as shown in Fig. 2.32. The traction remains constant during the entire simulation. This benchmark example is widely studied by Xu and Needleman (1994), Sharon et al. (1995), Ravi-Chandar (1998), Belytschko et al. (2003), Song et al. (2006) and Ha and Bobaru (2010), to name a few.

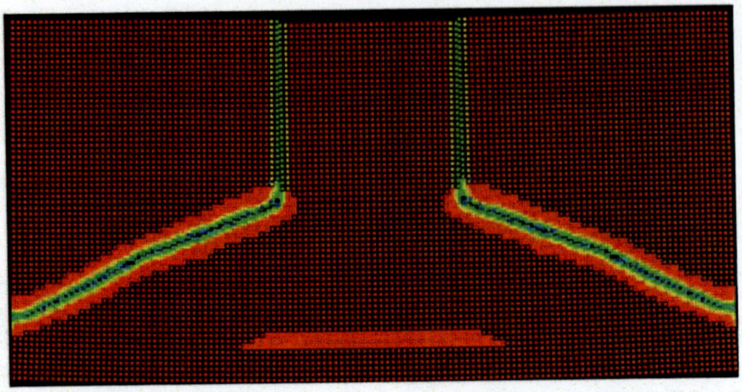

Fig. 2.30 The crack pattern of adaptive refined Kalthoff–Winkler experiment at step 925

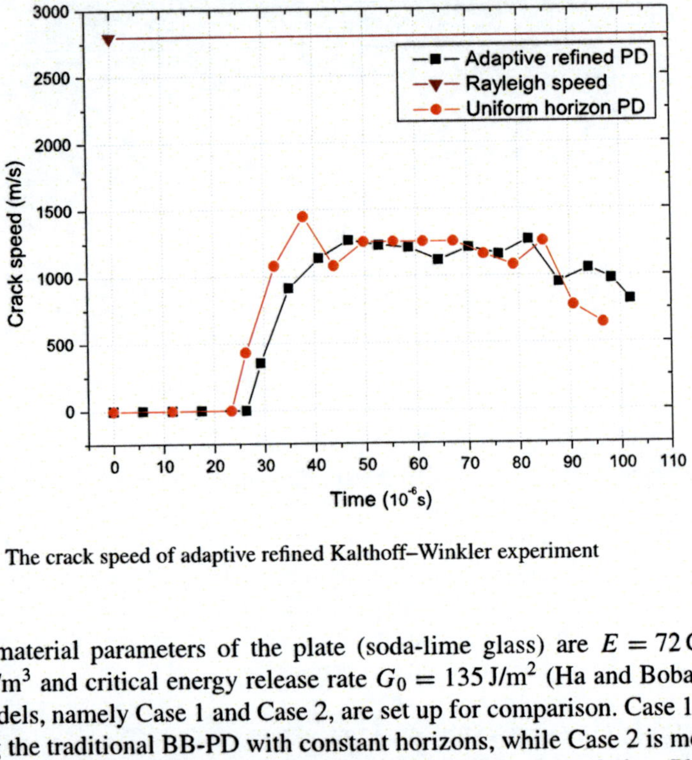

Fig. 2.31 The crack speed of adaptive refined Kalthoff–Winkler experiment

The material parameters of the plate (soda-lime glass) are $E = 72\,\text{GPa}$, $\rho = 2440\,\text{kg/m}^3$ and critical energy release rate $G_0 = 135\,\text{J/m}^2$ (Ha and Bobaru 2010). Two models, namely Case 1 and Case 2, are set up for comparison. Case 1 is solved by using the traditional BB-PD with constant horizons, while Case 2 is modeled by an adaptive refined BB-PD using the present dual-horizon formulation. Plane stress conditions are assumed, and all simulations are carried out in 2D. The particle sizes chosen are 5×10^{-4} m for Case 1 and 1×10^{-3} m for Case 2, respectively. Case 2 uses an adaptive refined model, allowing only 1 refinement step leading to 16,000 particles for Case 1 and 4000-6424 for Case 2, respectively. The final crack patterns are depicted in Figs. 2.33 and 2.34. The Rayleigh speed for the material is 3102 m/s according to Eq. (2.61). The maximum crack speed is 1881.4 m/s (60.6% of c_R) for

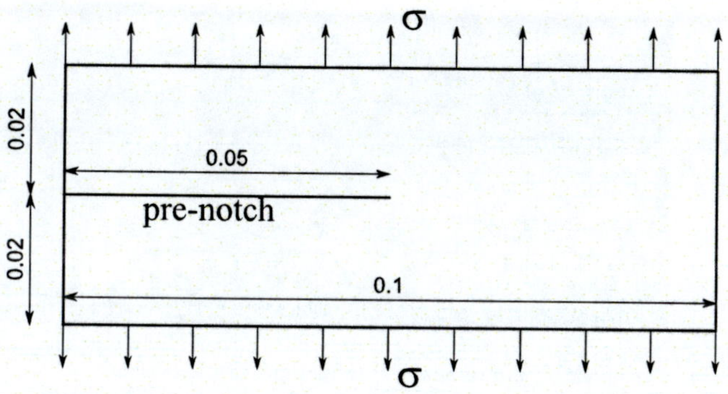

Fig. 2.32 Setup of the pre-cracked plate under traction load

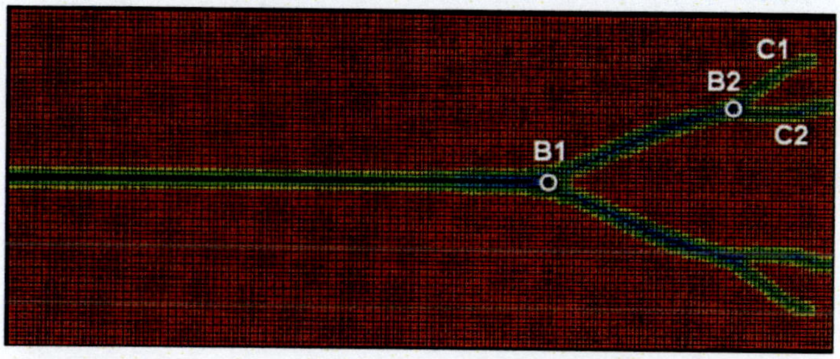

Fig. 2.33 The crack pattern of pre-cracked plate with uniform horizons

Case 1 and 2184.1 m/s (70.4% of c_R) for Case 2, respectively. One possible reason for the different maximum crack speeds is the deviation of the mapping method in adaptive refinement. Figure 2.35b compares the normalized crack speed of our DH-PD formulation with results obtained by the phantom node method (Song et al. 2006). It is well known that the dynamics crack speed is very sensitive for branching cracks. For this example, the results between the two methods generally agree, and both of them showed around 75% of the Rayleigh wave speed, which are physically reasonable.

For Case 1, the crack starts to propagate at 11.9 μs, and the first crack branching at point $B1$ takes place at 24.5 μs at the speed of 1043.6 m/s and the second crack branch point $B2$ is observed at 40.9 μs at the speed of 1147.1 m/s. Once branching occurs, the crack propagation speed decreases, since the energy released to create more fracture surfaces decelerates the propagation speed. Afterward, the crack speed increases again gradually. For Case 2, the initial crack starts to propagate at 10.3 μs; the first branching at point $B1$ occurs at 19.4 μs with a crack speed of 1247.6 m/s, while the second branching point $B2$ is observed at 34.2 μs with a crack propagation speed of 1330.6 m/s. The crack pattern is similar to Case 1.

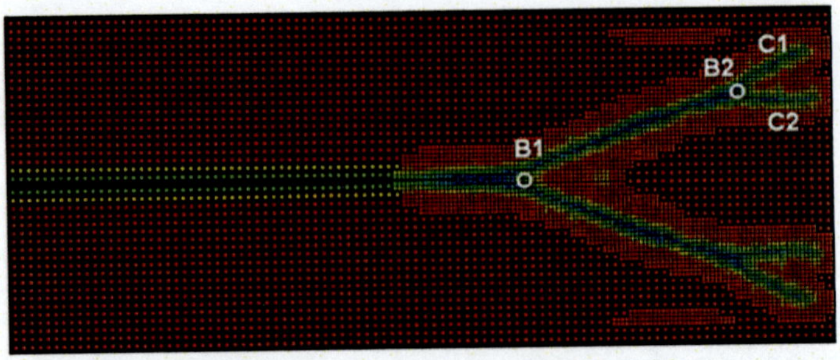

Fig. 2.34 The crack pattern of adaptive refined pre-cracked plate

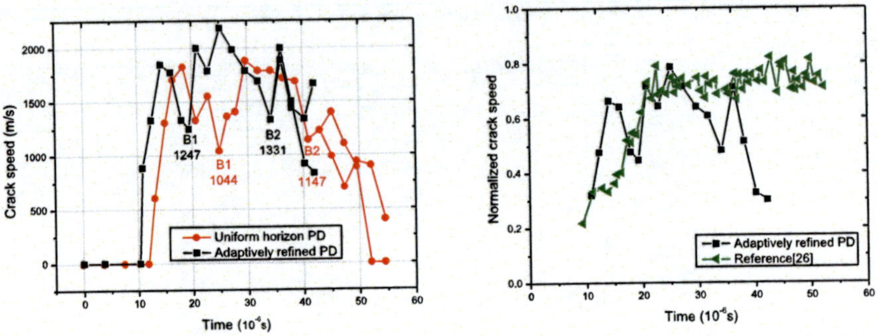

(a) Crack speed by DH-PD

(b) Normalized crack speed compared with the phantom node method[Song et al., 2006]

Fig. 2.35 The crack speed of pre-cracked plate under traction by the present DH-PD and the phantom node method (Song et al. 2006)

2.5.5 *Multiple Materials*

The material interface problem in peridynamics was formally discussed in Alali and Gunzburger (2015), where the authors analyzed the convergence of peridynamics for heterogeneous media, and found that the operator of linear peridynamics diverges in the presence of material interfaces. Some approaches have been proposed to deal with this issue. In reference Silling et al. (2007), the interfaces and mixtures between multiple materials are accounted for by using separate influence functions for each material. For example, if $\mathbf{x}$ is near an interface between two materials denoted by (1) and (2), two separate influence functions $\omega^{(1)}$ and $\omega^{(2)}$ can be defined such that each vanishes at points $\mathbf{x}+\boldsymbol{\xi}$ outside its associated material. The force state is given by

$$\mathbf{T}_{\mathbf{x}\mathbf{x}'} = \omega^{(1)}\langle\boldsymbol{\xi}\rangle P_{\mathbf{x}}^{(1)}\mathbf{K}^{-1}\boldsymbol{\xi} + \omega^{(2)}\langle\boldsymbol{\xi}\rangle P_{\mathbf{x}}^{(2)}\mathbf{K}^{-1}\boldsymbol{\xi} \tag{2.62}$$

where $P_{\mathbf{x}}^{(1)}$ and $P_{\mathbf{x}}^{(1)}$ are the first Piola–Kirchhoff stress from the respective constitution models and respective deformation gradient tensors $\mathbf{F}^{(1)}$ and $\mathbf{F}^{(2)}$, $\mathbf{K}$ being the shape tensor. In the DH-PD formulation, any point $\mathbf{x}$ has a force state $\mathbf{T}_{\mathbf{x}\square}$, which

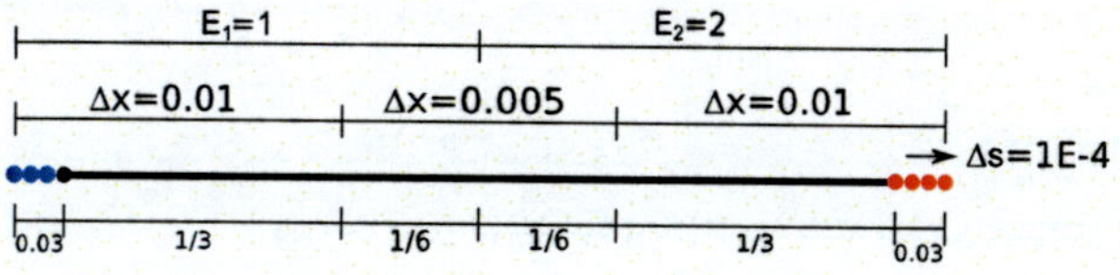

Fig. 2.36 The setup of 1D bar

acts like a force potential field applied to its nearby neighbors in $H_{\mathbf{x}}$. We assume $\mathbf{x}$ imposes reaction bond forces to its neighbors no matter what the neighbor's material type is. The bond force can be determined by one side or both sides. For example, in the BB-PD, microelastic modulus $C(\delta)$ and critical material stretch $s(\delta)$ are the primary parameters. For two points i and j with initial coefficient $[C(\delta_i), s(\delta_i)]$ (abbreviated as $[C_i, s_i]$) and $[C(\delta_j), s(\delta_j)]$, respectively, when taking into consideration the combination of the initial coefficients, the alternate forms can be formulated as the arithmetic average or geometric average of micro-elastic moduli or critical stretches.

For the OSB-PD and NOSB-PD in multiple materials, the force state $t_{\mathbf{xx}'}$ and $T_{\mathbf{xx}'}$ can be calculated similarly to the single homogeneous material and hence, it is not necessary to distinguish between homogeneous and heterogeneous materials.

1D bar tensile test

The length of the bar is 1 m, discretized with grid spacing $\Delta x = 0.01, 0.005$ m. The elastic modulus and density are $E_1 = 1$, $E_2 = 2$ and $\rho = 1$, respectively. Boundary conditions are applied at 3 material points at each end, i.e. fixed boundary condition at the left end (blue points in Fig. 2.36) and displacement boundary conditions of Δs =1e-4 m (red points in Fig. 2.36). The damping coefficient is $\alpha = $ 5e-3 m. Two cases were tested. Case I is based on the traditional constant-horizon BB-PD but without any treatment for multiple materials, while Case II on dual-horizon BB-PD with variable horizons. The equilibrium state is obtained at t =0.8 s. The bar internal sectional force is approximated by summing the bond force at one side given by

$$f_i = \delta_i \sum_{j \in H_i, j > i} C_i s_{ij} \Delta V_j \tag{2.63}$$

where s_{ij} is the bond stretch for bond ij, and δ_i is the radius of horizon H_i. Due to the error of discretization, Eq. (2.63) cannot sum the sectional force from material point i accurately; when the right half of H_i contains just 3 neighbors and volume correction is considered, there are 2.5 material points in Eq. (2.63), which corresponds to 5/6 of the theoretical sectional force.

For Case I, the force discontinuity in the interface of different point spacing is observed in Fig. 2.37. The force solution diverts from the analytical solution. The reason for the result is that the micro-elastic modulus is determined by one point. Consider a point i close to the interface of two materials with micro-elastic modulus

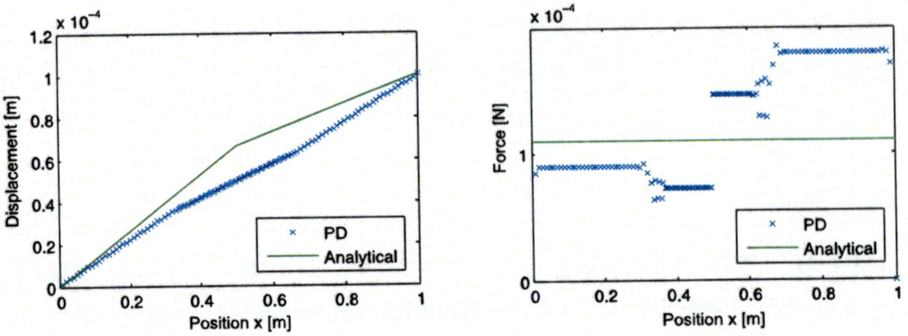

Fig. 2.37 The displacement and sectional force for Case 1

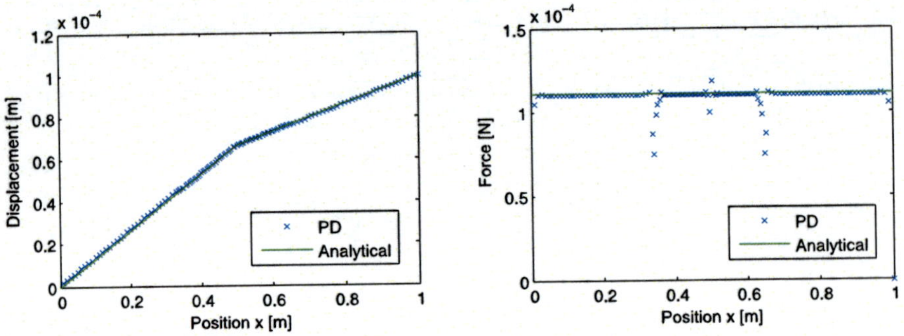

Fig. 2.38 The displacement and sectional force for Case 2

C_i; the equilibrium of i requires the total force from the left side to be equal to that from the right side. As a result, the displacement field in the left half and right half horizons are antisymmetrical with respect to i. Hence, there is no discontinuity in the strain field at the interface. At last, the spurious equilibrium displacement and the force field are obtained.

However, for Case II modeled with DH-PD, the force solution and displacement solution agree well with the analytical solutions, respectively; see Fig. 2.38. Both the material interface and point spacing interface have minimal influence on the final solution, though small variation is observed at the interfaces of material or grid spacing. The variation of the sectional force arises from the nonlocality in peridynamics. In PD, the force can transmit over a finite distance; the equilibrium of a point means the forces from horizon and dual-horizon are canceled. In this sense, the equilibrium is determined by the domains of horizon and dual-horizon; two nearby material points may have different domains, therefore, the jumping of internal sectional force is reasonable. This is essentially different from the ghost force which is unbalanced. In conclusion, we show for a simple 1D example that DH-PD has the potential to effectively deal with multiple materials.

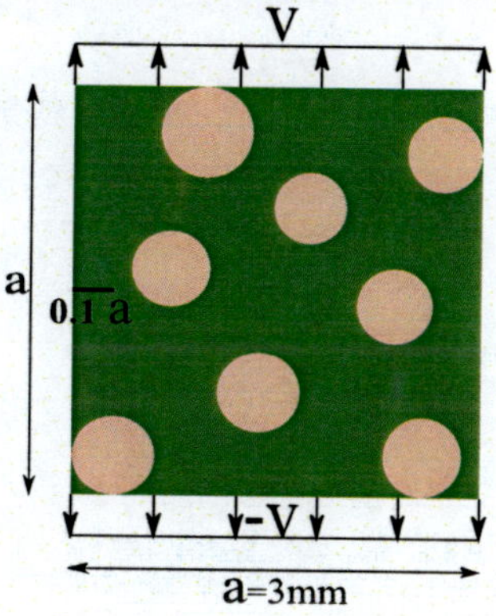

Fig. 2.39 The setup of the plate with inclusions

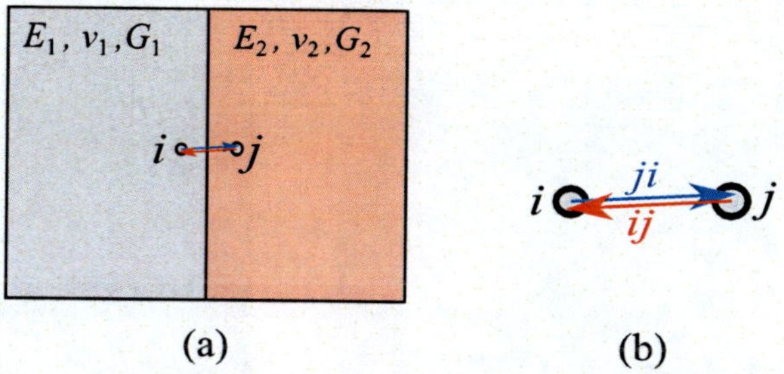

Fig. 2.40 The bonds in composite materials. Bond ji depends on i. When calculating i, all neighbors in H_i pretend to have the same parameters as i. Bond ij depends on j; when calculating j, all neighbors in H_j pretend to have the same parameters as j. Bonds ij and ji are broken separately

Crack propagation in heterogeneous materials

Consider a plate subjected to constant velocity traction at both the top and bottom edges, as shown in Fig. 2.39. The matrix is filled with many randomly distributed inclusions; all inclusions have the same material parameters. The material parameters for matrix include $E_1 = 72$ GPa, $\nu_1 = 1/3$ and $G_0 = 40$ J/m^2. The parameters for the inclusions are $E_2 = 144$ GPa, $\nu_2 = 1/3$ and $G_0 = 80$ J/m^2. The plate is discretized into 40,000 material points with point spacing of $\Delta x = 1.5$e-5m. The horizon radius is $\delta = 3.015\Delta x$. The material's critical bond stretch is $s_1 = s_2 \approx 4.64$e-3. Three cases are conducted to show the capability of DH-PD in dealing with composite materials. The velocity of traction on the boundary is $v_{bc} = 0.5$ m/s for Cases I, II and $v_{bc} = 2$ m/s for Case III. Two different heterogeneous structures are tested; the heterogeneous structures for Cases II and III coincide.

We use a simple damage rule to take into account the crack propagation in heterogeneous materials. In the framework of DH-PD, the bond and dual-bond are broken separately. For example, as shown in Fig. 2.40, bond ji is interpreted as i exerting force on j, and the strength of bond ji is assumed to depend only on i parameters, e.g. i's critical stretch; the same applies to bond ij. Figure 2.41b, c shows when the

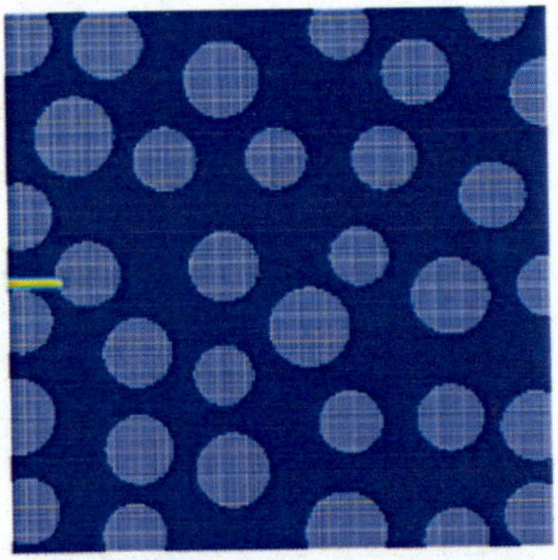

(a) Initial crack

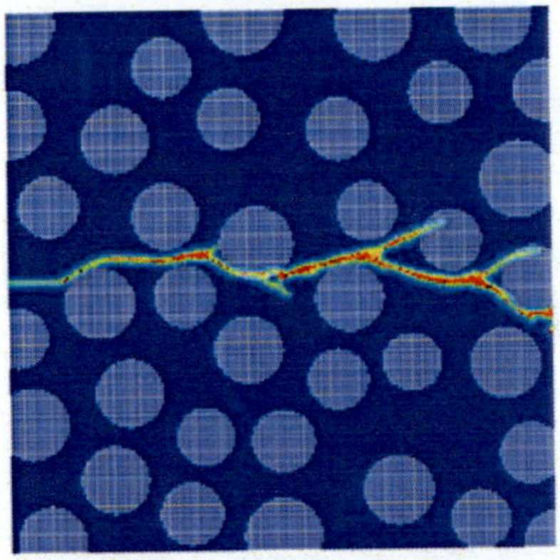

(b) Case I : $v_{bc} = 0.5$ m/s

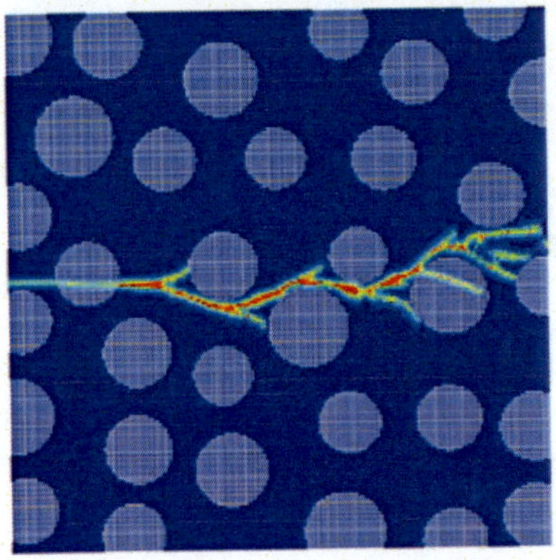

(c) Case II : $v_{bc} = 0.5$ m/s

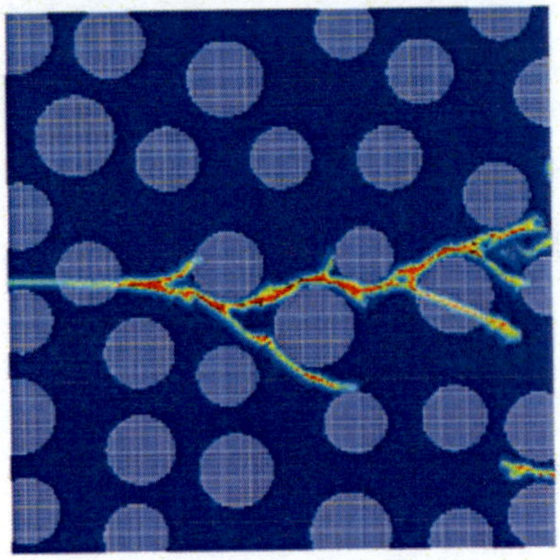

(d) Case III : $v_{bc} = 2$ m/s

Fig. 2.41 The crack patterns in multiple-material plate

initial crack tip is in the inclusion (Fig. 2.41c), the crack will firstly propagate in the inclusions; when the initial crack tip is in the matrix, the crack firstly propagates in the matrix. We also note two phenomena when the load rate is increased:

(1) The number of crack branches has increased. Such a phenomenon is well-known in dynamic fracture.
(2) The crack propagates through the inclusions. This tendency has been also observed experimentally, i.e. the number of aggregates/particle cracks in a weaker matrix material is increased with increasing strain rate/load rate.

2.6 Conclusions

This chapter contributes to the development of peridynamics with varying horizons. Therefore, the interactions between particles are based on two independent horizons, i.e. the direct force from the horizon and the reaction force from the dual-horizon have naturally eliminated spurious wave reflections and ghost forces present in the conventional peridynamics formulation. Based on the new concepts of horizons, the equation of motion of the dual-horizon peridynamics is derived. We show that both the balance of linear momentum and angular momentum are satisfied. The key difference between the present peridynamics formulation from the conventional one is the way the reaction forces are computed. Hence, the dual-horizon peridynamics can be implemented in any existing peridynamics code with minimal effort. Three numerical examples show the present method is free from spurious wave reflection, and the accuracy is retained along the interface where horizon sizes undergo sudden changes. The method also shows its capability for fracture problems including crack branching. A simple h-adaptive scheme is proposed to improve the crack path resolution, both for 2D and 3D cases. The present method is particularly promising for multiscale analysis where the models at different length scales can be bridged by using different horizon settings.

References

Alali B, Gunzburger M (2015) Peridynamics and material interfaces. J Elast 1–24

Askari E, Bobaru F, Lehoucq R, Parks M, Silling S, Weckner O (2008) Peridynamics for multiscale materials modeling. J Phys Conf Ser 125:012078. IOP Publishing. https://doi.org/10.1088/1742-6596/125/1/012078

Batra R, Ravinsankar M (2000) Three-dimensional numerical simulation of the Kalthoff experiment. Int J Fract 105(2):161–186

Belytschko T, Chen H, Xu J, Zi G (2003) Dynamic crack propagation based on loss of hyperbolicity and a new discontinuous enrichment. Int J Numer Meth Eng 58(12):1873–1905. https://doi.org/10.1002/nme.941

Bobaru F, Ha Y (2011) Adaptive refinement and multiscale modeling in 2D peridynamics. J Multiscale Comput Eng 9(635–659):635–660. https://doi.org/10.1615/intjmultcompeng.2011002793

Bobaru F, Yang M, Alves LF, Silling S, Askari E, Xu J (2009) Convergence, adaptive refinement, and scaling in 1D peridynamics. Int J Numer Methods Eng 77(6):852–877. https://doi.org/10.1002/nme.2439

Dipasquale D, Zaccariotto M, Galvanetto U (2014) Crack propagation with adaptive grid refinement in 2D peridynamics. Int J Fract 190(1–2):1–22. https://doi.org/10.1007/s10704-014-9970-4

Gerstle W, Sau N, Silling S (2007) Peridynamic modeling of concrete structures. Nucl Eng Des 237(12):1250–1258. https://doi.org/10.4028/www.scientific.net/amm.638-640.1725

Graff K (1975) Wave motion in elastic solids. Clarendon Oxford

Ha Y, Bobaru F (2010) Studies of dynamic crack propagation and crack branching with peridynamics. Int J Fract 162:229–244. https://doi.org/10.1007/978-90-481-9760-6_18

Li S, Liu W, Rosakis A, Belytschko T, Hao W (2002) Mesh-free Galerkin simulations of dynamic shear band propagation and failure mode transition. Int J Solids Struct 39(5):1213–1240. https://doi.org/10.1016/s0020-7683(01)00188-3

Madenci E, Oterkus E (2014) Peridynamic theory and its applications. Springer

Oterkus S, Madenci E, Agwai A (2014) Fully coupled peridynamic thermomechanics. J Mech Phys Solids 64:1–23

Rabczuk T, Areias PMA, Belytschko T (2007) A simplified mesh-free method for shear bands with cohesive surfaces. Int J Numer Meth Eng 69(5):993–1021. https://doi.org/10.1002/nme.1797

Rabczuk T, Zi G, Bordas S, Nguyen-Xuan H (2010) A simple and robust three-dimensional cracking-particle method without enrichment. Comput Meth Appl Mech Eng 199(37):2437–2455. https://doi.org/10.1016/j.cma.2010.03.031

Ravi-Chandar K (1998) Dynamic fracture of nominally brittle materials. Int J Fract 90(1):83–102. ISSN 1573-2673. https://doi.org/10.1023/A:1007432017290

Sharon E, Gross SP, Fineberg J (1995) Local crack branching as a mechanism for instability in dynamic fracture. Phys Rev Lett 74(25):5096. https://doi.org/10.1103/physrevlett.74.5096

Silling S (2000) Reformulation of elasticity theory for discontinuities and long-range forces. J Mech Phys Solids 48(1):175–209. https://doi.org/10.2172/1895

Silling S (2003) Dynamic fracture modeling with a meshfree peridynamic code. Comput Fluid Solid Mech 641–644. https://doi.org/10.1016/b978-008044046-0.50157-3

Silling S, Askari E (2005) A meshfree method based on the peridynamic model of solid mechanics. Comput Struct 83(17):1526–1535

Silling S, Lehoucq R (2010) Peridynamic theory of solid mechanics. Adv Appl Mech 44:73–168

Silling S, Epton M, Weckner O, Xu J, Askari E (2007) Peridynamic states and constitutive modeling. J Elast 88(2):151–184

Silling S, Littlewood D, Seleson P (2014) Variable horizon in a peridynamic medium. Technical report, Sandia National Laboratories (SNL-NM), Albuquerque, NM (United States)

Song JH, Areias PM, Belytschko T (2006) A method for dynamic crack and shear band propagation with phantom nodes. Int J Numer Meth Eng 67(6):868–893

Warren T, Silling S, Askari A, Weckner O, Epton M, Xu J (2008) A non-ordinary state-based peridynamic method to model solid material deformation and fracture. Int J Solids Struct 46(5):1186–1195. https://doi.org/10.1016/j.ijsolstr.2008.10.029

Xu X, Needleman A (1994) Numerical simulations of fast crack growth in brittle solids. J Mech Phys Solids 42(9):1397–1434. https://doi.org/10.1016/0022-5096(94)90003-5

Yu K, Xin X, Lease K (2011) A new adaptive integration method for the peridynamic theory. Model Simul Mater Sci Eng 19(4):045003. https://doi.org/10.1088/0965-0393/19/4/045003

Chapter 3
First-Order Nonlocal Operator Method

The present chapter aims to develop a nonlocal operator method for solving the weak forms of PDEs based on weighted residual method or variational principles. Though the nonlocal theory is more general than the local theory, we focus on solving the local problems with the nonlocal operator method. The nonlocal operator method constructs the nonlocal operator to represent the nonlocal interaction without shape function or its derivatives in traditional meshless or finite element methods. The content is outlined as follows. In Sect. 3.1, the concepts of support and dual-support are introduced. Based on the support, the general nonlocal operator and its variation in continuous form or discrete form are defined. Since the differential operator forms the basis of different energy functionals, we study the capabilities of the nonlocal operator based on variational principles in obtaining the strong forms or weak forms of different functionals in Sect. 3.2. In Sect. 3.3, we discuss the zero-energy mode in the nonlocal operator and propose a universal operator energy functional to remove the zero-energy mode. The higher order nonlocal operators and higher order energy functional are generalized and obtained in Sect. 3.4. Some numerical examples are presented to validate the method in Sect. 3.5.

3.1 Support, Dual-Support and Nonlocal Operators

Consider a domain as shown in Fig. 3.1a, let $\mathbf{x}$ be spatial coordinates in the domain Ω; $\mathbf{r} := \mathbf{x}' - \mathbf{x}$ is a Euclidean vector (or a spatial vector, or simply a vector) starts from $\mathbf{x}$ to $\mathbf{x}'$; $\mathbf{v} := \mathbf{v}(\mathbf{x}, t)$ and $\mathbf{v}' := \mathbf{v}(\mathbf{x}', t)$ are the field values for $\mathbf{x}$ and $\mathbf{x}'$, respectively; $\mathbf{v_r} := \mathbf{v}' - \mathbf{v}$ is the relative field vector for spatial vector $\mathbf{r}$.

Support $\mathcal{S}_\mathbf{x}$ of point $\mathbf{x}$ is the domain where any spatial point $\mathbf{x}'$ forms spatial vector $\mathbf{r}(= \mathbf{x}' - \mathbf{x})$ from $\mathbf{x}$ to $\mathbf{x}'$. Support $\mathcal{S}_\mathbf{x}$ specifies the range of nonlocal interaction that happened with respect to point $\mathbf{x}$. The main function of support is to define different nonlocal operators. Now we introduce a shape moment tensor $\mathbf{K}^n_\mathbf{x}$ which is a $n-$order tensor

T. Rabczuk et al., *Computational Methods Based on Peridynamics and Nonlocal Operators*, Computational Methods in Engineering & the Sciences,
https://doi.org/10.1007/978-3-031-20906-2_3

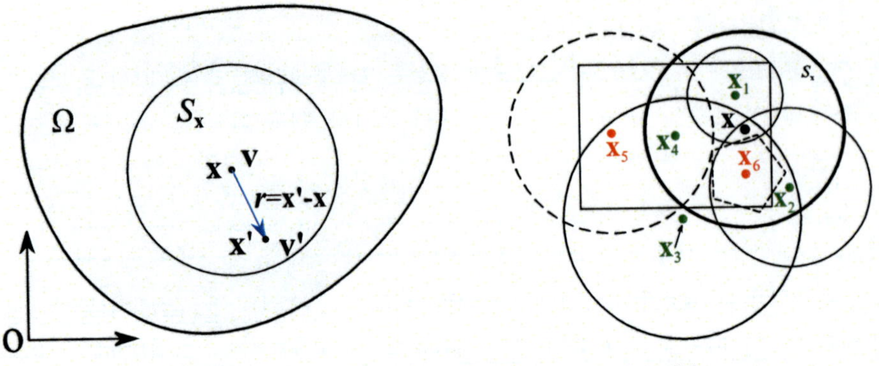

Fig. 3.1 **a** Domain and notations. **b** Schematic diagram for support and dual-support, all shapes above are supported, $\mathcal{S}_{\mathbf{x}} = \{\mathbf{x}_1, \mathbf{x}_2, \mathbf{x}_4, \mathbf{x}_6\}$, $\mathcal{S}'_{\mathbf{x}} = \{\mathbf{x}_1, \mathbf{x}_2, \mathbf{x}_3, \mathbf{x}_4\}$

$$\mathbf{K}^n_{\mathbf{x}} := \int_{\mathcal{S}_{\mathbf{x}}} w(\mathbf{r}) \underbrace{\mathbf{r} \otimes \mathbf{r} \otimes \cdots \otimes \mathbf{r}}_{\text{n terms}} dV_{\mathbf{x}'}, \tag{3.1}$$

where $w(\mathbf{r})$ is the weight function and can be, for example,

$$w(\mathbf{r}) = \frac{1}{r^3} \,. \tag{3.2}$$

Two special cases of the shape tensor for $\mathcal{S}_{\mathbf{x}}$ are the zeroth-order shape tensor (the weighted volume of the support) and the second-order shape tensor

$$\mathbf{K}_{\mathbf{x}} := \int_{\mathcal{S}_{\mathbf{x}}} w(\mathbf{r})\mathbf{r} \otimes \mathbf{r} dV_{\mathbf{x}'}. \tag{3.3}$$

Dual-support is defined as a union of the points whose supports include $\mathbf{x}$, denoted by

$$\mathcal{S}'_{\mathbf{x}} = \{\mathbf{x}' | \mathbf{x} \in \mathcal{S}_{\mathbf{x}'}\}. \tag{3.4}$$

The point $\mathbf{x}'$ forms dual-vector $\mathbf{r}'(= \mathbf{x} - \mathbf{x}' = -\mathbf{r})$ in $\mathcal{S}'_{\mathbf{x}}$. On the other hand, $\mathbf{r}'$ is the spatial vector formed in $\mathcal{S}_{\mathbf{x}'}$. One example to illustrate the support and dual-support is shown in Fig. 3.1b.

3.1.1 Nonlocal Operators in Support

The general operators in calculus include the gradient of the scalar and vector field, the curl and divergence of the vector field (Du et al. 2013). These operators have the corresponding nonlocal forms based on the Taylor series expansion. We use $\tilde{\nabla}$

to denote the nonlocal operator, while the local operators follow the conventional notations. The derivation adopts the conventions in classical calculus and linear algebra. The scalar, vector and two- or higher order tensor are denoted by small letter, small bold letter, capital bold letter, respectively. The vector is based on column form. For the purpose of being concise, the following notations are used simultaneously, i.e. $\mathbf{a}\cdot\mathbf{b}=\mathbf{a}^T\mathbf{b}, \mathbf{a}\otimes\mathbf{b}=\mathbf{a}\mathbf{b}^T, \mathbf{M}\cdot\mathbf{v}=\mathbf{M}\mathbf{v}, \mathbf{v}\cdot\mathbf{M}=\mathbf{v}^T\mathbf{M}, \mathbf{M}_1\cdot\mathbf{M}_2=\mathbf{M}_1\mathbf{M}_2, \mathbf{M}_1:\mathbf{M}_2=\text{tr}(\mathbf{M}_1\mathbf{M}_2^T)$, $\nabla=(\frac{\partial}{\partial x},\frac{\partial}{\partial y},\frac{\partial}{\partial z})^T$, and $\nabla\mathbf{v}=\nabla\otimes\mathbf{v}$, where superscript T refers to the transpose operation. The gradients of a scalar v and 3-vector $\mathbf{v}=(v_1,v_2,v_3)^T$ are denoted by, respectively.

$$\nabla v=\begin{bmatrix}\frac{\partial v}{\partial x}\\ \frac{\partial v}{\partial y}\\ \frac{\partial v}{\partial z}\end{bmatrix},\quad \nabla\mathbf{v}=\begin{bmatrix}\frac{\partial v_1}{\partial x} & \frac{\partial v_1}{\partial y} & \frac{\partial v_1}{\partial z}\\ \frac{\partial v_2}{\partial x} & \frac{\partial v_2}{\partial y} & \frac{\partial v_2}{\partial z}\\ \frac{\partial v_3}{\partial x} & \frac{\partial v_3}{\partial y} & \frac{\partial v_3}{\partial z}\end{bmatrix}.$$

Nonlocal gradient of a vector field $\mathbf{v}$ for point $\mathbf{x}$ in support $\mathcal{S}_\mathbf{x}$ is defined as

$$\tilde{\nabla}\otimes\mathbf{v}_\mathbf{x}:=\int_{\mathcal{S}_\mathbf{x}}w(\mathbf{r})\mathbf{v}_\mathbf{r}\otimes\mathbf{r}dV_{\mathbf{x}'}\cdot\mathbf{K}_\mathbf{x}^{-1}, \tag{3.5}$$

where $\mathbf{v}_\mathbf{r}=\mathbf{v}_{\mathbf{x}'}-\mathbf{v}_\mathbf{x}$, $\mathbf{K}_\mathbf{x}$ is the second-order shape tensor. One example of the nonlocal gradient is the nonlocal deformation gradient in Peridynamics (Silling et al. 2007).

In fact, the field value of nearby point $\mathbf{x}'$ in $\mathcal{S}_\mathbf{x}$ is obtained by Taylor series expansion as

$$\mathbf{v}_{\mathbf{x}'}=\mathbf{v}_\mathbf{x}+\nabla\otimes\mathbf{v}_\mathbf{x}\cdot\mathbf{r}+O(r^2), \tag{3.6}$$

where $O(r^2)$ represents order terms higher than one, and for a linear field $O(r^2)=0$. Insert Eq. 3.6 into RHS of Eq. 3.5 and integrate into support $\mathcal{S}_\mathbf{x}$, one verifies that the nonlocal operator converges to the local operator by the following derivation.

$$\begin{aligned}\tilde{\nabla}\otimes\mathbf{v}_\mathbf{x}-&\int_{\mathcal{S}_\mathbf{x}}w(\mathbf{r})\mathbf{v}_\mathbf{r}\otimes\mathbf{r}dV_{\mathbf{x}'}\cdot\mathbf{K}_\mathbf{x}^{-1}\\ =&\int_{\mathcal{S}_\mathbf{x}}w(\mathbf{r})(\mathbf{v}'_\mathbf{x}-\mathbf{v}_\mathbf{x})\otimes\mathbf{r}dV_{\mathbf{x}'}\cdot\mathbf{K}_\mathbf{x}^{-1}\\ =&\int_{\mathcal{S}_\mathbf{x}}w(\mathbf{r})\nabla\mathbf{v}_\mathbf{x}\cdot\mathbf{r}\otimes\mathbf{r}dV_{\mathbf{x}'}\cdot\mathbf{K}_\mathbf{x}^{-1}\\ =&\nabla\otimes\mathbf{v}_\mathbf{x}\cdot\int_{\mathcal{S}_\mathbf{x}}w(\mathbf{r})\mathbf{r}\otimes\mathbf{r}dV_{\mathbf{x}'}\cdot\mathbf{K}_\mathbf{x}^{-1}\\ =&\nabla\otimes\mathbf{v}_\mathbf{x}\cdot\mathbf{K}_\mathbf{x}\cdot\mathbf{K}_\mathbf{x}^{-1}\\ =&\nabla\otimes\mathbf{v}_\mathbf{x}.\end{aligned}$$

When $\mathbf{x}'$ is close enough to $\mathbf{x}$ or when support $\mathcal{S}_{\mathbf{x}}$ is small enough, the nonlocal operator can be considered as the linearization of the field. The nonlocal operator converges to the local operator in the continuous limit. On the other hand, the nonlocal operator defined by integral form, still holds in the case where strong discontinuity exists. The local operator can be viewed as a special case of the nonlocal operator.

Similarly, **nonlocal gradient** of a scalar field v for point $\mathbf{x}$ in support $\mathcal{S}_{\mathbf{x}}$ is defined as

$$\tilde{\nabla} v_{\mathbf{x}} := \int_{\mathcal{S}_{\mathbf{x}}} w(\mathbf{r}) v_{\mathbf{r}} \mathbf{r} dV_{\mathbf{x}'} \cdot \mathbf{K}_{\mathbf{x}}^{-1}, \tag{3.7}$$

where $v_{\mathbf{r}} = v_{\mathbf{x}'} - v_{\mathbf{x}}$.

Let $[\square]_\times$ denotes the conversion from 3×3 antisymmetric matrix to 3-vector,

$$\begin{bmatrix} 0 & -a_3 & a_2 \\ a_3 & 0 & -a_1 \\ -a_2 & a_1 & 0 \end{bmatrix}_\times \mapsto (a_1, a_2, a_3). \tag{3.8}$$

It is easy to verify that

$$\nabla \times \mathbf{a} = [\nabla \otimes \mathbf{a} - (\nabla \otimes \mathbf{a})^T]_\times \tag{3.9}$$

where superscript T refers to the transpose operation.

By analogy with Eq. 3.9, let

$$\begin{aligned}
\tilde{\nabla} \times \mathbf{v} &= [\tilde{\nabla} \otimes \mathbf{v} - (\tilde{\nabla} \otimes \mathbf{v})^T]_\times \\
&= \Big[\int_{\mathcal{S}_{\mathbf{x}}} w(\mathbf{r}) \mathbf{v}_{\mathbf{r}} \otimes (\mathbf{K}_{\mathbf{x}}^{-1} \cdot \mathbf{r}) dV_{\mathbf{x}'} - \int_{\mathcal{S}_{\mathbf{x}}} w(\mathbf{r}) (\mathbf{K}_{\mathbf{x}}^{-1} \cdot \mathbf{r}) \otimes \mathbf{v}_{\mathbf{r}} dV_{\mathbf{x}'}\Big]_\times \\
&= \Big[\int_{\mathcal{S}_{\mathbf{x}}} w(\mathbf{r}) \big(\mathbf{v}_{\mathbf{r}} \otimes (\mathbf{K}_{\mathbf{x}}^{-1} \cdot \mathbf{r}) - (\mathbf{K}_{\mathbf{x}}^{-1} \cdot \mathbf{r}) \otimes \mathbf{v}_{\mathbf{r}}\big) dV_{\mathbf{x}'}\Big]_\times \\
&= \int_{\mathcal{S}_{\mathbf{x}}} w(\mathbf{r}) \big[\mathbf{v}_{\mathbf{r}} \otimes (\mathbf{K}_{\mathbf{x}}^{-1} \cdot \mathbf{r}) - (\mathbf{K}_{\mathbf{x}}^{-1} \cdot \mathbf{r}) \otimes \mathbf{v}_{\mathbf{r}}\big]_\times dV_{\mathbf{x}'} \\
&= \int_{\mathcal{S}_{\mathbf{x}}} w(\mathbf{r}) (\mathbf{K}_{\mathbf{x}}^{-1} \cdot \mathbf{r}) \times \mathbf{v}_{\mathbf{r}} dV_{\mathbf{x}'}.
\end{aligned}$$

In the last step, $\mathbf{b} \times \mathbf{a} = \big[\mathbf{a} \otimes \mathbf{b} - \mathbf{b} \otimes \mathbf{a}\big]_\times$ is used.

Hence, **nonlocal curl** of a vector field $\mathbf{v}$ for point $\mathbf{x}$ in support $\mathcal{S}_{\mathbf{x}}$ is defined as

$$\tilde{\nabla} \times \mathbf{v}_{\mathbf{x}} := \int_{\mathcal{S}_{\mathbf{x}}} w(\mathbf{r}) (\mathbf{K}_{\mathbf{x}}^{-1} \cdot \mathbf{r}) \times \mathbf{v}_{\mathbf{r}} dV_{\mathbf{x}'}. \tag{3.10}$$

By analogy with $\nabla \cdot \mathbf{v} = \mathrm{tr}(\nabla \otimes \mathbf{v})$, where $\mathrm{tr}(\square)$ denotes the trace of the matrix, the nonlocal divergence of a vector field $\mathbf{v}$ for point $\mathbf{x}$ in support $\mathcal{S}_{\mathbf{x}}$ is derived as

$$
\begin{aligned}
\tilde{\nabla} \cdot \mathbf{v}_{\mathbf{x}} &= \mathrm{tr}(\tilde{\nabla} \otimes \mathbf{v}) \\
&= \mathrm{tr}\Big(\int_{\mathcal{S}_{\mathbf{x}}} w(\mathbf{r}) \mathbf{v}_{\mathbf{r}} \otimes (\mathbf{K}_{\mathbf{x}}^{-1} \cdot \mathbf{r}) \mathrm{d}V_{\mathbf{x}'}\Big) \\
&= \int_{\mathcal{S}_{\mathbf{x}}} w(\mathbf{r}) \mathrm{tr}\big(\mathbf{v}_{\mathbf{r}} \otimes (\mathbf{K}_{\mathbf{x}}^{-1} \cdot \mathbf{r})\big) \mathrm{d}V_{\mathbf{x}'} \\
&= \int_{\mathcal{S}_{\mathbf{x}}} w(\mathbf{r}) \mathbf{v}_{\mathbf{r}} \cdot (\mathbf{K}_{\mathbf{x}}^{-1} \cdot \mathbf{r}) \mathrm{d}V_{\mathbf{x}'}.
\end{aligned}
$$

In the third step, $\mathrm{tr}(\mathbf{a} \otimes \mathbf{b}) = \mathbf{a} \cdot \mathbf{b}$ is used.

Hence, **nonlocal divergence** is defined as

$$
\tilde{\nabla} \cdot \mathbf{v}_{\mathbf{x}} := \int_{\mathcal{S}_{\mathbf{x}}} w(\mathbf{r}) \mathbf{v}_{\mathbf{r}} \cdot (\mathbf{K}_{\mathbf{x}}^{-1} \cdot \mathbf{r}) \mathrm{d}V_{\mathbf{x}'}. \tag{3.11}
$$

In the continuous limit, based on the nonlocal gradient, the nonlocal curl and nonlocal divergence converge to the conventional curl and divergence operator, respectively.

3.1.2 Variation of the Nonlocal Operator

We present the variation of the general nonlocal operator in continuous form and discrete form. The discrete form is beneficial for the numerical implementation. Different nonlocal operators can be used to replace the local differential operators in PDEs, especially in the framework of weighted residual method and variational principles. We use the δ to denote the variation.

Nonlocal divergence operator

The variation of $\tilde{\nabla} \cdot \mathbf{F}_{\mathbf{x}}$ is given by

$$
\delta(\tilde{\nabla} \cdot \mathbf{F}_{\mathbf{x}}) = \tilde{\nabla} \cdot \delta \mathbf{F}_{\mathbf{x}} = \int_{\mathcal{S}_{\mathbf{x}}} w(\mathbf{r}) (\mathbf{K}_{\mathbf{x}}^{-1} \cdot \mathbf{r}) \cdot (\delta \mathbf{F}_{\mathbf{x}'} - \delta \mathbf{F}_{\mathbf{x}}) \mathrm{d}V_{\mathbf{x}'}, \tag{3.12}
$$

The number of dimensions of $\tilde{\nabla} \cdot \delta \mathbf{F}_{\mathbf{x}}$ is infinite, and discretization is required.

After discretization of the domain by particles, the whole domain is represented by

$$
\Omega = \sum_{j=1}^{N} \Delta V_j \tag{3.13}
$$

where j is the global index of volume ΔV_j, N is the number of particles in Ω.

The continuous support $\mathcal{S}$ at point I is represented by the discrete form

$$N_I = \{1, \ldots, k, \ldots, n_I\} \tag{3.14}$$

where $1, \ldots, k, \ldots, n_I$ are the local indices of neighbors of particle I. The discrete form of $\tilde{\nabla} \cdot \delta \mathbf{F}_I$ can be written as

$$\tilde{\nabla} \cdot \delta \mathbf{F}_I \simeq \sum_{k \in N_I} w(\mathbf{r}) \Delta V_k (\mathbf{K}_I^{-1} \mathbf{r}) \cdot (\delta \mathbf{F}_k - \delta \mathbf{F}_I) = \mathbf{B}_I^{div} \cdot \delta \mathbf{F}_{N_I}, \tag{3.15}$$

where $\simeq$ denotes discretization, $\delta \mathbf{F}_{N_I}$ is all the variations of the unknowns in support $\mathcal{S}_I$,

$$\delta \mathbf{F}_{N_I} = (\delta \mathbf{F}_I^T, \delta \mathbf{F}_1^T, \ldots, \delta \mathbf{F}_{n_I}^T)^T, \tag{3.16}$$

is a vector with length of $ndim \cdot (n_I + 1)$ and $ndim$ is the number of spatial dimensions. $\mathbf{B}_I^{div}$ is the coefficient vector with length of $3(n_I + 1)$ in 3D case,

$$\mathbf{B}_I^{div} = \begin{bmatrix} -\sum_{k=1}^{n_I} w(\mathbf{r}_{kI}) \Delta V_k \mathbf{K}_I^{-1} \mathbf{r}_{kI} \\ w(\mathbf{r}_{1I}) \Delta V_1 \mathbf{K}_I^{-1} \mathbf{r}_{1I} \\ \vdots \\ w(\mathbf{r}_{n_I I}) \Delta V_{n_I} \mathbf{K}_I^{-1} \mathbf{r}_{n_I I} \end{bmatrix}^T. \tag{3.17}$$

Nonlocal curl operator

The variation of $\tilde{\nabla} \times \mathbf{F}_I$ in discrete form reads

$$\tilde{\nabla} \times \delta \mathbf{F}_I \simeq \sum_{k \in N_I} w(\mathbf{r}) \Delta V_k (\mathbf{K}_I^{-1} \mathbf{r}) \times (\delta \mathbf{F}_k - \delta \mathbf{F}_I) = \mathbf{B}_I^{curl} \cdot \delta \mathbf{F}_{N_I}, \tag{3.18}$$

where ΔV_k is the volume for neighbor k. For the 3D case, $\mathbf{B}_I^{curl}$ is a $3 \times 3(N_I + 1)$ matrix, where N_I is the number of neighbors in $\mathcal{S}_I$, n_I is given by Eq. 3.14. $\delta \mathbf{F}_{N_I}$ is given in Eq. 3.16 and $\mathbf{B}_I^{curl}$ in 3D can be written as

$$\mathbf{B}_I^{curl} = \begin{bmatrix} 0 & \sum_{k=1}^{n_I} R_k^z & -\sum_{k=1}^{n_I} R_k^y & 0 & -R_1^z & R_1^y & \cdots & 0 & -R_{n_I}^z & R_{n_I}^y \\ -\sum_{k=1}^{n_I} R_k^z & 0 & \sum_{k=1}^{n_I} R_k^x & R_1^z & 0 & -R_1^x & \cdots & R_{n_I}^z & 0 & -R_{n_I}^x \\ \sum_{k=1}^{n_I} R_k^y & -\sum_{k=1}^{n_I} R_k^x & 0 & -R_1^y & R_1^x & 0 & \cdots & -R_{n_I}^y & R_{n_I}^x & 0 \end{bmatrix} \tag{3.19}$$

where $(R_k^x, R_k^y, R_k^z) = w(\mathbf{r}) \Delta V_k \mathbf{K}_I^{-1} \mathbf{r}_{kI}$.

Nonlocal gradient operator for vector field

Similarly, the variation of $\tilde{\nabla} \otimes \mathbf{F}_I$ in the discrete form reads

$$\tilde{\nabla} \otimes \delta \mathbf{F}_I \simeq \sum_{k \in N_I} w(\mathbf{r}) (\delta \mathbf{F}_k - \delta \mathbf{F}_I) \otimes (\mathbf{K}_I^{-1} \mathbf{r}) \Delta V_k = \mathbf{B}_I^{\mathbf{grad}} \cdot \delta \mathbf{F}_{N_I} \tag{3.20}$$

where ΔV_k is the volume for neighbor k. For the 3D case, $\mathbf{B}_I^{curl}$ is a $9 \times 3(n_I + 1)$ matrix, where n_I is the number of neighbors in $\mathcal{S}_I$, n_I is given by Eq. 3.14. $\delta\mathbf{F}_{N_I}$ is given in Eq. 3.16 and $\mathbf{B}_I^{\mathbf{grad}}$ in 3D can be written as

$$\mathbf{B}_I^{\mathbf{grad}} = \begin{bmatrix} -\sum_{k=1}^{n_I} R_k^x & 0 & 0 & R_1^x & 0 & 0 & \cdots & R_{n_I}^x & 0 & 0 \\ 0 & -\sum_{k=1}^{n_I} R_k^x & 0 & 0 & R_1^x & 0 & \cdots & 0 & R_{n_I}^x & 0 \\ 0 & 0 & -\sum_{k=1}^{n_I} R_k^x & 0 & 0 & R_1^x & \cdots & 0 & 0 & R_{n_I}^x \\ -\sum_{k=1}^{n_I} R_k^y & 0 & 0 & R_1^y & 0 & 0 & \cdots & R_{n_I}^y & 0 & 0 \\ 0 & -\sum_{k=1}^{n_I} R_k^y & 0 & 0 & R_1^y & 0 & \cdots & 0 & R_{n_I}^y & 0 \\ 0 & 0 & -\sum_{k=1}^{n_I} R_k^y & 0 & 0 & R_1^y & \cdots & 0 & 0 & R_{n_I}^y \\ -\sum_{k=1}^{n_I} R_k^z & 0 & 0 & R_1^z & 0 & 0 & \cdots & R_{n_I}^z & 0 & 0 \\ 0 & -\sum_{k=1}^{n_I} R_k^z & 0 & 0 & R_1^z & 0 & \cdots & 0 & R_{n_I}^z & 0 \\ 0 & 0 & -\sum_{k=1}^{n_I} R_k^z & 0 & 0 & R_1^z & \cdots & 0 & 0 & R_{n_I}^z \end{bmatrix} \tag{3.21}$$

where $(R_k^x, R_k^y, R_k^z) = w(\mathbf{r})\Delta V_k \mathbf{K}_I^{-1}\mathbf{r}_{kI}$.

Nonlocal gradient operator for scalar field

The variation of ∇v_I reads

$$\tilde{\nabla}\delta v_I \simeq \sum_{k \in N_I} w(\mathbf{r}_{kI})\Delta V_k \mathbf{K}_I^{-1}\mathbf{r}(\delta v_k - \delta v_I) = \mathbf{B}_I^{grad} \cdot \delta\mathbf{v}_{N_I}, \tag{3.22}$$

where $\delta\mathbf{v}_{N_I} = (\delta v_I, \delta v_1, \ldots, \delta v_{n_I})^T$, the dimensions of $\mathbf{B}_I^{grad}$ in 3D are $3 \times (N_I + 1)$,

$$\mathbf{B}_I^{grad} = \begin{bmatrix} -\sum_{k=1}^{n_I} R_k^x & R_1^x & \cdots & R_{n_I}^x \\ -\sum_{k=1}^{n_I} R_k^y & R_1^y & \cdots & R_{n_I}^y \\ -\sum_{k=1}^{n_I} R_k^z & R_1^z & \cdots & R_{n_I}^z \end{bmatrix}. \tag{3.23}$$

It can be seen that the basic element in nonlocal operator method is $w(\mathbf{r})\Delta V_k \mathbf{K}_I^{-1}\mathbf{r}$, which is quite similar to the derivative of shape function in meshless or finite element methods. For the nonlocal divergence, nonlocal curl and nonlocal gradient operators for scalar or vector fields, there exist corresponding nonlocal operator matrices such as Eqs. (3.17, 3.19, 3.21, 3.23). The operator matrix in any dimensions can be constructed based on $w(\mathbf{r})\Delta V_k \mathbf{K}_I^{-1}\mathbf{r}$ by following the similar procedure in Eqs. (3.17, 3.19, 3.21, 3.23).

3.2 The Variational Principles Based on the Nonlocal Operator

The problems based on variational principles start from the functional which describes the unknown functions defined in the domain and on the boundary. The residual is the gradient of the functional on the unknown vector, while the tangent stiffness matrix is the Hessian matrix of the functional on the unknown vector. The functional is usually expressed by the local operators such as divergence, curl and gradient. For simplicity, we assume the functional is a function on a single local operator. The boundary terms can be handled in the similar way.

Assuming four general functionals at a point which depend on the divergence of a vector field, the curl of a vector field, the gradient of a vector field and the gradient of a scalar field, respectively,

$$\mathcal{F}(\nabla \cdot \mathbf{v}), \mathcal{F}(\nabla \times \mathbf{v}), \mathcal{F}(\nabla \mathbf{v}), \mathcal{F}(\nabla v). \tag{3.24}$$

The examples are the strain energy functional in solid mechanics, the volume strain energy functional in solid mechanics, the wave vector form of electromagnetic field, and the thermal conduction, respectively.

3.2.1 Divergence Operator

The first- and second-order derivatives of the functional $\mathcal{F}(\nabla \cdot \mathbf{v})$ on operator $\nabla \cdot \mathbf{v}$ are, respectively,

$$p = \frac{\partial \mathcal{F}(\nabla \cdot \mathbf{v})}{\partial (\nabla \cdot \mathbf{v})}, \; d = \frac{\partial p}{\partial (\nabla \cdot \mathbf{v})} = \frac{\partial^2 \mathcal{F}(\nabla \cdot \mathbf{v})}{\partial (\nabla \cdot \mathbf{v}) \partial (\nabla \cdot \mathbf{v})} \tag{3.25}$$

where p is a scalar, d is a scalar.

The first and second variation of $\mathcal{F}$ at point I with discrete support $N_I = (1, \ldots, n_I)$ can be written as

$$\delta \mathcal{F}(\nabla \cdot \mathbf{v}_I) = \frac{\partial \mathcal{F}(\nabla \cdot \mathbf{v}_I)}{\partial (\nabla \cdot \mathbf{v}_I)} \tilde{\nabla} \cdot \delta \mathbf{v}_I = p \, \mathbf{B}_I^{div} \cdot \delta \mathbf{v}_{N_I}$$
$$\delta^2 \mathcal{F}(\nabla \cdot \mathbf{v}_I) = (\tilde{\nabla} \cdot \delta \mathbf{v}_I)^T d \, \tilde{\nabla} \cdot \delta \mathbf{v}_I = \delta \mathbf{v}_{N_I}^T (\mathbf{B}_I^{div})^T d \, \mathbf{B}_I^{div} \delta \mathbf{v}_{N_I},$$

and then residual and tangent stiffness matrix at a point can be extracted as

$$\mathbf{R}_I^{div} = p_I \, \mathbf{B}_I^{div}, \mathbf{K}_I^{div} = (\mathbf{B}_I^{div})^T d \, \mathbf{B}_I^{div} \tag{3.26}$$

where $\mathbf{B}_I^{div}$ is given by Eq. 3.17.

Let's consider the first variation of all particles, and let $p_{\mathbf{x}} = \frac{\partial \mathcal{F}(\nabla \cdot \mathbf{v}_{\mathbf{x}})}{\partial (\nabla \cdot \mathbf{v}_{\mathbf{x}})}$.

$$\delta\mathcal{F}(\nabla\cdot\mathbf{v})=\sum_{\Delta V_{\mathbf{x}}\in\Omega}\Delta V_{\mathbf{x}}\delta\mathcal{F}_{\mathbf{x}}=\sum_{\Delta V_{\mathbf{x}}\in\Omega}\Delta V_{\mathbf{x}}p_{\mathbf{x}}\cdot(\nabla\cdot\delta\mathbf{v}_{\mathbf{x}})$$
$$=\sum_{\Delta V_{\mathbf{x}}\in\Omega}\Delta V_{\mathbf{x}}p_{\mathbf{x}}\cdot\sum_{\mathcal{S}_{\mathbf{x}}}w(\mathbf{r})\Delta V_{\mathbf{x}'}\mathbf{K}_{\mathbf{x}}^{-1}\mathbf{r}\cdot(\delta\mathbf{v}_{\mathbf{x}'}-\delta\mathbf{v}_{\mathbf{x}})$$
$$=\sum_{\Delta V_{\mathbf{x}}\in\Omega}\Delta V_{\mathbf{x}}\Big(-\sum_{\mathcal{S}_{\mathbf{x}}}w(\mathbf{r})\Delta V_{\mathbf{x}'}\mathbf{K}_{\mathbf{x}}^{-1}\mathbf{r}\cdot\delta\mathbf{v}_{\mathbf{x}}\cdot p_{\mathbf{x}}+\sum_{\mathcal{S}'_{\mathbf{x}}}w(\mathbf{r}')\Delta V_{\mathbf{x}'}\mathbf{K}_{\mathbf{x}'}^{-1}\mathbf{r}'\cdot\delta\mathbf{v}_{\mathbf{x}}\cdot p_{\mathbf{x}'}\Big).$$

In the second and third step, the dual-support is considered as follows. In the second step, the term with $\delta\mathbf{v}_{\mathbf{x}'}$ is the vector from $\mathbf{x}$'s support, but is added to particle $\mathbf{x}'$; since $\mathbf{x}'\in\mathcal{S}_{\mathbf{x}}$, $\mathbf{x}$ belongs to the dual-support $\mathcal{S}'_{\mathbf{x}'}$ of $\mathbf{x}'$. In the third step, all the terms with $\delta\mathbf{v}_{\mathbf{x}}$ are collected from other particles whose supports contain $\mathbf{x}$ and therefore form the dual-support of $\mathbf{x}$. The terms with $\delta\mathbf{v}_{\mathbf{x}}$ in the first-order variation $\delta\mathcal{F}(\nabla\cdot\mathbf{v})=0$ are

$$-\sum_{\mathcal{S}_{\mathbf{x}}}w(\mathbf{r})\Delta V_{\mathbf{x}'}p_{\mathbf{x}}\mathbf{K}_{\mathbf{x}}^{-1}\mathbf{r}+\sum_{\mathcal{S}'_{\mathbf{x}}}w(\mathbf{r}')\Delta V_{\mathbf{x}'}p_{\mathbf{x}'}\mathbf{K}_{\mathbf{x}'}^{-1}\mathbf{r}'. \tag{3.27}$$

When any particle's volume $\Delta V_{\mathbf{x}'}\to 0$, the continuous form is

$$-\int_{\mathcal{S}_{\mathbf{x}}}w(\mathbf{r})p_{\mathbf{x}}\mathbf{K}_{\mathbf{x}}^{-1}\mathbf{r}\,\mathrm{d}V_{\mathbf{x}'}+\int_{\mathcal{S}'_{\mathbf{x}}}w(\mathbf{r}')p_{\mathbf{x}'}\mathbf{K}_{\mathbf{x}'}^{-1}\mathbf{r}'\,\mathrm{d}V_{\mathbf{x}'}. \tag{3.28}$$

Equation 3.28 is the nonlocal strong form for energy functional $\mathcal{F}(\nabla\cdot\mathbf{v})$ with the corresponding local strong form is $-\nabla(d\,\nabla\cdot\mathbf{v})$. The local strong form can be obtained by integration of part of the energy functional in the whole domain. Consider the variation of $\int_\Omega\mathcal{F}(\nabla\cdot\mathbf{v}_{\mathbf{x}})\mathrm{d}V_{\mathbf{x}}$

$$\delta\Big(\int_\Omega\mathcal{F}(\nabla\cdot\mathbf{v}_{\mathbf{x}})\mathrm{d}V_{\mathbf{x}}\Big)=\int_\Omega\delta\mathcal{F}(\nabla\cdot\mathbf{v}_{\mathbf{x}})\mathrm{d}V_{\mathbf{x}}$$
$$=\int_\Omega\frac{\partial\mathcal{F}(\nabla\cdot\mathbf{v}_{\mathbf{x}})}{\partial(\nabla\cdot\mathbf{v}_{\mathbf{x}})}\cdot\nabla\cdot\delta\mathbf{v}_{\mathbf{x}}\mathrm{d}V_{\mathbf{x}}$$
$$=\int_\Omega p_{\mathbf{x}}\nabla\cdot\delta\mathbf{v}_{\mathbf{x}}\mathrm{d}V_{\mathbf{x}}$$
$$=\int_{\partial\Omega}p_{\mathbf{x}}\mathbf{n}_{\mathbf{x}}\cdot\delta\mathbf{v}_{\mathbf{x}}\mathrm{d}S_{\mathbf{x}}-\int_\Omega(\nabla p_{\mathbf{x}})\cdot\delta\mathbf{v}_{\mathbf{x}}\mathrm{d}V_{\mathbf{x}}.$$

For any point in Ω, the term corresponding to $\mathbf{v}_{\mathbf{x}}$ is $-\nabla(d\,\nabla\cdot\mathbf{v}_{\mathbf{x}})$.

3.2.2 Curl Operator

The first- and second-order derivatives of the functional $\mathcal{F}(\nabla \times \mathbf{v})$ on operator $\nabla \times \mathbf{v}$ are, respectively,

$$\mathbf{p} = \frac{\partial \mathcal{F}(\nabla \times \mathbf{v})}{\partial(\nabla \times \mathbf{v})}, \ \mathbf{D} = \frac{\partial \mathbf{p}}{\partial(\nabla \times \mathbf{v})} = \frac{\partial^2 \mathcal{F}}{\partial(\nabla \times \mathbf{v})^T \partial(\nabla \times \mathbf{v})}. \tag{3.29}$$

In 3D $\mathbf{p}$ is a vector with length of 3, $\mathbf{D}$ is 3×3 matrices.

The first and second variation of $\mathcal{F}$ at point I with discrete support $N_I = (1, \dots, n_I)$ can be written as

$$\delta\mathcal{F}(\nabla \times \mathbf{v}_I) = \frac{\partial \mathcal{F}(\nabla \times \mathbf{v}_I)}{\partial(\nabla \times \mathbf{v}_I)} \tilde{\nabla} \times \delta\mathbf{v}_I = \mathbf{p}_I \mathbf{B}_I^{curl} \cdot \delta\mathbf{v}_{N_I}$$
$$\delta^2\mathcal{F}(\nabla \times \mathbf{v}_I) = (\tilde{\nabla} \times \delta\mathbf{v}_I)^T \mathbf{D}\, \tilde{\nabla} \times \delta\mathbf{v}_I = \delta\mathbf{v}_{N_I}^T (\mathbf{B}_I^{curl})^T \mathbf{D}\mathbf{B}_I^{curl} \delta\mathbf{v}_{N_I},$$

and then residual and tangent stiffness matrix at a point can be extracted as

$$\mathbf{R}_I^{curl} = \mathbf{p}_I^T \mathbf{B}_I^{curl}, \ \mathbf{K}_I^{curl} = (\mathbf{B}_I^{curl})^T \mathbf{D}\mathbf{B}_I^{curl} \tag{3.30}$$

where $\mathbf{B}_I^{curl}$ is given by Eq. 3.19.

Let's consider the first variation of all particles, and let $\mathbf{p}_\mathbf{x} = \frac{\partial \mathcal{F}(\nabla \times \mathbf{v}_\mathbf{x})}{\partial(\nabla \times \mathbf{v}_\mathbf{x})}$.

$$\begin{aligned}
\delta\mathcal{F}(\nabla \times \mathbf{v}) &= \sum_{\Delta V_\mathbf{x} \in \Omega} \Delta V_\mathbf{x} \delta\mathcal{F}_\mathbf{x} = \sum_{\Delta V_\mathbf{x} \in \Omega} \Delta V_\mathbf{x} (\nabla \times \delta\mathbf{v}_\mathbf{x}) \cdot \mathbf{p}_\mathbf{x} \\
&= \sum_{\Delta V_\mathbf{x} \in \Omega} \Delta V_\mathbf{x} \sum_{\mathcal{S}_\mathbf{x}} w(\mathbf{r}) \Delta V_{\mathbf{x}'} \mathbf{K}_\mathbf{x}^{-1} \mathbf{r} \times (\delta\mathbf{v}_{\mathbf{x}'} - \delta\mathbf{v}_\mathbf{x}) \cdot \mathbf{p}_\mathbf{x} \\
&= \sum_{\Delta V_\mathbf{x} \in \Omega} \Delta V_\mathbf{x} \sum_{\mathcal{S}_\mathbf{x}} w(\mathbf{r}) \Delta V_{\mathbf{x}'} \mathbf{p}_\mathbf{x} \times (\mathbf{K}_\mathbf{x}^{-1} \mathbf{r}) \cdot (\delta\mathbf{v}_{\mathbf{x}'} - \delta\mathbf{v}_\mathbf{x}) \\
&= \sum_{\Delta V_\mathbf{x} \in \Omega} \Delta V_\mathbf{x} \Big(-\sum_{\mathcal{S}_\mathbf{x}} w(\mathbf{r}) \Delta V_{\mathbf{x}'} \mathbf{p}_\mathbf{x} \times (\mathbf{K}_\mathbf{x}^{-1} \mathbf{r}) \cdot \delta\mathbf{v}_\mathbf{x} + \sum_{\mathcal{S}'_\mathbf{x}} w(\mathbf{r}') \Delta V_{\mathbf{x}'} \mathbf{p}_{\mathbf{x}'} \times (\mathbf{K}_{\mathbf{x}'}^{-1} \mathbf{r}') \cdot \delta\mathbf{v}_\mathbf{x} \Big).
\end{aligned}$$

The relation $\mathbf{a} \cdot (\mathbf{b} \times \mathbf{c}) = \mathbf{c} \cdot (\mathbf{a} \times \mathbf{b}) = \mathbf{b} \cdot (\mathbf{c} \times \mathbf{a})$ is used in the third step. In the third and fourth step, the dual-support is considered as follows. In the third step, the term with $\delta\mathbf{v}_{\mathbf{x}'}$ is the vector from $\mathbf{x}$'s support, but is added to particle $\mathbf{x}'$; since $\mathbf{x}' \in \mathcal{S}_\mathbf{x}$, $\mathbf{x}$ belongs to the dual-support $\mathcal{S}'_{\mathbf{x}'}$ of $\mathbf{x}'$. In the fourth step, all the terms with $\delta\mathbf{v}_\mathbf{x}$ are collected from other particles whose supports contain $\mathbf{x}$ and therefore form the dual-support of $\mathbf{x}$. The terms with $\delta\mathbf{v}_\mathbf{x}$ in the first-order variation $\delta\mathcal{F}(\nabla \times \mathbf{v}) = 0$ are

$$-\sum_{\mathcal{S}_\mathbf{x}} w(\mathbf{r}) \Delta V_{\mathbf{x}'} \mathbf{p}_\mathbf{x} \times (\mathbf{K}_\mathbf{x}^{-1} \mathbf{r}) + \sum_{\mathcal{S}'_\mathbf{x}} w(\mathbf{r}') \Delta V_{\mathbf{x}'} \mathbf{p}_{\mathbf{x}'} \times (\mathbf{K}_{\mathbf{x}'}^{-1} \mathbf{r}'). \tag{3.31}$$

When any particle's volume $\Delta V_{\mathbf{x}'} \to 0$, the continuous form is

$$-\int_{\mathcal{S}_\mathbf{x}} w(\mathbf{r})\mathbf{p}_\mathbf{x} \times (\mathbf{K}_\mathbf{x}^{-1}\mathbf{r})\, \mathrm{d}V_{\mathbf{x}'} + \int_{\mathcal{S}'_\mathbf{x}} w(\mathbf{r}')\mathbf{p}_{\mathbf{x}'} \times (\mathbf{K}_{\mathbf{x}'}^{-1}\mathbf{r}')\, \mathrm{d}V_{\mathbf{x}'}. \tag{3.32}$$

Equation 3.32 is the strong form for energy functional $\mathcal{F}(\nabla \times \mathbf{v})$, where the corresponding local strong from obtained by integration by part of the energy functional is $-\nabla \times (\mathbf{D}\nabla \times \mathbf{v})$, which is obtained as follows. In order to derive the local strong form of functional $\mathcal{F}(\nabla \times \mathbf{v})$, consider the variation of $\int_\Omega \mathcal{F}(\nabla \times \mathbf{v}_\mathbf{x})\mathrm{d}V_\mathbf{x}$

$$\begin{aligned}
\delta\Big(\int_\Omega \mathcal{F}(\nabla \times \mathbf{v}_\mathbf{x})\mathrm{d}V_\mathbf{x}\Big) &= \int_\Omega \delta\mathcal{F}(\nabla \times \mathbf{v}_\mathbf{x})\mathrm{d}V_\mathbf{x}\\
&= \int_\Omega \frac{\partial \mathcal{F}(\nabla \times \mathbf{v}_\mathbf{x})}{\partial(\nabla \times \mathbf{v}_\mathbf{x})} \cdot \nabla \times \delta\mathbf{v}_\mathbf{x}\mathrm{d}V_\mathbf{x}\\
&= \int_\Omega \mathbf{p}_\mathbf{x} \cdot \nabla \times \delta\mathbf{v}_\mathbf{x}\mathrm{d}V_\mathbf{x}\\
&= \int_{\partial\Omega} \mathbf{p}_\mathbf{x} \times \mathbf{n}_\mathbf{x} \cdot \delta\mathbf{v}_\mathbf{x}\mathrm{d}S_\mathbf{x} - \int_\Omega (\nabla \times \mathbf{p}_\mathbf{x}) \cdot \delta\mathbf{v}_\mathbf{x}\mathrm{d}V_\mathbf{x}.
\end{aligned}$$

For any point in Ω, the term corresponding to $\delta\mathbf{v}_\mathbf{x}$ is $-\nabla \times (\mathbf{D}\nabla \times \mathbf{v}_\mathbf{x})$.

3.2.3 *Gradient Operator of Vector Field*

The first- and second-order derivatives of the functional $\mathcal{F}(\nabla\mathbf{v})$ on the operator $\nabla\mathbf{v}$ are, respectively,

$$\mathbf{P} = \frac{\partial \mathcal{F}(\nabla\mathbf{v})}{\partial(\nabla\mathbf{v})},\ \mathbf{D} = \frac{\partial \mathbf{P}}{\partial(\nabla\mathbf{v})} = \frac{\partial^2 \mathcal{F}(\nabla\mathbf{v})}{\partial(\nabla\mathbf{v})^T\partial(\nabla\mathbf{v})} \tag{3.33}$$

$\mathbf{P}$ is a 3×3 tensor. When $\nabla\mathbf{v}$ is the deformation gradient with respect to the initial configuration, $\mathbf{P}$ is the first Piola–Kirchhoff stress. $\mathbf{D}$ is a $3 \times 3 \times 3 \times 3$ tensor. When $\nabla\mathbf{v}$ is the deformation gradient with respect to the initial configuration, $\mathbf{D}$ is the material tensor in solid mechanics. Let $\mathbf{D}_{9\times9}$ be the matrix form of the fourth-order tensor $\mathbf{D}$ and $\mathbf{P}_9$ be the vector form of $\mathbf{P}$.

The first and second variation of $\mathcal{F}$ at point I with discrete support $N_I = (1, \ldots, n_I)$ can be written as

$$\begin{aligned}
\delta\mathcal{F}(\nabla\mathbf{v}_I) &= \frac{\partial \mathcal{F}(\nabla\mathbf{v}_I)}{\partial(\nabla\mathbf{v}_I)}\tilde{\nabla}\delta\mathbf{v}_I = \mathbf{P}_{9_I}^T\,\mathbf{B}^{\mathbf{grad}} \cdot \delta\mathbf{v}_{N_I}\\
\delta^2\mathcal{F}(\nabla\mathbf{v}_I) &= (\tilde{\nabla}\delta\mathbf{v}_I)^T\mathbf{D}\,\tilde{\nabla}\delta\mathbf{v}_I = \delta\mathbf{v}_{N_I}^T(\mathbf{B}_I^{\mathbf{grad}})^T\mathbf{D}_{9\times9}\,\mathbf{B}_I^{\mathbf{grad}}\delta\mathbf{v}_{N_I},
\end{aligned}$$

and then residual and tangent stiffness matrix at a point can be extracted as

$$\mathbf{R}_I^{\mathbf{grad}} = \mathbf{P}_{9_I}^T \mathbf{B}^{\mathbf{grad}}, \mathbf{K}_I^{\mathbf{grad}} = (\mathbf{B}_I^{\mathbf{grad}})^T \mathbf{D}_{9\times 9} \mathbf{B}_I^{\mathbf{grad}} \tag{3.34}$$

where $\mathbf{B}_I^{\mathbf{grad}}$ is given by Eq. 3.21.

Let's consider the first variation of all particles, and let $\mathbf{P}_{\mathbf{x}} = \frac{\partial \mathcal{F}(\nabla \mathbf{v}_{\mathbf{x}})}{\partial(\nabla \mathbf{v}_{\mathbf{x}})}$.

$$\begin{aligned}
\delta\mathcal{F}(\nabla\mathbf{v}) &= \sum_{\Delta V_{\mathbf{x}}\in\Omega} \Delta V_{\mathbf{x}} \delta\mathcal{F}_{\mathbf{x}} = \sum_{\Delta V_{\mathbf{x}}\in\Omega} \Delta V_{\mathbf{x}} \mathbf{P}_{\mathbf{x}} \cdot (\nabla\delta\mathbf{v}_{\mathbf{x}}) \\
&= \sum_{\Delta V_{\mathbf{x}}\in\Omega} \Delta V_{\mathbf{x}} \mathbf{P}_{\mathbf{x}} \cdot \sum_{\mathcal{S}_{\mathbf{x}}} w(\mathbf{r}) \Delta V_{\mathbf{x}'} \mathbf{K}_{\mathbf{x}}^{-1} \mathbf{r} \otimes (\delta\mathbf{v}_{\mathbf{x}'} - \delta\mathbf{v}_{\mathbf{x}}) \\
&= \sum_{\Delta V_{\mathbf{x}}\in\Omega} \Delta V_{\mathbf{x}} \Big(- \sum_{\mathcal{S}_{\mathbf{x}}} w(\mathbf{r}) \Delta V_{\mathbf{x}'} \mathbf{K}_{\mathbf{x}}^{-1} \mathbf{r} \otimes \delta\mathbf{v}_{\mathbf{x}} \cdot \mathbf{P}_{\mathbf{x}} + \sum_{\mathcal{S}'_{\mathbf{x}}} w(\mathbf{r}') \Delta V_{\mathbf{x}'} \mathbf{K}_{\mathbf{x}'}^{-1} \mathbf{r}' \otimes \delta\mathbf{v}_{\mathbf{x}} \cdot \mathbf{P}_{\mathbf{x}'} \Big).
\end{aligned}$$

In the second and third step, the dual-support is considered as follows. In the second step, the term with $\delta\mathbf{v}_{\mathbf{x}'}$ is the vector from $\mathbf{x}$'s support, but is added to particle $\mathbf{x}'$; since $\mathbf{x}' \in \mathcal{S}_{\mathbf{x}}$, $\mathbf{x}$ belongs to the dual-support $\mathcal{S}'_{\mathbf{x}'}$ of $\mathbf{x}'$. In the third step, all the terms with $\delta\mathbf{v}_{\mathbf{x}}$ are collected from other particles whose supports contain $\mathbf{x}$ and therefore form the dual-support of $\mathbf{x}$. The terms with $\delta\mathbf{v}_{\mathbf{x}}$ in the first-order variation $\delta\mathcal{F}(\mathbf{v}) = 0$ are

$$-\sum_{\mathcal{S}_{\mathbf{x}}} w(\mathbf{r}) \Delta V_{\mathbf{x}'} \mathbf{P}_{\mathbf{x}} \cdot \mathbf{K}_{\mathbf{x}}^{-1} \mathbf{r} + \sum_{\mathcal{S}'_{\mathbf{x}}} w(\mathbf{r}') \Delta V_{\mathbf{x}'} \mathbf{P}_{\mathbf{x}'} \cdot \mathbf{K}_{\mathbf{x}'}^{-1} \mathbf{r}'. \tag{3.35}$$

When any particle's volume $\Delta V_{\mathbf{x}'} \to 0$, the continuous form is

$$-\int_{\mathcal{S}_{\mathbf{x}}} w(\mathbf{r}) \mathbf{P}_{\mathbf{x}} \cdot \mathbf{K}_{\mathbf{x}}^{-1} \mathbf{r} \, dV_{\mathbf{x}'} + \int_{\mathcal{S}'_{\mathbf{x}}} w(\mathbf{r}') \mathbf{P}_{\mathbf{x}'} \cdot \mathbf{K}_{\mathbf{x}'}^{-1} \mathbf{r}' \, dV_{\mathbf{x}'}. \tag{3.36}$$

Equation 3.36 is the nonlocal strong form of energy functional $\mathcal{F}_3$, where the local strong form obtained by integration by part of the energy functional is $-\nabla \cdot \mathbf{P}_{\mathbf{x}}$. If $\nabla\mathbf{v}$ denotes the deformation gradient with respect to the initial configuration and $\mathcal{F}(\nabla\mathbf{v})$ is the strain energy density, $\mathbf{P}_{\mathbf{x}}$ is the first Piola–Kirchhoff stress and the Eq. 3.36 is the key expression in the dual-horizon peridynamics (Ren et al. 2017, 2016).

The local strong form $-\nabla \cdot \mathbf{P}_{\mathbf{x}}$ is obtained as follows. In order to derive the local strong form of functional $\mathcal{F}(\nabla\mathbf{v})$, consider the variation of $\int_\Omega \mathcal{F}(\nabla\mathbf{v}_{\mathbf{x}}) dV_{\mathbf{x}}$

$$\begin{aligned}
\delta\Big(\int_\Omega \mathcal{F}(\nabla\mathbf{v}_{\mathbf{x}}) dV_{\mathbf{x}}\Big) &= \int_\Omega \delta\mathcal{F}(\nabla\mathbf{v}_{\mathbf{x}}) dV_{\mathbf{x}} \\
&= \int_\Omega \frac{\partial\mathcal{F}(\nabla\mathbf{v}_{\mathbf{x}})}{\partial(\nabla\mathbf{v}_{\mathbf{x}})} \cdot \nabla\delta\mathbf{v}_{\mathbf{x}} dV_{\mathbf{x}} \\
&= \int_\Omega \mathbf{P}_{\mathbf{x}} \cdot \nabla\delta\mathbf{v}_{\mathbf{x}} dV_{\mathbf{x}} \\
&= \int_{\partial\Omega} \mathbf{P}_{\mathbf{x}} \cdot \mathbf{n}_{\mathbf{x}} \cdot \delta\mathbf{v}_{\mathbf{x}} dS_{\mathbf{x}} - \int_\Omega (\nabla \cdot \mathbf{P}_{\mathbf{x}}) \cdot \delta\mathbf{v}_{\mathbf{x}} dV_{\mathbf{x}}.
\end{aligned}$$

For any point in Ω, the term corresponding to $\delta\mathbf{v}_{\mathbf{x}}$ is $-\nabla \cdot \mathbf{P}_{\mathbf{x}}$.

3.2.4 Gradient Operator of Scalar Field

The first- and second-order derivatives of the functional $\mathcal{F}(\nabla v)$ on the operator ∇v are, respectively,

$$\mathbf{p} = \frac{\partial \mathcal{F}(\nabla v)}{\partial(\nabla v)}, \mathbf{D} = \frac{\partial \mathbf{p}}{\partial(\nabla v)} = \frac{\partial^2 \mathcal{F}(\nabla v)}{\partial(\nabla v)^T \partial(\nabla v)}. \tag{3.37}$$

In 3D, $\mathbf{p}$ is a 3-vector. $\mathbf{D}$ are 3×3 matrices.

The first and second variation of $\mathcal{F}$ at point I with discrete support $N_I = (1, \ldots, n_I)$ can be written as

$$\delta\mathcal{F}(\nabla v_I) = \frac{\partial \mathcal{F}(\nabla v_I)}{\partial(\nabla v_I)} \tilde{\nabla}\delta v_I = \mathbf{p}_I^T \mathbf{B}_I^{grad} \cdot \delta\mathbf{v}_{N_I}$$
$$\delta^2\mathcal{F}(\nabla v_I) = (\tilde{\nabla}\delta v_I)^T \mathbf{D}\, \tilde{\nabla}\delta v_I = \delta\mathbf{v}_{N_I}^T (\mathbf{B}_I^{grad})^T \mathbf{D}\mathbf{B}_I^{grad} \delta\mathbf{v}_{N_I},$$

and then residual and tangent stiffness matrix at a point can be extracted as

$$\mathbf{R}_I^{grad} = \mathbf{p}_I^T \mathbf{B}_I^{grad}, \mathbf{K}_I^{grad} = (\mathbf{B}_I^{grad})^T \mathbf{D}\mathbf{B}_I^{grad} \tag{3.38}$$

where $\mathbf{B}_I^{grad}$ is given by Eq. 3.23.

In order to derive the nonlocal strong form, let's consider the first variation of all particles, and let $\mathbf{p}_\mathbf{x} = \frac{\partial \mathcal{F}(\nabla v_\mathbf{x})}{\partial(\nabla v_\mathbf{x})}$.

$$\delta\mathcal{F}(\nabla v) = \sum_{\Delta V_\mathbf{x} \in \Omega} \Delta V_\mathbf{x} \delta\mathcal{F}_\mathbf{x} = \sum_{\Delta V_\mathbf{x} \in \Omega} \Delta V_\mathbf{x} \nabla\delta v_\mathbf{x} \cdot \mathbf{p}_\mathbf{x}$$
$$= \sum_{\Delta V_\mathbf{x} \in \Omega} \Delta V_\mathbf{x} \Big(\sum_{\mathcal{S}_\mathbf{x}} w(\mathbf{r}) \Delta V_{\mathbf{x}'} \mathbf{K}_\mathbf{x}^{-1} \mathbf{r} (\delta v_{\mathbf{x}'} - \delta v_\mathbf{x}) \cdot \mathbf{p}_\mathbf{x} \Big)$$
$$= \sum_{\Delta V_\mathbf{x} \in \Omega} \Delta V_\mathbf{x} \Big(- \sum_{\mathcal{S}_\mathbf{x}} w(\mathbf{r}) \Delta V_{\mathbf{x}'} \mathbf{K}_\mathbf{x}^{-1} \mathbf{r} \delta v_\mathbf{x} \cdot \mathbf{p}_\mathbf{x} + \sum_{\mathcal{S}'_\mathbf{x}} w(\mathbf{r}') \Delta V_{\mathbf{x}'} \mathbf{K}_{\mathbf{x}'}^{-1} \mathbf{r}' \delta v_\mathbf{x} \cdot \mathbf{p}_{\mathbf{x}'} \Big).$$

In the second and third step, the dual-support is considered as follows. In the second step, the term with $\delta v_{\mathbf{x}'}$ is the vector from $\mathbf{x}$'s support, but is added to particle $\mathbf{x}'$; since $\mathbf{x}' \in \mathcal{S}_\mathbf{x}$, $\mathbf{x}$ belongs to the dual-support $\mathcal{S}'_{\mathbf{x}'}$ of $\mathbf{x}'$. In the third step, all the terms with $\delta v_\mathbf{x}$ are collected from other particles whose supports contain $\mathbf{x}$ and therefore form the dual-support of $\mathbf{x}$. The terms with $\delta v_\mathbf{x}$ in the first-order variation $\delta\mathcal{F}(\nabla v) = 0$ are

$$- \sum_{\mathcal{S}_\mathbf{x}} w(\mathbf{r}) \Delta V_{\mathbf{x}'} \mathbf{K}_\mathbf{x}^{-1} \mathbf{r} \cdot \mathbf{p}_\mathbf{x} + \sum_{\mathcal{S}'_\mathbf{x}} w(\mathbf{r}') \Delta V_{\mathbf{x}'} \mathbf{K}_{\mathbf{x}'}^{-1} \mathbf{r}' \cdot \mathbf{p}_{\mathbf{x}'}. \tag{3.39}$$

When any particle's volume $\Delta V_{\mathbf{x}'} \to 0$, the continuous form is

$$-\int_{S_{\mathbf{x}}} w(\mathbf{r})\mathbf{K}_{\mathbf{x}}^{-1}\mathbf{r}\cdot\mathbf{p}_{\mathbf{x}}\,\mathrm{d}V_{\mathbf{x}'}+\int_{S_{\mathbf{x}}'} w(\mathbf{r}')\mathbf{K}_{\mathbf{x}'}^{-1}\mathbf{r}'\cdot\mathbf{p}_{\mathbf{x}'}\,\mathrm{d}V_{\mathbf{x}'}. \tag{3.40}$$

Equation 3.40 is the nonlocal strong form for energy functional $\mathcal{F}(\nabla v)$.

The simplest example for this energy functional is

$$\mathcal{F}(\nabla T)=\frac{1}{2}\kappa\nabla T\cdot\nabla T,$$

where T is the temperature, κ is the thermal conductivity. The local strong form corresponding to Eq. 3.40 is $-\nabla\cdot\mathbf{p}$, where $\mathbf{p}=\kappa\nabla T$.

The local strong form $-\nabla\cdot\mathbf{p}_{\mathbf{x}}$ is obtained as follows. In order to derive the local strong form of functional $\mathcal{F}(\nabla T)$, consider the variation of $\int_\Omega \mathcal{F}(\nabla T)\mathrm{d}V_{\mathbf{x}}$

$$\begin{aligned}
&\delta\Big(\int_\Omega \frac{1}{2}\kappa\nabla T_{\mathbf{x}}\cdot\nabla T_{\mathbf{x}}\mathrm{d}V_{\mathbf{x}}\Big)\\
&=\int_\Omega \kappa\nabla T_{\mathbf{x}}\cdot\nabla\delta T_{\mathbf{x}}\mathrm{d}V_{\mathbf{x}}\\
&=\int_{\partial\Omega} \kappa\nabla T_{\mathbf{x}}\cdot\mathbf{n}_{\mathbf{x}}\cdot\delta T_{\mathbf{x}}\mathrm{d}S_{\mathbf{x}}-\int_\Omega(\nabla\cdot\kappa\nabla T_{\mathbf{x}})\cdot\delta T_{\mathbf{x}}\mathrm{d}V_{\mathbf{x}}.
\end{aligned}$$

For any point in Ω, the term corresponding to $\delta T_{\mathbf{x}}$ is $-\nabla\cdot(\kappa\nabla T_{\mathbf{x}})$.

When solving solid mechanics by finite element method, for an element with nodal unknown vector $\mathbf{a}^e$, the strain and stress can be expressed by the strain matrix $\mathbf{B}$ and material matrix $\mathbf{D}$, i.e.

$$\boldsymbol{\varepsilon}=\mathbf{B}\mathbf{a}^e,\quad \boldsymbol{\sigma}=\mathbf{D}\boldsymbol{\varepsilon}=\mathbf{D}\mathbf{B}\mathbf{a}^e. \tag{3.41}$$

The general expression for the stiffness matrix is

$$\mathbf{K}^e=\int_{V_e}\mathbf{B}^T\mathbf{D}\mathbf{B}\mathrm{d}V. \tag{3.42}$$

One of the critical steps in the tangent stiffness matrix is the calculation of the strain matrix $\mathbf{B}$.

For the Nonlocal Operator Method, $\mathbf{B}^{div}$, $\mathbf{B}^{curl}$, $\mathbf{B}^{grad}$, $\mathbf{B}^{\mathbf{grad}}$ in Sect. 3.1.1 are the operator matrices for divergence, curl, gradient of scalar, gradient of vector, respectively. NOM obtains the tangent stiffness matrix of the functional in a way quite similar to FEM. However, the nonlocal operator obtains the operator matrix directly, instead of the derivative of the shape function. Another difference is that the number of dimensions of the operator matrix varies with the number of particles in support.

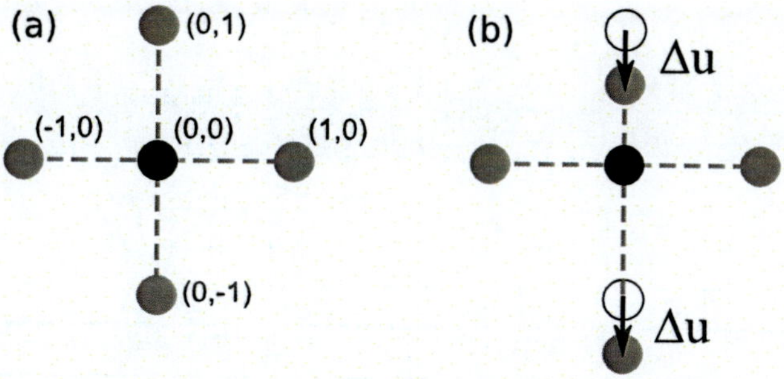

Fig. 3.2 Zero-energy mode demonstration. **a** initial configuration. **b** upper and lower particles with the same rigid translation Δu. The deformation gradients defined by (a) and (b) are the same

3.3 Operator Energy Functional

The same nonlocal operator can be defined by several configurations, for example, the initial configuration Fig. 3.2a with a rigid translation Δu for the upper and lower particles turns into Fig. 3.2b. It is easy to verify that the nonlocal gradient of **u** from Eq. 3.5 is zero, the same as that in the initial configuration. It can be seen that for the same field gradient, the configuration is not unique, where the extra deformation not accounted for by the gradient is called the zero-energy mode, which has zero energy contribution for the energy functional. The zero-energy mode in the nonlocal operator is due to the deformation vectors counteracting with each other during the summation.

In order to remove the zero-energy mode, we propose a penalty energy functional to achieve the linear field of the vector field. The penalty energy functional is defined as the weighted difference between the current value of a point and the value predicted by the field gradient. In fact, the vector field in the neighborhood of a particle is required to be linear. Therefore, it has to be exactly described by the gradient of the vector field, and the zero-energy modes are identified as that part of the vector field, which is not described by the vector field gradient. In practice, the difference of current deformed vector $\mathbf{v}_\mathbf{r}$ and predicted vector by field gradient ($\mathbf{F}_\mathbf{x}(=\nabla\mathbf{v})$ in Eq. 3.5) is $(\mathbf{Fr}-\mathbf{v}_\mathbf{r})$. We formulate the operator energy based on the difference in support as follows. Let $\alpha=\frac{\mu}{m_\mathbf{K}}$ be a coefficient for the operator energy, where $m_\mathbf{K}=\mathrm{tr}(\mathbf{K})$, μ is the penalty coefficient, the functional for zero-energy mode is

$$
\begin{aligned}
\mathcal{F}^{hg} &= \alpha \int_S w(\mathbf{r})(\mathbf{F}\mathbf{r} - \mathbf{v}_\mathbf{r})^T(\mathbf{F}\mathbf{r} - \mathbf{v}_\mathbf{r})\mathrm{d}V \\
&= \alpha \int_S w(\mathbf{r})\Big(\mathbf{r}^T\mathbf{F}^T\mathbf{F}\mathbf{r} + \mathbf{v}_\mathbf{r}^T\mathbf{v}_\mathbf{r} - 2\mathbf{v}_\mathbf{r}^T\mathbf{F}\mathbf{r}\Big)\mathrm{d}V \\
&= \alpha \int_S w(\mathbf{r})\Big(\mathbf{F}^T\mathbf{F} : \mathbf{r}\otimes\mathbf{r} + \mathbf{v}_\mathbf{r}^T\mathbf{v}_\mathbf{r} - 2\mathbf{F} : \mathbf{v}_\mathbf{r}\otimes\mathbf{r}\Big)\mathrm{d}V \\
&= \alpha\mathbf{F}^T\mathbf{F} : \int_S w(\mathbf{r})\mathbf{r}\otimes\mathbf{r}\mathrm{d}V + \alpha\int_S w(\mathbf{r})\mathbf{v}_\mathbf{r}^T\mathbf{v}_\mathbf{r}\mathrm{d}V - 2\alpha\mathbf{F} : \int_S w(\mathbf{r})\mathbf{v}_\mathbf{r}\otimes\mathbf{r}\mathrm{d}V \\
&= \alpha\mathbf{F}^T\mathbf{F} : \mathbf{K} + \alpha\int_S w(\mathbf{r})\mathbf{v}_\mathbf{r}\cdot\mathbf{v}_\mathbf{r}\mathrm{d}V - 2\alpha\mathbf{F} : (\mathbf{F}\mathbf{K}) \\
&= \frac{\mu}{m_\mathbf{K}}\Big(\int_S w(\mathbf{r})\mathbf{v}_\mathbf{r}\cdot\mathbf{v}_\mathbf{r}\mathrm{d}V - \mathbf{F} : \mathbf{F}\mathbf{K}\Big).
\end{aligned}
\tag{3.43}
$$

The above definition of operator energy is similar to the variance in probability theory and statistics. In above derivation, we used the relations: $\mathbf{F}^T\mathbf{F} : \mathbf{K} = \mathbf{F} : (\mathbf{F}\mathbf{K})$, $\mathbf{a}^T\mathbf{M}\mathbf{b} = \mathbf{M} : \mathbf{a}\otimes\mathbf{b}$, $\mathbf{A} : \mathbf{B} = \mathrm{tr}(\mathbf{A}\mathbf{B}^T)$, where capital letter denotes matrix and small letter is column vector. The purpose of $m_\mathbf{K}$ is to make the energy functional independent with the support since the shape tensor $\mathbf{K}$ is involved in $\mathbf{F}^T\mathbf{F} : \mathbf{K}$. It should be noted that the zero-energy functional is valid in any dimensions and there is no limitation on the shape of the support.

The variation of $\delta(\mathbf{F} : \mathbf{F}\mathbf{K})$ can be rewritten as

$$
\begin{aligned}
\delta(\mathbf{F} : \mathbf{F}\mathbf{K}) &= \delta(\mathbf{F}\mathbf{K} : \mathbf{F}) = 2\mathbf{F}\mathbf{K} : \delta\mathbf{F} \\
&= 2\mathbf{F}\mathbf{K} : \int_S w(\mathbf{r})\delta\mathbf{v}_\mathbf{r}\otimes(\mathbf{K}^{-1}\mathbf{r})\mathrm{d}V \\
&= 2\int_S w(\mathbf{r})\delta\mathbf{v}_\mathbf{r}^T\mathbf{F}\mathbf{K}(\mathbf{K}^{-1}\mathbf{r})\mathrm{d}V \\
&= 2\int_S w(\mathbf{r})\delta\mathbf{v}_\mathbf{r}^T(\mathbf{F}\mathbf{r})\mathrm{d}V \\
&= 2\int_S w(\mathbf{r})(\mathbf{F}\mathbf{r})\cdot\delta\mathbf{v}_\mathbf{r}\mathrm{d}V.
\end{aligned}
\tag{3.44}
$$

Then the variation of $\mathcal{F}^{hg}$ is

$$
\begin{aligned}
\delta\mathcal{F}^{hg} &= \frac{\mu}{m_\mathbf{K}}\Big(\int_S w(\mathbf{r})\delta(\mathbf{v}_\mathbf{r}\cdot\mathbf{v}_\mathbf{r})\mathrm{d}V - \delta(\mathbf{F} : \mathbf{F}\mathbf{K})\Big) \\
&= \frac{\mu}{m_\mathbf{K}}\Big(\int_S 2w(\mathbf{r})\mathbf{v}_\mathbf{r}\cdot\delta\mathbf{v}_\mathbf{r}\mathrm{d}V - 2\int_S w(\mathbf{r})(\mathbf{F}\mathbf{r})\cdot\delta\mathbf{v}_\mathbf{r}\mathrm{d}V\Big) \\
&= \frac{2\mu}{m_\mathbf{K}}\int_S w(\mathbf{r})(\mathbf{v}_\mathbf{r} - \mathbf{F}\mathbf{r})\cdot(\delta\mathbf{v}' - \delta\mathbf{v})\mathrm{d}V.
\end{aligned}
\tag{3.45}
$$

The term on $\delta\mathbf{v}$ is the operator-stabilized term from its support, while the terms on $\delta\mathbf{v}'$ are the operator-stabilized terms for the dual-support $\mathcal{S}'_{\mathbf{x}'}$ of point $\mathbf{x}'$. The operator-stabilized term for individual spatial vector $\mathbf{r}$ can be written as

$$\mathbf{T}_{\mathbf{r}}^{hg} = \partial_{\mathbf{v}}\mathcal{F}_{\mathbf{r}}^{hg} = \frac{2\mu}{m_{\mathbf{K}}} w(\mathbf{r})\big(\mathbf{v}_{\mathbf{r}} - \mathbf{F}_{\mathbf{x}}\mathbf{r}\big). \tag{3.46}$$

Equation 3.46 gives the explicit formula for the operator-stabilized force. The second variation of $\mathcal{F}^{hg}$ reads

$$\delta^2\mathcal{F}^{hg} = \frac{2\mu}{m_{\mathbf{K}}}\Big(\int_S w(\mathbf{r})(\delta\mathbf{v}' - \delta\mathbf{v})^T(\delta\mathbf{v}' - \delta\mathbf{v})\mathrm{d}V - \delta\mathbf{F}\mathbf{K} : \delta\mathbf{F}\Big). \tag{3.47}$$

After discretization of the domain into particles, $\delta\mathcal{F}^{hg}$, $\delta^2\mathcal{F}^{hg}$ for particle I with discrete support $N_I = (1, \ldots, n_I)$ can be written as

$$\delta\mathcal{F}_I^{hg} = \frac{2\mu}{m_{\mathbf{K}_I}} \sum_{k\in N_I} w(\mathbf{r}_{kI})\Delta V_k(\mathbf{v}_{\mathbf{r}} - \mathbf{F}\mathbf{r}) \cdot (\delta\mathbf{v}_k - \delta\mathbf{v}_I) = \mathbf{R}_I^{hg} \cdot \delta\mathbf{v}_{N_I} \tag{3.48}$$

$$\delta^2\mathcal{F}_I^{hg} = \frac{2\mu}{m_{\mathbf{K}_I}}\Big(\sum_{k\in N_I} w(\mathbf{r}_{kI})\Delta V_k(\delta\mathbf{v}_k - \delta\mathbf{v}_I)^T(\delta\mathbf{v}_k - \delta\mathbf{v}_I) - \tilde{\nabla}\delta\mathbf{v}_I\mathbf{K}_I : \tilde{\nabla}\delta\mathbf{v}_I\Big) = \delta\mathbf{v}_{N_I}^T\mathbf{K}_I^{hg}\delta\mathbf{v}_{N_I} \tag{3.49}$$

Let $(r_k^x, r_k^y, r_k^z) = w(\mathbf{r}_{kI})\Delta V_k(\mathbf{v}_{\mathbf{r}_{kI}} - \tilde{\nabla}\mathbf{v}_I \cdot \mathbf{r}_{kI})$. The residual $\mathbf{R}_I^{hg}$ and stiffness matrix $\mathbf{K}_I^{hg}$ can be extracted from the above equations.

$$\mathbf{R}_I^{hg} = \alpha(-\sum_{k=1}^{n_I} r_k^x, -\sum_{k=1}^{n_I} r_k^y, -\sum_{k=1}^{n_I} r_k^z, r_1^x, r_1^y, r_1^z, \ldots, r_{n_I}^x, r_{n_I}^y, r_{n_I}^z) \tag{3.50}$$

$$\mathbf{K}_I^{hg} = \alpha\begin{bmatrix} \sum_{k=1}^{n_I}\mathbf{I}_k & -\mathbf{I}_1 & \cdots & -\mathbf{I}_{n_I} \\ -\mathbf{I}_1 & \mathbf{I}_1 & \mathbf{0} & \mathbf{0} \\ \vdots & \mathbf{0} & \ddots & \mathbf{0} \\ -\mathbf{I}_{n_I} & \mathbf{0} & \mathbf{0} & \mathbf{I}_{n_I} \end{bmatrix} - \alpha(\mathbf{B}_I^{\mathbf{grad}})^T \begin{bmatrix} \mathbf{K}_I & 0 & 0 \\ 0 & \mathbf{K}_I & 0 \\ 0 & 0 & \mathbf{K}_I \end{bmatrix} \mathbf{B}_I^{\mathbf{grad}} \tag{3.51}$$

where $\mathbf{I}_k = w(\mathbf{r}_{kI})\Delta V_k \mathbf{1} \otimes \mathbf{1}^T$, $\mathbf{1} = (1, 1, 1)^T$.

When the unknown field is consistent with its gradient, both the operator energy functional and the operator energy residual are zeros. The constant operator tangent stiffness matrix can effectively remove the zero-energy mode, which will be demonstrated by the numerical examples. In this sense, operator energy functional contributes to the linear completeness of the nonlocal operators.

3.4 Higher Order Nonlocal Operators and Operator Energy Functional

3.4.1 Higher Order Operator Energy Functional

Consider a n-order operator energy functional defined by

$$\mathcal{F}_n^{hg}=\alpha\int_S w(\mathbf{r})\Big(\nabla u\cdot\mathbf{r}+\frac{1}{2!}\nabla^2 u:\mathbf{r}^2+\frac{1}{3!}\nabla^3 u\,\vdots\,\mathbf{r}^3+\cdots+\frac{1}{n!}\nabla^n u\cdot^{(n)}\mathbf{r}^n-u_{\mathbf{r}}\Big)^2\mathrm{d}V. \tag{3.52}$$

Let

$$S_{n-1}=u_{\mathbf{r}}-(\nabla u\cdot\mathbf{r}+\frac{1}{2!}\nabla^2 u:\mathbf{r}^2+\frac{1}{3!}\nabla^3 u\,\vdots\,\mathbf{r}^3+\cdots+\frac{1}{(n-1)!}\nabla^{n-1}u\cdot^{(n-1)}\mathbf{r}^{n-1}). \tag{3.53}$$

In the n-order Taylor series expansion of u, $S_{n-1}\approx\frac{1}{n!}\nabla^n u\cdot^{(n)}\mathbf{r}^n$.
On the other hand,

$$S_n=S_{n-1}-\frac{1}{n!}\nabla^n u\cdot^{(n)}\mathbf{r}^n. \tag{3.54}$$

E_n^{hg} can be simplified as

$$\begin{aligned}
\mathcal{F}_n^{hg}&=\alpha\int_S w(\mathbf{r})S_n^2\mathrm{d}V=\int_S w(\mathbf{r})(S_{n-1}-\frac{1}{n!}\nabla^n u\cdot^{(n)}\mathbf{r}^n)^2\mathrm{d}V\\
&=\alpha\int_S w(\mathbf{r})\big(S_{n-1}^2+(\frac{1}{n!}\nabla^n u\cdot^{(n)}\mathbf{r}^n)^2-2S_{n-1}(\frac{1}{n!}\nabla^n u\cdot^{(n)}\mathbf{r}^n)\big)\mathrm{d}V\\
&\approx\alpha\int_S w(\mathbf{r})\big(S_{n-1}^2+(\frac{1}{n!}\nabla^n u\cdot^{(n)}\mathbf{r}^n)^2-2(\frac{1}{n!}\nabla^n u\cdot^{(n)}\mathbf{r}^n)^2\big)\mathrm{d}V\\
&=\alpha\int_S w(\mathbf{r})\big(S_{n-1}^2-(\frac{1}{n!}\nabla^n u\cdot^{(n)}\mathbf{r}^n)^2\big)\mathrm{d}V\\
&=\mathcal{F}_{n-1}^{hg}-\alpha(\frac{1}{n!}\nabla^n u\cdot^{(n)})^2\cdot^{(2n)}\mathbf{K}_{2n}.
\end{aligned}$$

Hence, n-order operator energy functional can be written as

$$\mathcal{F}_n^{hg}=\alpha\Big(\int_S w(\mathbf{r})u_{\mathbf{r}}u_{\mathbf{r}}\mathrm{d}V-(\nabla u)^2:\mathbf{K}_2-(\frac{1}{2!}\nabla^2 u)^2\cdot^{(4)}\mathbf{K}_4-\cdots-(\frac{1}{n!}\nabla^n u)^2\cdot^{(2n)}\mathbf{K}_{2n}\Big). \tag{3.55}$$

The n-order operator energy functional depends on $(1-n)$-order nonlocal operators, and the operator residual and operator stiffness matrix can be obtained with ease by the variation of the nonlocal operator.

3.5 Applications

3.5.1 1D Beam and Bar Test

The energy functionals of cantilever beam and bar are, respectively,

$$\mathcal{F}_{beam}(u) = \frac{1}{2}\int_0^L (u_{,xx} E I u_{,xx} - uq)dx \tag{3.56}$$

$$\mathcal{F}_{bar}(u) = \frac{1}{2}\int_0^L (u_{,x} E A u_{,x} - uq)dx. \tag{3.57}$$

We consider the boundary conditions in Eq. 3.58 of the cantilever beam with a concentrated transverse load $P = 1$ applied on the end.

$$u(0) = 0, \frac{du}{dx}|_{x=0} = 0. \tag{3.58}$$

For the uniform bar, the left side is fixed and the other side is applied with a load $P = 1$. The theoretical solution for beam and bar are, respectively,

$$u(x) = \frac{P}{6EI}(3Lx^2 - x^3), u(x) = \frac{Px}{EA}, \tag{3.59}$$

where $EI = 1$, $EA = 1$ are the stiffness coefficient, $L = 1$ is the length of the beam. For the operators $u_{,xx}$, $u_{,x}$, the **B** matrix can be constructed based on the particles in support. The residual and tangent stiffness matrix of Eqs. 3.56 and 3.57 are obtained by the matrix multiplication of **B** matrices. The boundary conditions are enforced by a penalty method in the tangent stiffness matrix. The L2-norm is calculated by

$$\|\mathbf{u}\|_{L2} = \sqrt{\frac{\sum_j(\mathbf{u}_j - \mathbf{u}_j^{exact}) \cdot (\mathbf{u}_j - \mathbf{u}_j^{exact})\Delta V_j}{\sum_j \mathbf{u}_j^{exact} \cdot \mathbf{u}_j^{exact} \Delta V_j}}. \tag{3.60}$$

The convergence of the L2-norm for the displacement of bar under tension is shown in Fig. 3.3. The convergence of the L2-norm for the deflection of the cantilever beam is shown in Fig. 3.4. With the refinement in discretization, the numerical results converge to the theoretical solutions at a rate $r \approx 1$.

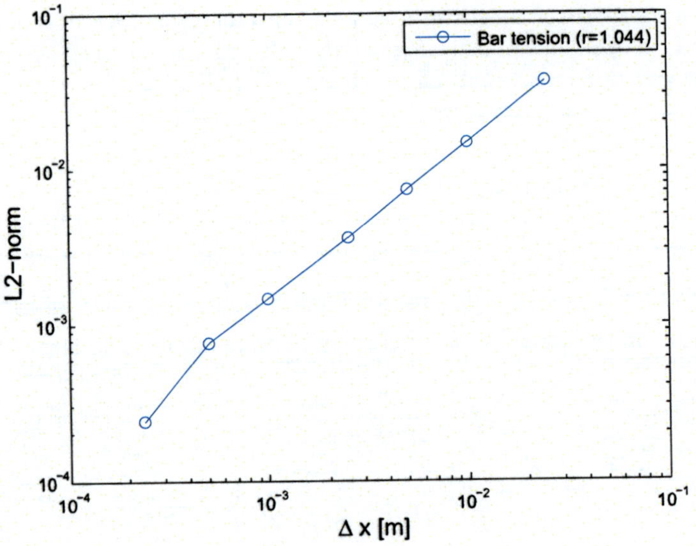

Fig. 3.3 Convergence of the L2-norm for the displacement of the bar under tension

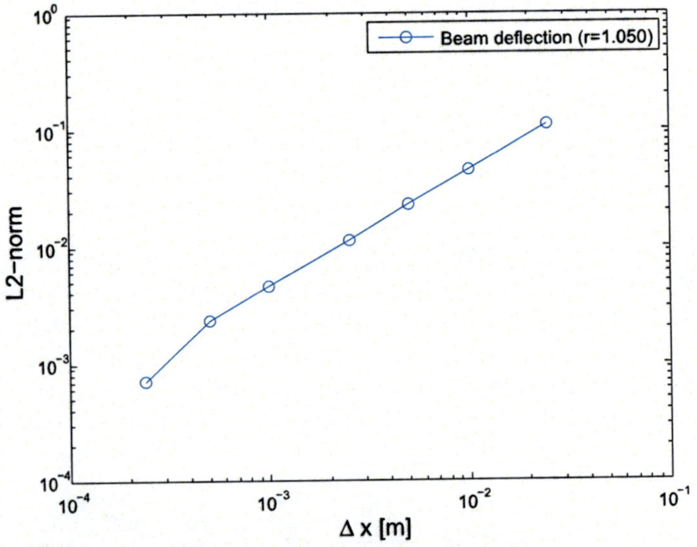

Fig. 3.4 Convergence of the L2-norm for the deflection of cantilever beam

3.5.2 Poisson Equation

In this section, we test the Poisson equation

$$\nabla^2 u = f(x, y), \quad (x, y) \in (0, 1) \times (0, 1), \tag{3.61}$$

where $f(x, y) = 2x(y-1)(y-2x+xy+2)e^{x-y}$, and the boundary conditions

$$u(x, 0) = u(x, 1) = 0, \; x \in [0, 1]$$
$$u(0, y) = u(1, y) = 0, \; y \in [0, 1].$$

The analytic solution is

$$u(x, y) = x(1-x)y(1-y)e^{x-y}. \tag{3.62}$$

The corresponding energy functional of Eq. 3.61 is

$$\mathcal{F} = \int_\Omega \big(-\frac{1}{2}\nabla u \cdot \nabla u - f(x, y)u\big)d\Omega. \tag{3.63}$$

The first and second variations of Π lead to the global residual and stiffness matrix

$$\mathbf{R}_g = \sum_{\Delta V_I \in \Omega} \Delta V_I \big(-\tilde{\nabla} u_I \cdot \mathbf{B}_{I2D} - f(x_I, y_I)\big) \tag{3.64}$$

$$\mathbf{K}_g = \sum_{\Delta V_I \in \Omega} \Delta V_I \big(-\mathbf{B}_{I2D}^T \cdot \mathbf{B}_{I2D}\big) \tag{3.65}$$

where $\mathbf{B}_{I2D}$ is the operator matrix for the 2D gradient of a scalar field. The Dirichlet boundary condition is enforced by a penalty method. The support radius is selected as $h = 1.2\Delta\mathbf{x}$. We test the convergence of the L2 error for u field under difference discretizations. The convergent plot is given in Fig. 3.5 with convergence rate of $r = 0.9567$. The contours of the u field with and without operator energy functional are shown in Fig. 3.6. It can be seen that the operator energy functional can stabilize the solution.

3.5.3 Nonlocal Theory for Linear Small Strain Elasticity

The elastic energy of a body V is given by the quadratic functional (Bažant and Jirásek 2002; Polizzotto 2001)

$$\mathcal{F} = \frac{1}{2}\int_V \int_V \boldsymbol{\varepsilon}^T(\mathbf{x})\mathbf{D}_e(\mathbf{x}, \mathbf{x}')\boldsymbol{\varepsilon}(\mathbf{x}')d\mathbf{x}'d\mathbf{x} = \frac{1}{2}\int_V \boldsymbol{\varepsilon}^T(\mathbf{x})\boldsymbol{\sigma}(\mathbf{x})d\mathbf{x}, \tag{3.66}$$

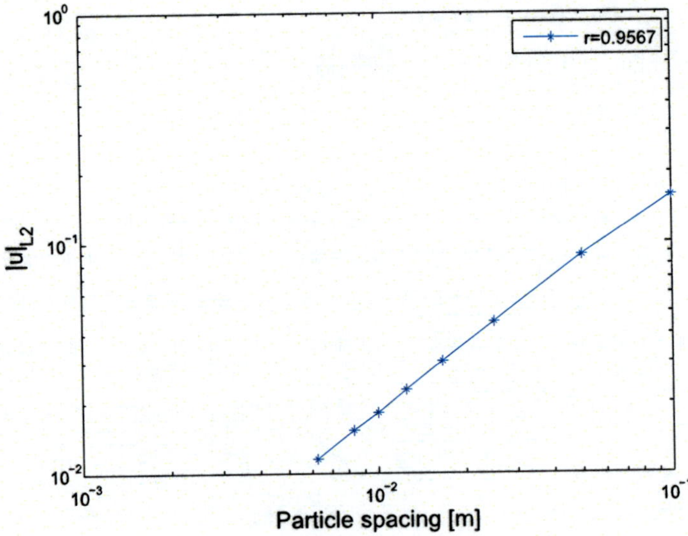

Fig. 3.5 Convergence of the L2 error of the displacement

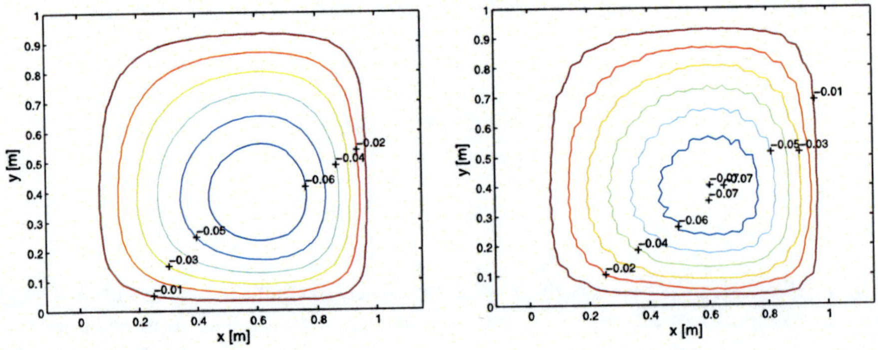

(a)

(b)

Fig. 3.6 Contour of u with operator-stabilized term $\mu = 0.1$ and without operator-stabilized term for discretization 40×40

where $\boldsymbol{\varepsilon}(\mathbf{x})$ = strain field, $\mathbf{D}_e(\mathbf{x}, \mathbf{x}')$ = generalized form of the elastic stiffness and

$$\boldsymbol{\sigma}(\mathbf{x}) = \int_V \mathbf{D}_e(\mathbf{x}, \mathbf{x}')\boldsymbol{\varepsilon}(\mathbf{x}')d\mathbf{x}' \tag{3.67}$$

is the stress dependent on the strain field in V. Only if $\mathbf{D}_e(\mathbf{x}, \mathbf{x}') = \mathbf{D}_e(\mathbf{x})\delta(\mathbf{x} - \mathbf{x}')$, Eq. 3.66 reduces to

$$\mathcal{F} = \frac{1}{2}\int_V \boldsymbol{\varepsilon}^T(\mathbf{x})\mathbf{D}_e(\mathbf{x})\boldsymbol{\varepsilon}(\mathbf{x})dx = \int_V \mathcal{F}[\boldsymbol{\varepsilon}(\mathbf{x}), \mathbf{x}]dx, \tag{3.68}$$

where $\mathcal{F}(\boldsymbol{\varepsilon}, \mathbf{x}) = \frac{1}{2}\boldsymbol{\varepsilon}^T\mathbf{D}_e(\mathbf{x})\boldsymbol{\varepsilon}$.

It can be assumed that the interaction effects decay with distance between the two points $\mathbf{x}$ and $\mathbf{x}'$, i.e.,

$$\mathbf{D}_e(\mathbf{x}, \mathbf{x}') = \mathbf{D}_e\alpha(\mathbf{x}, \mathbf{x}'), \tag{3.69}$$

where α is certain attenuation function satisfying the normalizing condition

$$\int_V \alpha(\mathbf{x}, \mathbf{x}')d\mathbf{x}' = 1. \tag{3.70}$$

α is also called the nonlocal weight function or the nonlocal averaging function, and is often assumed to have the form of Gauss distribution function

$$\alpha_\infty(r) = (l\sqrt{2\pi})^{-N_{dim}} \exp(-\frac{r^2}{2l^2}), \tag{3.71}$$

where l is the parameter with the dimension of length, N_{dim} the number of spatial dimensions. For reasons of computational efficiency, the attenuation function is often selected as the finite support, e.g., the polynomial bell-shaped function,

$$\alpha(\mathbf{r}) = c\big(\max(0, 1 - \frac{r^2}{R^2})\big)^2, \tag{3.72}$$

where c is determined by the normalizing condition Eq. 3.70.

Then the stress–strain law reads

$$\boldsymbol{\sigma}(\mathbf{x}) = \int_V \mathbf{D}_e\alpha(\mathbf{x}, \mathbf{x}')\boldsymbol{\varepsilon}(\mathbf{x}')d\mathbf{x}' = \mathbf{D}_e\int_V \alpha(\mathbf{x}, \mathbf{x}')\boldsymbol{\varepsilon}(\mathbf{x}')d\mathbf{x}' = \mathbf{D}_e\bar{\boldsymbol{\varepsilon}}(\mathbf{x}), \tag{3.73}$$

where

$$\bar{\boldsymbol{\varepsilon}}(\mathbf{x}) = \int_V \alpha(\mathbf{x}, \mathbf{x}')\boldsymbol{\varepsilon}(\mathbf{x}')d\mathbf{x}' \tag{3.74}$$

is the nonlocal strain.

With the aid of nonlocal operator, the strain tensor $\boldsymbol{\varepsilon}_I$ for particle I with discrete support $N_I = (1, \ldots, n_I)$ can be written as

$$\delta\boldsymbol{\varepsilon}_I = \frac{1}{2}(\tilde{\nabla}\delta\mathbf{u}_I + (\tilde{\nabla}\delta\mathbf{u}_I)^T) = \mathbf{B}_I \cdot \delta\mathbf{u}_{N_I} \tag{3.75}$$

where $\tilde{\nabla}\delta\mathbf{u}$ is the variation of nonlocal gradient operator in Eq. 3.5, $\mathbf{B}_I$ is the strain matrix for I. The residual and tangent stiffness matrix of nonlocal energy functional Eq. 3.66 are

$$\mathbf{R}_g = \sum_{\Delta V_I \in V} \sum_{\Delta V_J \in V} \boldsymbol{\varepsilon}_I^T \mathbf{D}_e(\mathbf{x}_I, \mathbf{x}_J)\mathbf{B}_J \Delta V_J \Delta V_I \tag{3.76}$$

$$\mathbf{K}_g = \sum_{\Delta V_I \in V} \sum_{\Delta V_J \in V} \mathbf{B}_I^T \mathbf{D}_e(\mathbf{x}_I, \mathbf{x}_J)\mathbf{B}_J \Delta V_J \Delta V_I. \tag{3.77}$$

It is found that the tangent stiffness matrix for nonlocal elasticity is equivalent to the matrix multiplication on the variational form of the nonlocal operator on each particle. The Neumann boundary conditions and Dirichlet boundary conditions can be applied directly on the residual and stiffness matrix.

3.5.4 Nonhomogeneous Biharmonic Equation

This example tests the nonlocal Hessian operator. The inhomogeneous biharmonic equation reads

$$\nabla^2\nabla^2 w = q_0, \quad (x, y) \in (0, 1) \times (-1/2, 1/2) \tag{3.78}$$

with boundary conditions

$$w(x, -1/2) = w(x, 1/2) = 0, \ x \in [0, 1]$$
$$w(0, y) = w(1, y) = 0, \ y \in [-1/2, 1/2].$$

This biharmonic equation corresponds to the simply supported square plate subjected to uniform load with parameters such as length $a = 1$ m, thickness $t = 0.01$ m, uniform pressure $q_0 = -100$ N, Poisson ratio $\nu = 0$, elastic modulus $E = 30$ GPa and $D_0 = \frac{Et^3}{12(1-\nu^2)}$.

The analytic solution for this plate is denoted by Timoshenko Woinowsky-Krieger (1959)

$$w = \frac{4q_0a^4}{\pi^5 D_0} \sum_{m=1,3,\ldots}^{\infty} \frac{1}{m^5}\Big(1 - \frac{\alpha_m \tanh\alpha_m + 2}{2\cosh\alpha_m}\cosh\frac{2\alpha_m y}{a} + \frac{\alpha_m}{2\cosh\alpha_m}\frac{2y}{a}\sinh\frac{2\alpha_m y}{a}\Big)\sin\frac{m\pi x}{a}, \tag{3.79}$$

where $\alpha_m = \frac{m\pi}{2}$.

The equivalent energy functional of Eq. 3.78 is

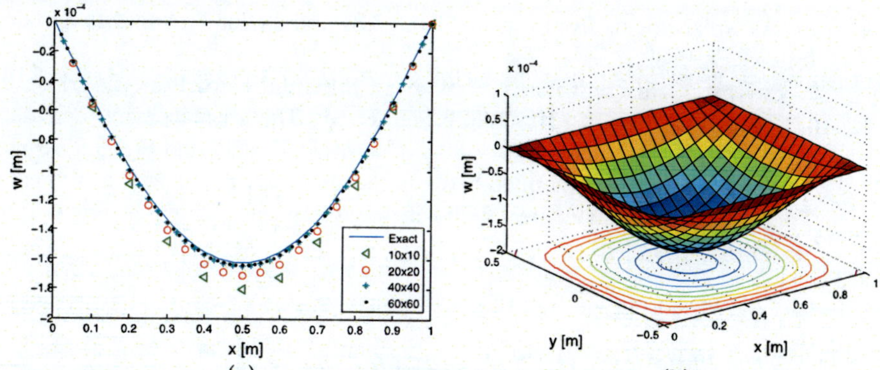

(a) (b)

Fig. 3.7 **a** Deflection of section $y = 0$ under different discretizations. **b** Contour of the deflection w for discretization of 20×20

$$\mathcal{F} = \int_\Omega \Big(\frac{1}{2}(\nabla^2 w)^T(\nabla^2 w) - q_0 w\Big) d\Omega.$$

With the aid of the operator matrix of nonlocal Hessian operators $\tilde{\nabla}^2 w$, the residual and tangent stiffness matrix can be obtained with ease. The plate is discretized uniformly and the support radius is selected as $h = 2.2\Delta\mathbf{x}$. The weight function is $w(\mathbf{r}) = \frac{1}{r^2}$. The second-order operator energy functional is used. The deflection curves for several discretizations are compared with the analytic solution in Fig. 3.7b. The contour of the deflection field for discretization of 20×20 is shown in Fig. 3.7b.

3.5.5 2D Solid Beam

A 2D cantilever beam loaded at the end with pure shear traction force is considered. The beam has dimensions of height of $D = 12$ m, length of $L = 48$ m and shear load of parabola distribution. The analytical solution for the beam is (Timoshenko and Goodier 1970; Zhuang and Augarde 2010)

$$u_x = \frac{Py}{6EI}\Big[(6L - 3x)x + (2 + \nu)(y^2 - \frac{D^2}{4})\Big] \tag{3.80}$$

$$u_y = -\frac{P}{6EI}\Big[3\nu y^2(L - x) + (4 + 5\nu)\frac{D^2 x}{4} + (3L - x)x^2\Big] \tag{3.81}$$

$$\sigma_{xx}(x, y) = \frac{P(L - x)y}{I}, \sigma_{yy}(x, y) = 0, \tau_{xy}(x, y) = -\frac{P}{2I}\Big(\frac{D^2}{4} - y^2\Big), \tag{3.82}$$

where $(x, y) \in [0, L] \times [-D/2, D/2]$, $P = -1000$ N, $I = \frac{D^3}{12}$ and material parameters $E = 10^8$ Pa, $\nu = 0.3$. The particles on the left boundary are constrained by the exact displacements from Eqs. 3.80 and 3.81 and the loading on the right boundary follows Eq. 3.82.

Several discretizations with different particle grids $\Delta x \in \{D/5, D/10, D/20, D/30, D/60\}$ are tested. The displacement and stress for discretization 10×40 and 20×80 are shown in Fig. 3.8 and Fig. 3.9, respectively. Good agreements are obtained between the numerical results and analytical results. The L2-norm of displacement field with approximately convergent rate of $r = 0.882$ is shown in Fig. 3.10.

3.5.6 Plate with Hole in Tension

This section solves the infinite plate with holes in tension and compares the numerical results by current method with that by analytical solutions. One quarter of the plate is modeled. For particles on $y = 0$ ($x = 0$) are fixed in y-direction (x-direction) by penalty method. The radius of the hole is $a = 1$ and the length of the plate is $L = 5$.

The stresses in Cartesian coordinates (Boresi et al. 2010) are

$$\begin{aligned}
\sigma_{xx}(r,\theta) &= T - T\frac{a^2}{r^2}(\frac{3}{2}\cos 2\theta + \cos 4\theta) + T\frac{3a^4}{2r^4}\cos 4\theta,\\
\sigma_{yy}(r,\theta) &= -T\frac{a^2}{r^2}(\frac{1}{2}\cos 2\theta - \cos 4\theta) - T\frac{3a^4}{2r^4}\cos 4\theta,\\
\tau_{xy}(r,\theta) &= -T\frac{a^2}{r^2}(\frac{1}{2}\sin 2\theta + \sin 4\theta) + T\frac{3a^4}{2r^4}\sin 4\theta.
\end{aligned} \tag{3.83}$$

For plane stress conditions, the displacement can be expressed as

$$\begin{aligned}
u_x(r,\theta) &= \frac{Ta}{8\mu}\left(\frac{r}{a}(\kappa+1)\cos\theta + \frac{2a}{r}((1+\kappa)\cos\theta + \cos 3\theta) - \frac{2a^3}{r^3}\cos 3\theta\right),\\
u_y(r,\theta) &= \frac{Ta}{8\mu}\left(\frac{r}{a}(\kappa-3)\sin\theta + \frac{2a}{r}((1-\kappa)\sin\theta + \sin 3\theta) - \frac{2a^3}{r^3}\sin 3\theta\right)
\end{aligned} \tag{3.84}$$

where $\mu = \frac{E}{2(1+\nu)}$, and $\kappa = \frac{3-\nu}{1+\nu}$.

For particles on $y = L (x = L)$ are applied with the surface traction force calculated by Eq. 3.83. The discretization with 11 nodes on the left edge is shown in Fig. 3.11b. It should be noted that only the nodes are used and the area associated with nodes are constructed from the element area. The material parameters are $E = 1000$, $\nu = 0.3$. Three cases with a total 525, 2050, 8019 nodes, respectively, are tested. The displacement and stress on polar coordinate $r = 2a$ are compared with the analytical solutions, as shown in Fig. 3.12a–c. The L2 norm of the displacement field by Eq. 3.60 are (0.0803, 0.0371, 0.0217) for three cases, respectively.

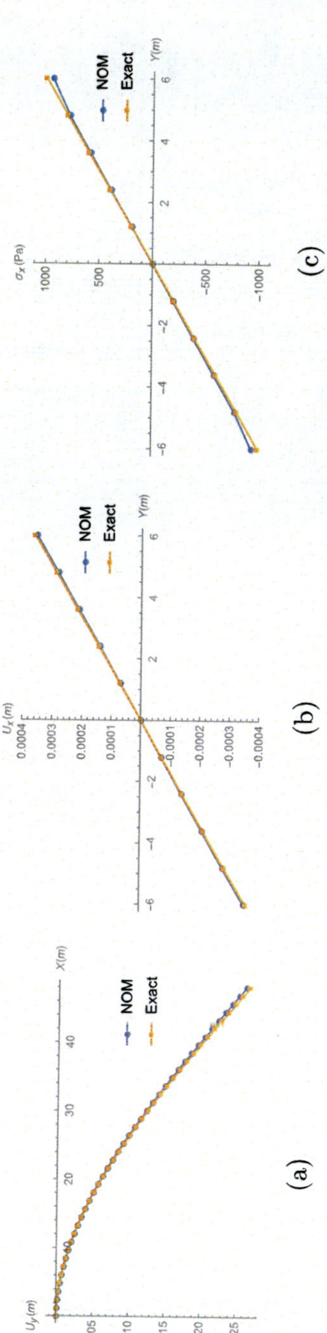

Fig. 3.8 Beam under discretization 10×40. **a** Displacement in y-direction for points on $x = 0$; **b** displacement in x-direction for points on $y = L/2$; **c** stress in x-direction for points on $y = L/2$

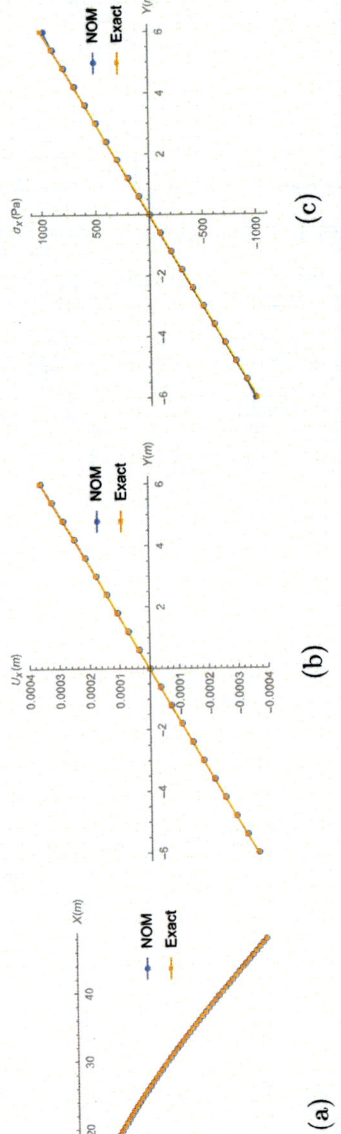

Fig. 3.9 Beam under discretization 20 × 80. **a** Displacement in y-direction for points on $x = 0$; **b** displacement in x-direction for points on $y = L/2$; **c** stress in x-direction for points on $y = L/2$

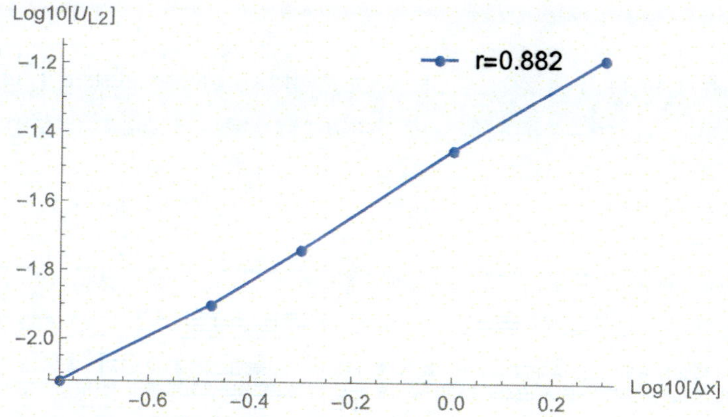

Fig. 3.10 Convergence of displacement on L_2-norm

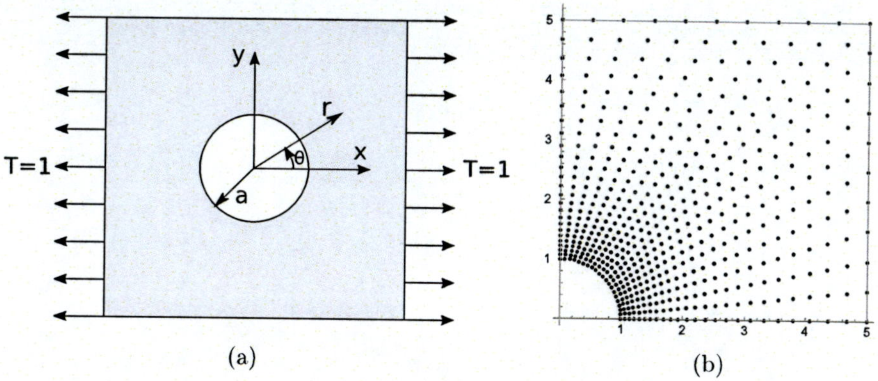

Fig. 3.11 **a** Setup of the plate with a hole; **b** discretization of the plate

3.6 Conclusions

We propose a nonlocal operator method for solving PDEs. The fundamental elements in nonlocal operator method include the support, dual-support, nonlocal operators and operator energy functional. The support is the basis to define the nonlocal operators. The nonlocal operator is a generalization of the conventional differential operators. Under certain conditions such as support decreasing to infinitesimal or linear field, the nonlocal operators converge to the local operators. On the other hand, the nonlocal operator is still valid in the case of fields involving discontinuity since the nonlocal operator is defined by integral form. The dual-support as the dual concept of support allows the inhomogeneous discretization of the computational domain. The dual-support contributes to deriving the nonlocal strong discrete or continuous forms of different functionals by means of variational principles.

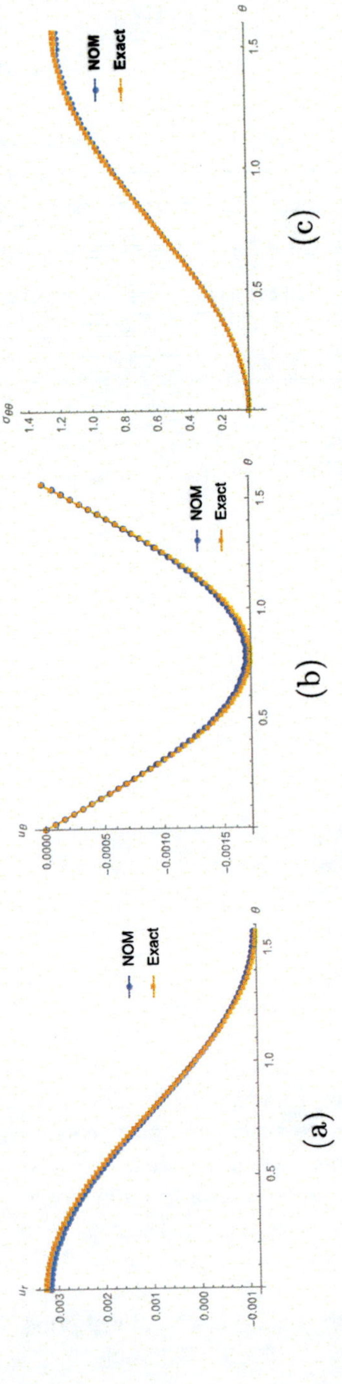

Fig. 3.12 Exact results versus numerical results. **a** u_r for points on $r = 2a$; **b** u_θ for points on $r = 2a$; **c** $\sigma_{\theta\theta}$ for points on $r = 2a$

The nonlocal operator is defined at one point but interacts with any other points in its support domain through nonlocal interactions. In this chapter, the continuous form is solved by discretizing the computational domain into particles, and finally results in a discrete system based on nodal integration. Nodal integration method suffers the rank deficiency and zero-energy mode. In order to remove the zero-energy mode, the operator energy functional is proposed, which can suppress the zero-energy modes in implicit/explicit analysis.

The nonlocal operator method is consistent with the weighted residual method and variational principle. The residual and tangent stiffness can be obtained with some matrix multiplication on common terms such as physical constitutions and nonlocal operators with variation. The nonlocal operator can be used to replace the traditional local operator of one-order or higher orders and thus obtains the discrete algebraic system of the PDEs with ease. In the example of nonlocal linear elasticity theory, the nonlocal operator method obtains the residual and tangent stiffness matrix concisely.

Several numerical examples include the deflection of cantilever beam and plate, the Poisson equation in 2D and eigenvalue problem are presented to illustrate the capabilities of the nonlocal operator method.

References

Bažant Z, Jirásek M (2002) Nonlocal integral formulations of plasticity and damage: survey of progress. J Eng Mech 128(11):1119–1149

Boresi AP, Chong K, Lee JD (2010) Elasticity in engineering mechanics. Wiley

Du Q, Gunzburger M, Lehoucq RB, Zhou K (2013) A nonlocal vector calculus, nonlocal volume-constrained problems, and nonlocal balance laws. Math Models Methods Appl Sci 23(03):493–540

Polizzotto C (2001) Nonlocal elasticity and related variational principles. Int J Solids Struct 38(42–43):7359–7380

Ren H, Zhuang X, Rabczuk T (2017) Dual-horizon peridynamics: a stable solution to varying horizons. Comput Methods Appl Mech Eng 318:762–782

Ren H, Zhuang X, Cai Y, Rabczuk T (2016) Dual-horizon peridynamics. Int J Numer Methods Eng

Silling S, Epton M, Weckner O, Xu J, Askari E (2007) Peridynamic states and constitutive modeling. J Elast 88(2):151–184

Timoshenko S, Goodier J (1970) Theory of elasticity, vol 412, 3rd edn. McGraw-Hill, New York

Timoshenko S, Woinowsky-Krieger S (1959) Theory of plates and shells. McGraw-Hill

Zhuang X, Augarde C (2010) Aspects of the use of orthogonal basis functions in the element-free Galerkin method. Int J Numer Methods Eng 81(3):366–380

Chapter 4
Nonlocal Operator Method for Computational Electromagnetic Field and Waveguide Problem

Electromagnetic analysis has been an indispensable part of many engineering and scientific study since Maxwell established a unified electromagnetic field theory—the Maxwell equations—in the 19th century. The Maxwell equations describing electromagnetic waves have numerous applications including radar, remote sensing, bioelectromagnetics, wireless communication and optics, just to name a few. Several computational methods have been developed for the solution of the Maxwell equations including the method of moments (Gibson 2007), finite element method (Jin 2015), time domain finite difference method (Taflove and Hagness 2005), ray theory (Deschamps 1972), meshless/meshfree methods (Ho et al. 2001; Zhuang et al. 2012), asymptotic-expansion methods (Bouche et al. 2012) and eigen expansion method (Chew et al. 1997). The purpose of this chapter is to apply the nonlocal operator method to the electromagnetic governing equations, by converting the local differential equation into nonlocal integral form. The specific content includes the definitions of the nonlocal curl and gradient operators, the matrix form of nonlocal operators, the consistent tangent stiffness, the hourglass energy functional and several benchmark tests.

4.1 Brief Review of Maxwell Equations

The general differential form of the Maxwell equations (Jin 2015) is given by

$$\nabla \times \mathbf{E} = -\frac{\partial \mathbf{B}}{\partial t} \qquad \text{(Faraday's law)} \tag{4.1}$$

$$\nabla \times \mathbf{H} = \frac{\partial \mathbf{D}}{\partial t} + \mathbf{J} \qquad \text{(Maxwell-Ampère law)} \tag{4.2}$$

T. Rabczuk et al., *Computational Methods Based on Peridynamics and Nonlocal Operators*, Computational Methods in Engineering & the Sciences,
https://doi.org/10.1007/978-3-031-20906-2_4

$$\nabla \cdot \mathbf{D} = \rho \qquad \text{(Gauss's law)} \tag{4.3}$$

$$\nabla \cdot \mathbf{B} = 0 \qquad \text{(Gauss's law-magnetic)} \tag{4.4}$$

$$\nabla \cdot \mathbf{J} = -\frac{\partial \rho}{\partial t}, \qquad \text{(equation of continuity)} \tag{4.5}$$

with
$\mathbf{E}$ = electric field intensity (V/m),
$\mathbf{D}$ = electric flux density (C/m^2),
$\mathbf{H}$ = magnetic field intensity (A/m),
$\mathbf{B}$ = magnetic flux density (Wb/m^2),
$\mathbf{J}$ = electric current density (A/m^2),
ρ = electric charge density (C/m^2).

The divergence-free requirement in Eq. (4.4) can be imposed for example with the penalty method (Rahman and Davies 1984), vector finite elements (Hano 1984; Nédélec 1980; Whitney 2012) or specially designed shape functions as presented in Evans and Hughes (2013), Kamensky et al. (2017). The constitutive relations can be written as

$$\mathbf{D} = \epsilon \mathbf{E} \tag{4.6}$$

$$\mathbf{B} = \mu \mathbf{H} \tag{4.7}$$

$$\mathbf{J} = \sigma \mathbf{E}, \tag{4.8}$$

where the constitutive parameters ϵ, μ and σ denote, respectively, the permittivity (F/m), permeability (H/m), and conductivity (S/m) of the medium. For the time-harmonic fields with a single frequency, the time dependent parts of Maxwell's equations can be written in simplified form as

$$\nabla \times \mathbf{E} = -j\omega \mathbf{B} \tag{4.9}$$

$$\nabla \times \mathbf{H} = j\omega \mathbf{D} + \mathbf{J} \tag{4.10}$$

$$\nabla \cdot \mathbf{J} = -j\omega \rho, \tag{4.11}$$

where j is the imaginary unit and ω is the angular frequency. The vector wave equations for $\mathbf{E}$ can be obtained by eliminating $\mathbf{H}$ from Eq. (4.10) and considering the constitutive relations Eqs. (4.6, 4.7, 4.8) to replace $\mathbf{H}$,

$$\nabla \times (\frac{1}{\mu} \nabla \times \mathbf{E}) - \omega^2 \epsilon \mathbf{E} = -j\omega \mathbf{J}. \tag{4.12}$$

The boundary conditions for equations based on $\mathbf{E}$ are

$$\mathbf{n} \times \mathbf{E} = \mathbf{P} \text{ on } S_1 \tag{4.13}$$

$$\frac{1}{\mu_r}\mathbf{n} \times (\nabla \times \mathbf{E}) = \mathbf{U} \text{ on } S_2. \tag{4.14}$$

The vector wave equations can also be formulated by using only $\mathbf{H}$. In this chapter, we will employ the vector wave equations based on electric fields.

4.2 Basic Concepts in Nonlocal Operator Method

Consider a domain as shown in Fig. 4.1a. Let $\mathbf{x}$ be the spatial coordinates in the domain Ω; $\mathbf{r} := \mathbf{x}' - \mathbf{x}$ is the spatial vector, the relative distance vector between $\mathbf{x}$ and $\mathbf{x}'$; $\mathbf{F} := \mathbf{F}(\mathbf{x}, t)$ and $\mathbf{F}' := \mathbf{F}(\mathbf{x}', t)$ are the electric field vectors for $\mathbf{x}$ and $\mathbf{x}'$, respectively; $\mathbf{F_r} := \mathbf{F}' - \mathbf{F}$ is the relative electric vector for vector $\mathbf{r}$.

Support $\mathcal{S}_{\mathbf{x}}$ is the domain where the nonlocal operator is defined, and any spatial point $\mathbf{x}'$ in support forms spatial vector $\mathbf{r}$. Support $\mathcal{S}_{\mathbf{x}}$ is usually presented by a spherical domain with a radius of $h_{\mathbf{x}}$. A point interacts with other points which fall inside the support of that point through nonlocal interactions.

In order to define the nonlocal operator, the shape tensor is defined as

$$\mathbf{K_x} = \int_{\mathcal{S}_{\mathbf{x}}} w(\mathbf{r})\mathbf{r} \otimes \mathbf{r} dV_{\mathbf{x}'}, \tag{4.15}$$

which is symmetric. The prerequisites of shape tensor are that it shall be invertible, which can be satisfied usually when enough particles fall inside the support. The

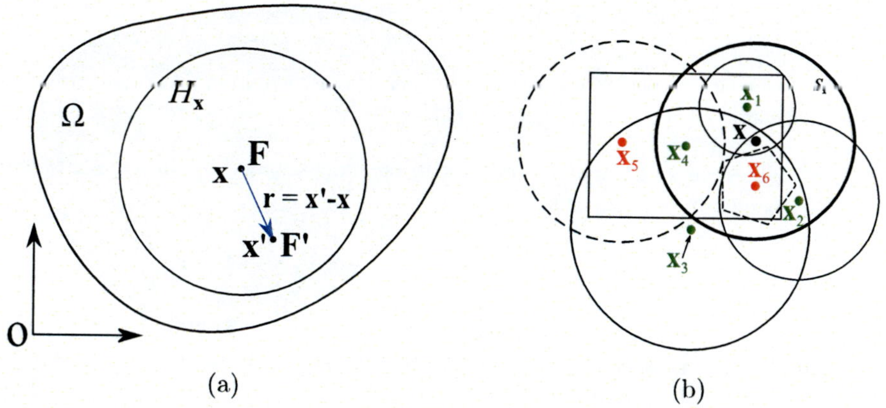

Fig. 4.1 **a** The electric field and notations. **b** Schematic diagram for support and dual-support, all circles above are support. $\mathcal{S}_{\mathbf{x}} = \{\mathbf{x}_1, \mathbf{x}_2, \mathbf{x}_4, \mathbf{x}_6\}$, $\mathcal{S}'_{\mathbf{x}} = \{\mathbf{x}_1, \mathbf{x}_2, \mathbf{x}_3, \mathbf{x}_4\}$

minimal number of neighbours in support are 2 and 3 for two-dimensional and three-dimensional problems, respectively. **Dual-support** is defined as the union of points whose supports include $\mathbf{x}$, denoted by

$$\mathcal{S}'_{\mathbf{x}} = \{\mathbf{x}' | \mathbf{x} \in \mathcal{S}_{\mathbf{x}'}\}. \tag{4.16}$$

Any point $\mathbf{x}'$ in $\mathcal{S}'_{\mathbf{x}}$ forms a dual-vector $\mathbf{r}'(= -\mathbf{r})$. On the other hand, $\mathbf{r}'$ is the spatial vector formed in $\mathcal{S}_{\mathbf{x}'}$. One example to illustrate the support and dual-support is shown in Fig. 4.1b.

4.2.1 Nonlocal Operators and Definitions Based on the Support

In the nonlocal operator method, key operators include the nonlocal operators for divergence, curl and gradient since they can be used to replace the local operators in the partial differential equations. We use $\tilde{\nabla}$ to denote the nonlocal operator, while the local operator is ∇. The nonlocal gradient of field $\mathbf{F}$ for point $\mathbf{x}$ in support $\mathcal{S}_{\mathbf{x}}$ is defined as

$$\tilde{\nabla}\mathbf{F}_{\mathbf{x}} = \int_{\mathcal{S}_{\mathbf{x}}} w(\mathbf{r})\mathbf{F}_{\mathbf{r}} \otimes \mathbf{r} dV_{\mathbf{x}'} \cdot \mathbf{K}_{\mathbf{x}}^{-1}, \tag{4.17}$$

with $\mathbf{F}_{\mathbf{r}} = \mathbf{F}_{\mathbf{x}'} - \mathbf{F}_{\mathbf{x}}$.

The nonlocal curl of field $\mathbf{F}$ for point $\mathbf{x}$ is defined as

$$\tilde{\nabla} \times \mathbf{F}_{\mathbf{x}} = \int_{\mathcal{S}_{\mathbf{x}}} w(\mathbf{r})(\mathbf{K}_{\mathbf{x}}^{-1} \cdot \mathbf{r}) \times \mathbf{F}_{\mathbf{r}} dV_{\mathbf{x}'}. \tag{4.18}$$

The nonlocal divergence of field $\mathbf{F}$ for point $\mathbf{x}$ is defined as

$$\tilde{\nabla} \cdot \mathbf{F}_{\mathbf{x}} = \int_{\mathcal{S}_{\mathbf{x}}} w(\mathbf{r})\mathbf{F}_{\mathbf{r}} \cdot (\mathbf{K}_{\mathbf{x}}^{-1} \cdot \mathbf{r}) dV_{\mathbf{x}'}. \tag{4.19}$$

The field value near a point $\mathbf{x}'$ can be approximated by Taylor series expansion by neglecting higher order terms as

$$\mathbf{F}_{\mathbf{x}'} = \mathbf{F}_{\mathbf{x}} + \nabla\mathbf{F}_{\mathbf{x}} \cdot \mathbf{r}, \text{ or } \mathbf{F}_{\mathbf{r}} = \nabla\mathbf{F}_{\mathbf{x}} \cdot \mathbf{r}. \tag{4.20}$$

Inserting Eq. (4.20) into the RHS of Eqs. (4.17), (4.18) and (4.19) and integrating in support $\mathcal{S}_{\mathbf{x}}$, it can be shown that the nonlocal operators are identical to the local operators. For example,

$$
\begin{aligned}
\tilde{\nabla}\mathbf{F}_{\mathbf{x}} &= \int_{\mathcal{S}_{\mathbf{x}}} w(\mathbf{r})\mathbf{F}_{\mathbf{r}} \otimes \mathbf{r} dV_{\mathbf{x}'} \cdot \mathbf{K}_{\mathbf{x}}^{-1} \\
&= \int_{\mathcal{S}_{\mathbf{x}}} w(\mathbf{r})(\mathbf{F}'_{\mathbf{x}} - \mathbf{F}_{\mathbf{x}}) \otimes \mathbf{r} dV_{\mathbf{x}'} \cdot \mathbf{K}_{\mathbf{x}}^{-1} \\
&= \int_{\mathcal{S}_{\mathbf{x}}} w(\mathbf{r})\nabla\mathbf{F}_{\mathbf{x}} \cdot \mathbf{r} \otimes \mathbf{r} dV_{\mathbf{x}'} \cdot \mathbf{K}_{\mathbf{x}}^{-1} \\
&= \nabla\mathbf{F}_{\mathbf{x}} \cdot \int_{\mathcal{S}_{\mathbf{x}}} w(\mathbf{r})\mathbf{r} \otimes \mathbf{r} dV_{\mathbf{x}'} \cdot \mathbf{K}_{\mathbf{x}}^{-1} \\
&= \nabla\mathbf{F}_{\mathbf{x}} \cdot \mathbf{K}_{\mathbf{x}} \cdot \mathbf{K}_{\mathbf{x}}^{-1} \\
&= \nabla\mathbf{F}_{\mathbf{x}}.
\end{aligned}
$$

When $\mathbf{x}'$ is close enough to $\mathbf{x}$ or when support $\mathbf{x}$ is small enough, the nonlocal operator can be considered as the linearization of the nonlinear field. The nonlocal operator converges to the local operator in the continuous limit. On the other hand, the nonlocal operator defined by integral form still holds in the case where strong discontinuity exists and the local operator cannot be defined. The local operator can be viewed as a special case of the nonlocal operator.

4.2.2 *Variation of Nonlocal Operators*

The nonlocal operators defined above are in vector or tensor form. The variation of the nonlocal operators leads to a higher-order tensor form, which is not convenient for implementation. We need to express the high-order tensor into vector or matrix form. Before we derive the variation of nonlocal operators, some notations to denote the variation and how the variations are related to the first- and second-order derivatives is to be discussed. Assuming a functional $\mathcal{F}(u, v)$, where $u := u(\mathbf{x})$, $v := v(\mathbf{x})$ are unknown functions in unknown vector $[u, v]$, the first and second variation can be expressed as

$$
\begin{aligned}
\delta\mathcal{F}(u, v) &= \partial_u\mathcal{F}\delta u + \partial_v\mathcal{F}\delta v = [\partial_u\mathcal{F}, \partial_v\mathcal{F}]\begin{bmatrix}\delta u\\ \delta v\end{bmatrix} \\
\delta^2\mathcal{F}(u, v) &= \partial_{uu}\mathcal{F}\delta u\delta u + \partial_{uv}\mathcal{F}\delta u\delta v + \partial_{vu}\mathcal{F}\delta v\delta u + \partial_{vv}F\delta v\delta v \\
&= \begin{bmatrix}\partial_{uu}\mathcal{F} & \partial_{uv}\mathcal{F}\\ \partial_{vu}\mathcal{F} & \partial_{vv}\mathcal{F}\end{bmatrix} : \begin{bmatrix}\delta u\delta u & \delta u\delta v\\ \delta v\delta u & \delta v\delta v\end{bmatrix}, \\
\text{where } \begin{bmatrix}\delta u\delta u & \delta u\delta v\\ \delta v\delta u & \delta v\delta v\end{bmatrix} &= \begin{bmatrix}\delta u\\ \delta v\end{bmatrix} \otimes \begin{bmatrix}\delta u & \delta v\end{bmatrix}.
\end{aligned}
$$

It can be seen that the second variation $\delta^2\mathcal{F}(u, v)$ is the double inner product of the Hessian matrix and the tensor formed by the variation of the unknowns, while the first variation $\delta\mathcal{F}(u, v)$ is inner product of the gradient vector and the variation of the unknowns. The gradient vector and the Hessian matrix represent the residual vector and tangent stiffness matrix of the functional, respectively, with unknown functions u, v being the independent variables,

$$\mathbf{R} = \nabla_{[u,v]}\mathcal{F}(u, v) = [\partial_u \mathcal{F}, \partial_v \mathcal{F}]$$
$$\mathbf{K} = \nabla^2_{[u,v]}\mathcal{F}(u, v) = \begin{bmatrix} \partial_{uu}\mathcal{F} & \partial_{uv}\mathcal{F} \\ \partial_{vu}\mathcal{F} & \partial_{vv}\mathcal{F} \end{bmatrix}.$$

The inner product or double inner product indicates that the location of an element in the residual or the tangent stiffness matrix corresponds to the location of the variation of the unknowns.

In this chapter, we use a special variation $\bar{\delta}$

$$\begin{aligned} \bar{\delta}\mathcal{F}(u, v) &= \partial_u \mathcal{F}\bar{\delta}u + \partial_v \mathcal{F}\bar{\delta}v = [\partial_u \mathcal{F}, \partial_v \mathcal{F}] \\ \bar{\delta}^2\mathcal{F}(u, v) &= \partial_{uu}\mathcal{F}\bar{\delta}u\bar{\delta}u + \partial_{uv}\mathcal{F}\bar{\delta}u\bar{\delta}v + \partial_{vu}\mathcal{F}\bar{\delta}v\bar{\delta}u + \partial_{vv}\mathcal{F}\bar{\delta}v\bar{\delta}v \\ &= \begin{bmatrix} \partial_{uu}\mathcal{F} & \partial_{uv}\mathcal{F} \\ \partial_{vu}\mathcal{F} & \partial_{vv}\mathcal{F} \end{bmatrix}, \end{aligned}$$

where $\bar{\delta}u$ denotes the index of $\partial_u \mathcal{F}$ in the residual vector by the index of u in the unknown vector. For example, the term $\partial_v \mathcal{F}\bar{\delta}v$ represents $\partial_v \mathcal{F}$ be in the second location of the residual vector since v is in the second position of $[u, v]$. The term $\partial_{uv}\mathcal{F}\bar{\delta}u\bar{\delta}v$ denotes that the location of $\partial_{uv}\mathcal{F}$ is (1,2), while the term $\partial_{vu}\mathcal{F}\bar{\delta}v\bar{\delta}u$ denotes that the location of $\partial_{vu}\mathcal{F}$ is (2,1).

Obviously

$$\delta\mathcal{F}(u, v) = \bar{\delta}\mathcal{F}(u, v)\begin{bmatrix} \delta u \\ \delta v \end{bmatrix}, \delta^2\mathcal{F}(u, v) = \bar{\delta}^2\mathcal{F}(u, v) : \begin{bmatrix} \delta u\delta u & \delta u\delta v \\ \delta v\delta u & \delta v\delta v \end{bmatrix} \tag{4.21}$$
$$\mathbf{K} = \bar{\delta}^2\mathcal{F}(u, v), \mathbf{R} = \bar{\delta}\mathcal{F}(u, v). \tag{4.22}$$

The special first-order and second-order variation of a functional lead to the residual and tangent stiffness matrix directly. The traditional variation can be recovered by the inner product of the special variation and the variation of the unknown vector.

The variation of $\tilde{\nabla} \cdot \mathbf{F_x}$ is given by

$$\tilde{\nabla} \cdot \delta\mathbf{F_x} = \int_{\mathcal{S}_\mathbf{x}} w(\mathbf{r})(\mathbf{K_x}^{-1}\mathbf{r}) \cdot (\delta\mathbf{F_{x'}} - \delta\mathbf{F_x}) dV_{\mathbf{x}'}. \tag{4.23}$$

The number of dimensions of $\tilde{\nabla} \cdot \delta\mathbf{F_x}$ is infinite, and discretization is required.

After discretization of the domain by particles, the whole domain is represented by

$$\Omega = \sum_{i=1}^{Nnode} \Delta V_i, \tag{4.24}$$

where i is the global index of volume ΔV_i, $Nnode$ is the number of particles in Ω.

Particles in $\mathcal{S}_i$ are represented by

$$N_i = \{i, j_1, ..., j_k, ..., j_{n_i}\}, \tag{4.25}$$

where $j_1, ..., j_k, ..., j_{n_i}$ are the global indices of neighbors of particle i. The discrete form of $\tilde{\nabla} \cdot \delta \mathbf{F}_i$ can be written as

$$\tilde{\nabla} \cdot \delta \mathbf{F}_i \simeq \sum_{j_k \in \mathcal{S}_i} w(\mathbf{r}) \Delta V_{j_k} (\mathbf{K}_i^{-1} \mathbf{r}) \cdot (\delta \mathbf{F}_{j_k} - \delta \mathbf{F}_i) = \tilde{\nabla} \cdot \bar{\delta} \mathbf{F}_i \cdot \delta \mathbf{F}_{N_i}. \tag{4.26}$$

where $\simeq$ denotes discretization, $\delta \mathbf{F}_{N_i}$ is all the variations of the unknowns in support $\mathcal{S}_i$,

$$\delta \mathbf{F}_{N_i} = (\delta \mathbf{F}_i, \delta \mathbf{F}_{j_1}, ..., \delta \mathbf{F}_{j_k}, ..., \delta \mathbf{F}_{j_{n_i}}), \tag{4.27}$$

$\tilde{\nabla} \cdot \bar{\delta} \mathbf{F}_i$ is the coefficient vector with a length of $3(n_i + 1)$ in 3D case,

$$\tilde{\nabla} \cdot \bar{\delta} \mathbf{F}_i = \sum_{j_k \in \mathcal{S}_i} w(\mathbf{r}) \Delta V_{j_k} (\mathbf{K}_i^{-1} \mathbf{r}) \cdot (\bar{\delta} \mathbf{F}_{j_k} - \bar{\delta} \mathbf{F}_i). \tag{4.28}$$

Based on the indices of $\bar{\delta} \mathbf{F}_{j_k}$ in $\delta \mathbf{F}_{N_i}$, $\tilde{\nabla} \cdot \bar{\delta} \mathbf{F}_i$ can be obtained by

$$\tilde{\nabla} \cdot \bar{\delta} \mathbf{F}_i[3k, 3k+1, 3k+2] = w(\mathbf{r}) \Delta V_{j_k} \mathbf{K}_i^{-1} \mathbf{r}, \; \tilde{\nabla} \cdot \bar{\delta} \mathbf{F}_i[0, 1, 2] = -\sum_{k=1}^{n_i} w(\mathbf{r}) \Delta V_{j_k} \mathbf{K}_i^{-1} \mathbf{r}, \tag{4.29}$$

where k is the index of particle j_k in N_i. The process to obtain $\tilde{\nabla} \cdot \bar{\delta} \mathbf{F}_i$ on nodal level is sometimes called the nodal assembly.

In the following section, we mainly discuss the special variation of the nonlocal operator and functional, while the actual variation can be recovered with ease. The variation of $\tilde{\nabla} \times \mathbf{F}_i$ in discrete form reads

$$\tilde{\nabla} \times \bar{\delta} \mathbf{F}_i \simeq \sum_{j_k \in \mathcal{S}_i} w(\mathbf{r}) \Delta V_{j_k} \mathbf{K}_i^{-1} \cdot \mathbf{r} \times (\bar{\delta} \mathbf{F}_{j_k} - \bar{\delta} \mathbf{F}_i), \tag{4.30}$$

where ΔV_{j_k} is the volume for particle j_k. For the 3D case, $\tilde{\nabla} \times \bar{\delta} \mathbf{F}_i$ is a $3 \times 3(n_i + 1)$ matrix, where n_i is the number of neighbors in $\mathcal{S}_i$, N_i is given by Eq. 4.25. For each particle j_k in N_i calculating $R_{j_k} = w(\mathbf{r}) \Delta V_{j_k} \mathbf{K}_i^{-1} \mathbf{r}$, we obtain

$$\tilde{\nabla}\times\bar{\delta}\mathbf{F}_i[1,3k]=R_{j_k}[2],\tilde{\nabla}\times\bar{\delta}\mathbf{F}_i[2,3k]=-R_{j_k}[1],\tilde{\nabla}\times\bar{\delta}\mathbf{F}_i[0,3k+1]=-R_{j_k}[2]$$
$$\tilde{\nabla}\times\bar{\delta}\mathbf{F}_i[2,3k+1]=R_{j_k}[0],\tilde{\nabla}\times\bar{\delta}\mathbf{F}_i[0,3k+2]=R_{j_k}[1],\tilde{\nabla}\times\bar{\delta}\mathbf{F}_i[1,3k+2]=-R_{j_k}[0]$$
$$\tilde{\nabla}\times\bar{\delta}\mathbf{F}_i[1,0]=-\sum_{k=1}^{n_i}R_{j_k}[2],\tilde{\nabla}\times\bar{\delta}\mathbf{F}_i[2,0]=\sum_{k=1}^{n_i}R_{j_k}[1],\tilde{\nabla}\times\bar{\delta}\mathbf{F}_i[0,1]=\sum_{k=1}^{n_i}R_{j_k}[2],$$
$$\tilde{\nabla}\times\bar{\delta}\mathbf{F}_i[2,1]=-\sum_{k=1}^{n_i}R_{j_k}[0],\tilde{\nabla}\times\bar{\delta}\mathbf{F}_i[0,2]=-\sum_{k=1}^{n_i}R_{j_k}[1],\tilde{\nabla}\times\bar{\delta}\mathbf{F}_i[1,2]=\sum_{k=1}^{n_i}R_{j_k}[0]. \quad (4.31)$$

where k is the index of particle j_k in N_i. The minus sign denotes the reaction from the dual-support, which guarantees the regularity of the stiffness matrix in the absence of external constraints. The nodal assembly for the variation of the vector cross product can be finally obtained by

$$\mathbf{F}^{\times}=\{R_0,R_1,R_2\}\times\{F_0,F_1,F_2\}=\{F_2R_1-F_1R_2,F_0R_2-F_2R_0,F_1R_0-F_0R_1\}, \quad (4.32)$$

while the gradient of $\mathbf{F}^{\times}$ on $\{F_0, F_1, F_2\}$ is given by

$$\begin{bmatrix}\frac{\partial F_0^{\times}}{\partial F_0} & \frac{\partial F_0^{\times}}{\partial F_1} & \frac{\partial F_0^{\times}}{\partial F_2}\\ \frac{\partial F_1^{\times}}{\partial F_0} & \frac{\partial F_1^{\times}}{\partial F_1} & \frac{\partial F_1^{\times}}{\partial F_2}\\ \frac{\partial F_2^{\times}}{\partial F_0} & \frac{\partial F_2^{\times}}{\partial F_1} & \frac{\partial F_2^{\times}}{\partial F_2}\end{bmatrix}=\begin{bmatrix}0 & -R_2 & R_1\\ R_2 & 0 & -R_0\\ -R_1 & R_0 & 0\end{bmatrix}. \quad (4.33)$$

The indices of R correspond to the locations in $\mathbf{F}^{\times}$.

Similarly, the variation of $\tilde{\nabla}\mathbf{F}_i$ in the discrete form reads

$$\tilde{\nabla}\bar{\delta}\mathbf{F}_i\simeq\sum_{j_k\in S_i}w(\mathbf{r})(\bar{\delta}\mathbf{F}_{j_k}-\bar{\delta}\mathbf{F}_i)\otimes(\mathbf{K}_i^{-1}\mathbf{r})\Delta V_{j_k}, \quad (4.34)$$

where ΔV_{j_k} is the volume for particle j_k. In 3D, $\tilde{\nabla}\bar{\delta}\mathbf{F}_i$ is a $9\times 3(n_i+1)$ matrix, where n_i is the number of neighbors in S_i, N_i is given by Eq. 4.25. For each particle in the neighbor list with $R_{j_k}=w(\mathbf{r})\Delta V_{j_k}\mathbf{K}_i^{-1}\mathbf{r}$, the terms in R_{j_k} can be added to the $\tilde{\nabla}\bar{\delta}\mathbf{F}_i$ as

$$\tilde{\nabla}\bar{\delta}\mathbf{F}_i[0,3k]=R_{j_k}[0],\tilde{\nabla}\bar{\delta}\mathbf{F}_i[3,3k]=R_{j_k}[1],\tilde{\nabla}\bar{\delta}\mathbf{F}_i[6,3k]=R_{j_k}[2],$$
$$\tilde{\nabla}\bar{\delta}\mathbf{F}_i[1,3k+1]=R_{j_k}[0],\tilde{\nabla}\bar{\delta}\mathbf{F}_i[4,3k+1]=R_{j_k}[1],\tilde{\nabla}\bar{\delta}\mathbf{F}_i[7,3k+1]=R_{j_k}[2],$$
$$\tilde{\nabla}\bar{\delta}\mathbf{F}_i[2,3k+2]=R_{j_k}[0],\tilde{\nabla}\bar{\delta}\mathbf{F}_i[5,3k+2]=R_{j_k}[1],\tilde{\nabla}\bar{\delta}\mathbf{F}_i[8,3k+2]=R_{j_k}[2],$$
$$\tilde{\nabla}\bar{\delta}\mathbf{F}_i[0,0]=-\sum_{k=1}^{n_i}R_{j_k}[0],\tilde{\nabla}\bar{\delta}\mathbf{F}_i[3,0]=-\sum_{k=1}^{n_i}R_{j_k}[1],\tilde{\nabla}\bar{\delta}\mathbf{F}_i[6,0]=-\sum_{k=1}^{n_i}R_{j_k}[2],$$
$$\tilde{\nabla}\bar{\delta}\mathbf{F}_i[1,1]=-\sum_{k=1}^{n_i}R_{j_k}[0],\tilde{\nabla}\bar{\delta}\mathbf{F}_i[4,1]=-\sum_{k=1}^{n_i}R_{j_k}[1],\tilde{\nabla}\bar{\delta}\mathbf{F}_i[7,1]=-\sum_{k=1}^{n_i}R_{j_k}[2],$$
$$\tilde{\nabla}\bar{\delta}\mathbf{F}_i[2,2]=-\sum_{k=1}^{n_i}R_{j_k}[0],\tilde{\nabla}\bar{\delta}\mathbf{F}_i[5,2]=-\sum_{k=1}^{n_i}R_{j_k}[1],\tilde{\nabla}\bar{\delta}\mathbf{F}_i[8,2]=-\sum_{k=1}^{n_i}R_{j_k}[2]. \quad (4.35)$$

where k is the index of particle j_k in N_i. The sub-index of $\tilde{\nabla}\delta\mathbf{F_x}$ can be obtained by the way similar to Eq. (4.33).

4.3 Waveguide

A waveguide is a structure that guides waves, such as electromagnetic waves or sound, with minimal loss of energy by restricting expansions to one or two dimensions. The study of waveguide requires the eigenmode of the wave propagation which in turn requires the tangent stiffness matrix of the field. In this section, we derive the tangent stiffness matrix in the framework of variational principle using the nonlocal operator method. The partial differential equation with boundary conditions for the waveguide problem can be expressed as

$$\nabla \times \frac{1}{\mu_r}\nabla \times \mathbf{E} - k_0^2\epsilon_r\mathbf{E} = 0, \text{ in } \Omega, \tag{4.36}$$

$$\nabla \cdot \mathbf{E} = 0, \text{ in } \Omega, \tag{4.37}$$

$$\mathbf{n} \times \mathbf{E} = 0 \text{ on } \Gamma_1, \tag{4.38}$$

$$\mathbf{n} \times \nabla \times \mathbf{E} = 0, \text{ on } \Gamma_2, \tag{4.39}$$

where Γ_1 is the electric boundary condition and Γ_2 is the magnetic boundary condition; $\epsilon_r(=\epsilon/\epsilon_0)$ and $\mu_r(=\mu/\mu_0)$ denote the relative permittivity and relative permeability, respectively while $k_0(=\omega\sqrt{\epsilon_0\mu_0})$ is the wavenumber in free space, $Z_0(=\sqrt{\mu_0/\epsilon_0})$ is the intrinsic impedance of free space and $\epsilon_0(=8.854 \times 10^{-12}$ F/m) and $\mu_0(=4\pi \times 10^{-7}$ H/m) are the permittivity and permeability of the free space, respectively.

Consider the inner product of Eq. (4.36) with arbitrary variation $\delta\mathbf{E}$ and integrate over the domain

$$\int_\Omega (\nabla \times (\frac{1}{\mu_r}\nabla \times \mathbf{E}) - \epsilon_r k_0^2\mathbf{E}) \cdot \delta\mathbf{E}dV = 0. \tag{4.40}$$

Applying the same procedure to Eq. (4.39) leads to

$$-\frac{1}{\mu_r}\int_{\Gamma_2} \mathbf{n} \times (\nabla \times \mathbf{E}) \cdot \delta\mathbf{E}d\Gamma = 0. \tag{4.41}$$

The sum of Eqs. (4.40) and (4.41) is

$$\int_\Omega (\nabla \times (\frac{1}{\mu_r}\nabla \times \mathbf{E}) - \epsilon_r k_0^2\mathbf{E}) \cdot \delta\mathbf{E}dV - \frac{1}{\mu_r}\int_{\Gamma_2} \mathbf{n} \times (\nabla \times \mathbf{E}) \cdot \delta\mathbf{E}d\Gamma = 0. \tag{4.42}$$

Applying second vector Green's theorem in Eq. (4.43)

$$\int_{\Omega} (u(\nabla \times \mathbf{a}) \cdot (\nabla \times \mathbf{b}) - \mathbf{a} \cdot (\nabla \times u \nabla \times \mathbf{b})) dV = \int_{\Gamma_2} u(\mathbf{a} \times \nabla \times \mathbf{b}) \cdot \mathbf{n} d\Gamma, \tag{4.43}$$

to Eq. (4.42) results in

$$\int_{\Omega} ((\nabla \times \delta \mathbf{E}) \cdot (\frac{1}{\mu_r} \nabla \times \mathbf{E}) - \epsilon_r k_0^2 \mathbf{E} \cdot \delta \mathbf{E}) dV = 0. \tag{4.44}$$

Equation (4.44) is equivalent to the variation of the functional $\mathcal{F}(\mathbf{E})$,

$$\mathcal{F}(\mathbf{E}) = \frac{1}{2} \int_{\Omega} ((\nabla \times \mathbf{E}) \cdot (\frac{1}{\mu_r} \nabla \times \mathbf{E}) - \epsilon_r k_0^2 \mathbf{E} \cdot \mathbf{E}) dV. \tag{4.45}$$

The divergence-free condition is enforced by the penalty method, so the functional becomes

$$\mathcal{F}(\mathbf{E}) = \frac{1}{2} \int_{\Omega} ((\nabla \times \mathbf{E}) \cdot (\frac{1}{\mu_r} \nabla \times \mathbf{E}) + \frac{p}{\mu_r} (\nabla \cdot \mathbf{E})^2 - \epsilon_r k_0^2 \mathbf{E} \cdot \mathbf{E}) dV, \tag{4.46}$$

where p is the penalty parameter which is set to 1 in our examples as suggested in Rahman and Davies (1984). Finally, the eigenvalue problem of the waveguide problem reads

$$\delta \mathcal{F}(\mathbf{E}) = 0, \tag{4.47}$$

$$\mathbf{n} \times \mathbf{E} = 0 \text{ on } \Gamma_1, \tag{4.48}$$

$$\mathbf{n} \cdot \mathbf{E} = 0 \text{ on } \Gamma_2. \tag{4.49}$$

Equation (4.48) is the electric wall and Eq. (4.49) is the magnetic wall, which is enforced for the sake of better accuracy and the elimination of some spurious solutions. For rectangular waveguide, the normal direction is parallel to a certain axis, for example

$$\mathbf{n} = (1, 0, 0), \mathbf{n} \times \mathbf{E} = 0 \Leftrightarrow (E_y = 0, E_z = 0).$$

$E_y = 0$, $E_z = 0$ can be applied the same as Dirichlet boundary conditions.

$F(\mathbf{E})$ on point $\mathbf{x}$ is

$$\mathcal{F}(\mathbf{E}_\mathbf{x}) = \frac{1}{2} (\nabla \times \mathbf{E}_\mathbf{x}) \cdot (\frac{1}{\mu_r} \nabla \times \mathbf{E}_\mathbf{x}) - \frac{1}{2} \epsilon_r k_0^2 \mathbf{E}_\mathbf{x} \cdot \mathbf{E}_\mathbf{x}, \tag{4.50}$$

and its first variation is written as

$$\delta\mathcal{F}(\mathbf{E_x}) = (\nabla\times\delta\mathbf{E_x})\cdot(\frac{1}{\mu_r}\nabla\times\mathbf{E_x}) + (\nabla\cdot\delta\mathbf{E_x})\cdot(\frac{p}{\mu_r}\nabla\cdot\mathbf{E_x}) - \epsilon_r k_0^2\mathbf{E_x}\cdot\delta\mathbf{E_x}. \tag{4.51}$$

Consider the first variation of all particles, and let $\mathbf{P_x} = (\frac{1}{\mu_r}\nabla\times\mathbf{E_x})$, we have

$$\begin{aligned}
\delta\mathcal{F}(\mathbf{E}) &= \sum_{\Delta V_\mathbf{x}\in\Omega}\Delta V_\mathbf{x}\delta F(\mathbf{E_x}) = \sum_{\Delta V_\mathbf{x}\in\Omega}\Delta V_\mathbf{x}\Big((\nabla\times\delta\mathbf{E_x})\cdot\mathbf{P_x} - \epsilon_r k_0^2\mathbf{E_x}\cdot\delta\mathbf{E_x}\Big)\\
&= \sum_{\Delta V_\mathbf{x}\in\Omega}\Delta V_\mathbf{x}\Big(\big(\sum_{\mathbf{x}'\in\mathcal{S}_\mathbf{x}} w(\mathbf{r})\Delta V_{\mathbf{x}'}\mathbf{K}_\mathbf{x}^{-1}\mathbf{r}\times(\delta\mathbf{E}_{\mathbf{x}'}-\delta\mathbf{E_x})\big)\cdot\mathbf{P_x} - \epsilon_r k_0^2\mathbf{E_x}\cdot\delta\mathbf{E_x}\Big)\\
&= \sum_{\Delta V_\mathbf{x}\in\Omega}\Delta V_\mathbf{x}\Big(\sum_{\mathbf{x}'\in\mathcal{S}_\mathbf{x}} w(\mathbf{r})\Delta V_{\mathbf{x}'}\mathbf{P_x}\times(\mathbf{K}_\mathbf{x}^{-1}\mathbf{r})\cdot(\delta\mathbf{E}_{\mathbf{x}'}-\delta\mathbf{E_x}) - \epsilon_r k_0^2\mathbf{E_x}\cdot\delta\mathbf{E_x}\Big)\\
&= \sum_{\Delta V_\mathbf{x}\in\Omega}\Delta V_\mathbf{x}\Big(-\sum_{\mathbf{x}'\in\mathcal{S}_\mathbf{x}} w(\mathbf{r})\Delta V_{\mathbf{x}'}\mathbf{P_x}\times(\mathbf{K}_\mathbf{x}^{-1}\mathbf{r})\cdot\delta\mathbf{E_x}+\\
&\sum_{\mathbf{x}'\in\mathcal{S}'_\mathbf{x}} w(-\mathbf{r})\Delta V_{\mathbf{x}'}\mathbf{P}_{\mathbf{x}'}\times(\mathbf{K}_{\mathbf{x}'}^{-1}(-\mathbf{r}))\cdot\delta\mathbf{E_x} - \epsilon_r k_0^2\mathbf{E_x}\cdot\delta\mathbf{E_x}\Big).
\end{aligned} \tag{4.52}$$

The relation $\mathbf{a}\cdot(\mathbf{b}\times\mathbf{c}) = \mathbf{c}\cdot(\mathbf{a}\times\mathbf{b}) = \mathbf{b}\cdot(\mathbf{c}\times\mathbf{a})$ is used in the third step. In the third and fourth steps of above derivation, the dual-support has been employed, i.e. in the third step, the term $\delta\mathbf{F}_{\mathbf{x}'}$ is the vector from $\mathbf{x}$'s support, but is added to particle $\mathbf{x}'$; since $\mathbf{x}'\in\mathcal{S}_\mathbf{x}$, $\mathbf{x}$ belongs to the dual-support $\mathcal{S}'_{\mathbf{x}'}$ of $\mathbf{x}'$. In the fourth step, all the terms with $\delta\mathbf{F_x}$ are collected from other particles whose supports contain $\mathbf{x}$ and therefore form the dual-support of $\mathbf{x}$. For any $\delta\mathbf{F_x}$, the first order variation $\delta\mathcal{F}(\mathbf{E}) = 0$ leads to the nonlocal form of the governing equation of the waveguide problem:

$$-\sum_{\mathbf{x}'\in\mathcal{S}_\mathbf{x}} w(\mathbf{r})\Delta V_{\mathbf{x}'}\mathbf{P_x}\times(\mathbf{K}_\mathbf{x}^{-1}\mathbf{r}) + \sum_{\mathbf{x}'\in\mathcal{S}'_\mathbf{x}} w(-\mathbf{r})\Delta V_{\mathbf{x}'}\mathbf{P}_{\mathbf{x}'}\times(\mathbf{K}_{\mathbf{x}'}^{-1}(-\mathbf{r})) - \epsilon_r k_0^2\mathbf{E_x} = 0. \tag{4.53}$$

When the particle's volume $\Delta V_{\mathbf{x}'}\to 0$, the continuous form is

$$-\int_{\mathcal{S}_\mathbf{x}} w(\mathbf{r})\mathbf{P_x}\times(\mathbf{K}_\mathbf{x}^{-1}\mathbf{r})dV_{\mathbf{x}'} + \int_{\mathcal{S}'_\mathbf{x}} w(-\mathbf{r})\mathbf{P}_{\mathbf{x}'}\times(-\mathbf{K}_{\mathbf{x}'}^{-1}\mathbf{r})dV_{\mathbf{x}'} - \epsilon_r k_0^2\mathbf{E_x} = 0. \tag{4.54}$$

Equation (4.54) is the nonlocal governing equation of the waveguide on the electric field.

For the eigenvalue problem, the stiffness matrix is required.

The special second variation of $\mathcal{F}(\mathbf{E_x})$ leads to the tangent stiffness matrix,

$$\begin{aligned}
\mathbf{K}(\mathbf{E_x}) &= \bar{\delta}^2\mathcal{F}(\mathbf{E_x}) \\
&= (\tilde{\nabla}\times\bar{\delta}\mathbf{E_x})\cdot(\frac{1}{\mu_r}\tilde{\nabla}\times\bar{\delta}\mathbf{E_x}) + (\tilde{\nabla}\cdot\bar{\delta}\mathbf{E_x})\cdot(\frac{p}{\mu_r}\tilde{\nabla}\cdot\bar{\delta}\mathbf{E_x}) - \epsilon_r k_0^2\bar{\delta}\mathbf{E_x}\cdot\bar{\delta}\mathbf{E_x} \\
&= \mathbf{K}_1(\mathbf{E_x}) + \mathbf{K}_2(\mathbf{E_x}) - k_0^2 M(\mathbf{E_x}),
\end{aligned} \tag{4.55}$$

where

$$\mathbf{K}_1(\mathbf{E_x}) = (\tilde{\nabla}\times\bar{\delta}\mathbf{E_x})\cdot(\frac{1}{\mu_r}\tilde{\nabla}\times\bar{\delta}\mathbf{E_x}), \tag{4.56}$$

$$\mathbf{K}_2(\mathbf{E_x}) = (\tilde{\nabla}\cdot\bar{\delta}\mathbf{E_x})\cdot(\frac{p}{\mu_r}\tilde{\nabla}\cdot\bar{\delta}\mathbf{E_x}), \tag{4.57}$$

$$\mathbf{M}(\mathbf{E_x}) = \epsilon_r\bar{\delta}\mathbf{E_x}\cdot\bar{\delta}\mathbf{E_x}. \tag{4.58}$$

It should be noted that the latter term $\epsilon_r k_0^2\bar{\delta}\mathbf{E_x}\cdot\bar{\delta}\mathbf{E_x}$ takes the local value at the particle and can be obtained easily, while the nonlocal term $\tilde{\nabla}\times\bar{\delta}\mathbf{E_x}$ can be evaluated by Eq. (4.31). Assembling the stiffness matrix of all particles, one gets the global stiffness matrix and global "mass" matrix.

$$\mathbf{K}_g = \sum_{\mathbf{x}\in\Omega}\Delta V_{\mathbf{x}}\big(\mathbf{K}_1(\mathbf{E_x}) + \mathbf{K}_2(\mathbf{E_x})\big), \tag{4.59}$$

$$\mathbf{M}_g = \sum_{\mathbf{x}\in\Omega}\Delta V_{\mathbf{x}}\mathbf{M}(\mathbf{E_x}), \tag{4.60}$$

leading to the generalized eigenvalue problem

$$(\mathbf{K}_g - k_0^2\mathbf{M}_g)\mathbf{E} = 0. \tag{4.61}$$

Note that the nodal integration of the above integrals results in hourglass modes which can be removed by introducing so-called hourglass energy, which will be addressed in the next section.

4.4 Hourglass Energy Functional

In order to remove the hourglass or zero-energy modes, a penalty term is added to achieve the linear completeness of the electric field, in which the penalty is proportional to the difference between the current value of a point and the value predicted by the field gradient (Ren et al. 2020).

The electric field in the neighborhood of a particle is required to be linear. Therefore, it has to be exactly described by the gradient of the electric field, and the hourglass modes are identified as that part of the electric field, which is not described by the electric gradient. The difference between the current vector $\mathbf{E_r}$ and the predicted vector by the field gradient ($\mathbf{F} := \tilde{\nabla}\mathbf{E}$ in Eq. (4.17)) is $(\mathbf{Fr} - \mathbf{E_r})$. We formulate the hourglass energy based on the difference in the support as follows: Let $\alpha = \frac{p^{hg}}{2\mu m_{\mathbf{K}}}$ be a coefficient for the hourglass energy, where $m_{\mathbf{K}} = \mathrm{tr}(\mathbf{K})$, μ is the magnetic coefficient, p^{hg} is the penalty which can be set to 1. Then, the functional for zero-energy mode is

$$\begin{aligned}
\mathcal{F}^{hg} &= \alpha \int_S w(\mathbf{r})(\mathbf{Fr} - \mathbf{E_r})^T(\mathbf{Fr} - \mathbf{E_r})dV \\
&= \alpha \int_S w(\mathbf{r})\Big(\mathbf{r}^T\mathbf{F}^T\mathbf{Fr} + \mathbf{E_r}^T\mathbf{E_r} - 2\mathbf{E_r}^T\mathbf{Fr}\Big)dV \\
&= \alpha \int_S w(\mathbf{r})\Big(\mathbf{F}^T\mathbf{F} : \mathbf{r} \otimes \mathbf{r} + \mathbf{E_r}^T\mathbf{E_r} - 2\mathbf{F} : \mathbf{E_r} \otimes \mathbf{r}\Big)dV \\
&= \alpha\mathbf{F}^T\mathbf{F} : \int_S w(\mathbf{r})\mathbf{r} \otimes \mathbf{r}dV + \alpha \int_S w(\mathbf{r})\mathbf{E_r}^T\mathbf{E_r}dV - 2\alpha\mathbf{F} : \int_S w(\mathbf{r})\mathbf{E_r} \otimes \mathbf{r}dV \\
&= \alpha\mathbf{F}^T\mathbf{F} : \mathbf{K} + \alpha \int_S w(\mathbf{r})\mathbf{E_r}^T\mathbf{E_r}dV - 2\alpha\mathbf{F} : (\mathbf{FK}) \\
&= \frac{p^{hg}}{2\mu m_{\mathbf{K}}}\Big(\int_S w(\mathbf{r})\mathbf{E_r}^T\mathbf{E_r}dV - \mathbf{F}^T\mathbf{F} : \mathbf{K}\Big).
\end{aligned} \tag{4.62}$$

The above definition of the hourglass energy is similar to the variance in probability theory and statistics. In the above derivation, we have used the relations, $\mathbf{F}^T\mathbf{F} : \mathbf{K} = \mathbf{F} : (\mathbf{FK})$, $\mathbf{a}^T\mathbf{Mb} = \mathbf{M} : \mathbf{a} \otimes \mathbf{b}$, $\mathbf{A} : \mathbf{B} = \mathrm{tr}(\mathbf{AB}^T)$, where the capital letter denotes the matrix and small letter the column vector. The purpose of $m_{\mathbf{K}}$ is to make the energy functional independent of the support since the shape tensor $\mathbf{K}$ is involved in $\mathbf{F}^T\mathbf{F} : \mathbf{K}$. It should be noted that the zero-energy functional is valid in any dimensions and there is no limitation on the shape of the support. Consider the variation of the zero-energy functional

$$\mathbf{R}^{hg} = \bar{\delta}\mathcal{F}^{hg} = \frac{p^{hg}}{\mu m_{\mathbf{K}}}\Big(\int_S w(\mathbf{r})(\mathbf{E}' - \mathbf{E})^T(\bar{\delta}\mathbf{E}' - \bar{\delta}\mathbf{E})dV - \mathbf{F} : \mathbf{K}\bar{\delta}\mathbf{F}\Big). \tag{4.63}$$

The residual of the hourglass mode is required in the explicit time integration method. In this chapter, we only discuss implicit analysis.

The electric flux of the hourglass mode for one vector $\mathbf{r}$ is given by

$$\mathbf{T}_{\mathbf{r}}^{hg} = \partial_{\mathbf{E}}\mathcal{F}^{hg}{}_{\mathbf{r}} = \frac{p^{hg}}{\mu m_{\mathbf{K}}} w(\mathbf{r})\big(\mathbf{E_r} - \mathbf{Fr}\big). \tag{4.64}$$

Equation (4.63) is an explicit formula for the hourglass flux. The term on $\delta\mathbf{E}$ is the hourglass term from its support, while the terms on $\delta\mathbf{E}'$ are the hourglass terms for the dual support $\mathcal{S}'_{\mathbf{x}'}$ of point $\mathbf{x}'$.

$$\mathbf{K}^{hg} = \bar{\delta}^2 \mathcal{F}^{hg} = \frac{p^{hg}}{\mu m_{\mathbf{K}}} \Big(\int_S w(\mathbf{r})(\bar{\delta}\mathbf{E}' - \bar{\delta}\mathbf{E})^T (\bar{\delta}\mathbf{E}' - \bar{\delta}\mathbf{E}) dV - \bar{\delta}\mathbf{F} : \mathbf{K}\bar{\delta}\mathbf{F} \Big). \quad (4.65)$$

The second variation of the zero-energy functional is its stiffness matrix on one point. The global tangent stiffness matrix for hourglass energy functional can be assembled by

$$\mathbf{K}_g^{hg} = \sum_{\mathbf{x} \in \Omega} \Delta V_{\mathbf{x}} \mathbf{K}_{\mathbf{x}}^{hg}. \quad (4.66)$$

The above equations indicate when the electric field is consistent with the field gradient, then the hourglass energy residual is zero. Once the hourglass mode is eliminated, the residual of the hourglass mode is zero. The stiffness matrix of the hourglass mode overcomes the rank deficiency of the matrix system when nodal integration is used. The generalized eigenvalue problem becomes

$$(\mathbf{K}_g + \mathbf{K}_g^{hg} - k_0^2 \mathbf{M}_g)\mathbf{E} = 0, \quad (4.67)$$

where $\mathbf{E}$ is the eigenvector for all unknowns.

4.5 Nonlocal Operator Method for Electromagnetic in the Time Domain

Consider a volume Ω bounded by the surface S. The electromagnetic field generated by an electric current density $\mathbf{J}_{\mathbf{x}}$ satisfies the Maxwell equations. Eliminating the magnetic field with the aid of the constitutive relations, the curl-curl equation for the electric field $\mathbf{E}$ is obtained by Jin (2015)

$$\nabla \times (\frac{1}{\mu} \nabla \times \mathbf{E}_{\mathbf{x}}) + \epsilon \ddot{\mathbf{E}}_{\mathbf{x}} + \sigma \dot{\mathbf{E}}_{\mathbf{x}} = -\dot{\mathbf{J}}_{\mathbf{x}}. \quad (4.68)$$

In order to obtain the equivalent nonlocal form of Eq. (4.68), let us consider Eq. (4.51)from the previous section. From the variational derivation of the waveguide problem, $\nabla \times (\frac{1}{\mu} \nabla \times \mathbf{E}_{\mathbf{x}})$ is equivalent to the functional

$$\mathcal{F}(\mathbf{E}_{\mathbf{x}}) = \frac{1}{2} (\nabla \times \mathbf{E}_{\mathbf{x}}) \cdot (\frac{1}{\mu_r} \nabla \times \mathbf{E}_{\mathbf{x}}). \quad (4.69)$$

Based on Eq. (4.54), one can get the nonlocal form of $\nabla \times (\frac{1}{\mu}\nabla \times \mathbf{E_x})$

$$\int_{S_\mathbf{x}} w(\mathbf{r})\mathbf{S_x} \times (\mathbf{K_x}^{-1}\mathbf{r})dV_{\mathbf{x}'} - \int_{S'_\mathbf{x}} w(-\mathbf{r})\mathbf{S_{x'}} \times (-\mathbf{K_{x'}}^{-1}\mathbf{r})dV_{\mathbf{x}'} + \epsilon\ddot{\mathbf{E}}_\mathbf{x} + \sigma\dot{\mathbf{E}}_\mathbf{x} = -\dot{\mathbf{J}}_\mathbf{x}, \tag{4.70}$$

where

$$\mathbf{S_x} = \frac{\partial\mathcal{F}(\nabla \times \mathbf{E_x})}{\partial(\nabla \times \mathbf{E_x})} = \frac{1}{\mu_r}\nabla \times \mathbf{E_x}. \tag{4.71}$$

The Dirichlet boundary conditions are

$$\mathbf{n} \times \mathbf{E_x} = \mathbf{P_x}, \quad \mathbf{x} \in S_1, \tag{4.72}$$

where $\mathbf{P_x}$ is the specified electric wall on point $\mathbf{x}$. The Neumann boundary condition on S_2 can be written as

$$\mathbf{n} \times (\frac{1}{\mu}\nabla \times \mathbf{E_x}) = \mathbf{U}(\mathbf{x}, t), \quad \mathbf{x} \in S_2. \tag{4.73}$$

Finally, the central difference scheme can be used for the time integration yielding

$$\frac{d\mathbf{E}}{dt} \approx \frac{\mathbf{E}^{n+1} - \mathbf{E}^{n-1}}{2\delta t}, \tag{4.74}$$

$$\frac{d^2\mathbf{E}}{dt^2} \approx \frac{\mathbf{E}^{n+1} - 2\mathbf{E}^n + \mathbf{E}^{n-1}}{(\delta t)^2}. \tag{4.75}$$

4.6 Numerical Examples

4.6.1 *The Schrödinger Equation in 1D*

In this section, we test the accuracy of the eigenvalue. The Schrödinger equation written in adimensional units for a one-dimensional harmonic oscillator is

$$[-\frac{1}{2}\frac{\partial^2}{\partial x^2} + V(x)]\phi(x) = \lambda\phi(x), \quad V(x) = \frac{1}{2}\omega^2 x^2. \tag{4.76}$$

For simplicity, we use $\omega = 1$. The particles are distributed with constant/variable spacing Δx on the region $[-10,10]$.

The exact wave functions and eigenvalues can be expressed as

$$\phi_n(x) = H_n(x)\exp(\pm\frac{x^2}{2}), \quad \lambda_n = n + \frac{1}{2}, \tag{4.77}$$

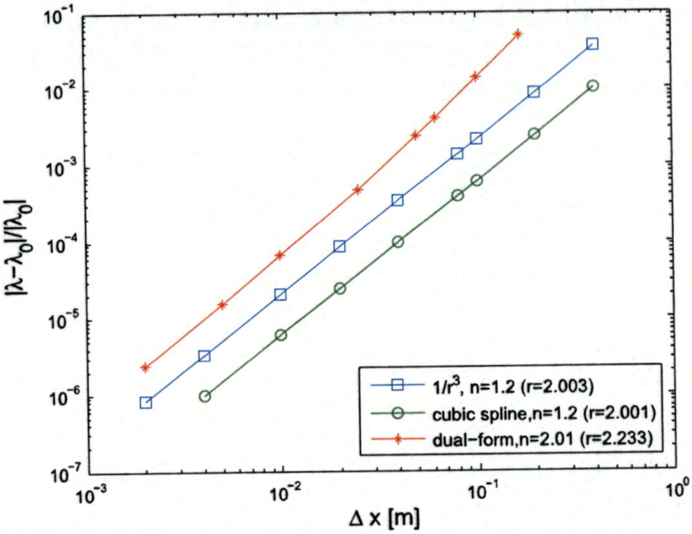

Fig. 4.2 Convergence of the lowest eigenvalue for a one-dimensional harmonic oscillator; $1/r^3$ is the influence function; support is selected as $h = n\Delta x$; dual-form with influence function $1/r^3$ uses an inhomogeneous discretization in Fig. 4.3; the particle spacing in dual-form is selected as the minimal particle spacing in the discretization

where n is a non-negative integer. $H_n(x)$ is the n-order Hermite polynomial.

The Schrödinger equation in 1D is reformulated in variational form as

$$\mathcal{F}(\phi) = \int_{-10}^{10} \frac{1}{2}\Big(\frac{\partial\phi}{\partial x}\frac{\partial\phi}{\partial x} + V(x)\phi(x)^2 - \lambda\phi(x)^2\Big)dx. \tag{4.78}$$

The tangent stiffness matrix is obtained as

$$\bar{\delta}^2\mathcal{F}(\phi) = \sum_{x\in[-10,10]} \Delta x\Big(\frac{\partial\bar{\delta}\phi}{\partial x}\frac{\partial\bar{\delta}\phi}{\partial x} + (V(x)-\lambda)\bar{\delta}\phi(x)\cdot\bar{\delta}\phi(x)\Big), \tag{4.79}$$

where $\frac{\partial\bar{\delta}\phi}{\partial x}$ is the nonlocal operator in 1D. The hourglass energy functional in 1D can be obtained with the procedure similar to that in Sect. 4.4.

We calculate the lowest eigenvalue and compare the numerical result with $\lambda_0 = 0.5$. The convergence plot of the error with different influence function and discretizations is shown in Fig. 4.2. It can be seen that the convergence rate is around 2. The influence function and inhomogeneous discretization have limited effect on the convergence. Good agreement is observed between the numerical result and the exact solution. The discretization of the dual-form is given in Fig. 4.3. The first three wave functions are given in Fig. 4.4.

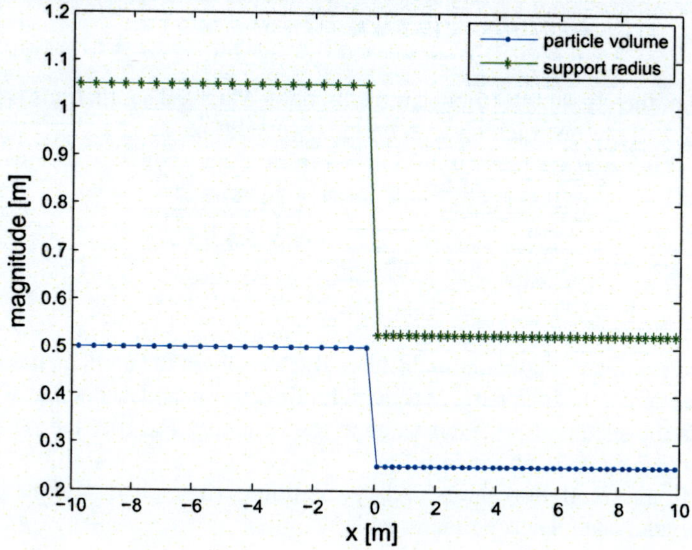

Fig. 4.3 The discretization of the dual form based on inhomogeneous discretization

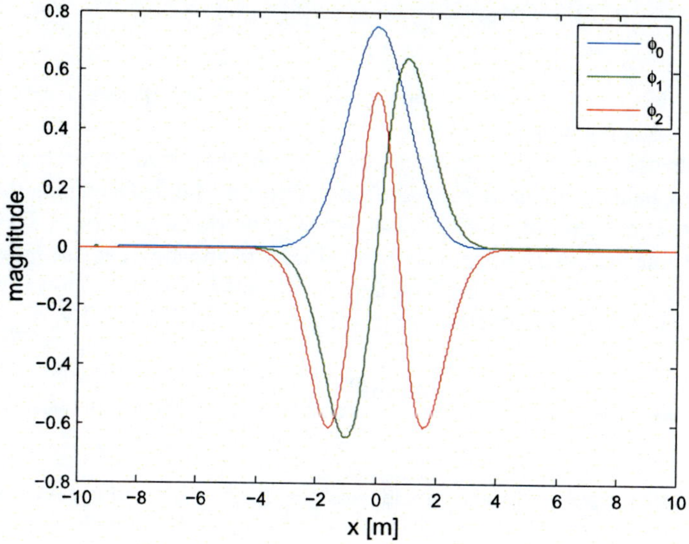

Fig. 4.4 The first three wave functions

4.6.2 Electrostatic Field Problems

When there is no electricity in the domain, Maxwell's equations can be simplified into the Poisson equation with boundary conditions as

$$\nabla^2\phi(\mathbf{x}) = \rho, \forall \mathbf{x} \in \Omega \tag{4.80}$$

$$\phi = \bar{\phi}, \forall \mathbf{x} \in \Gamma_\phi \tag{4.81}$$

$$\frac{\partial \phi}{\partial \mathbf{n}} = \bar{q}, \forall \mathbf{x} \in \Gamma_q, \tag{4.82}$$

where ϕ denotes potential, ρ is the charge density of the domain, $\mathbf{n}$ is outward normal direction of the boundary, Ω is the solution domain, and its boundary $\partial\Omega = \Gamma_\phi \cup \Gamma_q$, $\bar{\phi}$ is the specified potential value at the boundary Γ_ϕ, q is the potential derivative value given on the boundary Γ_q.

In the simulation, the boundary conditions are applied with the penalty method. The equivalent energy functional is

$$\mathcal{F}(\phi) = \int_\Omega \nabla\phi(\mathbf{x}) \cdot \nabla\phi(\mathbf{x}) dV_\mathbf{x} + \alpha \int_{\Gamma_\phi} (\phi - \bar{\phi})^2 d\Gamma_\mathbf{x} + \alpha \int_{\Gamma_q} (\frac{\partial \phi}{\partial \mathbf{n}} - \bar{q})^2 d\Gamma_\mathbf{x}, \tag{4.83}$$

where $\alpha = 1 \times 10^6$ is the penalty coefficient. The stiffness matrix can be obtained the similar way in Sect. 4.3.

In order to validate the accuracy of nonlocal operator formulation on the electrostatic field problem, we calculate the electrostatic field of a rectangular plate, which is a benchmark problem with the analytical solution given in Eq. 4.84. In this example, the potential on the upper and lower sides is 0, and the right side is $\phi = (a, y) = U_0$, the horizontal electric field strength on the left side is 0, as shown in Fig. 4.5. In the simulation, the parameters are $a = 1$ m, $b = 1$ m, $U_0 = 1.0$ V.

$$\phi(x, y) = \frac{4U_0}{\pi} \sum_{k=1}^{\infty} \frac{\cosh((2k-1)\pi x/b)}{(2k-1)\cosh((2k-1)\pi a/b)} \times \sin((2k-1)\pi y/b) \tag{4.84}$$

The support radius is selected as $\mathcal{S}_i = 1.2\Delta x_i$. The hourglass penalty of $p^{hg} = 1$ is used in all simulations. The plate is discretized with different mesh densities to test the convergence. The contour plot of the electric potential from the numerical solution and analytical solution are shown in Fig. 4.6 and Fig. 4.7, respectively. A satisfactory agreement is observed between Figs. 4.6 and 4.7. The L2 norm of the electric potential decreases with the refinement of the mesh with the convergence rate of $r = 0.8709$, as shown in Fig. 4.8.

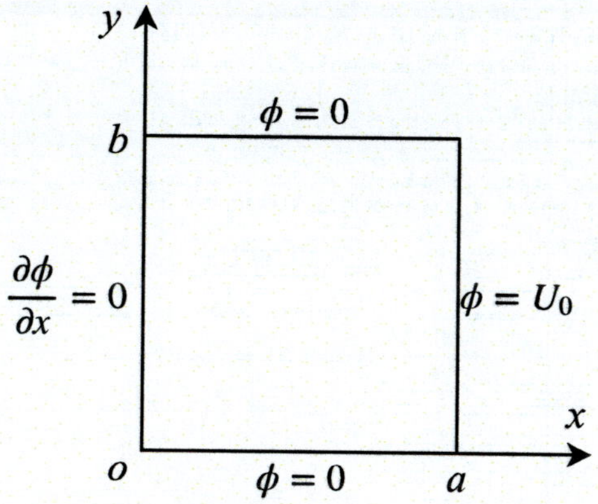

Fig. 4.5 Boundary condition of a rectangular plate

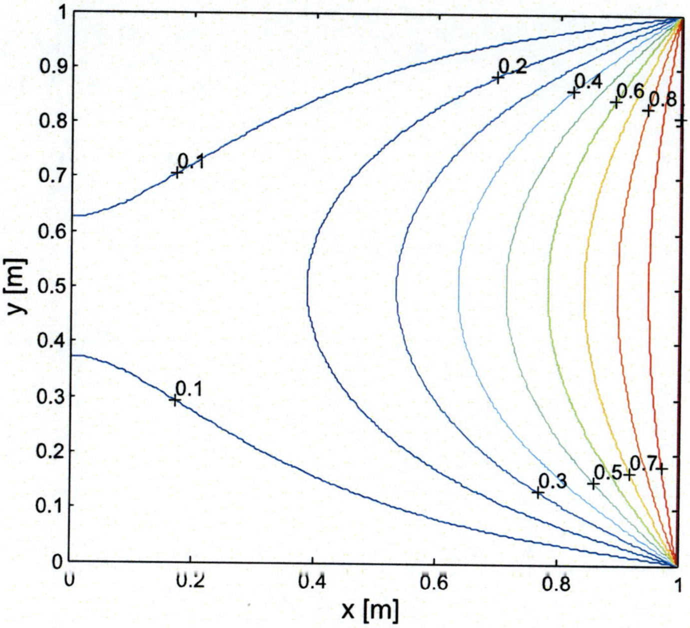

Fig. 4.6 Contour plot of numerical solution of electric potential under mesh 50 × 50

4.6.3 Rectangular Waveguide Problem

A hollow waveguide is a transmission line that looks like an empty metallic pipe. It supports the propagation of transverse electric (TE) and transverse magnetic (TM) modes, but not transverse electromagnetic (TEM) modes. There are an infinite number of modes that can propagate as long as the operating frequency is above the

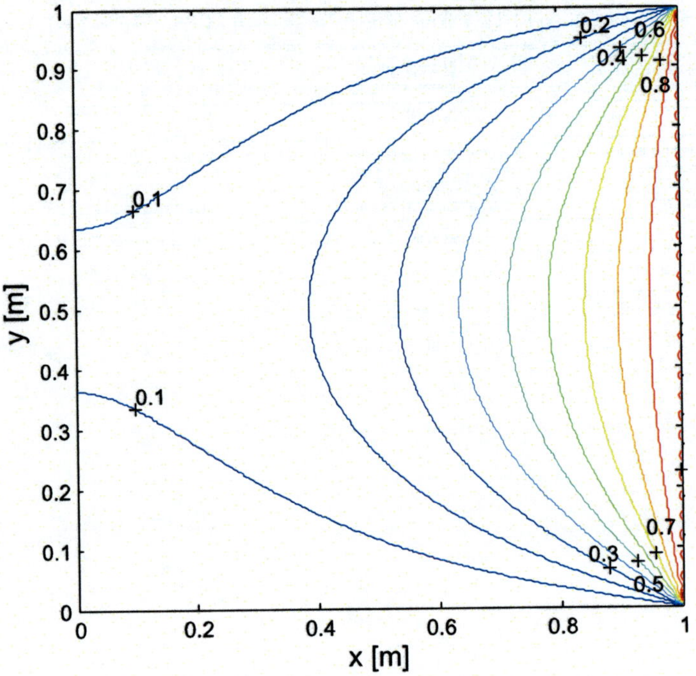

Fig. 4.7 Contour plot of analytical solution of electric potential

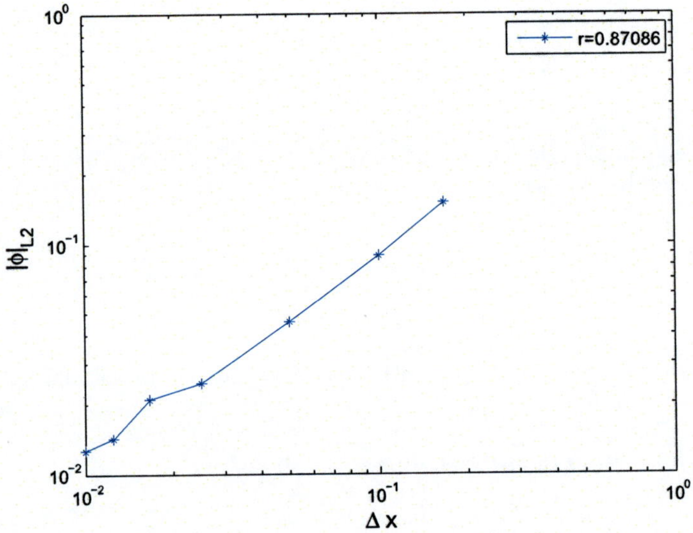

Fig. 4.8 Convergence plot of L_2 norm of electric potential

cutoff frequency of the mode. The notation TE_{mn} and TM_{mn} is commonly used to denote the type of the wave and its mode, where m and n are the mode number in the horizontal and vertical directions, respectively. The mode with the lowest cutoff frequency is called the fundamental mode or dominant mode. For a hollow rectangular waveguide, the dominant mode is TE_{10}. The analytical solution for the E-field in the TE mode is expressed as

$$E_x = A_{mn}\frac{n\pi}{b\varepsilon}\cos(\frac{m\pi x}{a})\sin(\frac{n\pi y}{b})e^{-jk_z z}, \tag{4.85}$$

$$E_y = A_{mn}\frac{m\pi}{a\varepsilon}\sin(\frac{m\pi x}{a})\cos(\frac{n\pi y}{b})e^{-jk_z z}, \tag{4.86}$$

$$E_z = 0. \tag{4.87}$$

The electromagnetic analysis of a rectangular waveguide is well known (Pozar 2009). Let us focus on the results used to verify our formulation, i.e.

$$f_{c_{mn}} = \frac{c}{2\pi}k_{c_{mn}}, \tag{4.88}$$

$$k_{c_{mn}} = \sqrt{(\frac{m\pi}{a})^2 + (\frac{n\pi}{b})^2}, \tag{4.89}$$

where $f_{c_{mn}}$ is the cutoff frequency of mode mn, $k_{c_{mn}}$ denotes the wavenumber corresponding to mode mn while a and b are the width and height of the waveguide, respectively.

A section of a rectangular waveguide is modeled with the proposed nonlocal operator formulation, and the first 3 modes are calculated and their field distributions analyzed. Since the background is set to a perfect electric conductor (PEC) material, we only need to model the vacuum inside the waveguide. The boundary conditions are "electric" in all directions, and the model is simulated using an eigenvalue solver in Matlab (1998). In this model the first 3 modes are calculated. The dimensions of the waveguide are set to $a = 22.86$ mm, $b = 10.16$ mm and $l = 40$ mm; the boundaries in blue illustrate the electric walls and the red boundary is the magnetic wall, see Fig. 4.9. The domain of the waveguide is discretized with two different particle spacings, as shown in Fig. 4.10. The support is selected as $h = 2.2\Delta x$ and influence function $w(r) = 1/r^2$.

The calculated frequencies are given in Table 4.1. The error in the frequency for Case 2 is less than 4%. The modes of the E field are shown in Fig. 4.11 for Case 1 and in Fig. 4.12 for Case 2. Good agreements are obtained between the numerical results and theoretical results.

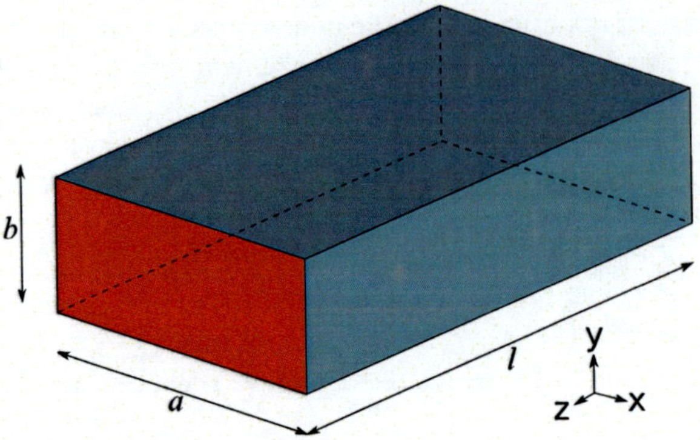

Fig. 4.9 Section of a rectangular waveguide, where a = 22.86 mm, b = 10.16 mm and l = 40 mm. Blue boundary denotes electric wall and red boundary is magnetic wall

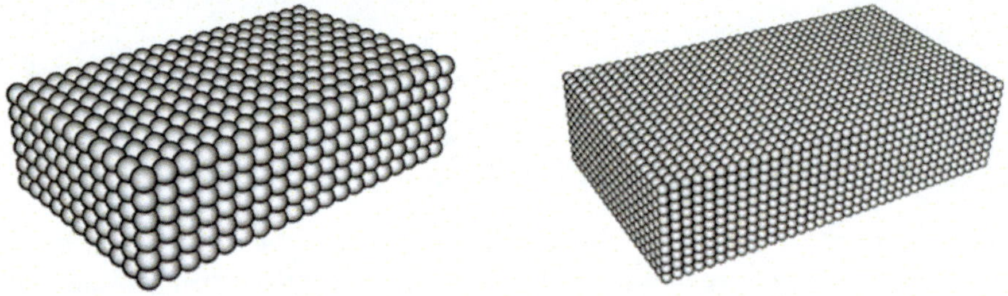

(a) Case 1:11 × 6 × 19 (b) Case 2:21 × 10 × 36

Fig. 4.10 The discretizations for two cases

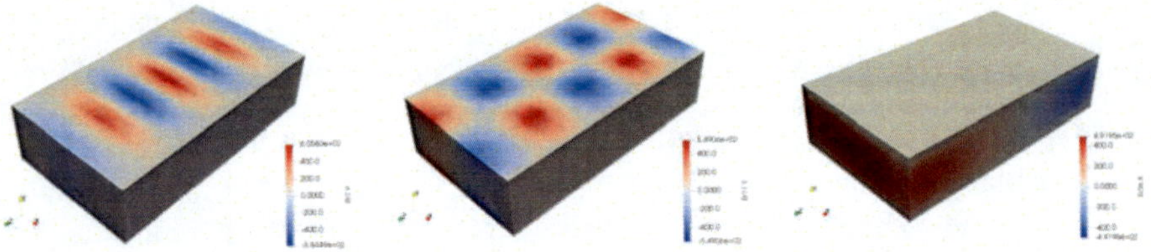

(a) TE10 (b) TE20 (c) TE01

Fig. 4.11 The TE modes for case 1

Table 4.1 Comparison of $f_{c_{mn}}$ between simulation and analytical results

Mode	TE_{10}(GHz)	TE_{20} (GHz)	TE_{01} (GHz)
Case 1	6.02(−8.29%)	12.33(−5.28%)	15.08(3.13%)
Case 2	6.30(−3.96%)	12.67(−2.67%)	14.91(1.88%)
Exact	6.56	13.02	14.63

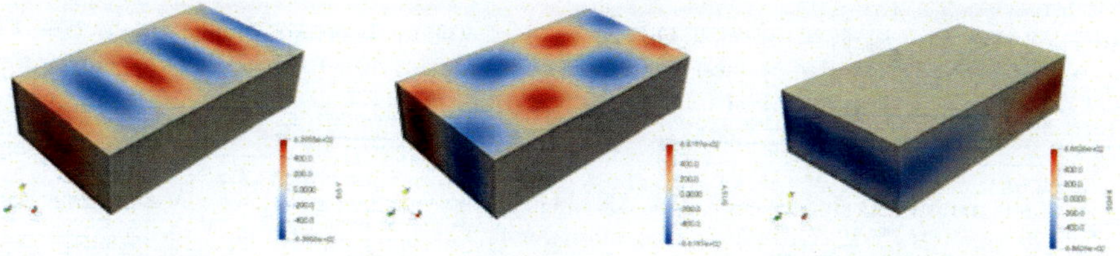
(a) TE10 (b) TE20 (c) TE01

Fig. 4.12 The TE modes for case 2

4.7 Conclusion

In this chapter, we proposed a nonlocal operator formulation for electromagnetic problems employing variational principles. The formulation is implicit and provides the tangent stiffness matrix, which is needed for the solution of the eigenvalue problem. We presented a scheme for assembling the global stiffness matrix based on nonlocal operators. The nonlocal form of the electromagnetic vector wave equations based on the electric field is obtained by means of the variational principles. Three numerical examples, including the Schrödinger equation in 1D, electrostatic field problem in 2D and waveguide in 3D are tested and show good agreement to available analytical solutions. In the future, we intend to also solve the transient problem and study problems involving strong discontinuities which is one of the key strengths of the nonlocal operator method.

References

Bouche D, Molinet F, Mittra R (2012) Asymptotic methods in electromagnetics. Springer

Chew WC, Jin J-M, Lu C-C, Michielssen E, Song JM (1997) Fast solution methods in electromagnetics. IEEE Trans Antennas Propag 45(3):533–543

Deschamps GA (1972) Ray techniques in electromagnetics. Proc IEEE 60(9):1022–1035

Evans JA, Hughes TJ (2013) Isogeometric divergence-conforming b-splines for the unsteady navier-stokes equations. J Comput Phys 241:141–167

Gibson WC (2007) The method of moments in electromagnetics. Chapman and Hall/CRC

Hano M (1984) Finite-element analysis of dielectric-loaded waveguides. IEEE Trans Microw Theory Tech 32(10):1275–1279

Ho S, Yang S, Machado JM, Wong H-CC (2001) Application of a meshless method in electromagnetics. IEEE Trans Magn 37(5):3198–3202

Jin J (2015) The finite element method in electromagnetics. Wiley

Kamensky D, Hsu M-C, Yu Y, Evans JA, Sacks MS, Hughes TJ (2017) Immersogeometric cardiovascular fluid-structure interaction analysis with divergence-conforming b-splines. Comput Methods Appl Mech Eng 314:408–472

Nédélec JC (1980) Mixed finite elements in R-3. Numerische Mathematik 35(3):315–341

Pozar DM (2009) Microwave engineering. Wiley

Rahman BA, Davies JB (1984) Penalty function improvement of waveguide solution by finite elements. IEEE Trans Microw Theory Tech 32(8):922–928

Ren H, Zhuang X, Rabczuk T (2020) A nonlocal operator method for solving partial differential equations. Comput Methods Appl Mech Eng 358:112621. ISSN 0045-7825. https://doi.org/10.1016/j.cma.2019.112621

Taflove A, Hagness SC (2005) Computational electrodynamics: the finite-difference time-domain method. Artech House

The mathworks. Inc., Natick, MA, 5:333 (1998)

Whitney H (2012) Geometric integration theory. Courier Corporation

Zhuang X, Augarde C, Mathisen K (2012) Fracture modelling using meshless methods and level sets in 3D: framework and modelling. Int J Numer Methods Eng 92(11):969–998

Chapter 5
Higher Order Nonlocal Operator Method

The low-order nonlocal operators (Ren et al. 2020) are suitable for solving low-order (not more than fourth-order) PDEs, but not higher order PDEs. The purpose of the chapter is to develop a higher order nonlocal operator method for solving higher order PDEs of multiple fields in multiple spatial dimensions. The nonlocal operator method obtains a set of partial derivatives of different orders at once. Combining with the weighted residual method and variational principles, the nonlocal operator method establishes the residual and tangent stiffness matrix for PDEs by some matrix operation on common terms, operator matrix. In contrast with the finite element method or meshless method with shape functions, the nonlocal operator method leads to the differential operators directly and adopts the nodal integration method. The remainder of the chapter is outlined as follows. In Sect. 5.1, the basic concepts such as support and dual-support, and the low-order nonlocal operators are reviewed and then the higher order nonlocal operator method based on Taylor series expansion of multiple variables is developed. We define a special quadratic functional to derive the nonlocal strong form for a $2n$-order PDEs based on the nonlocal operators in Sect. 5.2. We give several numerical examples to demonstrate the capabilities of this method in solving PDEs by the strong form in Sect. 5.3 and by the weak form in Sect. 5.4.

5.1 Nonlocal Operator Method

5.1.1 Basic Concepts

Consider a domain as shown in Fig. 5.1a; let $\mathbf{x}_i$ be spatial coordinates in the domain Ω; $\mathbf{r} := \mathbf{x}_j - \mathbf{x}_i$ is a spatial vector that starts from $\mathbf{x}_i$ to $\mathbf{x}_j$; $\mathbf{v}_i := \mathbf{v}(\mathbf{x}_i, t)$ and $\mathbf{v}_j := \mathbf{v}(\mathbf{x}_j, t)$ are the field values for $\mathbf{x}_i$ and $\mathbf{x}_j$, respectively; $\mathbf{v}_{ij} := \mathbf{v}_j - \mathbf{v}_i$ is the relative field vector for spatial vector $\mathbf{r}$.

T. Rabczuk et al., *Computational Methods Based on Peridynamics and Nonlocal Operators*, Computational Methods in Engineering & the Sciences,
https://doi.org/10.1007/978-3-031-20906-2_5

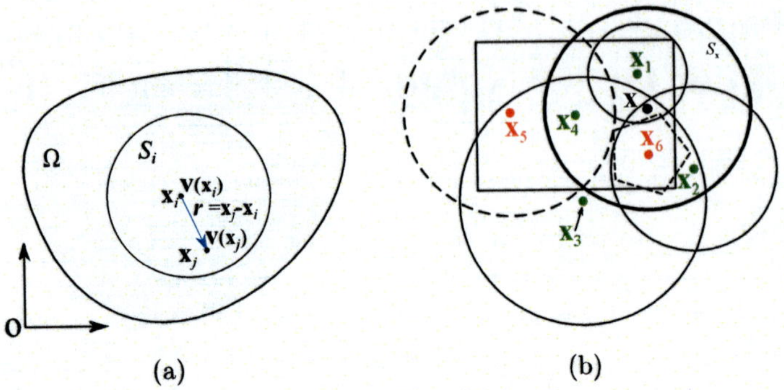

Fig. 5.1 **a** Domain and notation. **b** Schematic diagram for support and dual-support; all shapes above are supports, $\mathcal{S}_{\mathbf{x}} = \{\mathbf{x}_1, \mathbf{x}_2, \mathbf{x}_4, \mathbf{x}_6\}$, $\mathcal{S}'_{\mathbf{x}} = \{\mathbf{x}_1, \mathbf{x}_2, \mathbf{x}_3, \mathbf{x}_4\}$

Support $\mathcal{S}_i$ of point $\mathbf{x}_i$ is the domain where any spatial point $\mathbf{x}_j$ forms spatial vector $\mathbf{r}(= \mathbf{x}_j - \mathbf{x}_i)$ from $\mathbf{x}_i$ to $\mathbf{x}_j$. The support serves as the basis for the nonlocal operators. There is no restriction on the support shapes, which can be a spherical domain, cube, semi-spherical domain, triangle and so on.

Dual-support is defined as a union of the points whose supports include $\mathbf{x}$, denoted by

$$\mathcal{S}'_i = \{\mathbf{x}_j | \mathbf{x}_i \in \mathcal{S}_j\}. \tag{5.1}$$

Point $\mathbf{x}_j$ forms dual-vector $\mathbf{r}'(= \mathbf{x}_i - \mathbf{x}_j = -\mathbf{r})$ in $\mathcal{S}'_i$. On the other hand, $\mathbf{r}'$ is the spatial vector formed in $\mathcal{S}_j$. One example to illustrate the support and dual-support is shown in Fig. 5.1b.

The nonlocal operator method uses the basic nonlocal operators to replace the local operator in calculus such as the gradient, divergence, curl and Hessian operators. The functional formulated by the local differential operator can be used to construct the residual or tangent stiffness matrix by replacing the local operator with the corresponding nonlocal operator. However, the convergence rate of the original nonlocal operator is limited to 1 since the basic nonlocal operator is one-order.

The nonlocal gradient of a vector field $\mathbf{v}$ for point $\mathbf{x}_i$ in support $\mathcal{S}_i$ is defined as

$$\tilde{\nabla}\mathbf{v}_i := \int_{\mathcal{S}_i} w(\mathbf{r})\mathbf{v}_{ij} \otimes \mathbf{r} dV_j \cdot \Big(\int_{\mathcal{S}_i} w(\mathbf{r})\mathbf{r} \otimes \mathbf{r} dV_j \Big)^{-1}. \tag{5.2}$$

The nonlocal gradient operator and its variation in discrete form are

$$\tilde{\nabla}\mathbf{v}_i = \sum_{j\in\mathcal{S}_i} w(\mathbf{r}_j)\mathbf{v}_{ij} \otimes \mathbf{r}_j \Delta V_j \cdot \Big(\sum_{j\in\mathcal{S}_i} w(\mathbf{r})\mathbf{r} \otimes \mathbf{r}\Delta V_j \Big)^{-1}, \tag{5.3}$$

$$\tilde{\nabla}\delta\mathbf{v}_i = \sum_{j\in\mathcal{S}_i} w(\mathbf{r}_j)\delta\mathbf{v}_{ij} \otimes \mathbf{r}_j \Delta V_j \cdot \Big(\sum_{j\in\mathcal{S}_i} w(\mathbf{r})\mathbf{r} \otimes \mathbf{r}\Delta V_j \Big)^{-1}. \tag{5.4}$$

The operator energy functional for vector field at point $\mathbf{x}_i$ is

$$\mathcal{F}_i^{hg} = p^{hg} \int_{\mathcal{S}_i} w(\mathbf{r})(\tilde{\nabla}\mathbf{v}_i \cdot \mathbf{r} - \mathbf{v}_{ij}) \cdot (\tilde{\nabla}\mathbf{v}_i \cdot \mathbf{r} - \mathbf{v}_{ij}) \mathbf{d}V_j \tag{5.5}$$

where p^{hg} is the penalty coefficient. The residual and tangent stiffness matrix of $\mathcal{F}_i^{hg}$ can be obtained with ease; we refer to Ren et al. (2020) for more details.

5.1.2 *Taylor Series Expansion*

There are several formulations for the Taylor series expansion of the function of multiple variables. The conventional Taylor series of a function at origin can be written as Hormander (1983)

$$u(x_1, ..., x_d) = \sum_{n_1=0}^{\infty} ... \sum_{n_d=0}^{\infty} \frac{x_1^{n_1} ... x_d^{n_d}}{n_1! ... n_d!} \left(\frac{\partial^{n_1+\cdots+n_d} u}{\partial x_1^{n_1} ... \partial x_d^{n_d}} \right)(0, ..., 0) \tag{5.6}$$

$$\begin{aligned} &= u(0, ..., 0) + \sum_{j=1}^{d} \frac{\partial u(0, ..., 0)}{\partial x_j} x_j + \frac{1}{2!} \sum_{j=1}^{d} \sum_{k=1}^{d} \frac{\partial^2 u(0, ..., 0)}{\partial x_j \partial x_k} x_j x_k + \\ &+ \frac{1}{3!} \sum_{j=1}^{d} \sum_{k=1}^{d} \sum_{l=1}^{d} \frac{\partial^3 u(0, ..., 0)}{\partial x_j \partial x_k \partial x_l} x_j x_k x_l + \cdots . \end{aligned} \tag{5.7}$$

By using the generalization of the inner product, the Taylor series expansion is

$$u_j = u_i + \nabla u_i \cdot \mathbf{r} + \frac{1}{2!} \nabla^2 u_i : \mathbf{r}^2 + \cdots + \frac{1}{n!} \nabla^n u_i \cdot^{(n)} \mathbf{r}^n + \cdots \tag{5.8}$$

where $\mathbf{r} = \mathbf{x}_j - \mathbf{x}_i$, $\mathbf{r}^n = \mathbf{r} \otimes ... \otimes \mathbf{r}$, and $\cdot^{(n)}$ is the generalization of inner product, where two special cases are $\cdot^{(1)} = \cdot$, $\cdot^{(2)} =:$.

Or using the d-dimensional multi-index, the Taylor series expansion is

$$u_j = \sum_{(\alpha_1, ..., \alpha_d) \in \alpha} \frac{r_1^{\alpha_1} ... r_d^{\alpha_d}}{\alpha_1! ... \alpha_d!} u_{i, \alpha_1 ... \alpha_d} \tag{5.9}$$

where $\alpha = \{(\alpha_1, ..., \alpha_d) | \alpha_i \in \mathbb{N}^0, 1 \leq i \leq d\}$. In this chapter, Eq. 5.9 is adopted for Taylor series expansion.

5.1.3 Mathematica Code for Multi-index

For the multi-indices in

$$\alpha_d^n = \{(n_1, ..., n_d)|1 \leq \sum_{i=1}^{d} n_i \leq n,\ n_i \in \mathbb{N}^0, 1 \leq i \leq d\},$$

the Mathematica code with high efficiency is

```
MultiIndexList[d_,n_]:=Module[{a,b,c},a=Subsets[Range[d+n],{d}];
Do[c=a[[i]];b=c-1;b[[2;;]]-=c[[1;;-2]];a[[i]]=b,{i,Length[a]}];a[[2;;]]];
(*note: d=number of spatial dimensions, n=maximal order of derivative*)
```

The number of elements in α_d^n can be determined by counting the combination of positive integer k as a sum of d non-negative integers up to non-commutativity. Imagine a line of $d + k - 1$ positions, where each position can contain either a cat or a divider. If one has k (nameless) cats and $d - 1$ dividers, he can split the cats into d groups by choosing positions for the dividers: $C_{k+d-1}^{d-1} = C_{k+d-1}^{k}$, where C_n^d is a binomial coefficient and can be written as $C_n^d = \binom{n}{d} = \frac{n!}{d!(n-d)!}$. The size of each group of cats corresponds to one of the non-negative integers in the sum.

Therefore, in d-dimension space, the number of k-order derivatives by Eq. 5.8 is

$$N(\nabla^k u_i) = C_{k+d-1}^{k}. \tag{5.10}$$

The number of all derivatives with maximal order n in d dimensional space is

$$N_d^n = \sum_{k=1}^{n} C_{k+d-1}^{k} = C_{n+d}^{n} - 1.$$

In order to obtain the n-order derivatives, the number of neighbors in $\mathcal{S}_i$ must be not less than $C_{n+d}^{n} - 1$, so that the coefficient matrix for all derivatives is invertible. If more neighbors are in the support, the least squares method can be used to find the approximation. The minimal number of neighbors in support is listed in Table 5.1.

5.1.4 Higher Order Nonlocal Operator Method

Several formulations of the Taylor series expansion of a function of multiple variables are available in Sect. 5.1.2. A scalar field u_j at a point $j \in \mathcal{S}_i$ can be obtained by the Taylor series expansion at u_i in d dimensions with maximal derivative order not more than n,

Table 5.1 Minimal number of neighbors in support. d = number of spatial dimensions, and n = maximal order of derivatives

N_d^n	n = 1	n = 2	n = 3	n = 4	n = 5	n = 6
d = 1	1	2	3	4	5	6
d = 2	2	5	9	14	20	27
d = 3	3	9	19	34	55	83
d = 4	4	14	34	69	125	209
d = 5	5	20	55	125	251	461
d = 6	6	27	83	209	461	923

$$u_j = u_i + \sum_{(n_1,\ldots,n_d)\in\alpha_d^n} \frac{r_1^{n_1}\ldots r_d^{n_d}}{n_1!\ldots n_d!} u_{i,n_1\ldots n_d} + O(r^{|\alpha|+1}) \tag{5.11}$$

where

$$\mathbf{r} = (r_1, \ldots, r_d) = (x_{j1} - x_{i1}, \ldots, x_{jd} - x_{id}) \tag{5.12}$$

$$u_{i,n_1\ldots n_d} = \frac{\partial^{n_1+\cdots+n_d} u_i}{\partial x_{i1}^{n_1}\ldots\partial x_{id}^{n_d}} \tag{5.13}$$

$$|\alpha| = \max\,(n_1 + \cdots + n_d), \tag{5.14}$$

α_d^n is the list of flattened multi-indices, where d denotes the number of spatial dimensions and n is the maximal order of partial derivatives for one index. Two special multi-indices can be written as

$$\alpha_d^n = \{(n_1, \ldots, n_d)|1 \leq \sum_{i=1}^{d} n_i \leq n,\ n_i \in \mathbb{N}^0, 1 \leq i \leq d\} \tag{5.15}$$

$$\text{or } \alpha_d^n = \{(n_1, \ldots, n_d)|1 \leq \sum_{i=1}^{d} n_i,\ 0 \leq n_i \leq n, n_i \in \mathbb{N}^0, 1 \leq i \leq d\}, \tag{5.16}$$

where $\mathbb{N}^0 = \{0, 1, 2, 3, \ldots\}$. Equation 5.16 gives a multi-index with $(1+n)^d - 1$ elements and $|\alpha| = nd$, while the multi-index by Eq. 5.15 has $\frac{(n+d)!}{n!d!} - 1$ elements according to Combinatorics. In this chapter, we adopt the multi-index by Eq. 5.15 since it avoids the mixed higher order terms and has some benefit for numerical computation. The way to obtain all elements in α_d^n of Eq. 5.15 by Mathematica sees Sect. 5.1.3.

For any multi-index $(n_1, \ldots, n_d) \in \alpha_d^n$, the partial derivative and the polynomial are

$$u_{i,n_1\ldots n_d}, \quad \frac{r_1^{n_1}\ldots r_d^{n_d}}{n_1!\ldots n_d!}, \quad \forall (n_1, \ldots, n_d) \in \alpha_d^n. \tag{5.17}$$

However, the original form of the Taylor series expansion is very sensitive to the round-off error. For example,

$$r_1^{n_1}...r_d^{n_d} \propto h^{n_1+\dots+n_d}$$

where h is the characteristic length scale of the support. The higher order terms reduce to 0 quickly when $h < 1$, or explode as $h > 1$. It is expected to have a length scale h approaching 1. When length scale of support $\mathcal{S}_i$ at u_i is taken into account, Taylor series expansion by Eq. 5.11 can be written as

$$\begin{aligned} u_j &= u_i + \sum_{(n_1,\dots,n_d)\in\alpha_d^n} \frac{r_1^{n_1}...r_d^{n_d}}{h_i^{n_1+\dots+n_d}}\left(\frac{h_i^{n_1+\dots+n_d}}{n_1!...n_d!}u_{i,n_1...n_d}\right) + O(r^{n+1}) \\ &= u_i + \sum_{(n_1,\dots,n_d)\in\alpha_d^n} \frac{r_1^{n_1}...r_d^{n_d}}{h_i^{n_1+\dots+n_d}} u^h_{i,n_1...n_d} + O(r^{n+1}) \end{aligned} \tag{5.18}$$

where h_i is the characteristic length of $\mathcal{S}_i$, and

$$u^h_{i,n_1...n_d} = \frac{h_i^{n_1+\dots+n_d}}{n_1!...n_d!}u_{i,n_1...n_d}. \tag{5.19}$$

Let $\mathbf{p}^h_j$, $\partial^h_\alpha u_i$ and $\partial_\alpha u_i$ be the list of the flattened polynomials, scaled partial derivatives and partial derivatives, respectively, based on multi-index notation α_d^n in Eq. 5.15,

$$\mathbf{p}^h_j = (\frac{r_d}{h}, \dots, \frac{r_1^{n_1}...r_d^{n_d}}{h^{n_1+\dots+n_d}}, \dots, \frac{r_1^n}{h^n})^T \tag{5.20}$$

$$\partial^h_\alpha u_i = (u^h_{i,0...1}, \dots, u^h_{i,n_1...n_d}, \dots, u^h_{i,n...0})^T \tag{5.21}$$

$$\partial_\alpha u_i = (u_{i,0...1}, \dots, u_{i,n_1...n_d}, \dots, u_{i,n...0})^T. \tag{5.22}$$

Introducing h in Eq. 5.20 enables the terms in Eq. 5.20 being in the "same" characteristic length scale. The actual partial derivatives can be recovered by

$$\partial_\alpha u_i = \mathbf{H}_i^{-1}\partial^h_\alpha u_i \tag{5.23}$$

where

$$\mathbf{H}_i = \text{diag}\big[h_i, \dots, \frac{h_i^{n_1+\dots+n_d}}{n_1!...n_d!}, \dots, \frac{h_i^n}{n!}\big] \tag{5.24}$$

where diag$[a_1, \dots, a_n]$ denotes a diagonal matrix whose diagonal entries starting in the upper left corner are $a_1, \dots, a_n$.

Therefore, Taylor series expansion with u_i being moved to the left side of the equation can be written as

$$u_{ij} = (\partial_\alpha^h u_i)^T \mathbf{p}_j^h, \forall j \in \mathcal{S}_i \tag{5.25}$$

where $u_{ij} = u_j - u_i$.

Integrating u_{ij} with weighted coefficient $w(\mathbf{r})(\mathbf{p}_j^h)^T$ in support $\mathcal{S}_i$, we obtain

$$\begin{aligned}\int_{\mathcal{S}_i} w(\mathbf{r}) u_{ij} (\mathbf{p}_j^h)^T \mathbf{d}V_j &= (\partial_\alpha^h u_i)^T \int_{\mathcal{S}_i} w(\mathbf{r}) \mathbf{p}_j^h \otimes (\mathbf{p}_j^h)^T \mathbf{d}V_j \\ &= (\partial_\alpha u_i)^T \mathbf{H}_i \int_{\mathcal{S}_i} w(\mathbf{r}) \mathbf{p}_j^h \otimes (\mathbf{p}_j^h)^T \mathbf{d}V_j \end{aligned}\tag{5.26}$$

where $w(\mathbf{r})$ is the weight function.

Therefore, the nonlocal operator $\tilde{\partial}_\alpha u_i$ can be obtained as

$$\tilde{\partial}_\alpha u_i := \mathbf{H}_i^{-1} \Big(\int_{\mathcal{S}_i} w(\mathbf{r}) \mathbf{p}_j^h \otimes (\mathbf{p}_j^h)^T \mathbf{d}V_j \Big)^{-1} \int_{\mathcal{S}_i} w(\mathbf{r}) u_{ij} \mathbf{p}_j^h \mathbf{d}V_j = \mathbf{K}_i \cdot \int_{\mathcal{S}_i} w(\mathbf{r}) \mathbf{p}_j^h u_{ij} \mathbf{d}V_j \tag{5.27}$$

where

$$\mathbf{K}_i := \mathbf{H}_i^{-1} \Big(\int_{\mathcal{S}_i} w(\mathbf{r}) \mathbf{p}_j^h \otimes (\mathbf{p}_j^h)^T \mathbf{d}V_j \Big)^{-1}. \tag{5.28}$$

The reason to call Eq. 5.27 nonlocal operator is that it is defined in the support, in contrast with the local operator defined at a point. The nonlocal operator approximates the local operator with order up to $|\alpha|$. The traditional local operator is suitable for theoretical derivation but not for numerical analysis since its definition is limited to infinitesimal. The nonlocal operator can be viewed as a generalization of the conventional local operator.

The variation of $\tilde{\partial}_\alpha u_i$ is

$$\tilde{\partial}_\alpha \delta u_i := \mathbf{K}_i \cdot \int_{\mathcal{S}_i} w(\mathbf{r}) \mathbf{p}_j^h (\delta u_j - \delta u_i) \mathbf{d}V_j. \tag{5.29}$$

In the continuous form, the number of dimensions of $\partial \delta u_i$ is infinite and discretization is required. After discretization of the domain by particles, the whole domain is represented by

$$\Omega = \sum_{i=1}^{N} \Delta V_i \tag{5.30}$$

where i is the global index of volume ΔV_i, and N is the number of particles in Ω.

Particles in $\mathcal{S}_i$ are represented by

$$\mathcal{S}_i = \{j_1, ..., j_k, ..., j_{n_i}\} \tag{5.31}$$

where $j_1, ..., j_k, ..., j_{n_i}$ are the global indices of neighbors of particle i, n_i being the number of neighbors of i in $\mathcal{S}_i$.

The discrete form of Eq. 5.27 and its variation are

$$\tilde{\partial}_\alpha u_i = \mathbf{K}_i \cdot \sum_{j \in \mathcal{S}_i} u_{ij} w(\mathbf{r}_j) \mathbf{p}_j^h \Delta V_j = \mathbf{K}_i \mathbf{p}_{wi}^h \Delta \mathbf{u}_i \tag{5.32}$$

$$\tilde{\partial}_\alpha \delta u_i = \mathbf{K}_i \cdot \sum_{j \in \mathcal{S}_i} \delta u_{ij} w(\mathbf{r}_j) \mathbf{p}_j^h \Delta V_j = \mathbf{K}_i \mathbf{p}_{wi}^h \delta \Delta \mathbf{u}_i \tag{5.33}$$

where

$$\mathbf{K}_i = \mathbf{H}_i^{-1} \Big(\sum_{j \in \mathcal{S}_i} w(\mathbf{r}) \mathbf{p}_j^h \otimes (\mathbf{p}_j^h)^T \Delta V_j \Big)^{-1}, \tag{5.34}$$

$$\mathbf{p}_{wi}^h = \Big(w(\mathbf{r}_{j_1}) \mathbf{p}_{j_1}^h \Delta V_{j_1}, ..., w(\mathbf{r}_{j_{n_i}}) \mathbf{p}_{j_{n_i}}^h \Delta V_{j_{n_i}} \Big) \tag{5.35}$$

$$\Delta \mathbf{u}_i = (u_{ij_1}, ..., u_{ij_k}, ..., u_{ij_{n_i}})^T. \tag{5.36}$$

When the weight function $w(\mathbf{r})$ is selected as the reciprocal of the volume, Eqs. 5.34 and 5.35 can be simplified further. The nonlocal operator provides all the partial derivatives with maximal order for a single index up to n. The set of derivatives in PDEs of real applications is a subset of the nonlocal operator. It should be noted that when the number of points in support is the same as the length of multi-index α_d^n and the coefficient matrix from Eq. 5.25 for all points in support is well conditioned, the nonlocal operator can be obtained directly by the inverse of the coefficient matrix. In this case, the nonlocal operator serves as an efficient way to obtain the higher order finite difference scheme.

Each term in $\tilde{\partial}_\alpha u_i$ corresponds to the row of $\mathbf{K}_i \mathbf{p}_{wi}^h$ multiplying $\Delta \mathbf{u}_i$. Equation 5.32 can be used to replace the differential operators in PDEs to form the algebraic equations. This way is through the strong form of the PDEs. The other ways to solve the linear (nonlinear) PDEs are through the weak formulations (weighted residual method) or the variational formulations (i.e. Ren et al. (2020)). In these cases, the variation of $\partial_\alpha u_i$ in Eq. 5.33 is required.

Equation 5.32 can be written more concisely as

$$\tilde{\partial}_\alpha u_i = \mathbf{K}_i \mathbf{p}_{wi}^h \Delta \mathbf{u}_i = \mathbf{B}_{\alpha i} \mathbf{u}_i \tag{5.37}$$

with $\mathbf{B}_{\alpha i}$ being the operator matrix for point i based on multi-index α_d^n

$$\mathbf{B}_{\alpha i} = \begin{bmatrix} -(1, \cdots, 1)_{n_p} \mathbf{K}_i \mathbf{p}_{wi}^h \\ \mathbf{K}_i \mathbf{p}_{wi}^h \end{bmatrix} \tag{5.38}$$

$$\mathbf{u}_i = (u_i, u_{j_1}, u_{j_2}, \cdots, u_{j_{n_i}})^T \tag{5.39}$$

where $(1, \cdots, 1)_{n_p} \mathbf{K}_i \mathbf{p}^h_{wi}$ is the column sum of $\mathbf{K}_i \mathbf{p}^h_{wi}$, n_p being the length of α^n_d. The operator matrix obtains all the partial derivatives of maximal order less than $|\alpha| + 1$ by the nodal values in support. For real applications, one can select the specific rows in the operator matrix based on the partial derivatives contained in the specific PDEs.

The traditional differential operator and their combination of one order or higher order and the corresponding variations can be constructed from Eq. 5.32 and Eq. 5.33, respectively. For example, the multi-index, polynomials and partial derivatives in two dimensions with maximal second-order derivatives are

$$\begin{aligned} \alpha^2_2 &= (01, 02, 10, 11, 20) \\ \mathbf{p}^h_j &= (y/h, y^2/h^2, x/h, xy/h^2, x^2/h^2)^T \\ \tilde{\partial}_\alpha u_i &= (u_{,01}, u_{,02}, u_{,10}, u_{,11}, u_{,20})^T. \end{aligned} \tag{5.40}$$

For the case of the Poisson equation in 2D, $\nabla^2 u = f$. In the strong form, the operator $\nabla^2 u = \frac{\partial^2 u}{\partial x^2} + \frac{\partial^2 u}{\partial y^2}$ is required; one can select the $\partial_\alpha u_i[2]$ in Eq. 5.40 for $\frac{\partial^2 u}{\partial y^2}$ and $\partial_\alpha u_i[5]$ in Eq. 5.40 for $\frac{\partial^2 u}{\partial x^2}$. When solved in weak form, one can select the $\partial_\alpha u_i[1]$ in Eq. 5.40 for $\frac{\partial u}{\partial y}$ and $\partial_\alpha u_i[3]$ in Eq. 5.40 for $\frac{\partial u}{\partial x}$ to construct the tangent stiffness matrix.

In fact, the nonlocal operator $\partial_\alpha u_i$ in discrete form can be obtained by least squares. Consider the weighted square sum of the Taylor series expansion in $\mathcal{S}_i$,

$$\mathcal{F}_i(\mathbf{u}) = \sum_{j \in \mathcal{S}_i} w(\mathbf{r}) \big(u_{ij} - (\mathbf{p}^h_j)^T \tilde{\partial}^h_\alpha u_i\big)^2 \Delta V_j \tag{5.41}$$

$$\begin{aligned} &= \sum_{j \in \mathcal{S}_i} w(\mathbf{r}) \big(u^2_{ij} + \partial^h_\alpha u^T_i \mathbf{p}^h_j (\mathbf{p}^h_j)^T \partial^h_\alpha u_i - 2u_{ij} (\mathbf{p}^h_j)^T \partial^h_\alpha u_i\big) \Delta V_j \\ &= \sum_{j \in \mathcal{S}_i} w(\mathbf{r}) u^2_{ij} \Delta V_j + \partial^h_\alpha u^T_i \sum_{j \in \mathcal{S}_i} w(\mathbf{r}) \mathbf{p}^h_j (\mathbf{p}^h_j)^T \Delta V_j \; \partial^h_\alpha u_i - 2\Delta \mathbf{u}^T_i \mathbf{p}^T_{wi} \partial^h_\alpha u_i \\ &= \sum_{j \in \mathcal{S}_i} w(\mathbf{r}) u^2_{ij} \Delta V_j + \tilde{\partial}_\alpha u^T_i \mathbf{H}_i \sum_{j \in \mathcal{S}_i} w(\mathbf{r}) \mathbf{p}^h_j (\mathbf{p}^h_j)^T \Delta V_j \; \mathbf{H}_i \partial^h_\alpha u_i - 2\Delta \mathbf{u}^T_i (\mathbf{p}^h_{wi})^T \mathbf{H}_i \tilde{\partial}_\alpha u_i. \end{aligned} \tag{5.42}$$

$\frac{\partial \mathcal{F}_i(\mathbf{u})}{\partial(\tilde{\partial}_\alpha u_i)} = 0$ leads to

$$\tilde{\partial}_\alpha u_i = \mathbf{H}^{-1}_i \Big(\sum_{j \in \mathcal{S}_i} w(\mathbf{r}) \mathbf{p}^h_j (\mathbf{p}^h_j)^T \Delta V_j\Big)^{-1} \mathbf{p}^h_{wi} \Delta \mathbf{u}_i = \mathbf{K}_i \mathbf{p}^h_{wi} \Delta \mathbf{u}_i \tag{5.43}$$

which is the same as Eq. 5.32.

Meanwhile, Eq. 5.41 represents the operator energy functional in the nonlocal operator method, and can be used to construct the tangent stiffness matrix of operator energy functional. The operator energy functional is the quadratic functional of the Taylor series expansion. By Eq. 5.43, 5.42 can be simplified into

$$\begin{aligned}\mathcal{F}_i(\mathbf{u}) =&\sum_{j\in\mathcal{S}_i} w(\mathbf{r})u_{ij}^2\Delta V_j - \Delta\mathbf{u}_i^T(\mathbf{p}_{wi}^h)^T\Big(\sum_{j\in\mathcal{S}_i} w(\mathbf{r})\mathbf{p}_j^h(\mathbf{p}_j^h)^T\Delta V_j\Big)^{-1}\mathbf{p}_{wi}^h\Delta\mathbf{u}_i\\ =&\Delta\mathbf{u}_i^T\mathbf{W}_i\Delta\mathbf{u}_i - \Delta\mathbf{u}_i^T(\mathbf{p}_{wi}^h)^T\Big(\sum_{j\in\mathcal{S}_i} w(\mathbf{r})\mathbf{p}_j^h(\mathbf{p}_j^h)^T\Delta V_j\Big)^{-1}\mathbf{p}_{wi}^h\Delta\mathbf{u}_i\\ =&\Delta\mathbf{u}_i^T\mathbf{M}_i\Delta\mathbf{u}_i \end{aligned}\tag{5.44}$$

where

$$\mathbf{W}_i = \text{diag}\big[w(\mathbf{r}_{j_1})\Delta V_{j_1}, ..., w(\mathbf{r}_{j_{n_i}})\Delta V_{j_{n_i}}\big]\tag{5.45}$$

$$\mathbf{M}_i = \mathbf{W}_i - (\mathbf{p}_{wi}^h)^T\Big(\sum_{j\in\mathcal{S}_i} w(\mathbf{r})\mathbf{p}_j^h(\mathbf{p}_j^h)^T\Delta V_j\Big)^{-1}\mathbf{p}_{wi}^h.\tag{5.46}$$

The first and second variations of $\mathcal{F}_i(\mathbf{u})$ read

$$\delta\mathcal{F}_i(\mathbf{u}) = 2\Delta\mathbf{u}_i^T\mathbf{M}_i\delta\Delta\mathbf{u}_i.\tag{5.47}$$

$$\delta^2\mathcal{F}_i(\mathbf{u}) = 2\delta\Delta\mathbf{u}_i^T\mathbf{M}_i\delta\Delta\mathbf{u}_i.\tag{5.48}$$

Let $\mathbf{v}_i(j) = \sum_{k=1}^{n_i}\mathbf{M}_i(j,k)$ be the sum of row of matrix $\mathbf{M}_i$, then the tangent stiffness matrix can be extracted from Eq. 5.48,

$$\mathbf{K}_i^{hg} = \frac{p_{hg}}{m_i}\begin{bmatrix}\sum\mathbf{v}_i & -\mathbf{v}_i^T\\ -\mathbf{v}_i & \mathbf{M}_i\end{bmatrix}\tag{5.49}$$

where the first row (column) denotes the entries for point i, while the neighbors start from the second row (column), p_{hg} is the penalty coefficient and m_i the normalization coefficient

$$m_i = \sum_{j\in\mathcal{S}_i} w(\mathbf{r})\mathbf{r}\cdot\mathbf{r}\Delta V_j,\tag{5.50}$$

where $\mathbf{r}$ varies for each j.

Let n_i be the number of neighbors in $\mathcal{S}_i$ and n_p be the length of $\tilde{\partial}_\alpha^h u_i$. The dimensions of terms in $\mathbf{M}_i$ are

$$\begin{aligned}&\text{Dim}(\mathbf{W}_i) = n_i\times n_i, \text{Dim}(\mathbf{p}_{wi}^h) = n_p\times n_i,\\ &\text{Dim}\Big(\Big(\sum_{j\in\mathcal{S}_i} w(\mathbf{r})\mathbf{p}_j^h(\mathbf{p}_j^h)^T\Delta V_j\Big)\Big) = n_p\times n_p\\ &\text{Rank}\Big((\mathbf{p}_{wi}^h)^T\Big(\sum_{j\in\mathcal{S}_i} w(\mathbf{r})\mathbf{p}_j^h(\mathbf{p}_j^h)^T\Delta V_j\Big)^{-1}\mathbf{p}_{wi}^h\Big) \le \min(n_p, n_i).\end{aligned}$$

When $n_i < n_p$, $\sum_{j\in S_i} w(\mathbf{r})\mathbf{p}_j^h(\mathbf{p}_j^h)^T \Delta V_j$ is singular. It is required that $n_i \geq n_p$ so that $\mathbf{M}_i$ in Eq. 5.46 is well defined. The number of neighbors is selected as $5p + n_p$, where p denotes the order of the nonlocal operator. These extra nodes are used to overcome the rank deficiency in nodal integration.

The operator energy functional $\mathcal{F}(\mathbf{u})$ represents the topology of the nonlocal operator method. Any field derived from $\mathcal{F}(\mathbf{u})$ should try to satisfy $\mathcal{F}(\mathbf{u}) = 0$ at the first step, which is independent of the actual physical model to be solved.

5.2 Quadratic Functional

A very special functional has the form

$$\mathcal{F} = \frac{1}{2}\tilde{\partial} u^T \mathbf{D} \tilde{\partial} u \tag{5.51}$$

where $\mathbf{D}$ is an arbitrary symmetric matrix, $\tilde{\partial} u \subset \tilde{\partial}_\alpha u$ in Eq. 5.32. The operator matrix $\mathbf{B}$ is constructed from $\mathbf{B}_\alpha$ based on the index of terms ∂u in $\partial_\alpha u$. Some examples of Eq. 5.51 are given in Sect. 5.2.2.

When $\mathbf{D}$ is independent with the unknown functions u, the functional $\mathcal{F}(\tilde{\partial} u)$ is pure quadratic; the first and second variations of $\mathcal{F}(\tilde{\partial} u)$ at a point are

$$\delta\mathcal{F} = \frac{\partial \mathcal{F}}{\partial(\tilde{\partial} u)} = \tilde{\partial}\delta u^T \mathbf{D} \tilde{\partial} u = \delta \mathbf{u}^T \mathbf{B}^T \mathbf{D} \tilde{\partial} u \tag{5.52}$$

$$\delta^2\mathcal{F} = \frac{\partial^2 \mathcal{F}}{\partial(\tilde{\partial} u^T)\partial(\tilde{\partial} u)} = \tilde{\partial}\delta u^T \mathbf{D} \tilde{\partial}\delta u = \delta \mathbf{u}^T \mathbf{B}^T \mathbf{D}\mathbf{B}\delta\mathbf{u} \tag{5.53}$$

and the residual and tangent stiffness matrix at a point can be written as

$$\mathbf{R}(u) = \mathbf{B}^T \mathbf{D} \tilde{\partial} u, \ \mathbf{K}(u) = \mathbf{B}^T \mathbf{D}\mathbf{B}. \tag{5.54}$$

When $\mathbf{D} := \mathbf{D}(u, \partial u)$ is a nonlinear tensor, the functional can be converted into quadratic functional by linearization and the Newton–Raphson method can be employed to find the solution.

According to $\tilde{\partial} u_i \subset \tilde{\partial}_\alpha u_i$ in Eq. 5.27, we write $\tilde{\partial} u_i = \mathbf{K}_i' \cdot \int_{S_i} w(\mathbf{r})\mathbf{p}_j^h u_{ij} \mathbf{d}V_j$, where $\mathbf{K}_i' \subset \mathbf{K}_i$ in Eq. 5.27. Let $\boldsymbol{\sigma}_i = \mathbf{D}\tilde{\partial} u_i$ and consider the variation in domain

$$\begin{aligned}\delta \int_\Omega \mathcal{F} &= \delta\Big(\int_\Omega \frac{1}{2}\tilde{\partial} u_i^T \mathbf{D}\tilde{\partial} u_i \mathbf{d}V_i\Big) = \int_\Omega \tilde{\partial} u_i^T \mathbf{D}\tilde{\partial}\delta u_i \,\mathbf{d}V_i \\ &= \int_\Omega \boldsymbol{\sigma}_i^T \tilde{\partial}\delta u_i \mathbf{d}V_i = \int_\Omega \boldsymbol{\sigma}_i^T \mathbf{K}_i' \int_{S_i} w(\mathbf{r})\mathbf{p}_j^h \delta u_{ij} \mathbf{d}V_j \mathbf{d}V_i\end{aligned}$$

$$= \int_{\Omega}\int_{S_i} w(\mathbf{r})\boldsymbol{\sigma}_i^T \mathbf{K}'_i \mathbf{p}_j^h (\delta u_j - \delta u_i)\mathbf{d}V_j \mathbf{d}V_i$$
$$= \int_{\Omega}\Big(- \int_{S_i} w(\mathbf{r})\boldsymbol{\sigma}_i^T \mathbf{K}'_i \mathbf{p}_j^h \mathbf{d}V_j + \int_{S'_i} \boldsymbol{\sigma}_j^T \mathbf{K}'_j \mathbf{p}_i^h \mathbf{d}V_j \Big)\delta u_i \mathbf{d}V_i. \quad (5.55)$$

Note that $\mathbf{p}_i^h$ in S'_i varies for different j since $\mathbf{p}_i^h$ is computed in j's support S_j. The terms with δu_i in the first-order variation $\delta \int_{\Omega} \mathcal{F} = 0$ are

$$- \int_{S_i} w(\mathbf{r})\boldsymbol{\sigma}_i^T \mathbf{K}'_i \mathbf{p}_j^h \mathbf{d}V_j + \int_{S'_i} \boldsymbol{\sigma}_j^T \mathbf{K}'_j \mathbf{p}_i^h \mathbf{d}V_j, \quad (5.56)$$

with "equivalent" higher order partial differential term $-\partial^T(\mathbf{D}\partial u)$, where $\partial := (\ldots, \frac{\partial^{n_1+\cdots+n_d}}{\partial x_1^{n_1}\ldots\partial x_d^{n_d}}, \ldots)^T$ is the differential operator based on the subset of multi-index α_d^n in Eq. 5.15. The PDE given by $-\partial^T(\mathbf{D}\partial u)$ has a maximal differential order of $2n$. The nonlocal strong form by Eq. 5.56 can be solved directly by an explicit integration algorithm. It should be noted that $\boldsymbol{\sigma}$, $\tilde{\partial} u$ are in the form of column vectors for a scalar field u. The generalization of u to the vector field is straightforward.

Equation 5.56 alone may suffer numerical instabilities (zero-energy mode), and therefore the operator energy functional by Eq. 5.41 is required. Equation 5.56 with correction terms can be written as

$$\tilde{\partial}_{\alpha}^T \boldsymbol{\sigma}_i \approx \int_{S_i} (w(\mathbf{r})\boldsymbol{\sigma}_i^T \mathbf{K}'_i \mathbf{p}_j^h + T_{ij}^{hg})\mathbf{d}V_j - \int_{S'_i} (\boldsymbol{\sigma}_j^T \mathbf{K}'_j \mathbf{p}_i^h + T_{ji}^{hg})dV_j \quad (5.57)$$

$$T_{ij}^{hg} = w(\mathbf{r})\frac{p^{hg}}{m_i}((\mathbf{p}_j^h)^T \tilde{\partial}_{\alpha}^h u_i - u_{ij}). \quad (5.58)$$

5.2.1 Newton–Raphson Method for Nonlinear Functional

The core of NOM is the functional, which comprises the physical functional and the operator energy functional. The physical functional may contain the functional on the domain and the other functional on the boundaries. In all,

$$\mathcal{F}(\mathbf{u}) = \int_{\Omega} (\mathcal{F}_1^{ph}(\mathbf{u}) + \mathcal{F}^{hg}(\mathbf{u}))\mathbf{d}V + \int_{\partial\Omega} \mathcal{F}_2^{ph}(\mathbf{u})dS. \quad (5.59)$$

The first and second derivatives on all unknowns lead to the residual and the tangent stiffness matrix, respectively,

$$\mathbf{R} = \frac{\partial \mathcal{F}}{\partial \mathbf{u}} = \int_{\Omega} (\frac{\partial \mathcal{F}_1^{ph}}{\partial \mathbf{u}} + \frac{\partial \mathcal{F}^{hg}}{\partial \mathbf{u}})\mathbf{d}V + \int_{\partial\Omega} \frac{\partial \mathcal{F}_2^{ph}}{\partial \mathbf{u}} dS \quad (5.60)$$

$$\mathbf{K} = \frac{\partial \mathbf{R}}{\partial \mathbf{u}^T} = \frac{\partial^2 \mathcal{F}}{\partial \mathbf{u}\partial \mathbf{u}^T} = \int_{\Omega} (\frac{\partial^2 \mathcal{F}_1^{ph}}{\partial \mathbf{u}\partial \mathbf{u}^T} + \frac{\partial^2 \mathcal{F}^{hg}}{\partial \mathbf{u}\partial \mathbf{u}^T})\mathbf{d}V + \int_{\partial\Omega} \frac{\partial^2 \mathcal{F}_2^{ph}}{\partial \mathbf{u}\partial \mathbf{u}^T} dS. \quad (5.61)$$

When any term in $\mathcal{F}$ is a nonlinear functional, the Newton–Raphson method is required. The solution is updated by iteration in each step. In the n step, the residual $\mathbf{R}(\mathbf{u}_n) = 0$ is satisfied; $\mathbf{R}(\mathbf{u}_{n+1})$ in the next step can be approximated by Taylor series expansion

$$\mathbf{R}(\mathbf{u}_{n+1}) \approx \mathbf{R}(\mathbf{u}_n) + \frac{\partial \mathbf{R}}{\partial \mathbf{u}^T}|_{\mathbf{u}=\mathbf{u}_n} \cdot (\mathbf{u}_{n+1} - \mathbf{u}_n). \tag{5.62}$$

The solution in $n+1$ step can be obtained by the iterations

$$0 = \mathbf{R}(\mathbf{u}^{k+1}) \approx \mathbf{R}(\mathbf{u}^k) + \mathbf{K}(\mathbf{u}^k) \cdot \Delta\mathbf{u}^{k+1} \rightarrow \mathbf{K}(\mathbf{u}^k)\Delta\mathbf{u}^{k+1} = -\mathbf{R}(\mathbf{u}^k) \tag{5.63}$$

where k denotes the iteration number in $n+1$ step, $\mathbf{u}^0 = \mathbf{u}_n$ and $\mathbf{u}^{k+1} = \mathbf{u}^k + \Delta\mathbf{u}^{k+1}$. When

$$\frac{\|\Delta\mathbf{u}^{k+1}\|}{\|\sum_{i=1}^{k+1} \Delta\mathbf{u}^i\|} \leq \text{Tol},$$

the iteration converges.

5.2.2 *Elastic Solid Materials*

In this section, we give some examples of how to express the linear/nonlinear elastic mechanics in the form of a nonlocal operator method. The maximal derivative order in linear elastic mechanics is 2, and the corresponding weak form only requires first-order partial derivatives. The internal energy functional for plane stress, plane strain and 3D linear elastic solid at a point are

$$\mathcal{F}_{plane\,stress} = \frac{1}{2}\sigma : \varepsilon = \frac{1}{2}\tilde{\partial}\mathbf{u}_{2d}^T D_{plane\,stress}\tilde{\partial}\mathbf{u}_{2d} \tag{5.64}$$

$$\mathcal{F}_{plane\,strain} = \frac{1}{2}\sigma : \varepsilon = \frac{1}{2}\tilde{\partial}\mathbf{u}_{2d}^T D_{plane\,strain}\tilde{\partial}\mathbf{u}_{2d} \tag{5.65}$$

$$\mathcal{F}_{3d} = \frac{1}{2}\sigma : \varepsilon = \frac{1}{2}\tilde{\partial}\mathbf{u}_{3d}^T D_{3d}\tilde{\partial}\mathbf{u}_{3d}, \tag{5.66}$$

where

$$\tilde{\partial}\mathbf{u}_{2d} = (u_x, u_y, v_x, v_y)^T \tag{5.67}$$

$$\tilde{\partial}\mathbf{u}_{3d} = (u_x, u_y, u_z, v_x, v_y, v_z, w_x, w_y, w_z)^T \tag{5.68}$$

$$D_{plane\,stress} = \frac{E}{1-\nu^2}\begin{bmatrix} 1 & 0 & 0 & \nu \\ 0 & \frac{1-\nu}{2} & \frac{1-\nu}{2} & 0 \\ 0 & \frac{1-\nu}{2} & \frac{1-\nu}{2} & 0 \\ \nu & 0 & 0 & 1 \end{bmatrix} \tag{5.69}$$

$$D_{plane\,strain} = \frac{E}{(1-2\nu)(1+\nu)}\begin{bmatrix} 1-\nu & 0 & 0 & \nu \\ 0 & 1/2-\nu & 1/2-\nu & 0 \\ 0 & 1/2-\nu & 1/2-\nu & 0 \\ \nu & 0 & 0 & 1-\nu \end{bmatrix} \tag{5.70}$$

$$D_{3D} = \begin{bmatrix} \lambda+2\mu & 0 & 0 & 0 & \lambda & 0 & 0 & 0 & \lambda \\ 0 & \mu & 0 & \mu & 0 & 0 & 0 & 0 & 0 \\ 0 & 0 & \mu & 0 & 0 & 0 & \mu & 0 & 0 \\ 0 & \mu & 0 & \mu & 0 & 0 & 0 & 0 & 0 \\ \lambda & 0 & 0 & 0 & \lambda+2\mu & 0 & 0 & 0 & \lambda \\ 0 & 0 & 0 & 0 & 0 & \mu & 0 & \mu & 0 \\ 0 & 0 & \mu & 0 & 0 & 0 & \mu & 0 & 0 \\ 0 & 0 & 0 & 0 & 0 & \mu & 0 & \mu & 0 \\ \lambda & 0 & 0 & 0 & \lambda & 0 & 0 & 0 & \lambda+2\mu \end{bmatrix}. \tag{5.71}$$

The tangent stiffness matrix of that point can be extracted by performing the first- or second-order variation of the above functionals.

For nonlinear elastic material, the strain energy density is a function of the deformation gradient, i.e.

$$\mathcal{F}(\mathbf{F})$$

while $\mathbf{F}$ is consistent with the nonlocal operators in $\tilde{\partial}\mathbf{u}_{3d}$

$$\mathbf{F} = \begin{bmatrix} F_1 & F_2 & F_3 \\ F_4 & F_5 & F_6 \\ F_7 & F_8 & F_9 \end{bmatrix} = \begin{bmatrix} u_x+1 & u_y & u_z \\ v_x & v_y+1 & v_z \\ w_x & w_y & w_z+1 \end{bmatrix}. \tag{5.72}$$

Within the framework of total Lagrangian formulation, the first Piola–Kirchhoff stress is the direct derivative of the strain energy over the deformation gradient,

$$\mathbf{P} = \frac{\partial\mathcal{F}(\mathbf{F})}{\partial\mathbf{F}}. \tag{5.73}$$

Furthermore, the material tensor (stress–strain relation) which is required in the implicit analysis can be obtained with the derivative of the first Piola–Kirchhoff stress,

$$\mathbf{D}_4 = \frac{\partial \mathbf{P}}{\partial \mathbf{F}} = \frac{\partial^2 \mathcal{F}(\mathbf{F})}{\partial \mathbf{F}^T \partial \mathbf{F}}. \tag{5.74}$$

The fourth-order material tensor $\mathbf{D}_4$ can be expressed in matrix form when the deformation gradient is flattened:

$$\mathbf{D} = \begin{bmatrix} \frac{\partial P_1}{\partial F_1} & \frac{\partial P_1}{\partial F_2} & \cdots & \frac{\partial P_1}{\partial F_9} \\ \frac{\partial P_2}{\partial F_1} & \frac{\partial P_2}{\partial F_2} & \cdots & \frac{\partial P_2}{\partial F_9} \\ \vdots & \vdots & \ddots & \vdots \\ \frac{\partial P_9}{\partial F_1} & \frac{\partial P_1}{\partial F_2} & \cdots & \frac{\partial P_9}{\partial F_9} \end{bmatrix} = \begin{bmatrix} \frac{\partial^2 \mathcal{F}(F)}{\partial F_1^2} & \frac{\partial^2 \mathcal{F}(F)}{\partial F_1 \partial F_2} & \cdots & \frac{\partial^2 \mathcal{F}(F)}{\partial F_1 \partial F_9} \\ \frac{\partial^2 \mathcal{F}(F)}{\partial F_2 \partial F_1} & \frac{\partial^2 \mathcal{F}(F)}{\partial F_2 \partial F_2} & \cdots & \frac{\partial^2 \mathcal{F}(F)}{\partial F_2 \partial F_9} \\ \vdots & \vdots & \ddots & \vdots \\ \frac{\partial^2 \mathcal{F}(F)}{\partial F_9 \partial F_1} & \frac{\partial^2 \mathcal{F}(F)}{\partial F_9 \partial F_2} & \cdots & \frac{\partial^2 \mathcal{F}(F)}{\partial F_9 \partial F_9} \end{bmatrix}, \tag{5.75}$$

where the flattened deformation gradient and first Piola–Kirchhoff stress are

$$F = (F_1, F_2, F_3, F_4, F_5, F_6, F_7, F_8, F_9) \tag{5.76}$$

and

$$P = \frac{\partial \mathcal{F}(F)}{\partial F} = (\frac{\partial \mathcal{F}(F)}{\partial F_1}, \frac{\partial \mathcal{F}(F)}{\partial F_2}, \cdots, \frac{\partial \mathcal{F}(F)}{\partial F_9}). \tag{5.77}$$

For the case of nearly incompressible Neo-Hooke material (Reese et al. 2000), the strain energy can be expressed as

$$\mathcal{F}(\mathbf{F}) = \frac{1}{2}\kappa(J-1)^2 + \frac{1}{2}\mu(\mathbf{F}:\mathbf{F}-3) \tag{5.78}$$

where $J = \det \mathbf{F}$.

The first Piola–Kirchhoff stress is

$$\mathbf{P} = \frac{\partial \mathcal{F}(\mathbf{F})}{\partial \mathbf{F}} = \mu\mathbf{F} + (J-1)\kappa J_{,\mathbf{F}}. \tag{5.79}$$

With some derivation, the material tensor in matrix form can be written as

$$\mathbf{D} = \mu\mathbf{I}_{9\times 9} + (J-1)\kappa J_{,FF} + \kappa J_{,F} \otimes J_{,F} \tag{5.80}$$

where $J_{,F}$ is the vector form of $J_{,\mathbf{F}}$, and

$$J_{,FF} = \begin{bmatrix} 0 & 0 & 0 & 0 & F_9 & -F_8 & 0 & -F_6 & F_5 \\ 0 & 0 & 0 & -F_9 & 0 & F_7 & F_6 & 0 & -F_4 \\ 0 & 0 & 0 & F_8 & -F_7 & 0 & -F_5 & F_4 & 0 \\ 0 & -F_9 & F_8 & 0 & 0 & 0 & 0 & F_3 & -F_2 \\ F_9 & 0 & -F_7 & 0 & 0 & 0 & -F_3 & 0 & F_1 \\ -F_8 & F_7 & 0 & 0 & 0 & 0 & F_2 & -F_1 & 0 \\ 0 & F_6 & -F_5 & 0 & -F_3 & F_2 & 0 & 0 & 0 \\ -F_6 & 0 & F_4 & F_3 & 0 & -F_1 & 0 & 0 & 0 \\ F_5 & -F_4 & 0 & -F_2 & F_1 & 0 & 0 & 0 & 0 \end{bmatrix}. \quad (5.81)$$

The numerical example based on the material model Eq. 5.78 is given in Sect. 5.4.4.

5.3 Numerical Examples by Strong Form

The nonlocal operator defined in Eq. 5.32 can be used to replace the partial derivatives of different orders in the partial differential equation. In other words, we can use the nonlocal operator to solve the PDE by its strong form. In this sense, the nonlocal operator is similar to the finite difference method. However, finite difference schemes of different order are constructed on the regular grid, where the extension to higher dimensions or higher order derivatives require special treatment, while the nonlocal operator is established simply based on the neighbor list in the support. In this section, we test the accuracy of nonlocal operators in solving second-order ordinary differential equations (ODE) or PDE by strong form. Note that the operator energy functional is not required in solving PDE by strong form.

The first three numerical examples demonstrate the capabilities of the nonlocal operator method in obtaining a high-order finite difference scheme.

5.3.1 *Second-Order ODE*

The ODE with boundary condition is given by

$$\frac{d^2u(x)}{dx^2} = 20x^3 + \pi^2 \cos(\pi x), u(0) = 0, u(1) = 0, x \in [0, 1] \quad (5.82)$$

with analytic solution

$$u(x) = x^5 - 3x - \cos(\pi x) + 1.$$

Since the highest order derivative in the ODE is two, the order of derivatives in the nonlocal operator list should be $p \geq 2$. We test the nonlocal operator with $p = 2, 3, 4, 5, 6$ in solving the second-order ODE. The minimal number of neighbors in

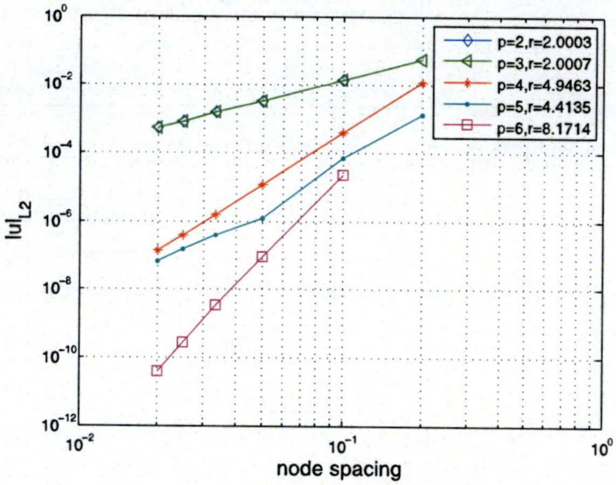

Fig. 5.2 Convergence of the L2-norm for u

the support is selected as the number of terms in the nonlocal operator. The difference between the numerical result and the theoretical solution is measured by the L2-norm, which is calculated by

$$\|\mathbf{u}\|_{L2} = \sqrt{\frac{\sum_j (\mathbf{u}_j - \mathbf{u}_j^{exact}) \cdot (\mathbf{u}_j - \mathbf{u}_j^{exact}) \Delta V_j}{\sum_j \mathbf{u}_j^{exact} \cdot \mathbf{u}_j^{exact} \Delta V_j}}. \quad (5.83)$$

The convergence of the L2-norm for u is shown in Fig. 5.2. It can be seen that with the increase of order in the nonlocal operator, the convergence rate increases greatly. $p = 2, 3$ have the same convergence rate.

5.3.2 1D Schrödinger Equation

This section tests the accuracy of the eigenvalue problem in 1D. The Schrödinger equation written in adimensional units for a 1D harmonic oscillator is

$$[-\frac{1}{2}\frac{\partial^2}{\partial x^2} + V(x)]\phi(x) = \lambda\phi(x), \quad V(x) = \frac{1}{2}\omega^2 x^2. \quad (5.84)$$

For simplicity, we use $\omega = 1$. The particles are uniformly distributed with constant spacing Δx on the region [−10,10].

The exact wave functions and eigenvalues can be expressed as

$$\phi_n(x) = H_n(x)\exp(\pm\frac{x^2}{2}), \quad \lambda_n = n + \frac{1}{2} \quad (5.85)$$

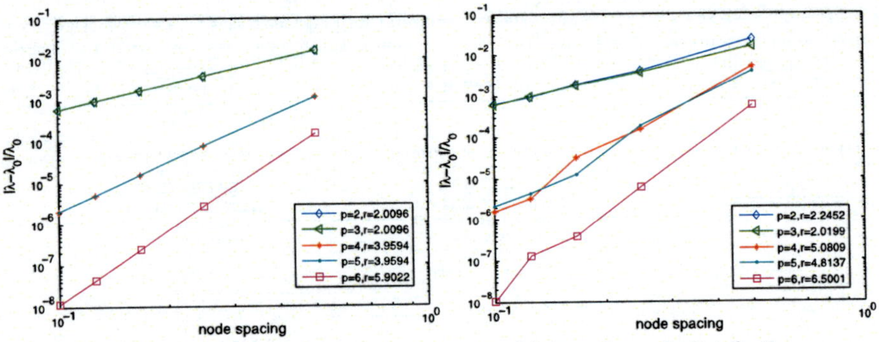

(a) regular node distribution (b) irregular node distribution

Fig. 5.3 Convergence of the lowest eigenvalue for a 1D harmonic oscillator

where n is a non-negative integer. $H_n(x)$ is the n-order Hermite polynomial. We calculate the lowest eigenvalue and compare the numerical result with $\lambda_0 = 0.5$. The convergence plot of the error is shown in Fig. 5.3.

5.3.3 *Poisson Equation*

In this section, we test the Poisson equation

$$\nabla^2 u = 2x(y-1)(y-2x+xy+2)e^{x-y}, \quad (x, y) \in (0, 1) \times (0, 1) \tag{5.86}$$

with the boundary conditions

$$\begin{aligned} u(x, 0) &= u(x, 1) = 0, \ x \in [0, 1] \\ u(0, y) &= u(1, y) = 0, \ y \in [0, 1]. \end{aligned}$$

The analytic solution is

$$u(x, y) = x(1-x)y(1-y)e^{x-y}. \tag{5.87}$$

The number of neighbors for each point is selected as the number of terms in the nonlocal operator. We test the convergence of the L2 error for the u field under uniform discretizations and nonuniform discretization in Fig. 5.4. The convergent plot is given in Fig. 5.5.

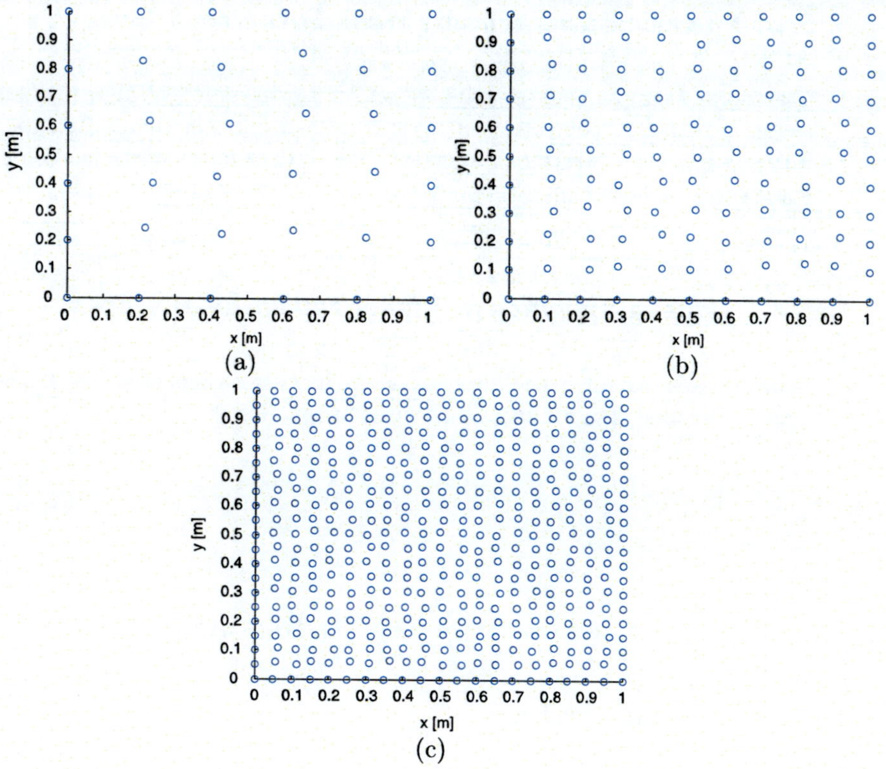

Fig. 5.4 Irregular nodal distributions

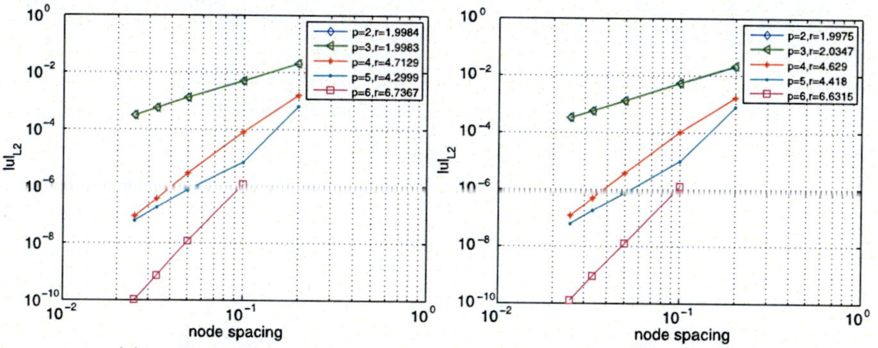

(a) regular node distribution (b) irregular node distribution

Fig. 5.5 The L_2 norm of different polynomial orders and node spacings for regular/irregular nodal distributions

5.4 Numerical Examples by Weak Form

The fourth example aims at solving the Poisson equation in higher dimensional space by both the "equivalent" integral form and operator energy functional. The fifth example is the biharmonic equation. The sixth example solves the Von Karman plate with simple support.

5.4.1 Poisson Equation in Higher Dimensional Space

In this section, we solve the Poisson equation in n dimensional space by the nonlocal operator method. The n dimensional Poisson equation is

$$\nabla^2 u = f(\mathbf{x}), \quad \mathbf{x} \in [0, 1]^n \tag{5.88}$$

with analytic solution

$$u(\mathbf{x}) = \exp\Big(\sum_{i=1}^{n}(-1)^{i-1}x_i\Big)\Pi_{i=1}^{n}x_i(1-x_i) \tag{5.89}$$

under the boundary conditions

$$u(x_1, ..., x_i = 0, ..., x_n) = u(x_1, ..., x_i = 1, ..., x_n) = 0,\ 1 \leq i \leq n \tag{5.90}$$

where $\mathbf{x} = (x_1, ..., x_n)$, $f(\mathbf{x}) = \nabla^2\big(\exp(\sum_{i=1}^{n}(-1)^{i-1}x_i)\Pi_{i=1}^{n}x_i(1-x_i)\big)$.

The equivalent integral functional for Eq. 5.88 is

$$\mathcal{F} = \int_\Omega (\frac{1}{2}\nabla u \cdot \nabla u - f(\mathbf{x})u)\mathbf{d}V. \tag{5.91}$$

The tangent stiffness is constructed from the operator matrix $\mathbf{B}$ for ∇u, e.g.

$$\mathbf{K}_g = \sum_{\Delta V_i \in \Omega} \Delta V_i \mathbf{B}_i^T \mathbf{B}_i. \tag{5.92}$$

The Dirichlet boundary conditions are applied by the penalty method.

The Poisson equation with dimensional number $n = (2, 3, 4, 5)$ under different discretization and order of nonlocal operators is tested, where the statistical results are shown in Tables 5.2, 5.3, 5.4 and 5.5.

Table 5.2 gives the statistical results for 2D Poisson equation under different discretizations. When the order of nonlocal operators increased from one to three, the L2-norm and error for u_{max} decreased gradually as shown in several cases. However, for fourth-order nonlocal operators, the result is not better than the third-order

Table 5.2 Statistical results for 2D Poisson equation

Node	Δx	L_2 norm	$\frac{u_{max}}{u_{max}^{exact}} - 1$	p-order	p^{hg}
1681	0.025	0.0485	−0.0281	1	1
1681	0.025	0.0262	0.01	2	1
1681	0.025	0.0139	−0.00256	3	1
1681	0.025	0.0175	−0.00308	4	1
6561	0.0125	0.0379	0.033	1	0
6561	0.0125	0.0179	0.0714	1	1
6561	0.0125	0.011	0.00505	2	1
25921	0.00625	0.0202	0.0221	1	1
25921	0.00625	0.00501	0.00266	2	1
25921	0.00625	0.00191	−0.000417	3	1
40401	0.005	0.00777	−0.00263	1	1
160801	0.0025	0.00291	0.0007	1	1

Table 5.3 Statistical results for 3D Poisson equation

Node	Δx	L_2 norm	$\frac{u_{max}}{u_{max}^{exact}} - 1$	p-order	p^{hg}
10648	0.04763	0.0907	−0.0406	1	1
29791	0.03333	0.0604	−0.0248	1	1
68921	0.025	0.0485	−0.02	1	1

Table 5.4 Statistical results for 4D Poisson equation

Node	Δx	L_2 norm	$\frac{u_{max}}{u_{max}^{exact}} - 1$	p-order	p^{hg}
14641	0.1	0.169	−0.0514	1	1
65536	0.0667	0.118	−0.0171	1	1
160000	0.0526	0.0983	−0.0203	1	1
810000	0.0345	0.0579	0.00304	1	1
2560000	0.0256	0.0454	0.00152	1	1

scheme. The thir-order scheme with 25921 nodes can achieve better results than the first-order scheme with 160801 nodes. The comparison between 5, 6 rows shows that the operator energy functional has a positive effect in improving the accuracy. In contrast with the scheme by strong form, the convergence property of weak form is slightly affected by the operator energy functional.

For the 3D Poisson equation, we tested three cases with discretization ranging from 22, 31 and 41 nodes in each direction. The statistical results are given in Table 5.4. The L2-norm and error for u_{max} decrease with the point grid space. When 41 nodes are used for each direction, the L2-norm is approximately 5%.

Table 5.5 Statistical results for 5D Poisson equation

Node	Δx	L_2 norm	$\frac{u_{max}}{u_{max}^{exact}} - 1$	p-order	p^{hg}
7776	0.2	0.229	−0.114	1	1
100000	0.111	0.181	−0.0944	1	1
1048576	0.0667	0.13	−0.0485	1	1
4084101	0.05	0.0985	−0.0352	1	1

For the 4D Poisson equation, we tested four cases with discretization ranging from 11, 16, 20, 30 and 40 nodes for each direction. The statistical results are given in Table 5.4. The L2-norm and error for u_{max} decrease with the point grid space. When 40 nodes are used for each direction, the L2-norm is approximately 5%.

For the 5D Poisson equation, when 16 nodes are assigned in each direction, the number of nodes reaches 1,048,576. More nodes in each direction will lead to the dimension disaster. The statistical results for different discretization are given in Table 5.5. The L2-norm and error for maximal u decrease with the node spacing. We tested maximal 21 nodes in each direction (the computational scale is restricted by the computational power of a desktop PC), the L2-norm is approximately 9.85% and the error for u_{max} with respect to the theoretical solution is less than 4%.

5.4.2 *Square Plate with Simple Support*

The plate equation reads

$$w_{,04} + 2w_{,22} + w_{,40} = \frac{q_0}{D_0}, \quad (x, y) \in (0, 1) \times (-1/2, 1/2) \tag{5.93}$$

where $D_0 = \frac{Et^3}{12(1-\nu^2)}$, with Dirichlet boundary conditions

$$w(x, -1/2) = w(x, 1/2) = 0, \; x \in [0, 1]$$
$$w(0, y) = w(1, y) = 0, \; y \in [-1/2, 1/2].$$

The analytic solution for the simply supported square plate subjected to uniform load is denoted by Timoshenko and Woinowsky-Krieger (1959)

$$w = \frac{4q_0a^4}{\pi^5 D_0} \sum_{m=1,3,...}^{\infty} \frac{1}{m^5}\Big(1 - \frac{\alpha_m \tanh \alpha_m + 2}{2\cosh \alpha_m} \cosh \frac{2\alpha_m y}{a} + \frac{\alpha_m}{2\cosh \alpha_m} \frac{2y}{a} \sinh \frac{2\alpha_m y}{a}\Big) \sin \frac{m\pi x}{a} \tag{5.94}$$

where $\alpha_m = \frac{m\pi}{2}$.

The "equivalent" integral form for Eq. 5.93 is

$$\mathcal{F}_{plate} = \frac{1}{2}\tilde{\partial} w^T D_{plate} \tilde{\partial} w \tag{5.95}$$

where

$$\tilde{\partial} w = (w_{yy}, w_{xx}, w_{xy})^T \tag{5.96}$$

$$\mathbf{D}_{plate} = \frac{Et^3}{12(1-\nu^2)} \begin{bmatrix} 1 & \nu & 0 \\ \nu & 1 & 0 \\ 0 & 0 & 2-2\nu \end{bmatrix}. \tag{5.97}$$

The parameters for the plate include length $a = 1$, thickness $t = 0.01$ m and uniform pressure $q_0 = -100$ N, Poisson ratio $\nu = 0.3$, elastic modulus $E = 30$ GPa and $D_0 = \frac{Et^3}{12(1-\nu^2)}$.

With the aid of nonlocal operator $\tilde{\partial} w$ and its operator matrix $\mathbf{B}$, the first and second variations of the energy functional are

$$\delta \mathcal{F}_{plate} = \sum_{\Delta V_i \in \Omega} \Delta V_i (\tilde{\partial} w_i^T \mathbf{D}_{plate} \tilde{\partial} \delta w_i - q_0 \delta w_i) = \sum_{\Delta V_i \in \Omega} \Delta V_i \tilde{\partial} w_i^T D_{plate} \mathbf{B}_i \delta \mathbf{w}_i - q_0 \delta w_i$$

$$\delta^2 \mathcal{F}_{plate} = \sum_{\Delta V_i \in \Omega} \Delta V_i \tilde{\partial} \delta w_i^T \mathbf{D}_{plate} \tilde{\partial} \delta w_i = \sum_{\Delta V_i \in \Omega} \Delta V_i \delta \mathbf{w}_i^T \mathbf{B}_i^T \mathbf{D}_{plate} \mathbf{B}_i \delta \mathbf{w}_i$$

where $\delta \mathbf{w}_i$ is the vector for all unknowns in support $\mathcal{S}_i$.

The plate is discretized uniformly and the number of neighbors for each point is selected as $n = 5p + length(\tilde{\partial} u)$, where p is the order of the nonlocal operator. The deflection curves for several discretizations are compared with the analytic solution in Fig. 5.6. The contour of the deflection field for discretization of 40×40 is shown in Fig. 5.7b. Compared with the original nonlocal operator method, the higher order NOM obtains the nonlocal operator in a simpler way.

5.4.3 Von Kármán Equations for a Thin Plate

The Von Kármán equations (Von Kármán 1910) are a set of nonlinear partial differential equations describing the large deflections of thin flat plates. The equations are based on Kirchhoff hypothesis: the surface normals to the plane of the plate remain perpendicular to the plate after deformation and in-plane (membrane) displacements are small and the change in thickness of the plate is negligible. These assumptions imply that the displacement field v in the plate can be expressed as (Ciarlet 1980)

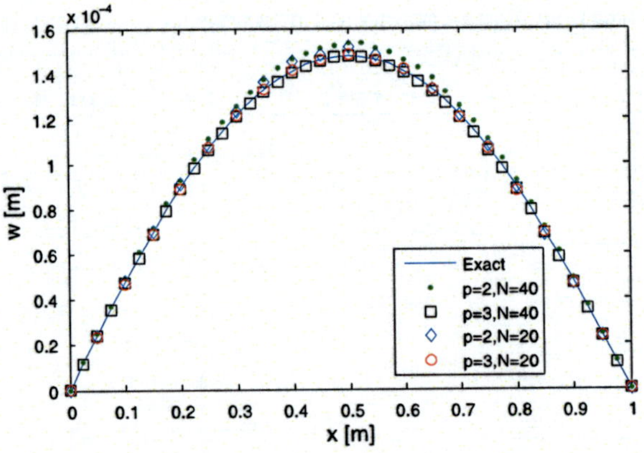

Fig. 5.6 Deflection of section $y = 0$ under different discretizations, where p denotes order of nonlocal operator, and N is the number of nodes in one direction

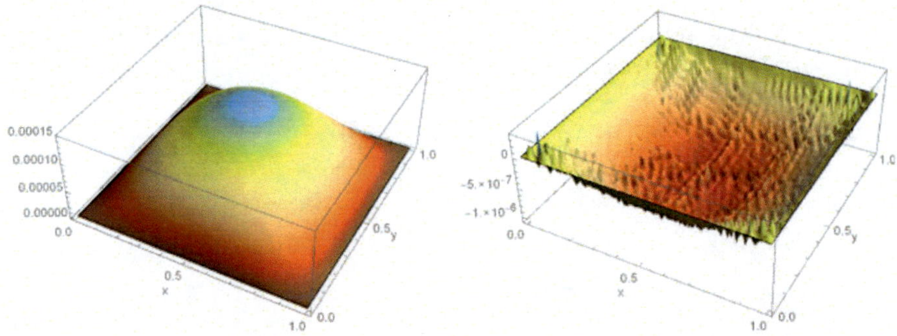

(a) (b)

Fig. 5.7 **a** Deflection. **b** Error of deflection w for discretization of 40×40 with respect to exact solution

$$
\begin{aligned}
v_1(x_1, x_2, x_3) &= u_1(x_1, x_2) - x_3 \frac{\partial w}{\partial x_1}, \\
v_2(x_1, x_2, x_3) &= u_2(x_1, x_2) - x_3 \frac{\partial w}{\partial x_2}, \\
v_3(x_1, x_2, x_3) &= w(x_1, x_2).
\end{aligned}
\tag{5.98}
$$

For a plate of a thickness h defined on the mid-surface (x_1, x_2), Von Kármán energy is given by Landau and Lifshitz (1986), Patrício da Silva and Krauth (1997)

$$
\iint_\Omega \{\frac{Eh^3}{24(1-\nu^2)}\{(\Delta w)^2 - 2(1-\nu)[w, w]\} + \frac{h}{2}\varepsilon_{ij}\sigma_{ij} - qw\}dx_1dx_2 \tag{5.99}
$$

where Laplace operator $\Delta w = \frac{\partial^2 w}{\partial x_1^2} + \frac{\partial^2 w}{\partial x_2^2}$ and

$$[w, w] = \frac{\partial^2 w}{\partial x_1^2}\frac{\partial^2 w}{\partial x_2^2} - (\frac{\partial^2 w}{\partial x_1 \partial x_2})^2, \tag{5.100}$$

ε_{ij} is the strain tensor with nonlinear terms in the deformations $u_1 = u_1(x_1, x_2)$, $u_2 = u_2(x_1, x_2)$, $w = w(x_1, x_2)$:

$$\varepsilon_{ij} = \frac{1}{2}(\frac{\partial u_i}{\partial x_j} + \frac{\partial u_j}{\partial x_i}) + \frac{1}{2}\frac{\partial w}{\partial x_i}\frac{\partial w}{\partial x_j} \tag{5.101}$$

where (u_1, u_2) is the lateral displacement field due to membrane effect, w is the deflection, σ_{ij} is the stress tensor, linearly proportional to ε_{ij}, $E = 30 \times 10^6$ Pa and $\nu = 0.3$ are the Young modulus and the Poisson ratio, respectively and $q = 1000$Pa is the external normal force per unit area of the plate. The dimensions of the plate are $1.0 \times 1.0 \times 0.01$ m^3. The energy functional in Eq. 5.99 leads to the governing equations

$$\frac{Eh^3}{12(1-\nu^2)}\nabla^4 w - h\frac{\partial}{\partial x_j}\left(\sigma_{ij}\frac{\partial w}{\partial x_i}\right) = q, \quad \frac{\partial \sigma_{ij}}{\partial x_j} = 0. \tag{5.102}$$

The Cauchy stress tensor in mid-plane can be written as

$$\sigma = \frac{E}{1-\nu^2}\left(\nu \mathrm{tr}\varepsilon \mathbf{I}_{2\times 2} + (1-\nu)\varepsilon\right) \tag{5.103}$$

$$\varepsilon = \begin{bmatrix} \varepsilon_{11} & \varepsilon_{12} \\ \varepsilon_{21} & \varepsilon_{22} \end{bmatrix}. \tag{5.104}$$

Herein, we write the moment and curvature by tensor form. The conventional vectorial form can be recovered with ease. The moment tensor and curvature tensor are

$$\mathbf{M} = \begin{bmatrix} M_{11} & M_{12} \\ M_{21} & M_{22} \end{bmatrix} = D_0\left(\nu \mathrm{tr}\boldsymbol{\kappa} \boldsymbol{I}_{2\times 2} + (1-\nu)\boldsymbol{\kappa}\right) \tag{5.105}$$

$$\boldsymbol{\kappa} = \nabla\nabla w = \begin{bmatrix} \frac{\partial^2 w}{\partial x_1^2} & \frac{\partial^2 w}{\partial x_1 \partial x_2} \\ \frac{\partial^2 w}{\partial x_2 \partial x_1} & \frac{\partial^2 w}{\partial x_2^2} \end{bmatrix} \tag{5.106}$$

where $D_0 = \frac{Eh^3}{12(1-\nu^2)}$. The moment tensor is similar to the stress tensor in the plane stress conditions.

The rotation in direction $\boldsymbol{n}$ is

$$\frac{\partial w}{\partial n} = \nabla w \cdot \boldsymbol{n}, \text{ where } \boldsymbol{n} = (n_1, n_2). \tag{5.107}$$

The curvature in direction $\boldsymbol{n}$ is

$$\kappa_n = \boldsymbol{n}^T \boldsymbol{\kappa} \boldsymbol{n}. \tag{5.108}$$

The momentum in direction $\boldsymbol{n}$ is

$$M_n = \boldsymbol{n}^T \mathbf{M} \boldsymbol{n}. \tag{5.109}$$

The nonlocal differential operators in Eq. 5.99 can be written as

$$\tilde{\partial} u = (u_{1,01}, u_{1,10}, u_{2,01}, u_{2,10}, w_{,01}, w_{,02}, w_{,10}, w_{,11}, w_{,20}). \tag{5.110}$$

The gradient of energy functional on $\tilde{\partial} u$ is

$$\frac{\partial \mathcal{F}}{\partial \tilde{\partial} \mathbf{u}} = D_0 \begin{bmatrix} \frac{(1-\nu)}{2}(u_{1,01} u_{2,10} + w_{,01} w_{,10}) \\ \frac{1}{2}\left(\nu(2u_{2,01} + w_{,01}^2) + 2u_{1,10} + w_{,10}^2\right) \\ \frac{1}{2}\left(\nu(2u_{1,10} + w_{,10}^2) + 2u_{2,01} + w_{,01}^2\right) \\ \frac{(1-\nu)}{2}(u_{1,01} u_{2,10} + w_{,01} w_{,10}) \\ \frac{1}{2}\left(w_{,01}(2\nu u_{1,10} + 2u_{2,01} + w_{,01}^2) + w_{,10}((1-\nu)(u_{1,01} u_{2,10}) + w_{,01} w_{,10})\right) \\ \frac{1}{12} h^2 (\nu w_{,20} + w_{,02}) \\ \frac{1}{2}\left((u_{1,01} u_{2,10})(w_{,01} - \nu w_{,01}) + w_{,10}(2\nu u_{2,01} + 2u_{1,10} + w_{,01}^2 + w_{,10}^2)\right) \\ \frac{1}{6} h^2 (1-\nu) w_{,11} \\ \frac{1}{12} h^2 (\nu w_{,02} + w_{,20}) \end{bmatrix}. \tag{5.111}$$

The Hessian matrix of $\mathcal{F}$ can be obtained with ease by computing $\frac{\partial^2 \mathcal{F}}{\partial \tilde{\partial} \mathbf{u}^2}$. The solution can be obtained when using the Newton–Raphson method in Sect. 5.2.1. For simplicity, we only consider the simple support boundary conditions.

The plate solved by NOM is discretized by 50×50 nodes. The reference results are calculated by S4R plate/shell element in ABAQUS (Hibbett et al. 1998). S4R element is a 4-node doubly curved thin or thick shell element with reduced integration, hourglass control and finite membrane strains. In ABAQUS, the flat thin plate with the same material parameters are discretized into 100×100 elements.

Displacement in membrane and deflection out-of-plane for nodes on $y = 0.5$ under different load levels are depicted in Figs. 5.8 and 5.9, respectively, where the lines represent the results by ABAQUS while the discrete symbols are the results by NOM. The displacement results agree well with that by ABAQUS.

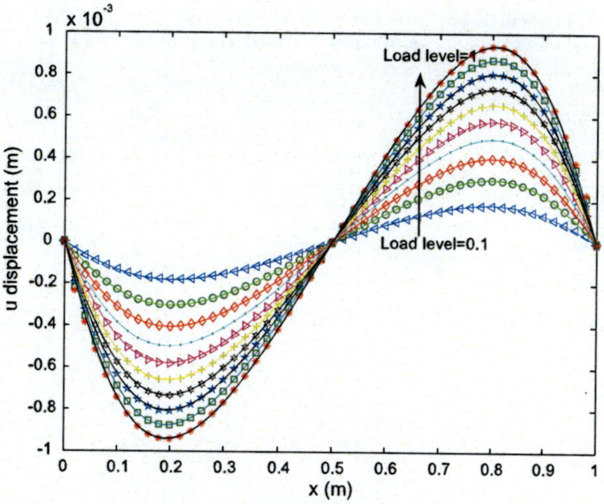

Fig. 5.8 Displacement in membrane for nodes in $y = L/2$ under load level from 0.1 to 1, where the lines represent the numerical results by S4R element in ABAQUS while the star, diamond, etc. symbols are the results by NOM

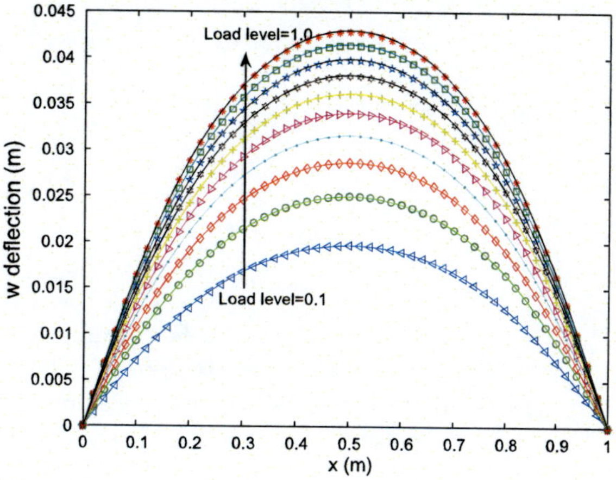

Fig. 5.9 Deflection for nodes in $y = L/2$ under load level from 0.1 to 1, where the lines represent the numerical results by S4R element in ABAQUS while the star, diamond, etc. symbols are the results by NOM

Maximal central deflection is plotted in Fig. 5.10, which shows the nonlinearity increase with load level significantly. It can be seen that the result by NOM matches well with the finite element method.

5.4.4 Nearly Incompressible Block

In this section, we model the nearly incompressible block of material constitution in Eq. 5.78 by nonlocal operator method with the Newton–Raphson iteration method. The nearly incompressible block of height $h = 50$ mm, length $2h$ and width $2h$ is loaded by an equally distributed pressure $p = 3$ MPa at its top center of area $h \times h$

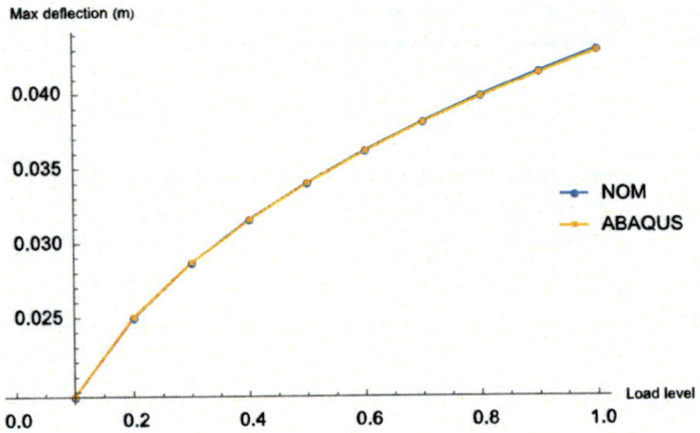

Fig. 5.10 Maximal deflection for node in $(L/2, L/2)$ under load level from 0.1 to 1

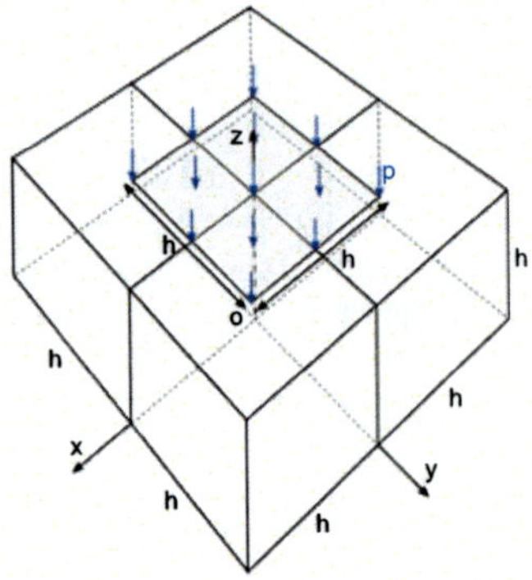

Fig. 5.11 Setup of the block

mm^2, as shown in Fig. 5.11. For symmetry reasons, only a quarter of the block is modeled. The bottom face is fixed in z-direction, while the nodes on plane $y = 0$ are fixed in y-direction and the nodes on plane $x = 0$ are fixed in x-direction, as similarly presented in reference Reese et al. (2000). The material parameters are $\kappa = 499.92568$ MPa and $\mu = 1.61148$ MPa.

The deformed block at final load level is depicted in Fig. 5.12, where good agreement is obtained between the finite element method and the NOM. The maximal displacements in z-direction by linear hexahedral element (H1), quadratic hexahedral elements (H2) and nonlocal operator method are given in Table 5.6.

5.4.5 *Fracture Modeling by Phase Field Method*

In this section, we implement the phase field fracture method by the nonlocal operator scheme. The phase field fracture methods (Karma et al. 2001; Hakim and Karma 2009; Miehe et al. 2010b; Borden et al. 2012, 2014; Wu 2017) model the fracture by introducing an additional field to describe the damage status of a material point, which offers great flexibilities to model complex fractures. In the literature, the phase field

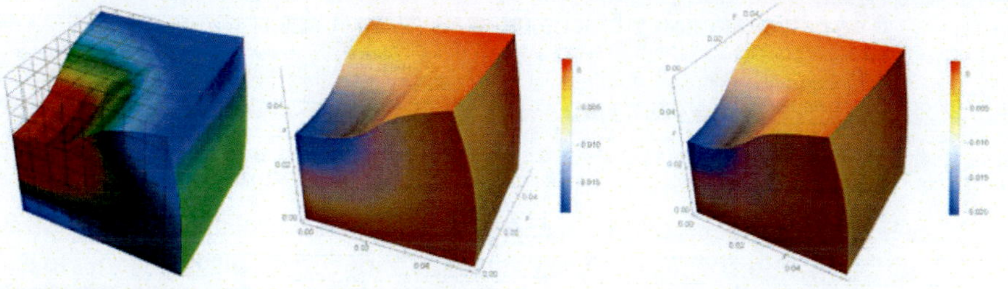

(a) FEM[Korelc and Wriggers, 2016] with 8^3 mesh (b) NOM with 11^3 nodes (c) NOM with 21^3 nodes

Fig. 5.12 z-direction displacement in deformed configuration at final load level

Table 5.6 Nearly incompressible block: displacement w_{max} (mm)

H1 element	13.17 (8^3 mesh)	19.52 (32^3 mesh)
H2 element	19.54 (8^3 mesh)	20.01 (32^3 mesh)
NOM	19.14 (11^3 nodes)	20.43 (21^3 nodes)

method is implemented by finite element methods or isogeometric analysis methods. Since NOM obtains all specified partial derivatives at once without resorting to shape functions, NOM can be viewed as an alternative to FEM to implement the phase field method.

The regularized crack functional is given as

$$\Gamma(s) \approx \int_{\Omega} (\frac{s^2}{2l} + \frac{l}{2} \nabla s \cdot \nabla s) \mathbf{d}\Omega, \tag{5.112}$$

where phase field $s = 1$ denotes full damage of the material, while $s = 0$ represents zero damage in the material and $\Gamma(s)$ is the "diffused" crack surface (Miehe et al. 2010b). When the intrinsic length scale l approaches 0, the sharp discontinuous interface of the crack is recovered, as is indicated by the Γ-limit.

The surface energy due to the formation of cracks is given by

$$\int_{\Gamma} g_c \mathbf{d}\Gamma \approx \int_{\Omega} g_c (\frac{s^2}{2l} + \frac{l}{2} \nabla s \cdot \nabla s) \mathbf{d}\Omega, \tag{5.113}$$

where g_c is the critical energy release rate.

The potential energy functional of the system reads

$$\Pi_l(\boldsymbol{u}, s) = \int_\Omega \psi_e(\boldsymbol{\varepsilon}(\nabla \boldsymbol{u}), s)\mathrm{d}\Omega - \int_{\partial\Omega} \boldsymbol{t}^* \cdot \boldsymbol{u}\mathrm{d}\Gamma - \int_\Omega \mathbf{b} \cdot \boldsymbol{u}\mathrm{d}\Omega + \int_\Omega g_c(\frac{s^2}{2l} + \frac{l}{2}\nabla s \cdot \nabla s)\mathrm{d}\Omega, \tag{5.114}$$

where ψ_e represents the strain energy density, $\boldsymbol{u}$ the displacement, $\boldsymbol{\varepsilon}$ the strain, $\boldsymbol{t}^*$ the surface traction on the boundary and $\mathbf{b}$ the body force density. In order to distinguish the different tensile and compressive strengths of the material, the strain energy density comprises two parts, the tensile part affected by the phase field and the compressive part independent with the phase field,

$$\psi_e(\boldsymbol{\varepsilon}(\nabla \boldsymbol{u}), s) = (1-s)^2 \psi_e^+(\boldsymbol{\varepsilon}(\nabla \boldsymbol{u})) + \psi_e^-(\boldsymbol{\varepsilon}(\nabla \boldsymbol{u})). \tag{5.115}$$

The variation $\delta \Pi_l = 0$ leads to mechanical and phase field governing equations

$$\nabla \cdot \boldsymbol{\sigma}^s + \mathbf{b} = 0, \tag{5.116}$$

$$g_c(s - l^2 \nabla^2 s) = 2l(1-s)H, \tag{5.117}$$

where $H(x, T) := \max_{t \in [0,T]} \psi^+(\boldsymbol{\varepsilon}(\boldsymbol{x}, t))$, the local history field of the maximum positive reference energy (Miehe et al. 2010b), which guarantees the irreversibility of the fracture; $\boldsymbol{\sigma}^s$ is the Cauchy stress

$$\boldsymbol{\sigma}^s = \frac{\partial \psi_e(\boldsymbol{\varepsilon}, s)}{\partial \boldsymbol{\varepsilon}} = (1-s)^2 \frac{\partial \psi_e^+}{\partial \boldsymbol{\varepsilon}} + \frac{\partial \psi_e^-}{\partial \boldsymbol{\varepsilon}} = (1-s)^2 \boldsymbol{\sigma}^+ + \boldsymbol{\sigma}^-, \tag{5.118}$$

where $\boldsymbol{\sigma}^\pm$ are the Cauchy stresses for the tensile and compressive parts given by

$$\boldsymbol{\sigma}^\pm = \frac{\partial \psi_e^\pm}{\partial \boldsymbol{\varepsilon}}. \tag{5.119}$$

The strain energy based on the eigendecomposition of the strain tensor (Miehe et al. 2010b) is expressed as

$$\psi_e^\pm(\boldsymbol{\varepsilon}) := \lambda \langle \varepsilon_1 + \varepsilon_2 + \varepsilon_3 \rangle_\pm^2 / 2 + \mu(\langle \varepsilon_1 \rangle_\pm^2 + \langle \varepsilon_2 \rangle_\pm^2 + \langle \varepsilon_3 \rangle_\pm^2), \tag{5.120}$$

where λ, μ are the Lamé constants, $\langle x \rangle_\pm := (x \pm |x|)/2$. The "positive" and "negative" stresses are derived as

$$\boldsymbol{\sigma}^\pm = \lambda \langle \varepsilon_1 + \varepsilon_2 + \varepsilon_3 \rangle_\pm I_{3\times 3} + 2\mu(\langle \varepsilon_1 \rangle_\pm \boldsymbol{n}_1 \otimes \boldsymbol{n}_1 + \langle \varepsilon_2 \rangle_\pm \boldsymbol{n}_2 \otimes \boldsymbol{n}_2 + \langle \varepsilon_3 \rangle_\pm \boldsymbol{n}_3 \otimes \boldsymbol{n}_3), \tag{5.121}$$

where $I_{3\times 3}$ is the identity matrix, and $\boldsymbol{n}_1, \boldsymbol{n}_2, \boldsymbol{n}_3$ are eigenvectors for principal strain $\varepsilon_1, \varepsilon_2, \varepsilon_3$ of $\boldsymbol{\varepsilon}(= \sum_{i=1}^3 \varepsilon_i \boldsymbol{n}_i \otimes \boldsymbol{n}_i)$, respectively.

The material tensor is constructed by

$$\mathbf{D} = \frac{\partial \sigma}{\partial \varepsilon} = (1-s)^2 \frac{\partial \sigma^+}{\partial \varepsilon} + \frac{\partial \sigma^-}{\partial \varepsilon} \tag{5.122}$$

where $\frac{\partial \sigma^+}{\partial \varepsilon}$, $\frac{\partial \sigma^-}{\partial \varepsilon}$ can be obtained by two fourth-order projection tensors

$$\mathbb{P}^+ := \partial_\varepsilon \left[\varepsilon^+(\varepsilon)\right] \quad \text{and} \quad \mathbb{P}^- := \mathbb{I} - \mathbb{P}^+. \tag{5.123}$$

For more details of the computation of these objects, we refer to Miehe and Lambrecht (2001), Zhou et al. (2018). The staggered scheme (Miehe et al. 2010a) for phase field fracture is employed.

The implementation of the phase field by NOM is the same as that by FEM, except that the derivatives of the shape function are replaced by nonlocal differential operators. The maximal derivative order in weak form is 1, therefore the first-order NOM scheme is used. The Dirichlet boundary conditions for phase field and displacement fields are enforced by a penalty method. The penalties in the operator functional are selected as 0.1 and μ for the phase field and displacement field, respectively.

In order to test the feasibility of NOM, a notched plate as shown in Fig. 5.13 is considered, with the tension boundary condition in Case (a) and shear boundary condition in Case (b) being solved separately. The elastic bulk modulus is chosen to $\lambda = 121.15\,\text{kN/mm}^2$, the shear modulus to $\mu = 80.77\,\text{kN/mm}^2$ and the critical energy release rate to $g_c = 2.7 \times 10^{-3}$ kN/mm. Plane strain condition is assumed. The plate is discretized by particle spacing of $\Delta x = 10^{-2}$ mm into 10, 000 particles. Four layers of particles closest to the initial crack are removed to form the crack. For each particle, 8 nearest neighbors are selected to form the support, where the characteristic length scale is determined by the farthest particle in the support. The internal length of the phase field is selected as $l = 4\Delta x$. For the tension test, a constant displacement increment of $\Delta u = 1 \times 10^{-5}$ mm is applied for 600 steps. For the shear test, the first 500 steps employ displacement increment of $\Delta u = 1 \times 10^{-5}$ mm and the remaining time steps use $\Delta u = 1 \times 10^{-6}$ mm. The final displacement field in y direction and phase field for tension test are given in Fig. 5.14a, b. The load for tension test is depicted in Fig. 5.15a. The load curves by NOM agree well with the reference result in Miehe et al. (2010b).

The final displacement field in x direction and phase field for the shear test are given in Fig. 5.14c, d. The load for shear test is depicted in Fig. 5.15b.

5.5 Concluding Remarks

We have proposed a higher order nonlocal operator method for solving higher order PDEs based on the strong form or the equivalent integral formulated by the weighted residual method or variational principles.

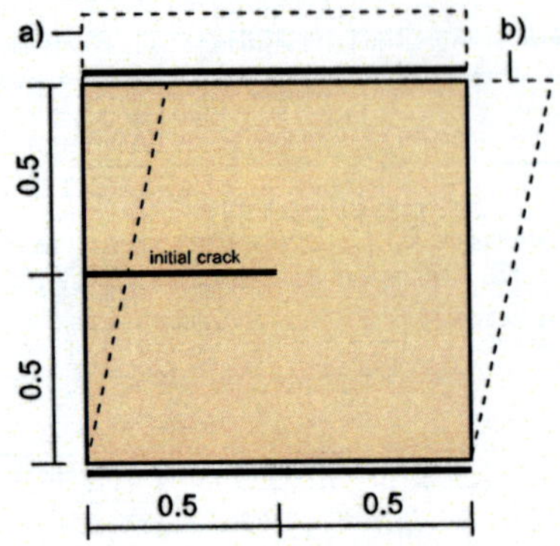

Fig. 5.13 Single edge notched specimen. Geometry and boundary conditions: **a** tension test; **b** shear test

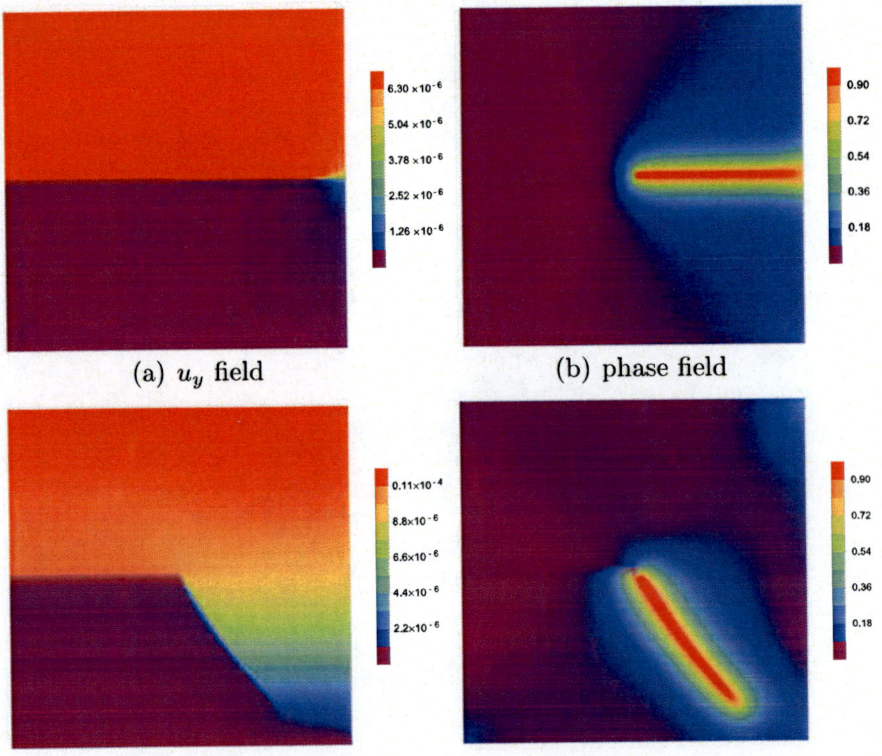

(a) u_y field (b) phase field

(c) u_x field (d) phase field

Fig. 5.14 Displacement fields and phase fields for tension and shear tests. **a** u_y displacement field. **b** Phase field for tension test with $l = 4 \times 10^{-2}$ mm. **c** The u_x displacement field. **d** Phase field for shear test with $l = 4 \times 10^{-2}$ mm

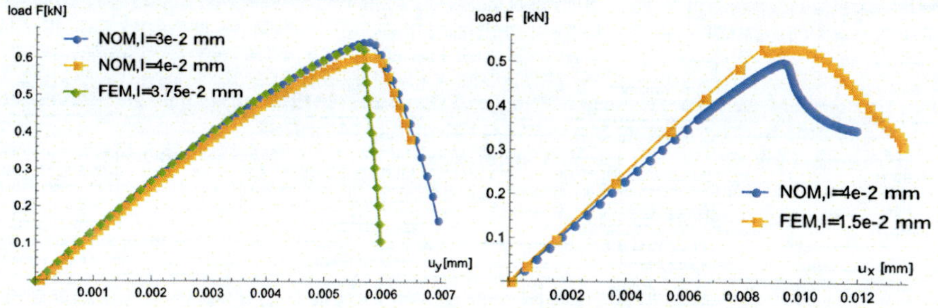

(a) (b)

Fig. 5.15 Load curves for **a** tension test, **b** shear test, where FEM results are taken from Fig. 12a in Miehe et al. (2010b), and Fig. 9a in Miehe et al. (2010a), respectively

The relation of the nonlocal operator and local operator is that the local operator is defined on a point, while the nonlocal operator method is defined on the support with a finite characteristic length scale. When the support decreases to a point, the nonlocal operator degenerates into the local operator. A nonlocal operator is constructed from the Taylor series expansion and approximates the local derivative with orders up to n. In order to establish the nonlocal operator, only finite points in support are required. Nonlocal operators can be viewed as a generalization of the local operator. Most rules applied to the local operator can be adopted directly by the nonlocal operator method.

In certain cases such as the regular grid, the nonlocal operator method is similar to the finite difference. One difference with the finite difference method is that the finite difference method requires a regular grid. When handling multiple fields, the finite difference method should adopt a staggered grid for the reason of numerical stability, which complicates the numerical implementation. For the nonlocal operator method, all the nodes have the same functions, in contrast with the finite difference method with a staggered grid, where different nodes represent different fields. In terms of numerical stability, the nonlocal operator method monitors and enhances the robustness of the derivative estimation by the operator energy functional, the quadratic functional of the Taylor series expansion. When adding the quadratic functional of the Taylor series expansion to the functional of a physical problem, the numerical stability can be enhanced and traced.

Taylor series expansion of multiple variables based on the multi-index notation is powerful in deriving various partial derivatives of different orders. Multi-index notation α_d^n in Sect. 5.1.3 can obtain automatically all the partial derivatives with orders up to n in d spatial dimensions. In addition, the characteristic length scale is introduced for high precision of the derivative estimation. With all the partial derivatives available, all linear PDEs up to $2n$ orders can be described with ease. By replacing the differential operator with the nonlocal one, the nonlocal operator method converts the PDEs into algebraic equations directly. The nonlocal operator method can be viewed as a tool to study the higher order PDEs.

References

Borden M, Verhoosel C, Scott M, Hughes T, Landis C (2012) A phase-field description of dynamic brittle fracture. Comput Methods Appl Mech Eng 217:77–95

Borden MJ, Hughes TJ, Landis CM, Verhoosel CV (2014) A higher-order phase-field model for brittle fracture: formulation and analysis within the isogeometric analysis framework. Comput Methods Appl Mech Eng 273:100–118

Ciarlet PG (1980) A justification of the von kármán equations. Arch Ration Mech Anal 73(4):349–389

Hakim V, Karma A (2009) Laws of crack motion and phase-field models of fracture. J Mech Phys Solids 57(2):342–368

Hibbett, Karlsson, and Sorensen (1998) ABAQUS/standard: user's manual, vol 1

Hormander L (1983) The analysis of partial differential operators. Springer

Karma A, Kessler D, Levine H (2001) Phase-field model of mode III dynamic fracture. Phys Rev Lett 87(4):045501

Korelc J, Wriggers P (2016) Automation of finite element methods. Springer

Landau LD, Lifshitz E (1986) Theory of elasticity, vol 7. Course of theoretical physics, vol 3, p 109

Miehe C, Lambrecht M (2001) Algorithms for computation of stresses and elasticity moduli in terms of Seth-Hill's family of generalized strain tensors. Commun Numer Methods Eng 17(5):337–353

Miehe C, Hofacker M, Welschinger F (2010a) A phase field model for rate-independent crack propagation: Robust algorithmic implementation based on operator splits. Comput Methods Appl Mech Eng 199(45–48):2765–2778

Miehe C, Welschinger F, Hofacker M (2010b) Thermodynamically consistent phase-field models of fracture: variational principles and multi-field fe implementations. Int J Numer Methods Eng 83(10):1273–1311

Patrício da Silva P, Krauth W (1997) Numerical solutions of the von Karman equations for a thin plate. Int J Mod Phys C 8(02):427–434

Reese S, Wriggers P, Reddy B (2000) A new locking-free brick element technique for large deformation problems in elasticity. Comput Struct 75(3):291–304

Ren H, Zhuang X, Rabczuk T (2020) A nonlocal operator method for solving partial differential equations. Comput Methods Appl Mech Eng 358:112621. ISSN 0045-7825. https://doi.org/10.1016/j.cma.2019.112621

Timoshenko S, Woinowsky-Krieger S (1959) Theory of plates and shells. McGraw-Hill

Von Kármán T (1910) Festigkeitsprobleme im maschinenbau. Teubner

Wu J (2017) A unified phase-field theory for the mechanics of damage and quasi-brittle failure. J Mech Phys Solids 103:72–99

Zhou S, Rabczuk T, Zhuang X (2018) Phase field modeling of quasi-static and dynamic crack propagation: COMSOL implementation and case studies. Adv Eng Softw 122:31–49

Chapter 6
Nonlocal Operator Method with Numerical Integration for Gradient Solid

NOM as a particle-based method (Ren et al. 2020) has difficulties in accurately enforcing boundary conditions yielding to a deterioration of the convergence rate. Moreover, particle-based methods require stabilization which can be achieved for instance by an operator energy functional as proposed for NOM in (Ren et al. 2020). However, the operator energy functional involves a penalty parameter and results might be sensitive w.r.t. the choice of this parameter.

One purpose of the chapter is to develop a nonlocal operator method with so-called approximation properties. The approximation property means that the numerical method can approximate the field value and its derivatives at any point of the domain. Similar to many mesh-free methods, we construct a background mesh and employ Gauss quadrature. Based on the shape functions in the background mesh, the coordinates of the integration points and weights can be calculated. Based on the node set in the support, the nonlocal operators of different orders can be defined. The NOM proposed in the chapter will make use of the Lagrange multipliers or modified variational principle to enforce the boundary conditions. To show the capability of the proposed NOM, we focus on problems in gradient elasticity. Gradient elasticity theories can describe some phenomena such as size effects and edge and skin effects in materials, which cannot be handled by conventional continuum mechanics. Gradient elasticity theory introduces an internal length scale to account for size effects at micro- or nano-scale.

The remainder of the chapter is outlined as follows. In Sect. 6.1, the basic concepts such as support and dual-support, and the low-order nonlocal operators are reviewed. In Sect. 6.2, we develop the higher order nonlocal operator method with approximation property based on the Taylor series expansion of functions with multiple variables. In order to validate the current method, the linear gradient elasticity theory is described and a modified energy functional for linear gradient elasticity is formulated in Sect. 6.3. We present several numerical examples to demonstrate the capabilities of this method in Sect. 6.4.

T. Rabczuk et al., *Computational Methods Based on Peridynamics and Nonlocal Operators*, Computational Methods in Engineering & the Sciences,
https://doi.org/10.1007/978-3-031-20906-2_6

6.1 Review of Nonlocal Operator Method

Consider a domain as shown in Fig. 6.1a. Let $\mathbf{x}_i$ be spatial coordinates in the domain $\mathbf{\Omega}$; $\mathbf{r}_{ij} := \mathbf{x}_j - \mathbf{x}_i$ is a spatial vector from $\mathbf{x}_i$ to $\mathbf{x}_j$; $\mathbf{v}_i := \mathbf{v}(\mathbf{x}_i)$ and $\mathbf{v}_j := \mathbf{v}(\mathbf{x}_j)$ are the field values for $\mathbf{x}_i$ and $\mathbf{x}_j$, respectively; $\mathbf{v}_{ij} := \mathbf{v}_j - \mathbf{v}_i$ is the relative field vector for the spatial vector $\mathbf{r}_{ij}$. NOM is based on the concept of support and dual-support.

The **support** $\mathcal{S}_i$ of a point $\mathbf{x}_i$ is the domain in which any spatial point $\mathbf{x}_j$ forms a spatial vector $\mathbf{r}_{ij}(= \mathbf{x}_j - \mathbf{x}_i)$ from $\mathbf{x}_i$ to $\mathbf{x}_j$. The support serves as the basis for the nonlocal operators.

The **dual-support** is defined as a union of points whose supports include $\mathbf{x}$ and is denoted by

$$\mathcal{S}'_i = \{\mathbf{x}_j | \mathbf{x}_i \in \mathcal{S}_j\}. \tag{6.1}$$

Point $\mathbf{x}_j$ forms a dual-vector $\mathbf{r}_{ji}(= \mathbf{x}_i - \mathbf{x}_j = -\mathbf{r}_{ij})$ in $\mathcal{S}'_i$. On the other hand, $\mathbf{r}_{ji}$ is the spatial vector formed in $\mathcal{S}_j$. One example to illustrate the support and dual-support is shown in Fig. 6.1b.

NOM uses the basic nonlocal operators to replace the local operators in calculus such as the gradient, divergence, curl and Hessian operators. The functional formulated by the local differential operator can be used to construct the residual or tangent stiffness matrix by replacing the local operator with the corresponding nonlocal operator. It can be shown (Ren et al. 2020) that the nonlocal gradient of a vector field $\mathbf{v}$ for point $\mathbf{x}_i$ in support $\mathcal{S}_i$ is defined as

$$\tilde{\nabla}\mathbf{v}_i := \int_{\mathcal{S}_i} w(\mathbf{r}_{ij})\mathbf{v}_{ij} \otimes \mathbf{r}_{ij} dV_j \cdot \mathbf{K}_i^{-1}, \tag{6.2}$$

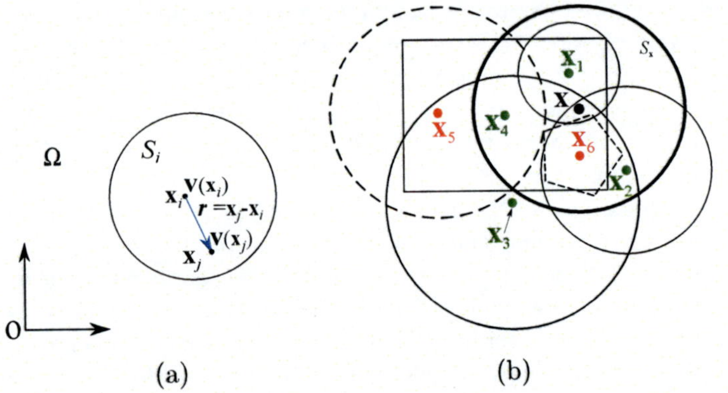

Fig. 6.1 **a** Domain and notation. **b** Schematic diagram for support and dual-support, all shapes above are supports. $\mathcal{S}_{\mathbf{x}} = \{\mathbf{x}_1, \mathbf{x}_2, \mathbf{x}_4, \mathbf{x}_6\}$, $\mathcal{S}'_{\mathbf{x}} = \{\mathbf{x}_1, \mathbf{x}_2, \mathbf{x}_3, \mathbf{x}_4\}$

where $\mathbf{v}_{ij} = \mathbf{v}_j - \mathbf{v}_i$, $\mathbf{K}_i = \int_{\mathcal{S}_i} w(\mathbf{r}_{ij})\mathbf{r}_{ij} \otimes \mathbf{r}_{ij} \mathrm{d}V_j$ is the shape tensor. Using nodal integration, the nonlocal gradient operator and its variation in the discrete form are

$$\tilde{\nabla}\mathbf{v}_i = \sum_{j \in \mathcal{S}_i} w(\mathbf{r}_{ij})\mathbf{v}_{ij} \otimes \mathbf{r}_{ij} \Delta V_j \cdot \mathbf{K}_i^{-1}, \tag{6.3}$$

$$\tilde{\nabla}\delta\mathbf{v}_i = \sum_{j \in \mathcal{S}_i} w(\mathbf{r}_{ij})\delta\mathbf{v}_{ij} \otimes \mathbf{r}_{ij} \Delta V_j \cdot \mathbf{K}_i^{-1}. \tag{6.4}$$

Besides the physical energy functional, NOM employs an operator energy functional for nonlocal gradients at point $\mathbf{x}_i$ in order to avoid instabilities, which occur due to under integration (Ren et al. 2020). The operator energy functional can be formulated as

$$\mathcal{F}_i^{hg} = \frac{p^{hg}}{m_i} \int_{\mathcal{S}_i} w(\mathbf{r}_{ij})(\tilde{\nabla}\mathbf{v}_i \cdot \mathbf{r}_{ij} - \mathbf{v}_{ij}) \cdot (\tilde{\nabla}\mathbf{v}_i \cdot \mathbf{r}_{ij} - \mathbf{v}_{ij}) \mathrm{d}V_j \tag{6.5}$$

where p^{hg} is the penalty coefficient, $m_i = \mathrm{tr}[\mathbf{K}_i]$. For solid mechanics, p^{hg} can be selected as the shear modulus. When the vector field (i.e. linear vector field) is consistent with its gradient, the value of the operator energy functional is zero. The second derivative of the operator energy functional leads to a tangent stiffness matrix, which needs to be superimposed with the tangent stiffness matrix of the physical energy functional. The rank of the operator tangent stiffness matrix is not zero, thus increasing the rank of the tangent stiffness matrix. Compared with the artificial viscosity (Monaghan 1992), stress point method (Dyka and Ingel 1995; Randles and Libersky 1996) and least square residual method (Beissel and Belytschko 1996), the operator tangent stiffness matrix has the most compact form since it is derived from the stability of the Taylor series expansion and is independent of the physical model to be solved. However, the operator tangent stiffness matrix requires a penalty coefficient. In addition, since the nodal integration is used, the imposition of exact boundary conditions remains problematic. In order to overcome the above issues, we develop the numerical integration scheme for the NOM in the subsequent section.

6.2 Nonlocal Operator Approximation Scheme

The Taylor series expansion of a function of multiple variables is presented in Sect. 5.1.2. There are three descriptions: multiple summation form, tensor form and vector form for the Taylor series expansion. For the sake of simplicity, the vector form of Taylor series expansion is employed for the subsequent derivation of the NOM. A scalar field u_j at a point $j \in \mathcal{S}_i$ can be obtained by the Taylor series expansion at u_i in d dimensions with maximal derivative order not more than n,

$$u_j = \sum_{(n_1,...,n_d)\in\alpha_d^n} \frac{r_1^{n_1}...r_d^{n_d}}{n_1!...n_d!} u_{i,n_1...n_d} + O(r^{|\alpha|+1}) \tag{6.6}$$

where

$$\mathbf{r}_{ij} = (r_1, ..., r_d) = (x_{j1} - x_{i1}, ..., x_{jd} - x_{id}) \tag{6.7}$$

$$u_{i,n_1...n_d} = \frac{\partial^{n_1+...+n_d} u_i}{\partial x_{i1}^{n_1}...\partial x_{id}^{n_d}} \tag{6.8}$$

$$|\alpha| = \max(n_1 + ... + n_d) \tag{6.9}$$

α_d^n is the list of multi-indices in d-dimensional space with maximal n-order of partial derivative for one index, which can be written as

$$\alpha_d^n = \{(n_1, ..., n_d)|0 \le \sum_{i=1}^{d} n_i \le n,\ n_i \in \mathbb{N}^0, 1 \le i \le d\} \tag{6.10}$$

where $\mathbb{N}^0 = \{0, 1, 2, 3, ...\}$. According to combinatorics, there are $\frac{(n+d)!}{n!d!}$ elements in α_d^n, which can be obtained by the simple code presented in Sect. 5.1.3. For each multi-index $(n_1, ..., n_d) \in \alpha_d^n$, the partial derivative and the polynomial are

$$u_{i,n_1...n_d}, \quad \frac{r_1^{n_1}...r_d^{n_d}}{n_1!...n_d!}, \quad \forall (n_1, ..., n_d) \in \alpha_d^n. \tag{6.11}$$

The original form of Taylor series expansion depends on $\|\mathbf{r}_{ij}\|$ and thus is sensitive to the round-off error, which however can be overcome by introducing the length scale of support $\mathcal{S}_i$ at u_i. The modified Taylor series expansion of Eq. 6.6 can be written as

$$\begin{aligned} u_j &= \sum_{(n_1,...,n_d)\in\alpha_d^n} \frac{r_1^{n_1}...r_d^{n_d}}{h_i^{n_1+...+n_d}} \left(\frac{h_i^{n_1+...+n_d}}{n_1!...n_d!} u_{i,n_1...n_d} \right) + O(r^{n+1}) \\ &= \sum_{(n_1,...,n_d)\in\alpha_d^n} \frac{r_1^{n_1}...r_d^{n_d}}{h_i^{n_1+...+n_d}} u_{i,n_1...n_d}^h + O(r^{n+1}) \end{aligned} \tag{6.12}$$

where h_i is the characteristic length of $\mathcal{S}_i$, and

$$u_{i,n_1...n_d}^h = \frac{h_i^{n_1+...+n_d}}{n_1!...n_d!} u_{i,n_1...n_d}. \tag{6.13}$$

The length scale of the support is similar to the scaling or shifting technique in meshless methods (Huerta et al. 2018). Let $\mathbf{p}_j^h$, $\partial_\alpha^h \mathbf{u}_i$ and $\partial \mathbf{u}_i$ be the list of the polynomials, scaled partial derivatives and partial derivatives, respectively, based on multi-index notation α_d^n in Eq. 6.10, which can be obtained by

$$\mathbf{p}_j^h = \left(1, \frac{r_d}{h}, ..., \frac{r_1^{n_1}...r_d^{n_d}}{h^{n_1+...+n_d}}, ..., \frac{r_1^n}{h^n}\right)^T \tag{6.14}$$

$$\partial_\alpha^h \mathbf{u}_i = (u_i^h, u_{i,0...1}^h, ..., u_{i,n_1...n_d}^h, ..., u_{i,n...0}^h)^T \tag{6.15}$$

$$\partial_\alpha \mathbf{u}_i = (u_i, u_{i,0...1}, ..., u_{i,n_1...n_d}, ..., u_{i,n...0})^T. \tag{6.16}$$

Here, u_i at point i is interpolated by the unknown nodal values in the support. It should be noted that particle-based NOM does not include $1, u_i^h, u_i$ in $\mathbf{p}_j^h, \partial_\alpha^h \mathbf{u}_i, \partial_\alpha \mathbf{u}_i$, respectively. The actual partial derivatives can be recovered by

$$\boldsymbol{H}_i \partial_\alpha \mathbf{u}_i = \partial_\alpha^h \mathbf{u}_i \tag{6.17}$$

where

$$\boldsymbol{H}_i = \mathrm{diag}\left[1, h_i, ..., \frac{h_i^{n_1+...+n_d}}{n_1!...n_d!}, ..., \frac{h_i^n}{n!}\right] \tag{6.18}$$

where $\mathrm{diag}[a_1, ..., a_n]$ denotes a diagonal matrix with diagonal entries starting in the upper left corner are $a_1, ..., a_n$. Therefore, the modified Taylor series expansion can be written as

$$u_j = (\partial_\alpha \mathbf{u}_i)^T \boldsymbol{H}_i \mathbf{p}_j^h, \forall j \in \mathcal{S}_i. \tag{6.19}$$

Summing over u_j with weighted coefficient $w(\mathbf{r}_{ij})(\mathbf{p}_j^h)^T$ for all nodes in support $\mathcal{S}_i$, we obtain

$$\sum_{\mathcal{S}_i} w(\mathbf{r}_{ij}) u_j (\mathbf{p}_j^h)^T = (\partial_\alpha \mathbf{u}_i)^T \boldsymbol{H}_i \sum_{\mathcal{S}_i} w(\mathbf{r}_{ij}) \mathbf{p}_j^h \otimes (\mathbf{p}_j^h)^T \tag{6.20}$$

where $w(\mathbf{r}_{ij})$ is the weighting function. Therefore, the nonlocal operator $\tilde{\partial}_\alpha \mathbf{u}_i$ at point i can be obtained as

$$\tilde{\partial}_\alpha \mathbf{u}_i := \mathbf{K}_i \sum_{\mathcal{S}_i} w(\mathbf{r}_{ij}) \mathbf{p}_j^h u_j \tag{6.21}$$

where

$$\mathbf{K}_i = \boldsymbol{H}_i^{-1} \left(\sum_{\mathcal{S}_i} w(\mathbf{r}_{ij}) \mathbf{p}_j^h \otimes (\mathbf{p}_j^h)^T\right)^{-1}. \tag{6.22}$$

We write the nodes in $\mathcal{S}_i$ as

$$\mathcal{S}_i = \{j_1, ..., j_k, ..., j_{n_i}\} \tag{6.23}$$

where $j_1, ..., j_k, ..., j_{n_i}$ are the global indices of neighbors of point i, n_i is the number of nodes in $\mathcal{S}_i$. The matrix form of Eq. 6.21 can be written as

$$\partial_\alpha \mathbf{u}_i = \mathbf{K}_i \mathbf{p}^h_{wi} \Delta \mathbf{u}_i \tag{6.24}$$

where

$$\mathbf{K}_i = \boldsymbol{H}_i^{-1}\Big(\sum_{j\in\mathcal{S}_i} w(\mathbf{r}_{ij})\mathbf{p}^h_j \otimes (\mathbf{p}^h_j)^T\Big)^{-1}, \tag{6.25}$$

$$\mathbf{p}^h_{wi} = \Big(w(\mathbf{r}_{ij_1})\mathbf{p}^h_{ij_1}, ..., w(\mathbf{r}_{ij_{n_i}})\mathbf{p}^h_{ij_{n_i}}\Big) \tag{6.26}$$

$$\Delta \mathbf{u}_i = (u_{j_1}, ..., u_{j_k}, ..., u_{j_{n_i}})^T. \tag{6.27}$$

The nonlocal operator provides all the partial derivatives with maximal order for single index up to n. For convenience, we may write the operator matrix as

$$\mathbf{B}_i = \mathbf{K}_i \mathbf{p}^h_{wi}. \tag{6.28}$$

Each term in $\partial_\alpha \mathbf{u}_i$ corresponds to the row of $\boldsymbol{K}_i \mathbf{p}^h_{wi}$ multiplied with $\Delta \mathbf{u}_i$. Equation 6.24 can be used to replace the differential operators in PDEs to form the algebraic system of equations. This is equivalent to solving the strong form of the PDEs. Alternatively, the weak form of the PDE can be solved by exploiting the weighted residual method or variational formulations as in Ren et al. (2020). This in turn requires the variation of $\partial_\alpha \mathbf{u}_i$ in Eq. 6.24. On the other hand, the nonlocal operator $\partial_\alpha \mathbf{u}_i$ in discrete form can be obtained by least squares. Consider the weighted square sum of the Taylor series expansion in $\mathcal{S}_i$,

$$\mathcal{F}_i(\mathbf{u}) = \sum_{j\in\mathcal{S}_i} w(\mathbf{r}_{ij})\big(u_j - (\mathbf{p}^h_j)^T \partial^h_\alpha \mathbf{u}_i\big)^2 \tag{6.29}$$

$$\begin{aligned} &= \sum_{j\in\mathcal{S}_i} w(\mathbf{r}_{ij})\big(u_j^2 + \partial^h_\alpha \mathbf{u}_i^T \mathbf{p}^h_j (\mathbf{p}^h_j)^T \partial^h_\alpha \mathbf{u}_i - 2u_j (\mathbf{p}^h_j)^T \partial^h_\alpha \mathbf{u}_i\big) \\ &= \sum_{j\in\mathcal{S}_i} w(\mathbf{r}_{ij})u_j^2 + \partial_\alpha \mathbf{u}_i^T \boldsymbol{H}_i\Big(\sum_{j\in\mathcal{S}_i} w(\mathbf{r}_{ij})\mathbf{p}^h_j(\mathbf{p}^h_j)^T\Big)\boldsymbol{H}_i \partial_\alpha \mathbf{u}_i - 2\Delta \mathbf{u}_i^T (\mathbf{p}^h_{wi})^T \boldsymbol{H}_i \partial_\alpha \mathbf{u}_i \end{aligned} \tag{6.30}$$

$\frac{\partial \mathcal{F}_i(\mathbf{u})}{\partial(\partial_\alpha \mathbf{u}_i)} = 0$ leads to

$$\partial_\alpha \mathbf{u}_i = \boldsymbol{H}_i^{-1}\Big(\sum_{j\in\mathcal{S}_i} w(\mathbf{r}_{ij})\mathbf{p}^h_j(\mathbf{p}^h_j)^T\Big)^{-1} \mathbf{p}^h_{wi} \Delta \mathbf{u}_i = \mathbf{K}_i \mathbf{p}^h_{wi} \Delta \mathbf{u}_i \tag{6.31}$$

which is identical to Eq. 6.24. It should be noted that the approximation scheme of NOM does not satisfy the Kronecker-delta property, which causes some difficulties for the enforcement of Dirichlet boundary conditions. In meshless methods, Dirichlet boundary conditions can be imposed in several ways. Among the most popular schemes are the penalty method, the Lagrange multiplier method and the method based on the modified variational principle. Herein, we adopt the modified variational principle to enforce the boundary conditions. Therefore, we decompose the functional into two parts related to the domain and the boundary.

$$\mathcal{F} = \int_{\Omega} \phi \mathrm{d}\Omega + \sum_{i} \int_{\Gamma_i} \bar{\phi}_i \mathrm{d}\Gamma \tag{6.32}$$

where ϕ, $\bar{\phi}_i$ are the functionals defined in the domain and on the boundaries, respectively. The first and second variational derivatives of $\mathcal{F}$ on the unknown functions can be derived using mathematical software, e.g. Mathematica (Wolfram 1999). Combined with the nonlocal operator matrix, the residual vector and tangent stiffness matrix can be obtained.

The main steps to find the stationary point of functional in Eq. 6.32 based on NOM are as follows:

1. Depending on the used integration scheme, calculate the weights and coordinates of all integration points in the domain and on the boundaries.
2. For each integration point, find the support list by nodes, as shown in Fig. 6.2. Use the support list to construct the operator matrix in Eq. 6.24.
3. Based on the partial derivatives in the functional, select the required operator matrix and construct the residual and tangent stiffness matrix.
4. Solve the algebraic system of equations.

We employ Gauss quadrature and employ available FE mesh generators to construct the background mesh. Let us consider for instance a tetrahedral background mesh. The integration point of the linear tetrahedron is expressed in terms of (ξ, η, ζ, w_c), where (ξ, η, ζ) are the local coordinates and w_c is the weighted coefficient. The global coordinate of the integration point can be calculated as

$$(x, y, z) = \sum_{j=1}^{4} N_j(\xi, \eta, \zeta)(x_j, y_j, z_j) \tag{6.33}$$

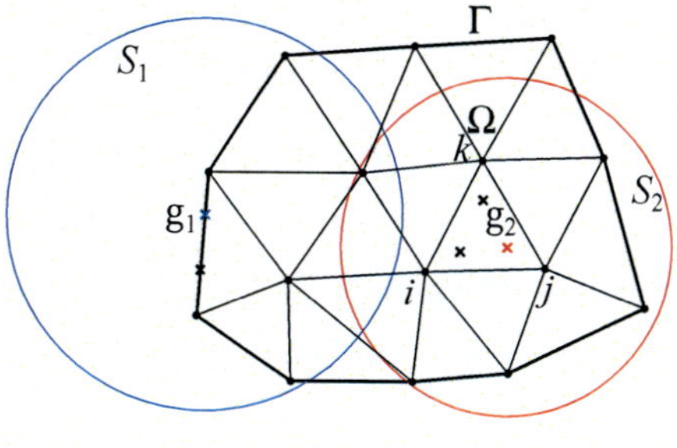

Fig. 6.2 Supports for integration points on the boundary and in the domain

where (x_j, y_j, z_j) are the jth nodal coordinates in the element and N_j is the shape function of the 4-node tetrahedron. The weight of the integration point is

$$w = w_c|\boldsymbol{J}| = w_c \begin{vmatrix} \frac{\partial x}{\partial \xi} & \frac{\partial y}{\partial \xi} & \frac{\partial z}{\partial \xi} \\ \frac{\partial x}{\partial \eta} & \frac{\partial y}{\partial \eta} & \frac{\partial z}{\partial \eta} \\ \frac{\partial x}{\partial \zeta} & \frac{\partial y}{\partial \zeta} & \frac{\partial z}{\partial \zeta} \end{vmatrix}. \tag{6.34}$$

The weight of the integration point on the boundary element with $\xi = 1$ is

$$w = w_c A = w_c \left[\left(\frac{\partial y}{\partial \eta}\frac{\partial z}{\partial \zeta} - \frac{\partial y}{\partial \zeta}\frac{\partial z}{\partial \eta} \right)^2 + \left(\frac{\partial z}{\partial \eta}\frac{\partial x}{\partial \zeta} - \frac{\partial z}{\partial \xi}\frac{\partial x}{\partial \eta} \right)^2 + \left(\frac{\partial x}{\partial \eta}\frac{\partial y}{\partial \xi} - \frac{\partial x}{\partial \xi}\frac{\partial y}{\partial \eta} \right)^2 . \right]^{\frac{1}{2}} \tag{6.35}$$

The weight on $\eta = 1, \zeta = 1$ can be obtained by coordinate permutation. Note that the background mesh is only required for integration and NOM does not require the derivatives of the FE shape functions. Therefore, the shape functions of low-order elements (i.e. triangular/tetrahedral or quadrilateral/hexahedral element) are sufficient for the solution of higher order PDEs. For simplicity, we employ linear triangle/tetrahedral or quadrilateral elements in the chapter.

It can be seen that the support domain of each integration point is constructed by the nearby nodes, other than the integration points. The node set serves as an approximation scheme for each Gauss point. The original form of NOM by nodal integration scheme becomes a special case of the current method when the integration points are located at the nodes. On the other hand, dual-support is not required for integration points since the integration points do not support any other nodes. We point out again that NOM obtains all partial derivatives without shape functions.

6.3 Gradient Solid Theory

6.3.1 *Linear Gradient Elasticity*

In order to demonstrate the capability of the current method, we apply the modified variational principle to solve the linear gradient solid problem. The internal energy functional of an isotropic linear gradient elastic solid (Aravas 2011) can be written as

$$\mathcal{F}_{int} = \int_\Omega \phi(\boldsymbol{\varepsilon}, \nabla\boldsymbol{\varepsilon})\,\mathrm{d}\Omega, \;\; \phi(\boldsymbol{\varepsilon}, \nabla\boldsymbol{\varepsilon}) = \frac{1}{2}\boldsymbol{\sigma} : \boldsymbol{\varepsilon} + \frac{l^2}{2}\nabla\boldsymbol{\sigma} \vdots \nabla\boldsymbol{\varepsilon} \tag{6.36}$$

where l is the gradient coefficient, $\boldsymbol{\varepsilon} = \frac{1}{2}\left(\nabla \boldsymbol{u} + (\nabla \boldsymbol{u})^T\right)$, $\mathrm{tr}\boldsymbol{\varepsilon} = \nabla \cdot \boldsymbol{u}$ and $\boldsymbol{\sigma}$ is the Cauchy stress. The Cauchy stress $\boldsymbol{\sigma}$, work-conjugate to $\boldsymbol{\varepsilon}$, and the double stress $\boldsymbol{\mu}$, work-conjugate to $\nabla\boldsymbol{\varepsilon}$, are derived from the partial derivatives of the strain energy density with respect to $\boldsymbol{\varepsilon}$ and $\nabla\boldsymbol{\varepsilon}$ (Khakalo and Niiranen 2017), respectively, as

$$\boldsymbol{\sigma} = \partial_{\boldsymbol{\varepsilon}}\phi = 2\mu\varepsilon + \lambda \mathrm{tr}\varepsilon\boldsymbol{I},\ \boldsymbol{\mu} = \partial_{\nabla\boldsymbol{\varepsilon}}\phi = l^2\nabla\boldsymbol{\sigma} \tag{6.37}$$

where λ, μ are the Lame constants and $\boldsymbol{I}$ is the identity matrix. The external work is

$$\begin{aligned}\mathcal{F}_{ext} = &\int_{\Omega} \boldsymbol{b}\cdot\boldsymbol{u}\mathrm{d}\Omega + \int_{\partial\Omega_P} \bar{\boldsymbol{P}}\cdot\boldsymbol{u}\mathrm{d}\Gamma + \int_{\partial\Omega_R} \bar{\boldsymbol{R}}\cdot(\boldsymbol{n}\cdot\nabla\boldsymbol{u})\mathrm{d}\Gamma + \oint_{\partial\partial\Omega_E} \bar{\boldsymbol{E}}\cdot\boldsymbol{u}\mathrm{d}l\\ &+ \int_{\partial\Omega_u} \boldsymbol{P}\cdot(\boldsymbol{u}-\bar{\boldsymbol{u}})\mathrm{d}\Gamma + \int_{\partial\Omega_{\nabla\mathbf{u}}} \boldsymbol{R}\cdot(\boldsymbol{n}\cdot\nabla\boldsymbol{u} - \frac{\partial\bar{\boldsymbol{u}}}{\partial\boldsymbol{n}})\mathrm{d}\Gamma\end{aligned} \tag{6.38}$$

where $\boldsymbol{b}$ stands for the body force per unit volume, $\bar{\boldsymbol{P}}$ the traction force on $\partial\Omega_P$, $\bar{\boldsymbol{R}}$ the double traction force on $\partial\Omega_R$, $\bar{\boldsymbol{E}}$ the external line loading at possible corners of the wedges $\partial\partial\Omega_E$, $\boldsymbol{P}$ the work-conjugate to the fixed displacement $\bar{\boldsymbol{u}}$ on $\partial\Omega_u$, $\boldsymbol{R}$ the work-conjugate to the fixed normal displacement gradient $\frac{\partial\bar{\boldsymbol{u}}}{\partial\boldsymbol{n}} := \overline{\boldsymbol{n}\cdot\nabla\boldsymbol{u}}$ on $\partial\Omega_{\nabla\mathbf{u}}$.

$$\boldsymbol{P} = \boldsymbol{n}\cdot\left(\boldsymbol{\sigma} - l^2\Delta\boldsymbol{\sigma}\right) - l^2\nabla_s\cdot(\boldsymbol{n}\cdot\nabla\boldsymbol{\sigma}) + l^2\left(\nabla_s\cdot\boldsymbol{n}\right)\boldsymbol{n}\otimes\boldsymbol{n} : \nabla\boldsymbol{\sigma} \text{ on } \partial\Omega_P \tag{6.39}$$

$$\boldsymbol{R} = l^2\boldsymbol{n}\otimes\boldsymbol{n} : \nabla\boldsymbol{\sigma} \text{ on } \partial\Omega_R \tag{6.40}$$

where $\Delta = \nabla^2$ denotes the Laplacian, ∇_s is the surface nabla-operator: $\nabla_s = \nabla - \boldsymbol{n}\otimes\boldsymbol{n}\cdot\nabla$. For simplicity, we ignore the external loadings at possible corners of the wedges in the formulation above; $\delta\mathcal{F}_{int} - \delta\mathcal{F}_{ext} = 0$ and integration by part in the body Ω and on the surface $\partial\Omega$ lead to the governing equation and boundary conditions,

$$\lambda l^2\nabla(\nabla\cdot\nabla\mathrm{tr}\varepsilon) + 2\mu l^2\nabla\cdot(\nabla\cdot(\nabla\varepsilon)) - \nabla\cdot\boldsymbol{\sigma} - \boldsymbol{b} = 0 \text{ in } \Omega \tag{6.41}$$

$$\boldsymbol{P} = \bar{\boldsymbol{P}} \text{ on } \partial\Omega_P \tag{6.42}$$

$$\boldsymbol{R} = \bar{\boldsymbol{R}} \text{ on } \partial\Omega_R \tag{6.43}$$

$$\boldsymbol{u} = \bar{\boldsymbol{u}} \text{ on } \partial\Omega_u \tag{6.44}$$

$$\boldsymbol{n}\cdot\nabla\boldsymbol{u} = \frac{\partial\bar{\boldsymbol{u}}}{\partial\boldsymbol{n}} \text{ on } \partial\Omega_{\nabla u}. \tag{6.45}$$

Therefore, the modified energy functional for linear elasticity solid is

$$\begin{aligned}\mathcal{F} = &\int_{\Omega} \frac{1}{2}\sigma : \varepsilon + \frac{l^2}{2}\nabla\sigma \vdots \nabla\varepsilon\, \mathrm{d}\Omega - \int_{\Omega} \boldsymbol{b}\cdot\boldsymbol{u}\mathrm{d}V - \int_{\partial\Omega_P} \bar{\boldsymbol{P}}\cdot\boldsymbol{u}\mathrm{d}\Gamma - \int_{\partial\Omega_R} \bar{\boldsymbol{R}}\cdot(\boldsymbol{n}\cdot\nabla\boldsymbol{u})\mathrm{d}\Gamma\\ &- \int_{\partial\Omega_u} \boldsymbol{P}\cdot(\boldsymbol{u}-\bar{\boldsymbol{u}})\mathrm{d}\Gamma - \int_{\partial\Omega_{\nabla\mathbf{u}}} \boldsymbol{R}\cdot(\boldsymbol{n}\cdot\nabla\boldsymbol{u} - \frac{\partial\bar{\boldsymbol{u}}}{\partial\boldsymbol{n}})\mathrm{d}\Gamma.\end{aligned} \tag{6.46}$$

The maximum order of partial derivatives in Eq. 6.46 is 3; therefore, the order of nonlocal operators in Eq. 6.24 should be at least 3. The variational derivative of the functional for the unknown functions can be easily obtained by software such as Mathematica. The numerical integration is performed both in the domain and on the boundaries. The support of the integration points on the boundary is identical to that in the domain, except that the integral weights are calculated based on the boundary elements.

6.3.2 Numerical Implementation

Let us consider a 2D problem with independent variables (u, v). The functions in Eq. 6.46 include

$$
\begin{aligned}
&u, u_{,x}, u_{,y}, u_{,xy}, u_{,xx}, u_{,yy}, u_{,xxx}, u_{,xxy}, u_{,xyy}, u_{,yyy}\\
&v, v_{,x}, v_{,y}, v_{,xy}, v_{,xx}, v_{,yy}, v_{,xxx}, v_{,xxy}, v_{,xyy}, v_{,yyy}.
\end{aligned} \tag{6.47}
$$

Let Ω be represented by a series of Gauss points. The ith Gauss point in Ω is associated to the integration weight w_i and global coordinate vector $\mathbf{x}_i$. Likewise, for boundary domain $\partial\Omega$, the weights and coordinates can be obtained by the boundary elements accordingly.

Let

$$
\begin{aligned}
\phi_1(\mathbf{v}_1) &= \frac{1}{2}\boldsymbol{\sigma} : \boldsymbol{\varepsilon}, \text{ in } \Omega, \text{ with } \mathbf{v}_1 = (u_{,x}, u_{,y}, v_{,x}, v_{,y})^T = \mathbf{B}_1 \cdot \boldsymbol{U}\\
\phi_2(\mathbf{v}_2) &= \frac{l^2}{2}\nabla\boldsymbol{\sigma}\vdots\nabla\boldsymbol{\varepsilon}, \text{ in } \Omega, \text{ with } \mathbf{v}_2 = (u_{,xx}, u_{,xy}, u_{,yy}, v_{,xx}, v_{,xy}, v_{,yy})^T = \mathbf{B}_2 \cdot \boldsymbol{U}\\
\phi_3(\mathbf{v}_3) &= -\boldsymbol{b} \cdot \boldsymbol{u} \text{ in } \Omega, \text{ with } \mathbf{v}_3 = (u, v)^T = \mathbf{B}_3 \cdot \boldsymbol{U}\\
\phi_4(\mathbf{v}_4) &= -\bar{\boldsymbol{P}} \cdot \boldsymbol{u} \text{ in } \partial\Omega_P, \text{ with } \mathbf{v}_4 = (u, v)^T = \mathbf{B}_4 \cdot \boldsymbol{U}\\
\phi_5(\mathbf{v}_5) &= -\bar{\boldsymbol{R}} \cdot (\boldsymbol{n} \cdot \nabla\boldsymbol{u}) \text{ in } \partial\Omega_R, \text{ with } \mathbf{v}_5 = (u_{,x}, u_{,y}, v_{,x}, v_{,y})^T = \mathbf{B}_5 \cdot \boldsymbol{U}
\end{aligned} \tag{6.48}
$$

$$
\phi_6(\mathbf{v}_6) = -\boldsymbol{P} \cdot (\boldsymbol{u} - \bar{\boldsymbol{u}}) \text{ in } \partial\Omega_u \tag{6.49}
$$

$$
\begin{aligned}
\text{with } \mathbf{v}_6 = (&u, u_{,x}, u_{,y}, u_{,xx}, u_{,xy}, u_{,yy}, u_{,xxx}, u_{,xxy}, u_{,xyy}, u_{,yyy},\\
&v, v_{,x}, v_{,y}, v_{,xx}, v_{,xy}, v_{,yy}, v_{,xxx}, v_{,xxy}, v_{,xyy}, v_{,yyy})^T = \mathbf{B}_6 \cdot \boldsymbol{U}
\end{aligned}
$$

$$
\phi_7(\mathbf{v}_7) = -\boldsymbol{R} \cdot (\boldsymbol{n} \cdot \nabla\boldsymbol{u} - \frac{\partial\bar{\boldsymbol{u}}}{\partial\boldsymbol{n}}) \text{ in } \partial\Omega_{\nabla u} \tag{6.50}
$$

$$
\text{with } \mathbf{v}_7 = (u_{,x}, u_{,y}, u_{,xx}, u_{,xy}, u_{,yy}, v_{,x}, v_{,y}, v_{,xx}, v_{,xy}, v_{,yy})^T = \mathbf{B}_7 \cdot \boldsymbol{U} \tag{6.51}
$$

where $\boldsymbol{U} = (\Delta\boldsymbol{u}^T, \Delta\boldsymbol{v}^T)^T$ is the vector containing the unknown nodal values in the support of the integration point; $\mathbf{B}_j$ is the operator sub-matrix constructed from $\mathbf{B}$

in Eq. 6.28 based on the index of element of $\mathbf{v}_i$ in $\partial_\alpha \boldsymbol{u}$. Equation 6.46 based on the integration point in the domain and boundaries can be written as

$$\mathcal{F} = \sum_{w_i \in \Omega} w_i(\phi_{i1} + \phi_{i2} + \phi_{i3}) + \sum_{w_i \in \partial\Omega_P} w_i \phi_{i4} + \sum_{w_i \in \partial\Omega_R} w_i \phi_{i5} + \sum_{w_i \in \partial\Omega_u} w_i \phi_{i6} + \sum_{w_i \in \partial\Omega_{\nabla u}} w_i \phi_{i7}. \quad (6.52)$$

The first variation leads to

$$\begin{aligned}
\delta\mathcal{F} &= \sum_{w_i \in \Omega} w_i \left(\frac{\partial \phi_{i1}}{\partial \mathbf{v}_{i1}} \delta \mathbf{v}_{i1} + \frac{\partial \phi_{i2}}{\partial \mathbf{v}_{i2}} \delta \mathbf{v}_{i2} + \frac{\partial \phi_{i3}}{\partial \mathbf{v}_{i3}} \delta \mathbf{v}_{i3} \right) + \sum_{w_i \in \partial\Omega_P} w_i \frac{\partial \phi_{i4}}{\partial \mathbf{v}_{i4}} \delta \mathbf{v}_{i4} \\
&+ \sum_{w_i \in \partial\Omega_R} \frac{\partial \phi_{i5}}{\partial \mathbf{v}_{i5}} \delta \mathbf{v}_{i5} + \sum_{w_i \in \partial\Omega_u} w_i \frac{\partial \phi_{i6}}{\partial \mathbf{v}_{i6}} \delta \mathbf{v}_{i6} + \sum_{w_i \in \partial\Omega_{\nabla u}} w_i \frac{\partial \phi_{i7}}{\partial \mathbf{v}_{i7}} \delta \mathbf{v}_{i7} \\
&= \sum_{w_i \in \Omega} w_i \left(\frac{\partial \phi_{i1}}{\partial \mathbf{v}_{i1}} \mathbf{B}_{i1} + \frac{\partial \phi_{i2}}{\partial \mathbf{v}_{i2}} \mathbf{B}_{i2} + \frac{\partial \phi_{i3}}{\partial \mathbf{v}_{i3}} \mathbf{B}_{i3} \right) \delta U_i + \sum_{w_i \in \partial\Omega_P} w_i \frac{\partial \phi_{i4}}{\partial \mathbf{v}_{i4}} \mathbf{B}_{i4} \delta U_i \\
&+ \sum_{w_i \in \partial\Omega_R} \frac{\partial \phi_{i5}}{\partial \mathbf{v}_{i5}} \mathbf{B}_{i5} \delta U_i + \sum_{w_i \in \partial\Omega_u} w_i \frac{\partial \phi_{i6}}{\partial \mathbf{v}_{i6}} \mathbf{B}_{i6} \delta U_i + \sum_{w_i \in \partial\Omega_{\nabla u}} w_i \frac{\partial \phi_{i7}}{\partial \mathbf{v}_{i7}} \mathbf{B}_{i7} \delta U_i
\end{aligned} \quad (6.53)$$

which can be used to construct the residual of the functional. The second variation yields

$$\begin{aligned}
\delta^2\mathcal{F} &= \sum_{w_i \in \Omega} w_i \left(\delta \mathbf{v}_{i1}^T \frac{\partial^2 \phi_{i1}}{\partial \mathbf{v}_{i1}^2} \delta \mathbf{v}_{i1} + \delta \mathbf{v}_{i2}^T \frac{\partial^2 \phi_{i2}}{\partial \mathbf{v}_{i2}^2} \delta \mathbf{v}_{i2} + \delta \mathbf{v}_{i3}^T \frac{\partial^2 \phi_{i3}}{\partial \mathbf{v}_{i3}^2} \delta \mathbf{v}_{i3} \right) + \sum_{w_i \in \partial\Omega_P} w_i \delta \mathbf{v}_{i4}^T \frac{\partial^2 \phi_{i4}}{\partial \mathbf{v}_{i4}^2} \delta \mathbf{v}_{i4} \\
&+ \sum_{w_i \in \partial\Omega_R} w_i \delta \mathbf{v}_{i5}^T \frac{\partial^2 \phi_{i5}}{\partial \mathbf{v}_{i5}^2} \delta \mathbf{v}_{i5} + \sum_{w_i \in \partial\Omega_u} w_i \delta \mathbf{v}_{i6}^T \frac{\partial^2 \phi_{i6}}{\partial \mathbf{v}_{i6}^2} \delta \mathbf{v}_{i6} + \sum_{w_i \in \partial\Omega_{\nabla u}} w_i \delta \mathbf{v}_{i7}^T \frac{\partial^2 \phi_{i7}}{\partial \mathbf{v}_{i7}^2} \delta \mathbf{v}_{i7} \\
&= \sum_{w_i \in \Omega} \delta U_i^T w_i \left(\mathbf{B}_{i1}^T \frac{\partial^2 \phi_{i1}}{\partial \mathbf{v}_{i1}^2} \mathbf{B}_{i1} + \mathbf{B}_{i2}^T \frac{\partial^2 \phi_{i2}}{\partial \mathbf{v}_{i2}^2} \mathbf{B}_{i2} + \mathbf{B}_{i3}^T \frac{\partial^2 \phi_{i3}}{\partial \mathbf{v}_{i3}^2} \mathbf{B}_{i3} \right) \delta U_i \\
&+ \sum_{w_i \in \partial\Omega_P} \delta U_i^T w_i \mathbf{B}_{i4}^T \frac{\partial^2 \phi_{i4}}{\partial \mathbf{v}_{i4}^2} \mathbf{B}_{i4} \delta U_i + \sum_{w_i \in \partial\Omega_R} \delta U_i^T w_i \mathbf{B}_{i5}^T \frac{\partial^2 \phi_{i5}}{\partial \mathbf{v}_{i5}^2} \mathbf{B}_{i5} \delta U_i \\
&+ \sum_{w_i \in \partial\Omega_u} \delta U_i^T w_i \mathbf{B}_{i6}^T \frac{\partial^2 \phi_{i6}}{\partial \mathbf{v}_{i6}^2} \mathbf{B}_{i6} \delta U_i + \sum_{w_i \in \partial\Omega_{\nabla u}} \delta U_i^T w_i \mathbf{B}_{i7}^T \frac{\partial^2 \phi_{i7}}{\partial \mathbf{v}_{i7}^2} \mathbf{B}_{i7} \delta U_i
\end{aligned} \quad (6.54)$$

where $w_i \mathbf{B}_{ik}^T \frac{\partial^2 \phi_{ik}}{\partial \mathbf{v}_{ik}^2} \mathbf{B}_{ik}$, $1 \leq k \leq 7$ is the tangent stiffness matrix of the integration point in the respective domain.

6.4 Numerical Examples

In this section, four numerical examples in 2D or 3D are given to validate the proposed NOM formulation. The weighting function $W(\mathbf{r}_{ij}) = 1/r_{ij}^2$ is used in all numerical examples.

6.4.1 *Static Rod in Tension*

Consider a straight prismatic rod of constant section area A and Length L. This example has been examined in Balobanov et al. 2016 and Papargyri-Beskou and Beskos 2010. The material parameters are the elastic modulus E and gradient coefficient l. The governing equation reads

$$AE\left(u'' - l^2 u''''\right) = 0 \text{ in } [0, L] \tag{6.55}$$

with boundary conditions

$$u|_{x=0} = 0, \quad P|_{x=L} \equiv AE\left(u' - l^2 u'''\right)\Big|_{x=L} = P_0 \tag{6.56}$$

$$u'\Big|_{x=L} = 0, \quad R|_{x=0} \equiv AEl^2\, u''\Big|_{x=0} = 0. \tag{6.57}$$

The analytical solution is

$$u(x) = \frac{P_0}{EA}\left(x - l\frac{e^{x/l} - e^{-x/l}}{e^{L/l} + e^{-L/l}}\right). \tag{6.58}$$

In order to compare the numerical result with the exact solution, we model the rod by 2D plate with zero Poisson ratio. The dimensions of the plate are $[0, 5] \times [0, 1]\,\text{m}^2$ and the discretization is depicted in Fig. 6.3. We assume $E = 210$ GPa and $P_0 = 10^6$ Pa and test different values for l. The displacement field is shown in Fig. 6.4. With increasing l, the maximum displacement decreases due to the growing higher order strain energy in the energy functional. When $l = 0$, conventional, i.e. local, elasticity theory is recovered. The slope of the displacement u at $x = L$ is zero indicating that the derivative boundary condition is correctly enforced by the modified energy functional.

The relative error in displacement norm is defined as follows:

$$Re_d = \sqrt{\sum_{i=1}^{ndof}(u_i^h - u_i^{exact})^2 / \sum_{i=1}^{ndof}(u_i^{exact})^2}, \tag{6.59}$$

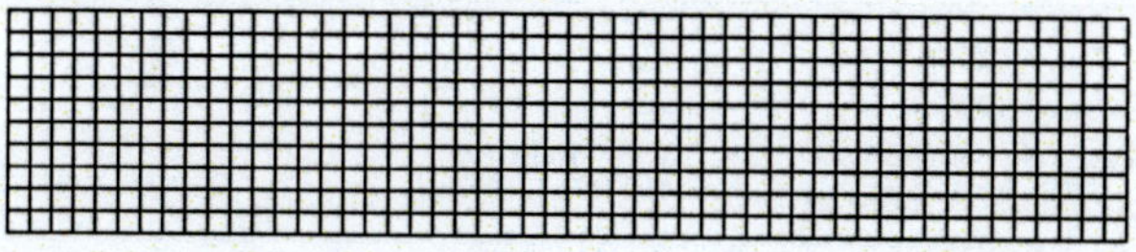

Fig. 6.3 Discretization of the plate. Quadrilateral element in the domain employs 4 Gauss points, line element on the boundary used 3 Gauss points

Fig. 6.4 Displacement u_x of the plate

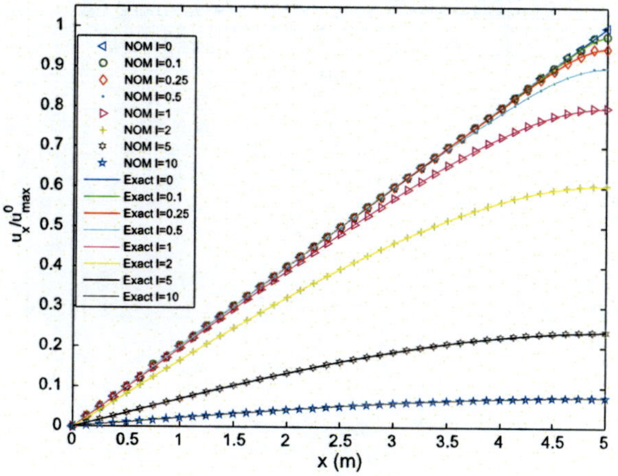

Fig. 6.5 The relative error of displacement for gradient coefficient $l = \{0.1, 0.5, 1, 2\}$

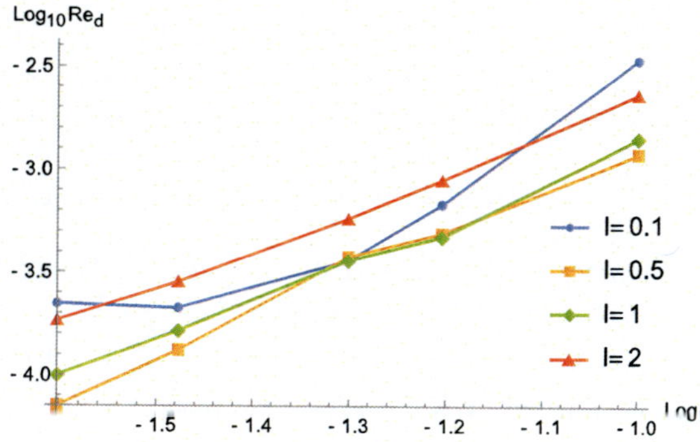

where $ndof$ is the number of degrees of freedom. The relative error Re_d of the displacement norm versus the mesh size $h \in \{0.05, 0.025, 0.0125\}$ for $l \in \{0.5, 1\}$ is depicted in Fig. 6.5.

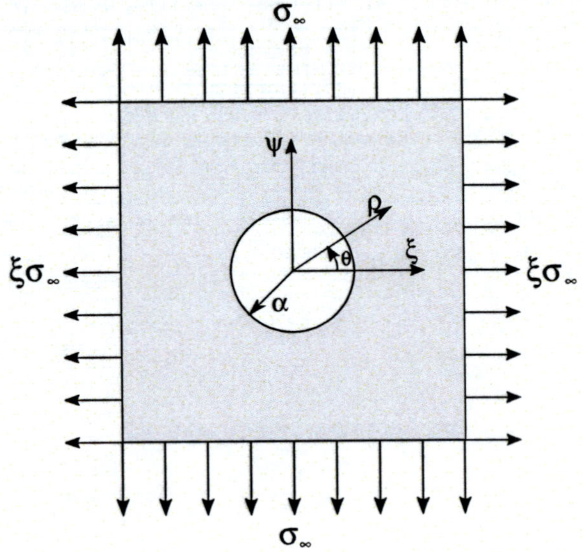

Fig. 6.6 Infinite plate with a hole under distributed uni-axial force at infinity; ξ is a coefficient representing the load in x-direction

6.4.2 Infinite Plate with Hole

In this section, a plate with holes based on gradient linear elasticity theory is solved and compared with the exact solution (Khakalo and Niiranen 2017). The boundary condition of the plate is given in Fig. 6.6. The theoretical solution can be written as

$$\boldsymbol{u}(r,\theta) = (\tilde{u}_1(r) + \hat{u}_1(r)\cos 2\theta)\boldsymbol{e}_r + (\tilde{u}_2(r) + \hat{u}_2(r)\sin 2\theta)\boldsymbol{e}_\theta \tag{6.60}$$

where $\boldsymbol{e}_r, \boldsymbol{e}_\theta$ are the unit basis vector in the polar coordinates, and $\tilde{u}_1(r), \tilde{u}_2(r), \hat{u}_1(r), \hat{u}_2(r)$ are

$$\tilde{u}_1(r) = A_1 r + \frac{A_2}{r} + A_4 K_1\left(\frac{r}{l}\right) \tag{6.61}$$

$$\tilde{u}_2(r) = 0 \tag{6.62}$$

$$\hat{u}_1(r) = -C_1 r + C_7\left[-\alpha_1\frac{1+2\gamma}{\gamma}\frac{a}{r} + \alpha_3\frac{a^3}{r^3} - K_1\left(\frac{r}{l}\right)\right] + C_8\left[-\alpha_2\frac{1+2\gamma}{\gamma}\frac{a}{r} + \alpha_4\frac{a^3}{r^3} + K_1\left(\frac{r}{l}\right)\left(1+8\frac{l^2}{r^2}\right) + 4\frac{l}{r}K_0\left(\frac{r}{l}\right)\right] \tag{6.63}$$

$$\hat{u}_2(r) = C_1 r + C_7\left[\alpha_1\frac{a}{r} + \alpha_3\frac{a^3}{r^3} + K_1\left(\frac{r}{l}\right)\right] + C_8\left[\alpha_2\frac{a}{r} + \alpha_4\frac{a^3}{r^3} + K_3\left(\frac{r}{l}\right)\right] \tag{6.64}$$

where I_i and $K_i, i = 0, 1, 3$, are the modified Bessel functions of the first and second kind, respectively, of order 0,1 and 3; the coefficients A_i, C_i, α_i are given as below.

$$A_1 = \frac{\sigma_\infty}{4(\mu+\lambda)}(1+\xi) \tag{6.65}$$

$$A_2 = \frac{\sigma_\infty}{4\mu}\frac{a^2(1+\xi)\big((1+2\gamma)h^2K_3(h)-2(2+3\gamma)hK_2(h)\big)}{4(1+\gamma)K_1(h)-2(3+5\gamma)hK_2(h)+(1+2\gamma)\big(h^2-12\big)K_3(h)+(2+4\gamma)hK_4(h)} \tag{6.66}$$

$$A_4 = -\frac{\sigma_\infty}{\lambda}\frac{a(1+\xi)}{4(1+\gamma)K_1(h)-2(3+5\gamma)hK_2(h)+(1+2\gamma)\big(h^2-12\big)K_3(h)+(2+4\gamma)hK_4(h)} \tag{6.67}$$

$$C_1 = -\frac{\sigma_\infty}{4\mu}(1-\xi) \tag{6.68}$$

$$C_7 = -\frac{\sigma_\infty}{2\lambda}a(1-\xi)\frac{\beta_2+\beta_4}{\beta_1\beta_4-\beta_2\beta_3} \tag{6.69}$$

$$C_8 = \frac{\sigma_\infty}{2\lambda}a(1-\xi)\frac{\beta_1+\beta_3}{\beta_1\beta_4-\beta_2\beta_3} \tag{6.70}$$

$$\begin{aligned}\beta_1 = {} & -2\gamma K_1(h) + \left[(1+3\gamma)\frac{12}{h} + (5+17\gamma)\frac{h}{2}\right]K_2(h) \\ & - (3+8\gamma)\frac{h^2}{4}K_3(h) - (1+2\gamma)hK_4(h)\end{aligned} \tag{6.71}$$

$$\begin{aligned}\beta_2 = {} & -4\gamma\left(\frac{1}{h}+\frac{24}{h^3}\right)K_0(h) - 2\left[1+2\gamma+(1+6\gamma)\frac{4}{h^2}+\gamma\frac{96}{h^4}\right]K_1(h) \\ & - \left[(2+3\gamma)\frac{h}{2}+(3+5\gamma)\frac{4}{h}+(1+2\gamma)\frac{32}{h^3}\right]K_2(h) \\ & + \left[3(1+2\gamma)\frac{h^2}{4}+13+14\gamma+(7+8\gamma)\frac{8}{h^2}\right]K_3(h) \\ & + \left[(2+7\gamma)\frac{4}{h}+(10\gamma-1)\frac{h}{2}\right]K_4(h) - \left[2(1+2\gamma)+\gamma\frac{h^2}{2}\right]K_5(h)\end{aligned} \tag{6.72}$$

$$\begin{aligned}\beta_3 = {} & 2\gamma K_1(h) + \left[(1+3\gamma)\frac{12}{h}+(1+\gamma)\frac{h}{2}\right]K_2(h) \\ & - \left[6(1+3\gamma)+(1+2\gamma)\frac{h^2}{4}\right]K_3(h) + \gamma hK_4(h)\end{aligned} \tag{6.73}$$

$$\begin{aligned}\beta_4 = {} & -4\gamma\left(\frac{1}{h}+\frac{24}{h^3}\right)K_0(h) - 8\gamma\left(\frac{5}{h^2}+\frac{24}{h^4}\right)K_1(h) \\ & + \frac{\gamma}{2}\left(h-\frac{64}{h^3}\right)K_2(h) + \left[9+16\gamma+(1+2\gamma)\frac{h^2}{4}+(9+10\gamma)\frac{8}{h^2}\right]K_3(h) \\ & - \left[(1-2\gamma)\frac{12}{h}+(1+2\gamma)\frac{h}{2}\right]K_4(h) - 14\gamma K_5(h) + \gamma hK_6(h)\end{aligned} \tag{6.74}$$

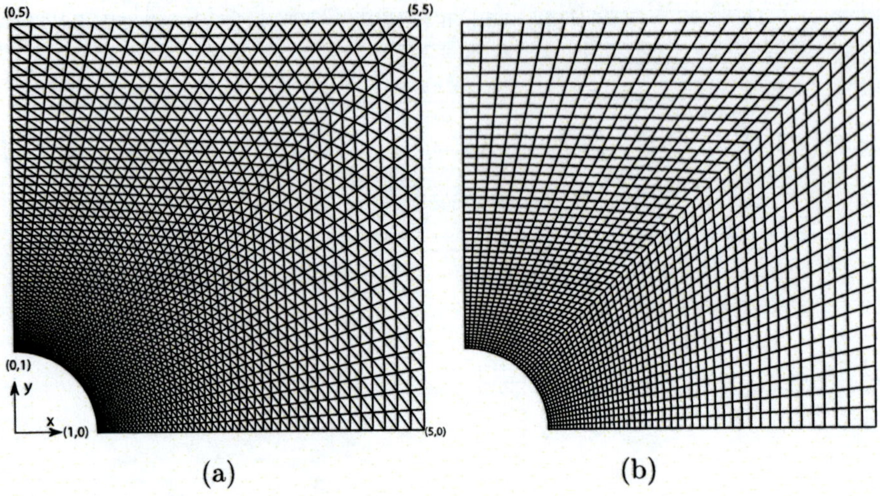

Fig. 6.7 Discritization of the plate

$$\alpha_1 = \frac{1+5\gamma}{2(1+\gamma)} h K_2(h) - \frac{1+3\gamma}{4(1+\gamma)} h^2 K_3(h) \tag{6.75}$$

$$\alpha_2 = \left(\frac{1+2\gamma}{4(1+\gamma)} h^2 + 3\right) K_3(h) + \frac{3\gamma-1}{2(1+\gamma)} h K_4(h) - \frac{\gamma}{4(1+\gamma)} h^2 K_5(h) \tag{6.76}$$

$$\alpha_3 = \frac{1+7\gamma}{12\gamma} h K_2(h) - \frac{1+4\gamma}{24\gamma} h^2 K_3(h) \tag{6.77}$$

$$\alpha_4 = -\frac{2}{3h} K_0(h) - \frac{1}{3}\left(1+\frac{4}{h^2}\right) K_1(h) - \frac{1}{12}\left(h+\frac{8}{h}\right) K_2(h) \\ + \frac{1}{24}\left(\frac{1+2\gamma}{\gamma} h^2 + 4\frac{3+2\gamma}{\gamma}\right) K_3(h) + \frac{6\gamma-1}{12\gamma} h K_4(h) - \frac{h^2}{12} K_5(h) \tag{6.78}$$

where $\gamma = \mu/\lambda$ and $h = a/l$.

One quarter of the plate is modeled as shown in Fig. 6.7. For simplicity, the coefficient $\xi = 0$ is used to denote the zero traction force in x-direction. The fixed displacement (the gradient of the displacement) for the Dirichlet boundary conditions is determined by the exact solution (the gradient of the exact solution) on the truncated sections of the plate. The material parameters include Young's modulus $E = 210 \times 10^9$ Pa, Poisson's ratio $\nu = 0.3$ and gradient coefficient l.

The fixed displacements by Eq. 6.60 on left boundary where $x = 0$ are

$$\bar{\mathbf{u}}|_{x=0} = \Big(0, A_4 K_1\left(\tfrac{y}{l}\right) + A_1 y + \frac{A_2}{y} + C_1 y - C_7(-K_1\left(\tfrac{y}{l}\right) + \frac{\alpha_3}{y^3} - \frac{\alpha_1(2\gamma+1)}{\gamma y}) \\ - C_8((\frac{8l^2}{y^2}+1)K_1\left(\tfrac{y}{l}\right) + \frac{4lK_0\left(\frac{y}{l}\right)}{y} + \frac{\alpha_4}{y^3} - \frac{\alpha_2(2\gamma+1)}{\gamma y})\Big), \tag{6.79}$$

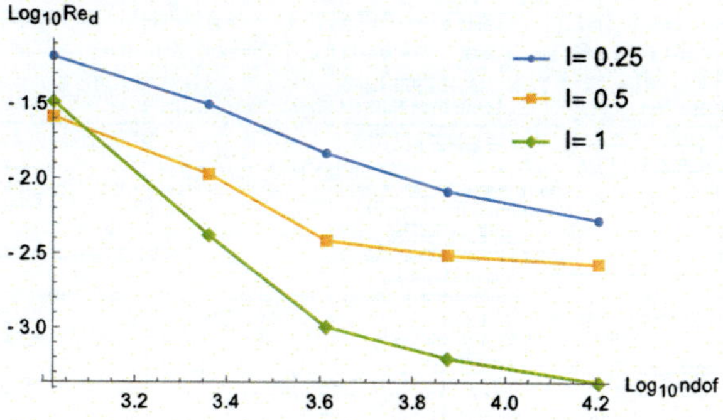

Fig. 6.8 The relative error of displacement norm for triangle element with 3 Gauss points and line element with 3 Gauss points

and on bottom boundary where $y = 0$ are

$$\bar{u}|_{y=0} = \Big(A_4K_1\left(\frac{x}{l}\right) + A_1x + \frac{A_2}{x} - C_1x + C_7(-K_1\left(\frac{x}{l}\right) + \frac{\alpha_3}{x^3} - \frac{\alpha_1(2\gamma+1)}{\gamma x}) + C_8((\frac{8l^2}{x^2} + 1)K_1\left(\frac{x}{l}\right) + \frac{4lK_0\left(\frac{x}{l}\right)}{x} + \frac{\alpha_4}{x^3} - \frac{\alpha_2(2\gamma+1)}{\gamma x}), 0\Big). \tag{6.80}$$

The displacement gradient calculated from Eq. 6.60 on bottom boundary where $y = 0$ is

$$\frac{\partial \bar{u}}{\partial n}|_{y=0} = \frac{1}{x}\Big(0, -A_4K_1\left(\frac{x}{l}\right) - A_1x - \frac{A_2}{x} - \frac{8C_8l^2K_1\left(\frac{x}{l}\right)}{x^2} - \frac{4C_8lK_0\left(\frac{x}{l}\right)}{x} - C_7K_1\left(\frac{x}{l}\right) - C_8K_1\left(\frac{x}{l}\right) - 2C_8K_3\left(\frac{x}{l}\right) - \frac{3\alpha_3C_7}{x^3} - \frac{3\alpha_4C_8}{x^3} + \frac{\alpha_1C_7}{\gamma x} + \frac{\alpha_2C_8}{\gamma x} - C_1x\Big). \tag{6.81}$$

The displacements and its gradient on line $x = 5$ and $y = 5$ can be obtained in the same manner. The values given by $\bar{u}|_{x=0}$, $\bar{u}|_{y=0}$, $\frac{\partial \bar{u}}{\partial n}|_{y=0}$ are used to formulate the boundary functionals in Eqs. 6.49 and 6.50.

The Re_d norm of the displacement field versus the number of degrees of freedom for different gradient coefficients is depicted in Fig. 6.8. The Re_d error of the displacement decreases with increasing gradient coefficient under the same discretization.

The hoop displacement curves are given in Fig. 6.9. With increasing l, the displacements for $r = 1$ decrease. Contour plots of the displacement field for $l \in \{0, 0.5, 1, 2\}$ are depicted in Figs. 6.11, 6.12, 6.13 and 6.14, respectively. In order to test the influence of numerical integration, the relative error of the displacement norm for 8 numerical integration schemes depicted in Fig. 6.10 is calculated.

The integration scheme for 3-node triangular elements with x Gauss points is abbreviated as $T3_x$. Similarly, $Q4_x$ denote 4-node quadrilateral elements with x Gauss points. Let us consider one quarter of the plate and a gradient coefficient $l = 1$. The plate is discretized with 2050 nodes and either triangular or quadrilateral

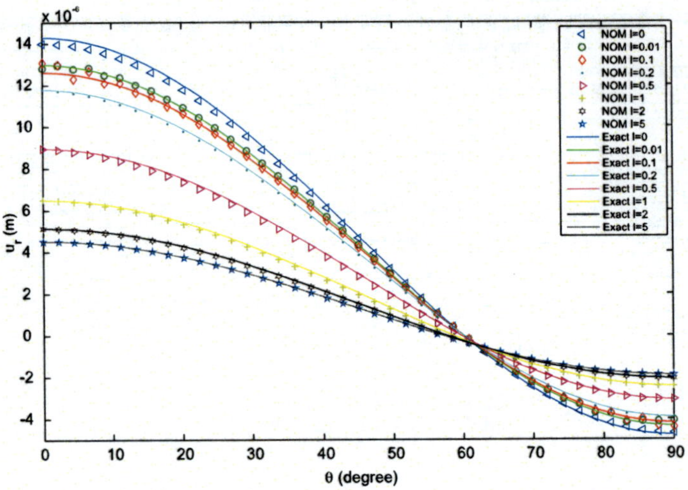

(a) u_r on $r = a$

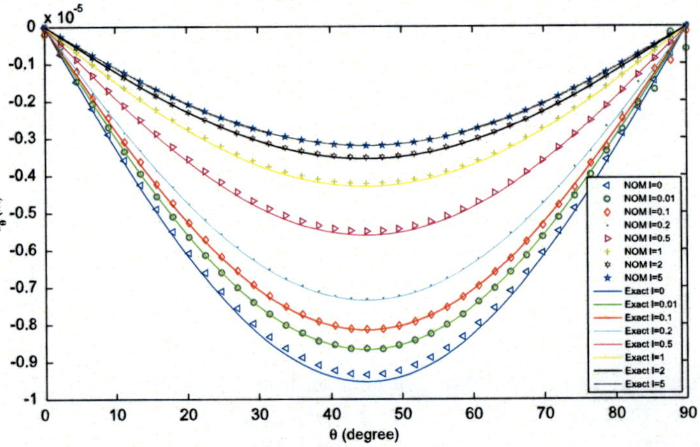

(b) u_θ on $r = a$

Fig. 6.9 Displacement on $r = a$ for different gradient coefficient l

Fig. 6.10 The numerical integration scheme for triangle element and quadrilateral element, where the dots denote the integration points in the element

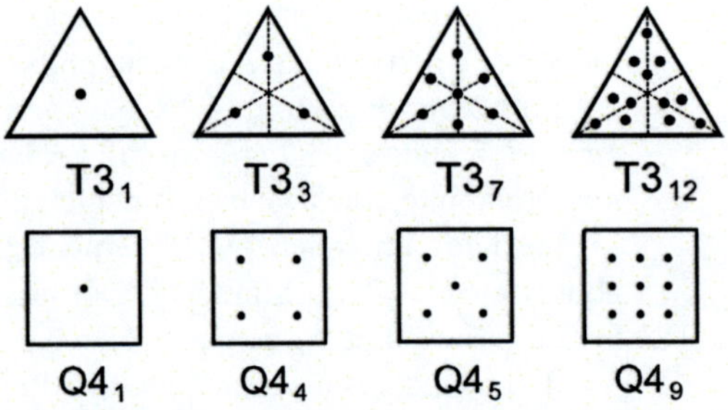

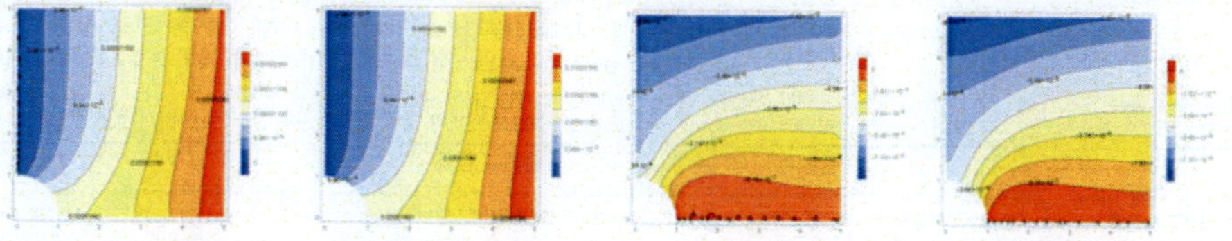

(a) NOM u_x (b) Exact u_x (c) NOM u_y (d) Exact u_y

Fig. 6.11 Contours of displacement field for $l = 0$

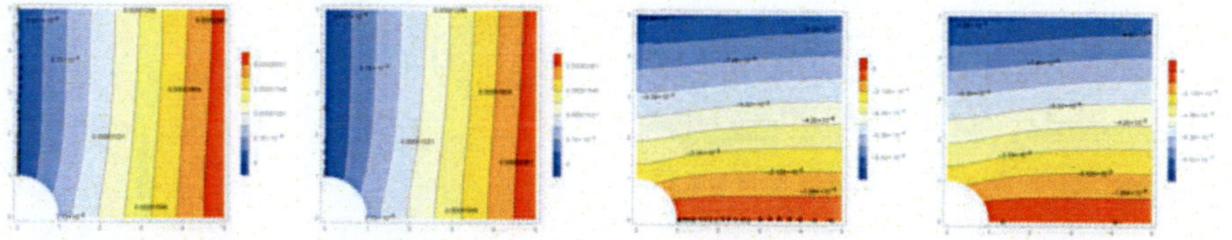

(a) NOM u_x (b) Exact u_x (c) NOM u_y (d) Exact u_y

Fig. 6.12 Contours of displacement field for $l = 0.5$

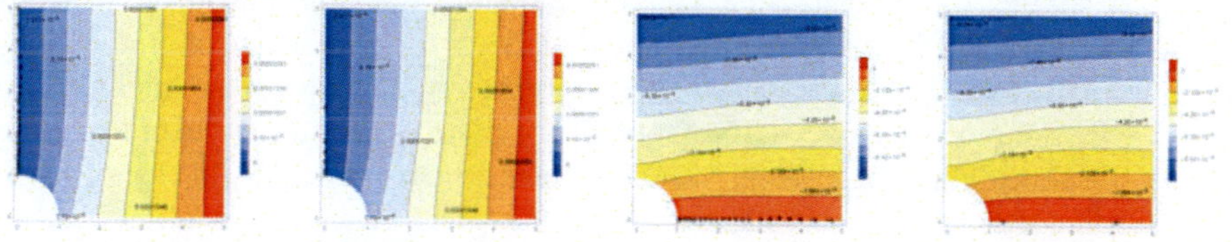

(a) NOM u_x (b) Exact u_x (c) NOM u_y (d) Exact u_y

Fig. 6.13 Contours of displacement field for $l = 1$

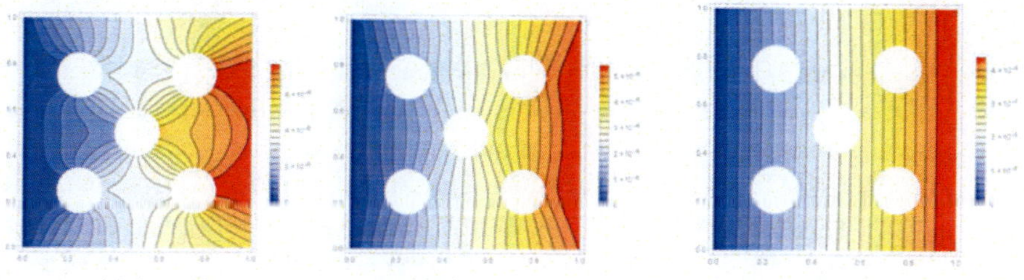

(a) $l = 0.01$ (b) $l = 0.1$ (c) $l = 1$

Fig. 6.14 Contours of displacement field for $l = 2$

elements. The relative errors in the displacement norm for the triangle elements can be found in Tables 6.1 and 6.2 showing that increasing the number of Gauss points improves the accuracy of the results.

Table 6.1 Relative error of displacement norm for discretization by triangle element

Integral scheme	$T3_1$	$T3_3$	$T3_7$	$T3_{12}$
Re_d	1.02e-2	1.48e-3	9.87e-4	5.67e-4

Table 6.2 Relative error of displacement norm for discretization by quadrilateral element

Integral scheme	$Q4_1$	$Q4_4$	$Q4_9$
Re_d	1.71e-3	6.05e-4	1.32e-4

6.4.3 2D Plate with Holes

In this section, we test a plate containing several holes as illustrated in Fig. 6.15. The material parameters are Young's modulus $E = 210$ GPa and Poisson's ratio $\nu = 0.3$. Three gradient coefficients $l = \{0.01, 0.1, 1\}$ are investigated. The nodes on the left side are fixed in all directions and a normal traction force of 10^6 Pa/m is applied on the right side. Contour plots of the displacement in x-direction (u_x) are given in Fig. 6.16. The associated plots for the displacement gradient $\frac{\partial u}{\partial x}$ and $\frac{\partial v}{\partial y}$ are depicted in Figs. 6.17 and 6.18, respectively. Increasing the gradient coefficient yields a more uniform deformation of the plate, where the influence of the holes in the plate diminishes. When $l = 0.01$, the maximum value of $\frac{\partial u}{\partial x}$ is located at the hole, and a stress concentration is expected.

6.4.4 Bending of 3D Block

The last example is a 3D block under bending. One end of the block is fixed in all directions while the other end is subjected to a uniform shear force $(0, 10^6, 0)$ Pa. The material parameters are Young's modulus $E = 210$ GPa and Poisson's ratio $\nu = 0.3$. The block has a dimension of $[0, 5] \times [0, 1] \times [0, 1]$ m^3. We employ 2560 linear hexahedral elements as background mesh for integration. Different gradient coefficients $l = \{0, 0.1, 0.2, 0.5, 1\}$ are tested. The displacement field for $l = 0.2$ is shown in Fig. 6.19. The displacement for nodes on line ($y = 0, z = 1$) is depicted in Fig. 6.20. For the case of gradient coefficient $l = 0$, the displacement curve by NOM agrees well with that by Abaqus.

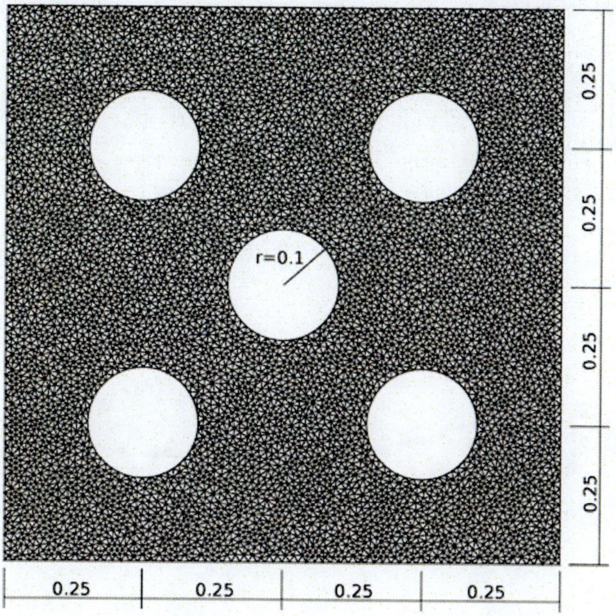

Fig. 6.15 Setup and mesh of the plate

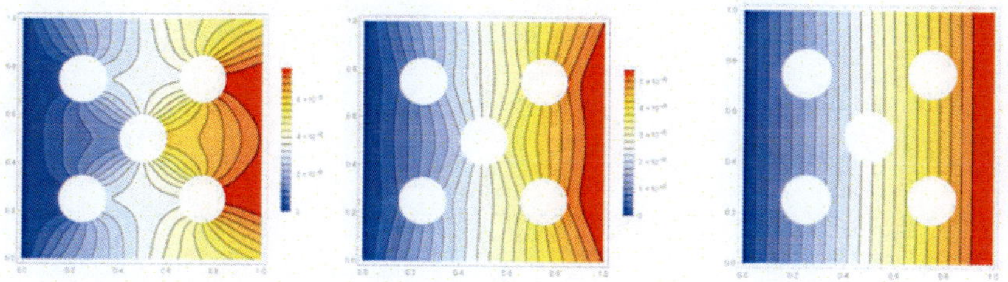

(a) $l = 0.01$ (b) $l = 0.1$ (c) $l = 1$

Fig. 6.16 Contours of displacement in x-direction

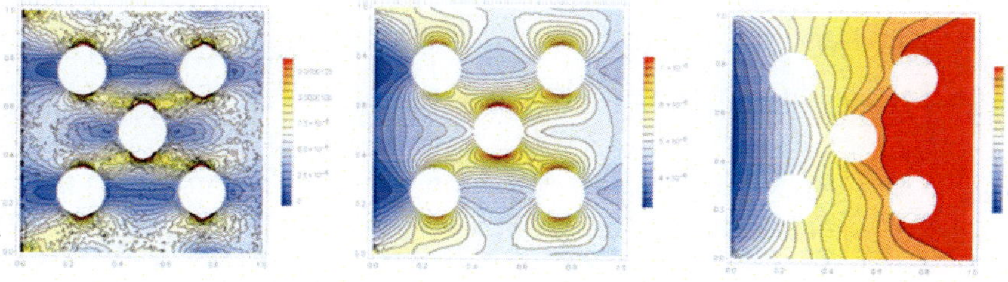

(a) $l = 0.01$ (b) $l = 0.1$ (c) $l = 1$

Fig. 6.17 Contours of $\frac{\partial u}{\partial x}$

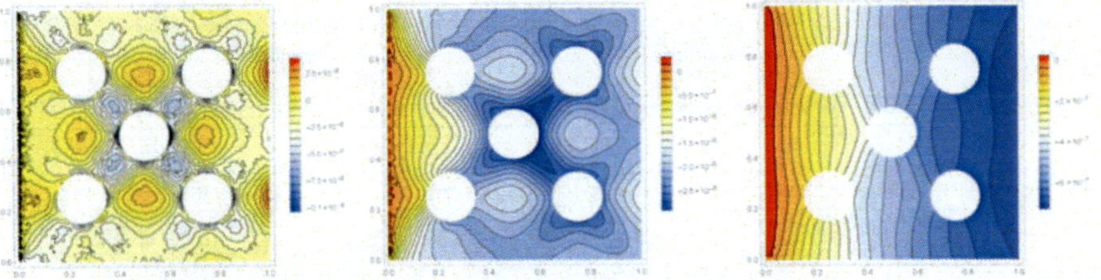

(a) $l = 0.01$ (b) $l = 0.1$ (c) $l = 1$

Fig. 6.18 Contours of $\frac{\partial v}{\partial y}$

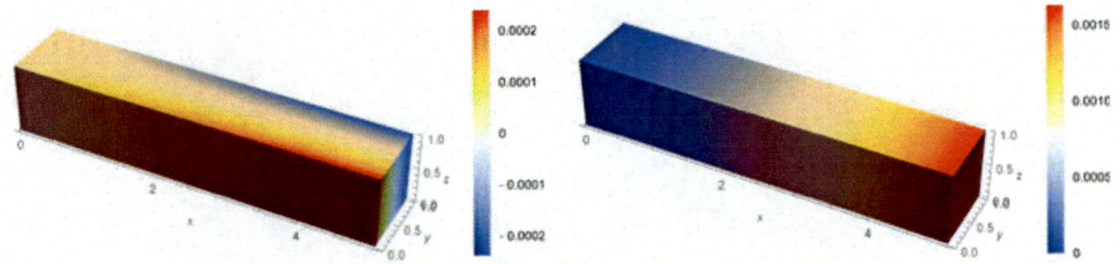

(a) u_x (b) u_y

Fig. 6.19 Displacement for $l = 0.2$

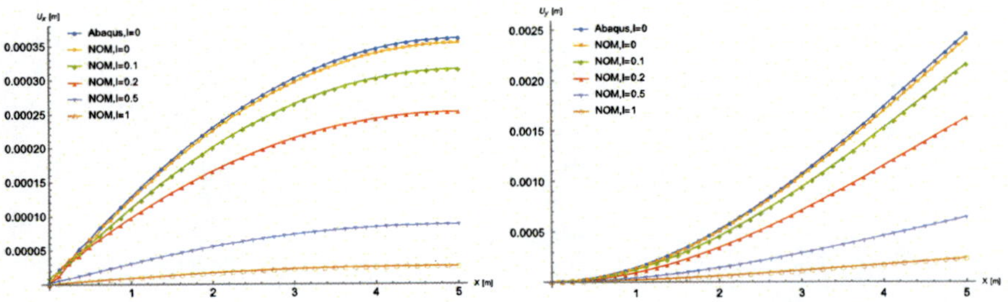

(a) u_x (b) u_y

Fig. 6.20 Displacement for nodes on line ($y = 0, z = 1$). The result by Abaqus is calculated with a very fine mesh of $160 \times 32 \times 32$ linear hexahedron elements

6.5 Concluding Remarks

We have developed a nonlocal operator method with approximation property and applied it to gradient solid problems taking advantage of the modified variational principle. The nonlocal differential operators of various orders are derived and a background mesh is constructed for numerical integration. The nonlocal differential operators are obtained via the Taylor series expansion of a multivariate function in the support. They are independent of the mesh's topology and connectivity for integration. Note also that NOM obtains all the partial derivatives with given maximal order at once without the need of partial derivatives of the FE shape functions.

The NOM approximation scheme enables an accurate numerical integration of the physical problems and various boundary conditions are enforced with ease based on the modified variational principle, which is used to formulate the energy functional.

In order to demonstrate the capability of the new NOM formulation, four numerical examples in gradient linear elasticity theory are presented. The governing equations and boundary conditions are reformulated by the modified variational principles. The functional is solved by a Galerkin formulation. The rod tension in 2D and infinite plate under tension are studied and compared with the exact solutions, where good agreement is observed. The study on the gradient solid shows that the gradient solid has a smaller deformation response compared with the classical elasticity.

References

Aravas N (2011) Plane-strain problems for a class of gradient elasticity models-stress function approach. J Elast 104(1–2):45–70

Balobanov V, Khakalo S, Niiranen J (2016) Isogeometric analysis of gradient-elastic 1D and 2D problems. In: Generalized continua as models for classical and advanced materials, pp 37–45. Springer

Beissel S, Belytschko T (1996) Nodal integration of the element-free galerkin method. Comput Methods Appl Mech Eng 139(1–4):49–74

Dyka C, Ingel R (1995) An approach for tension instability in smoothed particle hydrodynamics (sph). Comput Struct 57(4):573–580

Huerta A, Belytschko T, Fernández-Méndez S, Rabczuk T, Zhuang X, Arroyo M (2018) Meshfree methods. In: Encyclopedia of computational mechanics, 2nd edn, pp 1–38

Khakalo S, Niiranen J (2017) Gradient-elastic stress analysis near cylindrical holes in a plane under bi-axial tension fields. Int J Solids Struct 110:351–366

Monaghan J (1992) Smoothed particle hydrodynamics. Ann Rev Astron Astrophys 30(1):543–574

Papargyri-Beskou S, Beskos D (2010) Static analysis of gradient elastic bars, beams, plates and shells. Open Mech J 4:65–73

Randles P, Libersky L (1996) Smoothed particle hydrodynamics: some recent improvements and applications. Comput Methods Appl Mech Eng 139(1–4):375–408

Ren H, Zhuang X, Rabczuk T (2020) A higher order nonlocal operator method for solving partial differential equations. Comput Methods Appl Mech Eng 367:113132

Ren H, Zhuang X, Rabczuk T (2020) A nonlocal operator method for solving partial differential equations. Comput Methods Appl Mech Eng 358:112621. ISSN: 0045-7825. https://doi.org/10.1016/j.cma.2019.112621

Wolfram S (1999) The mathematica book. Assembly Automation

Chapter 7
Dual-Support Smoothed Particle Hydrodynamics in Solid: Variational Principle and Implicit Formulation

7.1 Introduction

Smoothed particle hydrodynamics (SPH) was introduced by Lucy (Lucy 1977b) and Gingold and Monaghan (Gingold and Monaghan 1977) to solve astrophysical problems such as the formation of stars and the evolution of dust clouds. Due to its flexibility, SPH has been extended to solve various engineering problems, i. e. free-surface flowing (Violeau and Rogers 2016), metal cutting (Limido et al. 2007), impacting simulation (Liu et al. 2006; Zhou et al. 2007), brittle/ductile fractures (Batra and Zhang 2004), plate and shell (Caleyron et al. 2012; Maurel and Combescure 2008), for more complete review of SPH, we refer to (Liu and Liu 2010; Monaghan et al. 2005). One of the key features of SPH is that the kernel approximation can convert the PDEs into simple algebraic equations, on which the solutions of the underlying PDEs are obtained. In contrast with finite element methods (Bonet and Wood 1997; Zienkiewicz et al. 1977) and boundary element methods (Fu et al. 2018a, b), SPH method discretize the continuous domain into a set of particles, each particle is associated with physical quantities such as mass, internal energy and velocity. Since no mesh is required, the SPH is considered as one of the oldest meshless methods. Though some advantages over finite element method (FEM) in arbitrarily large deformations and discontinuity modeling such as fractures, SPH is less accurate and robust than mesh-based methods due to the tensile instabilities and rank-deficiency in the nodal integration approach. A number of different schemes are devised to enhance the stability of SPH, such as artificial viscosity (Monaghan 1992), XSPH time integration scheme (Monaghan 1989), stress points method (Dyka and Ingel 1995; Randles and Libersky 1996) for rank-deficiency problem, Lagrange kernel (Rabczuk et al. 2004a) for tensile instabilities, hourglass force method for zero-energy mode (Ganzenmüller 2015). Meanwhile, various techniques have been developed through the years to alleviate these problems, among which include Corrected Smoothed Particle Method (CSPM) (Chen et al. 1999), Reproducing Kernel Particle Method (RKPM) (Liu et al. 1995), Symmetric Smoothed Particle

T. Rabczuk et al., *Computational Methods Based on Peridynamics and Nonlocal Operators*, Computational Methods in Engineering & the Sciences,
https://doi.org/10.1007/978-3-031-20906-2_7

Hydrodynamics (SSPH) (Zhang and Batra 2009), Optimal Transportation Meshless method (OTM) (Li et al. 2010) and so on.

The purpose of this chapter is to derive by variational principles the dual-support SPH and furthermore construct the tangent stiffness matrix for implicit analysis without zero-energy mode. The content of this chapter is outlined as follows. In Sect. 7.1, we review the basic concepts of support and dual-support and derive the dual-support SPH based on variational principles. In order to remove the hourglass mode, we introduce the hourglass energy functional, based on which the hourglass force, the hourglass residual and tangent stiffness matrix are derived in Sect. 7.2. The implementation and material constitutions are provided in Sect. 7.3. With the aid of the variation of the deformation gradient tensor, the nodal tangent stiffness matrix is simply the matrix multiplication of common terms. In order to verify the implicit scheme, we give in Sect. 7.4 four numerical examples in 2D/3D. The numerical results are compared with the theoretical solutions or finite element results as reference solutions and the good agreements are obtained. The performance of hourglass control is also analyzed in the same section.

7.2 Variational Derivation of Dual-Support SPH

Consider a solid in the initial and current configuration as shown in Fig. 8.1a. Let $\mathbf{X}_i$ be material coordinates in the initial configuration $_0$. A function ϕ mapping any point $\mathbf{X}$ in the reference coordinates to the current coordinate $\boldsymbol{x}$ at time t,

$$\boldsymbol{x} = \phi(\mathbf{X}, t). \tag{7.1}$$

Let $\boldsymbol{x}_i := \phi(\mathbf{X}_i, t)$ and $\boldsymbol{x}_j := \phi(\mathbf{X}_j, t)$ be the spatial coordinates in the current configuration $_t$ of the corresponding particles; $\mathbf{X}_{ij} := \mathbf{X}_j - \mathbf{X}_i$ is initial spatial vector, the relative distance vector between $\mathbf{X}_i$ and $\mathbf{X}_j$; $\mathbf{u}_i := \boldsymbol{x}_i - \mathbf{X}_i$ and $\mathbf{u}_j := \boldsymbol{x}_j - \mathbf{X}_j$ are the displacement vectors for $\mathbf{X}_i$ and $\mathbf{X}_j$, respectively; $\mathbf{u}_{ij} := \mathbf{u}_j - \mathbf{u}_i$ is the relative displacement vector for spatial vector $\mathbf{X}_{ij}$; $\boldsymbol{x}_{ij} := \phi(\mathbf{X}_j, t) - \phi(\mathbf{X}_i, t) = \mathbf{X}_{ij} + \mathbf{u}_{ij}$ is the current spatial vector for $\mathbf{X}_{ij}$.

The governing equations for SPH solid in Lagrangian description include

$$\rho_0 = \rho \det(\mathbf{F}) \tag{7.2}$$

$$\rho_0 \ddot{\boldsymbol{x}} = \nabla_{\mathbf{X}} \cdot \mathbf{P} + \mathbf{f} \tag{7.3}$$

$$\rho_0 \dot{e} = \mathbf{P} : \dot{\mathbf{F}}, \tag{7.4}$$

where $\mathbf{F}$ is the deformation gradient, $\mathbf{P}$ is the first Piola–Kirchhoff stress, e is the internal energy density. In the case of pure elastic solid, the continuity equation and the energy equation can be ignored and only the equation of motion is required.

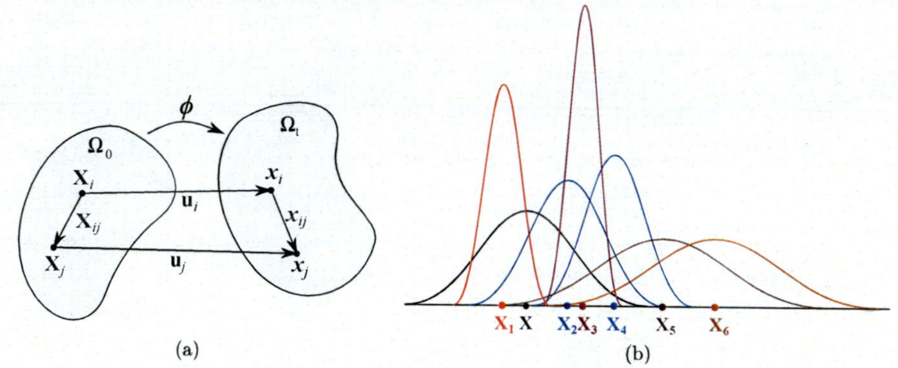

Fig. 7.1 **a** Configuration for deformed body. **b** Schematic diagram for support and dual-support in one dimension with cubic kernel function. $\mathcal{S}_{\mathbf{X}} = \{\mathbf{X}_1, \mathbf{X}_2, \mathbf{X}_3, \mathbf{X}_4\}$, $\mathcal{S}'_{\mathbf{X}} = \{\mathbf{X}_1, \mathbf{X}_2, \mathbf{X}_5\}$

Support $\mathcal{S}_i$ is the domain where any particle $\mathbf{X}_j$ with $X_{ij} = |\mathbf{X}_{ij}| \leq h_i$, where h_i is the smoothing length for particle i. The support $\mathcal{S}_i$ is usually presented by a spherical domain with radius of h_i,

$$\mathcal{S}_i = \{\mathbf{X}_j | X_{ij} \leq h_i\}. \tag{7.5}$$

Dual-support is defined as a union of the points whose supports include $\mathbf{X}_i$, denoted by

$$\mathcal{S}'_i = \{\mathbf{X}_j | \mathbf{X}_i \in \mathcal{S}_j\} = \{\mathbf{X}_j | X_{ij} \leq h_j\}. \tag{7.6}$$

One example to illustrate the support and dual-support is shown in Fig. 7.1b.

SPH approximation for a scalar function in the reference of material configuration can be written as

$$f(\mathbf{X}_i) = \sum_{\mathcal{S}_i} V_j f(\mathbf{X}_j) W_i(\mathbf{X}_{ij}), \tag{7.7}$$

where $W_i(\mathbf{X}_{ij})$ is the SPH kernel function for particle $\mathbf{X}_i$, which only depends on the distance vector between $\mathbf{X}_i$ and $\mathbf{X}_j$. V_j is the volume associated with the particle $\mathbf{X}_j$ in the initial configuration.

The symmetric SPH approximation of a derivative of scalar function f is obtained by the gradient operator on the kernel function,

$$\text{Grad}(f(\mathbf{X}_i)) = \sum_{\mathcal{S}_i} V_j \big(f(\mathbf{X}_j) - f(\mathbf{X}_i)\big) \nabla W_i(\mathbf{X}_{ij}), \tag{7.8}$$

where Grad denotes the gradient operator based on the initial configuration, the gradient of the kernel function is calculated by

$$\nabla W_i(\mathbf{X}_{ij}) = \frac{\mathrm{d}W_i(\mathbf{X}_{ij})}{\mathrm{d}X_{ij}} \frac{\mathbf{X}_{ij}}{X_{ij}}. \tag{7.9}$$

For the condition of zeroth-order and first-order completeness, the corrected kernel gradient is defined as

$$\tilde{\nabla} W_i(\mathbf{X}_{ij}) = \mathbf{K}_i^{-1} \nabla W_i(\mathbf{X}_{ij}), \tag{7.10}$$

where the correction matrix $\mathbf{K}_i$ is defined as

$$\mathbf{K}_i = \sum_{\mathcal{S}_i} V_j \nabla W_i(\mathbf{X}_{ij}) \otimes \mathbf{X}_{ij}. \tag{7.11}$$

The deformation gradient $\mathbf{F}$ for $\mathbf{X}_i$ in SPH is defined as

$$\mathbf{F}_i = \frac{\partial \boldsymbol{x}_i}{\partial \mathbf{X}_i} = \sum_{\mathcal{S}_i} \boldsymbol{x}_{ij} \otimes \tilde{\nabla} W_i(\mathbf{X}_{ij}) V_j. \tag{7.12}$$

The variation of the deformation gradient

$$\delta \mathbf{F}_i = \sum_{\mathcal{S}_i} \delta \boldsymbol{x}_{ij} \otimes \tilde{\nabla} W_i(\mathbf{X}_{ij}) V_j = \sum_{\mathcal{S}_i} (\delta \boldsymbol{x}_j - \delta \boldsymbol{x}_i) \otimes \tilde{\nabla} W_i(\mathbf{X}_{ij}) V_j. \tag{7.13}$$

Let $\mathcal{F}(\mathbf{F}_i)$ be the strain energy density functional that only depends on the deformation gradient. The first-order variation of strain energy $\mathcal{F}(\mathbf{F}_i)$ in $\mathcal{S}_i$ of point $\mathbf{X}_i$ is

$$\begin{aligned} \delta \mathcal{F}(\mathbf{F}_i) &= \partial_{\mathbf{F}_i} \mathcal{F} \cdot \delta \mathbf{F}_i = \partial_{\mathbf{F}_i} \mathcal{F} \cdot \sum_{\mathcal{S}_i} (\delta \boldsymbol{x}_j - \delta \boldsymbol{x}_i) \otimes \tilde{\nabla} W_i(\mathbf{X}_{ij}) V_j \\ &= \mathbf{P}_i \cdot \sum_{\mathcal{S}_i} (\delta \boldsymbol{x}_j - \delta \boldsymbol{x}_i) \otimes \tilde{\nabla} W_i(\mathbf{X}_{ij}) V_j. \end{aligned} \tag{7.14}$$

The first Piola–Kirchhoff stress $\mathbf{P}$ related to the deformation gradient is then given by

$$\mathbf{P} = \partial_{\mathbf{F}} \mathcal{F}(\mathbf{F}). \tag{7.15}$$

The Lagrangian for the system includes the kinetic energy, potential energy (strain energy, the body force energy and external work), and can be expressed as

$$L(\dot{\boldsymbol{x}}, \boldsymbol{x}) = \sum_{V_i \in \Omega_0} \left(\frac{1}{2} \rho \dot{\boldsymbol{x}}_i \cdot \dot{\boldsymbol{x}}_i - \mathcal{F}(\mathbf{F}_i) + \mathbf{b}_0 \cdot (\boldsymbol{x}_i - \mathbf{X}_i) \right) V_i + \int_{\Gamma_0} \mathbf{f}_0 \cdot (\boldsymbol{x} - \mathbf{X}) \mathrm{d}\Gamma_0. \tag{7.16}$$

The external work in time interval $[t_1, t_2]$ is $W^{ext} = \int_{t_1}^{t_2}\int_{\Gamma_0} \mathbf{f}_0 \cdot (\boldsymbol{x} - \mathbf{X})\mathrm{d}\Gamma_0 \mathrm{d}t$. The integral of the Lagrangian L between two instants of time t_1 and t_2 is $S = \int_{t_1}^{t_2} L(\dot{\boldsymbol{x}}, \boldsymbol{x})\mathrm{d}t$. In order to derive the internal force between particles, we neglect the external work for simplicity. Applying the principle of least action, we have

$$
\begin{aligned}
\delta S &= \int_{t_1}^{t_2} \sum_{V_i \in \Omega_0} \Big(\rho \dot{\boldsymbol{x}}_i \cdot \delta \dot{\boldsymbol{x}}_i - \delta \mathcal{F}(\mathbf{F}_i) + \mathbf{b}_0 \cdot \delta(\boldsymbol{x}_i - \mathbf{X}_i)\Big) V_i \mathrm{d}t \\
&= \int_{t_1}^{t_2} \sum_{V_i \in \Omega_0} \Big(- \rho \ddot{\boldsymbol{x}}_i \cdot \delta \boldsymbol{x}_i - \sum_{\mathcal{S}_i} \mathbf{P}_i \cdot (\delta \boldsymbol{x}_j - \delta \boldsymbol{x}_i) \otimes \tilde{\nabla} W_i(\mathbf{X}_{ij}) V_j + \mathbf{b}_0 \cdot \delta \boldsymbol{x}_i \Big) V_i \mathrm{d}t \\
&= \int_{t_1}^{t_2} \sum_{V_i \in \Omega_0} \Big((-\rho \ddot{\boldsymbol{x}}_i + \sum_{\mathcal{S}_i} \mathbf{P}_i \cdot \tilde{\nabla} W_i(\mathbf{X}_{ij}) V_j - \sum_{\mathcal{S}'_i} \mathbf{P}_j \cdot \tilde{\nabla} W_j(\mathbf{X}_{ji}) V_j + \mathbf{b}_0) \cdot \delta \boldsymbol{x}_i \Big) V_i \mathrm{d}t.
\end{aligned}
\tag{7.17}
$$

The derivation considers the boundary condition $\delta \boldsymbol{x}(t_1) = 0,\ \delta \boldsymbol{x}(t_2) = 0$. In the second and third step, the dual-support is considered as follows. In the second step, the term with $\delta \boldsymbol{x}_j$ is the force vector from $\mathbf{X}_i$'s support, but is added to particle $\mathbf{X}_j$; since $\mathbf{X}_j \in \mathcal{S}_i$, $\mathbf{X}_i$ belongs to the dual-support $\mathcal{S}'_j$ of $\mathbf{X}_j$. In the third step, all terms with $\delta \boldsymbol{x}_i$ are collected from other particles whose supports contain $\mathbf{X}_i$ and therefore form the dual-support of $\mathbf{X}_i$. For any $\delta \boldsymbol{x}_i$, the first-order variation $\delta S = 0$ leading to

$$
\rho \ddot{\boldsymbol{x}}_i = \sum_{\mathcal{S}_i} \mathbf{P}_i \cdot \tilde{\nabla} W_i(\mathbf{X}_{ij}) V_j - \sum_{\mathcal{S}'_i} \mathbf{P}_j \cdot \tilde{\nabla} W_j(\mathbf{X}_{ji}) V_j + \mathbf{b}_0, \forall \mathbf{X}_i \in \Omega_0. \tag{7.18}
$$

In the chapter, we use the kernel function of quintic spline

$$
W(\mathbf{r}) = \alpha_d \Big((1 - r/h)_+^5 - 6(\frac{2}{3} - r/h)_+^5 + 15(\frac{1}{3} - r/h)_+^5 \Big) \tag{7.19}
$$

where $r = \|\mathbf{r}\|$, h is the smoothing length scale; $\alpha_d = (3^5/40,\ 3^7 7/478\pi,\ 3^7/40\pi)$ for 1,2,3 dimensional spaces, respectively; $x_+ = \max(0, x)$. For more kernel functions with discussions on their properties, we refer to (Dehnen and Aly 2012). Based on our numerical test, the selection of kernel functions has very small influence on the final result.

7.3 Functional of Hourglass Energy

In order to remove the hourglass mode (zero-energy mode), the conventional SPH adds a penalty term to the force state, in which the penalty force is proportional to the difference between current location of a point and the position predicted by the deformation gradient (Ganzenmüller 2015):

$$\mathbf{T}_i^{hg} \propto \sum_{\mathcal{S}_i} \frac{(\mathbf{F}_i\mathbf{X}_{ij} - \boldsymbol{x}_{ij}) \cdot \boldsymbol{x}_{ij}}{x_{ij}}. \tag{7.20}$$

However, the above formulation is only feasible in the explicit formulation since x_{ij} exists in the denominator.

The displacement field in the neighborhood of a particle is required to be linear. Therefore, it has to be exactly described by the deformation gradient, and the hourglass modes are identified as that part of the displacement field, which is not described by the deformation gradient (Ganzenmüller 2015). In practice, the difference between the current deformed vector $\boldsymbol{x}_{ij}$ and the predicted vector by deformation gradient is $(\mathbf{F}_i\mathbf{X}_{ij} - \boldsymbol{x}_{ij})$. We formulate the hourglass energy based on the difference in the support as follows. Let $\alpha = \frac{\mu}{m_{\mathbf{K}_i}}$ be a coefficient for the hourglass energy, where $m_{\mathbf{K}_i} = \mathrm{tr}(\mathbf{K})$, μ is the shear modulus, the functional for zero-energy mode is defined as

$$\begin{aligned}
\mathcal{F}_i^{hg} &= \alpha \sum_{\mathcal{S}_i} \frac{1}{X_{ij}} \frac{\mathrm{d}W_i(\mathbf{X}_{ij})}{\mathrm{d}X_{ij}} (\mathbf{F}_i\mathbf{X}_{ij} - \boldsymbol{x}_{ij})^T (\mathbf{F}_i\mathbf{X}_{ij} - \boldsymbol{x}_{ij}) V_j \\
&= \alpha \sum_{\mathcal{S}_i} \frac{1}{X_{ij}} \frac{\mathrm{d}W_i(\mathbf{X}_{ij})}{\mathrm{d}X_{ij}} \Big(\mathbf{X}_{ij}^T \mathbf{F}_i^T \mathbf{F}_i \mathbf{X}_{ij} + \boldsymbol{x}_{ij}^T \boldsymbol{x}_{ij} - 2\boldsymbol{x}_{ij}^T \mathbf{F}_i \mathbf{X}_{ij}\Big) V_j \\
&= \alpha \sum_{\mathcal{S}_i} \frac{1}{X_{ij}} \frac{\mathrm{d}W_i(\mathbf{X}_{ij})}{\mathrm{d}X_{ij}} \Big(\mathbf{F}_i^T \mathbf{F}_i : \mathbf{X}_{ij} \otimes \mathbf{X}_{ij} + \boldsymbol{x}_{ij}^T \boldsymbol{x}_{ij} - 2\mathbf{F}_i : \boldsymbol{x}_{ij} \otimes \mathbf{X}_{ij}\Big) V_j \\
&= \alpha \mathbf{F}_i^T \mathbf{F}_i : \sum_{\mathcal{S}_i} \frac{1}{X_{ij}} \frac{\mathrm{d}W_i(\mathbf{X}_{ij})}{\mathrm{d}X_{ij}} \mathbf{X}_{ij} \otimes \mathbf{X}_{ij} V_j + \alpha \sum_{\mathcal{S}_i} \frac{1}{X_{ij}} \frac{\mathrm{d}W_i(\mathbf{X}_{ij})}{\mathrm{d}X_{ij}} \boldsymbol{x}_{ij}^T \boldsymbol{x}_{ij} V_j \\
&\quad - 2\alpha \mathbf{F}_i : \sum_{\mathcal{S}_i} \frac{1}{X_{ij}} \frac{\mathrm{d}W_i(\mathbf{X}_{ij})}{\mathrm{d}X_{ij}} \boldsymbol{x}_{ij} \otimes \mathbf{X}_{ij} V_j \\
&= \alpha \mathbf{F}_i^T \mathbf{F}_i : \sum_{\mathcal{S}_i} \mathbf{X}_{ij} \otimes \nabla W_i(\mathbf{X}_{ij}) V_j + \alpha \sum_{\mathcal{S}_i} \frac{\mathrm{d}W_i(\mathbf{X}_{ij})}{\mathrm{d}X_{ij}} \frac{\boldsymbol{x}_{ij}^T}{X_{ij}} \boldsymbol{x}_{ij} V_j - 2\alpha \mathbf{F}_i : \sum_{\mathcal{S}_i} \boldsymbol{x}_{ij} \otimes \nabla W_i(\mathbf{X}_{ij}) V_j \\
&= \alpha \mathbf{F}_i^T \mathbf{F}_i : \mathbf{K}_i + \alpha \sum_{\mathcal{S}_i} \frac{1}{X_{ij}} \frac{\mathrm{d}W_i(\mathbf{X}_{ij})}{\mathrm{d}X_{ij}} \boldsymbol{x}_{ij}^T \boldsymbol{x}_{ij} V_j - 2\alpha \mathbf{F}_i : (\mathbf{F}_i \mathbf{K}_i) \\
&= \frac{\mu}{m_{\mathbf{K}_i}} \Big(\sum_{\mathcal{S}_i} \frac{\mathrm{d}W_i(\mathbf{X}_{ij})}{\mathrm{d}X_{ij}} \frac{\boldsymbol{x}_{ij}^T \boldsymbol{x}_{ij}}{X_{ij}} V_j - \mathbf{F}_i : (\mathbf{F}_i \mathbf{K}_i) \Big).
\end{aligned} \tag{7.21}$$

The above definition of hourglass energy is similar to the variance in probability theory and statistics. In above derivation, we used the relations: $\mathbf{F}^T\mathbf{F} : \mathbf{K} = \mathbf{F} : (\mathbf{F}\mathbf{K})$, $\mathbf{a}^T\mathbf{M}\mathbf{b} = \mathbf{M} : \mathbf{a} \otimes \mathbf{b}$, $\mathbf{A} : \mathbf{B} = \mathrm{tr}(\mathbf{A}\mathbf{B}^T)$, where capital letter denotes matrix and small letter is column vector. The purpose of $m_{\mathbf{K}}$ is to make the energy functional independent with the support since $\mathbf{K}$ is involved in $\mathbf{F}^T\mathbf{F} : \mathbf{K}$.

In order to derive the residual and tangent stiffness matrix directly, some notation to denote the variation and how the variations are related to the residual and stiffness matrix are introduced subsequently. Assume a functional $\mathcal{F}(\mathbf{u})$, where $\mathbf{u}$ are unknown function vector, the first and second variations can be expressed as

$$\delta\mathcal{F}(\mathbf{u}) = \partial_{\mathbf{u}}\mathcal{F}(\mathbf{u}) \cdot \delta\mathbf{u} = \bar{\delta}\mathcal{F}(\mathbf{u}) \cdot \delta\mathbf{u}$$
$$\delta^2\mathcal{F}(\mathbf{u}) = \partial_{\mathbf{uu}}\mathcal{F}(\mathbf{u}) \cdot \delta\mathbf{u}\delta\mathbf{u} = \bar{\delta}^2\mathcal{F}(\mathbf{u}) \cdot \delta\mathbf{u}\delta\mathbf{u} \tag{7.22}$$

where the special variation $\bar{\delta}\mathcal{F}(\mathbf{u})$ and $\bar{\delta}^2\mathcal{F}(\mathbf{u})$ are defined as

$$\bar{\delta}\mathcal{F}(\mathbf{u}) := \partial_{\mathbf{u}}\mathcal{F}(\mathbf{u}) \tag{7.23}$$
$$\bar{\delta}^2\mathcal{F}(\mathbf{u}) := \partial_{\mathbf{uu}}\mathcal{F}(\mathbf{u}). \tag{7.24}$$

The gradient vector and Hessian matrix represent the residual vector and tangent stiffness matrix of the functional, respectively, with unknown functions $\mathbf{u}$ being the independent variables. Hence,

$$\mathbf{R}(\mathbf{u}) = \bar{\delta}\mathcal{F}(\mathbf{u})$$
$$\mathbf{K}(\mathbf{u}) = \bar{\delta}^2\mathcal{F}(\mathbf{u}).$$

For example, when $\mathbf{u} = [u, v]$, the special variations of functional $\mathcal{F}(u, v)$ are given as

$$\begin{aligned}
\bar{\delta}\mathcal{F}(u, v) &= \partial_u\mathcal{F}\bar{\delta}u + \partial_v\mathcal{F}\bar{\delta}v = [\partial_u\mathcal{F}, \partial_v\mathcal{F}] \\
\bar{\delta}^2\mathcal{F}(u, v) &= \partial_{uu}\mathcal{F}\bar{\delta}u\bar{\delta}u + \partial_{uv}\mathcal{F}\bar{\delta}u\bar{\delta}v + \partial_{vu}\mathcal{F}\bar{\delta}v\bar{\delta}u + \partial_{vv}\mathcal{F}\bar{\delta}v\bar{\delta}v \\
&= \begin{bmatrix} \partial_{uu}\mathcal{F} & \partial_{uv}\mathcal{F} \\ \partial_{vu}\mathcal{F} & \partial_{vv}\mathcal{F} \end{bmatrix}
\end{aligned}$$

where $\bar{\delta}u$ has no other meaning but denotes the index of $\partial_u\mathcal{F}$ in the residual vector by the index of u in the unknown vector. Namely, the term $\partial_v\mathcal{F}\bar{\delta}v$ represents $\partial_v\mathcal{F}$ be in the second location of the residual vector since v is in the second position of $[u, v]$. The term $\partial_{uv}\mathcal{F}\bar{\delta}u\bar{\delta}v$ denotes that the location of $\partial_{uv}\mathcal{F}$ is (1,2), while the term $\partial_{vu}\mathcal{F}\bar{\delta}v\bar{\delta}u$ denotes that the location of $\partial_{vu}\mathcal{F}$ is (2,1). The special first-order and second-order variations of a functional lead to the residual and tangent stiffness matrix directly. The traditional variation can be recovered by the inner product of the special variation and the variation of the unknown vector.

Therefore, the variation of $\mathbf{F}_i : \mathbf{F}_i\mathbf{K}_i$ can be rewritten as

$$\begin{aligned}
\bar{\delta}(\mathbf{F}_i : \mathbf{F}_i\mathbf{K}_i) &= \bar{\delta}(\mathbf{F}_i\mathbf{K}_i : \mathbf{F}_i) = 2\mathbf{F}_i\mathbf{K}_i : \bar{\delta}\mathbf{F}_i \\
&= 2\mathbf{F}_i\mathbf{K}_i : \sum_{\mathcal{S}_i} \bar{\delta}\boldsymbol{x}_{ij} \otimes \tilde{\nabla}W_i(\mathbf{X}_{ij})V_j \\
&= 2\sum_{\mathcal{S}_i} \bar{\delta}\boldsymbol{x}_{ij}^T\mathbf{F}_i\mathbf{K}_i\tilde{\nabla}W_i(\mathbf{X}_{ij})V_j \\
&= 2\sum_{\mathcal{S}_i} \bar{\delta}\boldsymbol{x}_{ij}^T(\mathbf{F}_i\nabla W_i(\mathbf{X}_{ij}))V_j \\
&= 2\sum_{\mathcal{S}_i} (\mathbf{F}_i\nabla W_i(\mathbf{X}_{ij})) \cdot \bar{\delta}\boldsymbol{x}_{ij}V_j.
\end{aligned} \tag{7.25}$$

Then the variation of $\mathcal{F}^{hg}$ is

$$\begin{aligned}\mathbf{R}_i^{hg} &= \bar{\delta}\mathcal{F}_i^{hg}\\ &= \frac{\mu}{m_{\mathbf{K}_i}}\Big(\sum_{S_i}\frac{1}{X_{ij}}\frac{dW_i(\mathbf{X}_{ij})}{dX_{ij}}\bar{\delta}(\boldsymbol{x}_{ij}\cdot\boldsymbol{x}_{ij})V_j - \bar{\delta}(\mathbf{F}_i:\mathbf{F}_i\mathbf{K}_i)\Big)\\ &= \frac{\mu}{m_{\mathbf{K}_i}}\Big(\sum_{S_i}2\frac{1}{X_{ij}}\frac{dW_i(\mathbf{X}_{ij})}{dX_{ij}}\boldsymbol{x}_{ij}\cdot\bar{\delta}\boldsymbol{x}_{ij}V_j - 2\sum_{S_i}(\mathbf{F}_i\nabla W_i(\mathbf{X}_{ij}))\cdot\bar{\delta}\boldsymbol{x}_{ij}V_j\Big)\\ &= \frac{2\mu}{m_{\mathbf{K}_i}}\sum_{S_i}\frac{1}{X_{ij}}\frac{dW_i(\mathbf{X}_{ij})}{dX_{ij}}(\boldsymbol{x}_{ij}-\mathbf{F}_i\mathbf{X}_{ij})\cdot(\bar{\delta}\boldsymbol{x}_j-\bar{\delta}\boldsymbol{x}_i)V_j\end{aligned} \tag{7.26}$$

$\mathbf{R}_i^{hg}$ is the residual for hourglass energy. Equation 7.26 gives the explicit formula for the hourglass force. The term on $\bar{\delta}\boldsymbol{x}_i$ is the hourglass force from its support, while the terms on $\bar{\delta}\mathbf{u}'$ are the hourglass forces for the dual support $\mathcal{S}_j'$ of point $\mathbf{X}_j$. When the displacement field is consistent with the deformation gradient, then the hourglass energy residual is zero. For individual vector $\mathbf{X}_{ij}$, the hourglass force vector can be obtained the same way as Eq. 7.17,

$$\mathbf{T}_{ij}^{hg} = -\left(\delta\mathcal{F}^{hg}\right)_{ij} = -\frac{2\mu}{m_{\mathbf{K}_i}}\frac{1}{X_{ij}}\frac{\mathrm{d}W_i(\mathbf{X}_{ij})}{\mathrm{d}X_{ij}}(\boldsymbol{x}_{ij}-\mathbf{F}_i\mathbf{X}_{ij}) = \frac{2\mu}{m_{\mathbf{K}_i}}\Big(\mathbf{F}_i\nabla W_i(\mathbf{X}_{ij}) - \frac{\boldsymbol{x}_{ij}}{X_{ij}}\frac{\mathrm{d}W_i(\mathbf{X}_{ij})}{\mathrm{d}X_{ij}}\Big). \tag{7.27}$$

The governing equation with hourglass force is

$$\rho\ddot{\mathbf{u}}_i = \sum_{S_i}\left(\mathbf{P}_i\cdot\tilde{\nabla}W_i(\mathbf{X}_{ij}) + \mathbf{T}_{ij}^{hg}\right)V_j - \sum_{S_i'}\left(\mathbf{P}_j\cdot\tilde{\nabla}W_j(\mathbf{X}_{ji}) + \mathbf{T}_{ji}^{hg}\right)V_j + \mathbf{b}_0. \tag{7.28}$$

One can see from Eq. 7.28 that one's particle's hourglass forces are divided into two groups, these from the support and the others from the dual-support. The derivation of the hourglass force is similar to the variational derivation on strain energy functional, thus is consistent with dual-support configuration. The hourglass forces from the support can be viewed as the direct forces, while the hourglass forces from the dual-support are the reaction forces. Therefore, the hourglass forces follow Newton's third law, the same as internal forces. When variable smoothing lengths are used, the hourglass forces have no influence on the conservation of linear momentum and angular momentum.

The variation of $\bar{\delta}\mathcal{F}^{hg}$ leads to the hourglass tangent stiffness matrix,

$$\mathbf{K}_i^{hg} = \bar{\delta}^2\mathcal{F}_i^{hg} = \frac{\mu}{m_{\mathbf{K}_i}}\Big(\sum_{S_i}\frac{\mathrm{d}W_i(\mathbf{X}_{ij})}{\mathrm{d}X_{ij}}\frac{1}{X_{ij}}(\bar{\delta}\boldsymbol{x}_j-\bar{\delta}\boldsymbol{x}_i)^T(\bar{\delta}\boldsymbol{x}_j-\bar{\delta}\boldsymbol{x}_i)V_j - \bar{\delta}\mathbf{F}_i\mathbf{K}_i:\bar{\delta}\mathbf{F}_i\Big). \tag{7.29}$$

Similarly, the hourglass correction for scalar field is

$$T_{ij}^{hg} = -\left(\bar{\delta}\mathcal{F}^{hg}\right)_{ij} = \frac{2\mu}{m_{\mathbf{K}_i}}\left(\nabla s_i \cdot \nabla W_i(\mathbf{X}_{ij}) - \frac{s_{ij}}{X_{ij}}\frac{\mathrm{d}W_i(\mathbf{X}_{ij})}{\mathrm{d}X_{ij}}\right). \tag{7.30}$$

where $s_{ij} = s_j - s_i$.

7.4 Numerical Implementation

For elastic material, the strain energy density is a function of the deformation gradient. For the total Lagrange formulation, it is convenient to use the first Piola–Kirchhoff stress, which is the direct derivative of the strain energy over the deformation gradient,

$$\mathbf{P} = \frac{\partial\psi(\mathbf{F})}{\partial\mathbf{F}}, \tag{7.31}$$

where

$$\mathbf{F} = \begin{bmatrix} F_1 & F_2 & F_3 \\ F_4 & F_5 & F_6 \\ F_7 & F_8 & F_9 \end{bmatrix}. \tag{7.32}$$

Furthermore, the material tensor (stress-strain relation) which is required in the implicit analysis can be obtained with the derivative of the first Piola–Kirchhoff stress,

$$\mathbf{D}_4 = \frac{\partial\mathbf{P}}{\partial\mathbf{F}} = \partial_{\mathbf{FF}}\psi(\mathbf{F}). \tag{7.33}$$

The fourth-order material tensor $\mathbf{D}_4$ can be expressed in matrix form when the deformation gradient is flattened.

$$\mathbf{D} = \begin{bmatrix} \frac{\partial P_1}{\partial F_1} & \frac{\partial P_1}{\partial F_2} & \cdots & \frac{\partial P_1}{\partial F_9} \\ \frac{\partial P_2}{\partial F_1} & \frac{\partial P_2}{\partial F_2} & \cdots & \frac{\partial P_2}{\partial F_9} \\ \vdots & \vdots & \ddots & \vdots \\ \frac{\partial P_9}{\partial F_1} & \frac{\partial P_1}{\partial F_2} & \cdots & \frac{\partial P_9}{\partial F_9} \end{bmatrix} = \begin{bmatrix} \frac{\partial^2\psi(F)}{\partial F_1^2} & \frac{\partial^2\psi(F)}{\partial F_1 \partial F_2} & \cdots & \frac{\partial^2\psi(F)}{\partial F_1 \partial F_9} \\ \frac{\partial^2\psi(F)}{\partial F_2 \partial F_1} & \frac{\partial^2\psi(F)}{\partial F_2 \partial F_2} & \cdots & \frac{\partial^2\psi(F)}{\partial F_2 \partial F_9} \\ \vdots & \vdots & \ddots & \vdots \\ \frac{\partial^2\psi(F)}{\partial F_9 \partial F_1} & \frac{\partial^2\psi(F)}{\partial F_9 \partial F_2} & \cdots & \frac{\partial^2\psi(F)}{\partial F_9 \partial F_9} \end{bmatrix}, \tag{7.34}$$

where the flattened deformation gradient and first Piola–Kirchhoff stress are

$$F = (F_1, F_2, F_3, F_4, F_5, F_6, F_7, F_8, F_9), \tag{7.35}$$

and

$$P = \frac{\partial \psi(F)}{\partial F} = (\frac{\partial \psi(F)}{\partial F_1}, \frac{\partial \psi(F)}{\partial F_2}, \cdots, \frac{\partial \psi(F)}{\partial F_9}). \tag{7.36}$$

The derivative of the determinant of deformation gradient on $\mathbf{F}$ is

$$J = \det(\mathbf{F}),\ J_{,\mathbf{F}} = \begin{bmatrix} F_5F_9 - F_6F_8 & F_6F_7 - F_4F_9 & F_4F_8 - F_5F_7 \\ F_3F_8 - F_2F_9 & F_1F_9 - F_3F_7 & F_2F_7 - F_1F_8 \\ F_2F_6 - F_3F_5 & F_3F_4 - F_1F_6 & F_1F_5 - F_2F_4 \end{bmatrix}. \tag{7.37}$$

Since the strain energy is formulated on the particles, the total discrete strain energy is the sum of all strain energy on the particles,

$$\mathcal{F} = \sum_{i=1}^{N} V_i \psi(F_i), \tag{7.38}$$

where V_i is the volume associated with particle i, N is the number of particles, F_i is the flattened deformation tensor. The first variation of $\mathcal{F}$ is the global residual

$$\mathbf{R}_g = \bar{\delta}\mathcal{F} = \sum_{i=1}^{N} V_i \frac{\partial \psi(F_i)}{\partial F_i} \bar{\delta} F_i = \sum_{i=1}^{N} V_i P_i \bar{\delta} F_i = \sum_{i=1}^{N} \mathbf{R}_i. \tag{7.39}$$

The variation of $\mathbf{R}_g$ is the global stiffness tangent matrix

$$\mathbf{K}_g = \bar{\delta}\mathbf{R}_g = \bar{\delta}^2\mathcal{F} = \sum_{i=1}^{N} V_i \bar{\delta} F_i^T \frac{\partial^2 \psi(F_i)}{\partial F_i^T \partial F_i} \bar{\delta} F_i = \sum_{i=1}^{N} V_i \bar{\delta} F_i^T \mathbf{D} \bar{\delta} F_i = \sum_{i=1}^{N} \mathbf{K}_i, \tag{7.40}$$

where V_i is the initial nodal volume; $\mathbf{R}_i$, $\mathbf{K}_i$ are the nodal residual and nodal tangent stiffness matrix, respectively:

$$\mathbf{R}_i = V_i P_i \bar{\delta} F_i,\ \mathbf{K}_i = V_i \bar{\delta} F_i^T \mathbf{D} \bar{\delta} F_i. \tag{7.41}$$

The summation of all particles is the global assembling, which is the same as the finite element method. The remaining work is on how to assemble the nodal residual and nodal stiffness matrix. Equation 7.41 shows that nodal residual and nodal stiffness are some matrix operations on $\bar{\delta} F$. In the framework of SPH, we have

$$\mathbf{F}_i = \sum_{\mathcal{S}_i} (\boldsymbol{x}_j - \boldsymbol{x}_i) \otimes \tilde{\nabla} W_i(\mathbf{X}_{ij}) V_j. \tag{7.42}$$

The variation of $\bar{\delta}\mathbf{F}_i$ reads

$$\bar{\delta}\mathbf{F}_i = \sum_{\mathcal{S}_i}(\bar{\delta}\boldsymbol{x}_j - \bar{\delta}\boldsymbol{x}_i) \otimes \tilde{\nabla} W_i(\mathbf{X}_{ij})V_j. \tag{7.43}$$

where V_j is the volume for particle $\mathbf{X}_j$. For the purpose of numerical implementation, $\bar{\delta}\mathbf{F}_i$ in 3D can be written as a matrix $\bar{\delta}F_i$ with the dimensions of $9 \times 3n_{\mathbf{X}_i}$, which can be assembled with the following order, where $n_{\mathbf{X}_i}$ is the number of particles in $\mathcal{S}_i$ ($\mathbf{X}_i$ is also included). The assembling process on the nodal level is called nodal assembly.

Assume particle $\mathbf{X}_i$'s neighbors $N_{\mathbf{X}_i} = \{j_0, j_1, ..., j_k, ..., j_{n_i-1}\}$, the first particle j_0 denotes the particle $\mathbf{X}_i$. Here the convention for index starts from 0. For each particle in the neighbor list, we use $R = \tilde{\nabla} W_i(\mathbf{X}_{ij})V_j$, the terms in R can be added to the $\bar{\delta}F$ as

$$\begin{aligned}
\bar{\delta}F_{0,3k} &= R_0, & \bar{\delta}F_{0,0} &= \bar{\delta}F_{0,0} - R_0\\
\bar{\delta}F_{3,3k} &= R_1, & \bar{\delta}F_{3,0} &= \bar{\delta}F_{3,0} - R_1\\
\bar{\delta}F_{6,3k} &= R_2, & \bar{\delta}F_{6,0} &= \bar{\delta}F_{6,0} - R_2\\
\bar{\delta}F_{1,3k+1} &= R_0, & \bar{\delta}F_{1,1} &= \bar{\delta}F_{1,1} - R_0\\
\bar{\delta}F_{4,3k+1} &= R_1, & \bar{\delta}F_{4,1} &= \bar{\delta}F_{4,1} - R_1\\
\bar{\delta}F_{7,3k+1} &= R_2, & \bar{\delta}F_{7,1} &= \bar{\delta}F_{7,1} - R_2\\
\bar{\delta}F_{2,3k+2} &= R_0, & \bar{\delta}F_{2,2} &= \bar{\delta}F_{2,2} - R_0\\
\bar{\delta}F_{5,3k+2} &= R_1, & \bar{\delta}F_{5,2} &= \bar{\delta}F_{5,2} - R_1\\
\bar{\delta}F_{8,3k+2} &= R_2, & \bar{\delta}F_{8,2} &= \bar{\delta}F_{8,2} - R_2,
\end{aligned}$$

where k is the index of particle $\mathbf{X}_j$ in $N_{\mathbf{X}_i}$. It should be noted that the above derivation is independent of the actual material constitutions, which can serve as a general framework for the implicit analysis using SPH for many materials.

With $\bar{\delta}F$ and $\mathbf{D}$ available for any particle i, the tangent stiffness matrix $\mathbf{K}_i$ at a point i in Eq. 7.41 is $V_i\bar{\delta}F_i^T\, \mathbf{D}_i\, \bar{\delta}F_i$, where $\bar{\delta}F_i$ is a matrix of $9 \times 3n_i$. The variation of the deformation gradient enables the construction of tangent stiffness matrix being simply the multiplication of some matrices.

The residual of hourglass energy functional in Eq. 7.26 and hourglass tangent stiffness matrix in Eq. 7.29 can be obtained with a similar procedure. The Dirichlet and Neumann boundary conditions can be applied on the particles, the same as the finite element method. After assembling the global stiffness matrix and residual, the solution is obtainable when solving the linear algebra system

$$(\mathbf{K}_g + \mathbf{K}^{hg})\mathbf{u} = \mathbf{R}_g,$$

where $\mathbf{K}_g$, $\mathbf{R}_g$ are the global stiffness matrix and global residual vector, respectively.

7.5 Material Constitutions

In this section, we consider only the elastic materials, including the linear elastic material, two hyperelastic materials. The other elastic materials can be formulated with the similar procedure.

The elastic energy density for linear isotropic material is

$$\psi(\boldsymbol{\varepsilon}) = \frac{1}{2}\lambda(\mathrm{tr}\,\boldsymbol{\varepsilon})^2 + \mu\boldsymbol{\varepsilon} : \boldsymbol{\varepsilon} \tag{7.44}$$

where $\boldsymbol{\varepsilon} = \frac{1}{2}(\mathbf{F}^T + \mathbf{F}) - \mathbf{I}$, λ, μ are the lamé constants for isotropic elastic material.

The material tensor $\mathbf{D}_4 = \frac{\partial^2\psi(\mathbf{F}_i)}{\partial\mathbf{F}_i^T\partial\mathbf{F}_i}$ can be written as matrix form $\mathbf{D}$

$$\mathbf{D} = \begin{bmatrix} \lambda+2\mu & 0\;0\;0 & \lambda & 0\;0\;0 & \lambda \\ 0 & \mu\;0\;\mu & 0 & 0\;0\;0 & 0 \\ 0 & 0\;\mu\;0 & 0 & 0\;\mu\;0 & 0 \\ 0 & \mu\;0\;\mu & 0 & 0\;0\;0 & 0 \\ \lambda & 0\;0\;0 & \lambda+2\mu & 0\;0\;0 & \lambda \\ 0 & 0\;0\;0 & 0 & \mu\;0\;\mu & 0 \\ 0 & 0\;\mu\;0 & 0 & 0\;\mu\;0 & 0 \\ 0 & 0\;0\;0 & 0 & \mu\;0\;\mu & 0 \\ \lambda & 0\;0\;0 & \lambda & 0\;0\;0 & \lambda+2\mu \end{bmatrix}. \tag{7.45}$$

For the case of Neo-Hooke material (Simo et al. 1985), the strain energy can be expressed as

$$\psi(\mathbf{F}) = \frac{1}{2}\kappa(J-1)^2 + \frac{1}{2}\mu(J^{-2/3}\mathbf{F} : \mathbf{F} - 3). \tag{7.46}$$

The first Piola–Kirchoff stress is

$$\mathbf{P} = \frac{\partial\psi(\mathbf{F})}{\partial\mathbf{F}} = \frac{\mu}{J^{2/3}}\mathbf{F} + \left((J-1)\kappa - \mu\frac{\mathbf{F} : \mathbf{F}}{3J^{5/3}}\right)J_{,\mathbf{F}}. \tag{7.47}$$

The material tensor can be written as

$$\mathbf{D} = \partial_F P = \frac{\mu}{J^{2/3}}\mathbf{I}_{9\times 9} - \frac{2\mu}{3J^{5/3}}(F \otimes J_{,F} + J_{,F} \otimes F) +$$
$$(\kappa + \frac{5\mu}{9J^{8/3}}\mathbf{F} : \mathbf{F})J_{,F} \otimes J_{,F} + \left(\kappa(J-1) - \frac{\mu}{3J^{5/3}}\mathbf{F} : \mathbf{F}\right)J_{,FF} \tag{7.48}$$

where $J_{,F}$ is the vector form of $J_{,\mathbf{F}}$, and

$$J_{,FF} = \begin{bmatrix} 0 & 0 & 0 & 0 & F_9 & -F_8 & 0 & -F_6 & F_5 \\ 0 & 0 & 0 & -F_9 & 0 & F_7 & F_6 & 0 & -F_4 \\ 0 & 0 & 0 & F_8 & -F_7 & 0 & -F_5 & F_4 & 0 \\ 0 & -F_9 & F_8 & 0 & 0 & 0 & 0 & F_3 & -F_2 \\ F_9 & 0 & -F_7 & 0 & 0 & 0 & -F_3 & 0 & F_1 \\ -F_8 & F_7 & 0 & 0 & 0 & 0 & F_2 & -F_1 & 0 \\ 0 & F_6 & -F_5 & 0 & -F_3 & F_2 & 0 & 0 & 0 \\ -F_6 & 0 & F_4 & F_3 & 0 & -F_1 & 0 & 0 & 0 \\ F_5 & -F_4 & 0 & -F_2 & F_1 & 0 & 0 & 0 & 0 \end{bmatrix}. \quad (7.49)$$

Another energy density functional for the compressible neo-Hookean material (Bonet and Wood 1997) is

$$\psi = \frac{\mu}{2}(\mathbf{F} : \mathbf{F} - 3) - \mu \ln J + \frac{\lambda}{2}(\ln J)^2. \quad (7.50)$$

The first Piola–Kirchhoff stress and the material matrix are

$$\mathbf{P} = \mu\mathbf{F} - \mu\frac{J_{,\mathbf{F}}}{J} + \lambda\frac{\ln J}{J}J_{,\mathbf{F}} \quad (7.51)$$

$$\mathbf{D} = \mu\mathbf{I}_{9\times9} + \lambda\frac{\ln J}{J}J_{,FF} + \frac{1}{J^2}(\lambda - \lambda\ln J + \mu)J_{,F} \otimes J_{,F}. \quad (7.52)$$

Equation 7.50 is used to model the rubber in Sect. 7.6.5.

7.6 Numerical Examples

We give six numerical examples to validate the implicit formulation of dual-support SPH and test the performance of the hourglass control. The numerical results are compared with the theoretical solutions or that by finite element method. Traditionally, SPH is solved by explicit integration methods, such as Velocity-Verlet algorithm, Leapfrog integration. Explicit integration method is limited by the maximal time increment for the reason of numerical stability. The maximal time increment depends on the minimal particle size Δx_{min} and the wave sound speed (C) in the media, i. e. $\Delta t_{max} \leq \Delta x_{min}/C$. The computer cost is economical for short-duration models but is very expensive for long-term models. Implicit algorithm is unconditionally stable for any time increment, thus has advantage for solving static or quasi-static problems. Traditional SPH can't be solved implicitly because the tangent stiffness matrix is not available. The variational formulation of SPH in this chapter obtains the residual and tangent stiffness matrix with ease, and thus provides great feasibility for implicit analysis by SPH for static/dynamic problems. For dynamic problems, Hilber–Hughes–Taylor (HHT) integration (Hilber et al. 1977), Newmark method (Newmark et al. 1959) can be readily used. For simplicity, we assume the acceleration in Eq. 7.18 to be zero and study a series of static problems.

7.6.1 3D Cantilever Loaded at the End

A 3D cantilever beam loaded at the end with pure shear traction force is considered. The beam with dimensions of height of $D = 3$ m, length of $L = 8$ m and thickness of $t = 2$ m and shear load of parabola distribution is shown in Fig. 7.2. The analytical solution for the beam is (Timoshenko and Goodier 1970; Zhuang and Augarde 2010)

$$u_x = \frac{Py}{6EI}[(6L - 3x)x + (2 + \nu)(y^2 - \frac{D^2}{4})] \tag{7.53}$$

$$u_y = -\frac{P}{6EI}\left[3\nu y^2(L - x) + (4 + 5\nu)\frac{D^2 x}{4} + (3L - x)x^2\right] \tag{7.54}$$

$$\sigma_{xx}(x, y) = \frac{P(L - x)y}{I}, \sigma_{yy}(x, y) = 0, \tau_{xy}(x, y) = -\frac{P}{2I}(\frac{D^2}{4} - y^2), \tag{7.55}$$

where $P = -1000$ N, $I = \frac{D^3}{12}$. The material parameters are taken as $E = 30$ GPa, $\nu = 0.3$. The particles on the left boundary are constrained by the exact displacements from Eqs. 7.53 and 7.54 and the loading on the right boundary follows Eq. 7.55. The error norm in displacement for particle i is calculated by

$$\|u\|_{error} = \sqrt{\frac{\sum_{i=1}^{N}(\mathbf{u}_i - \mathbf{u}_i^h) \cdot (\mathbf{u}_i - \mathbf{u}_i^h)V_i}{\sum_{i=1}^{N}\mathbf{u}_i \cdot \mathbf{u}_i V_i}} \tag{7.56}$$

The exact strain energy and numerical strain energy are computed by

$$\begin{aligned} E_{numerical} &= \frac{1}{2}\sum_{i=1}^{N} \varepsilon_i^T \mathbf{D} \varepsilon_i V_i \\ E_{exact} &= \frac{1}{2}\int_{\Omega} \varepsilon_i^T \mathbf{D} \varepsilon_i dV_i \\ E_{error} &= \frac{E_{numerical}}{E_{exact}} - 1, \end{aligned} \tag{7.57}$$

where $\mathbf{D}$ is the material tensor.

We tested four cases with different discretizations. The statistics of the particle number, the supports and the displacement error and energy are given in Table 7.1. It can be seen that the numerical results converge to the theoretical solution with the increase of the number of particles. The y-displacements of particles on the red line in Fig. 7.2 are plotted with good agreement to the theoretical solution in Fig. 7.3.

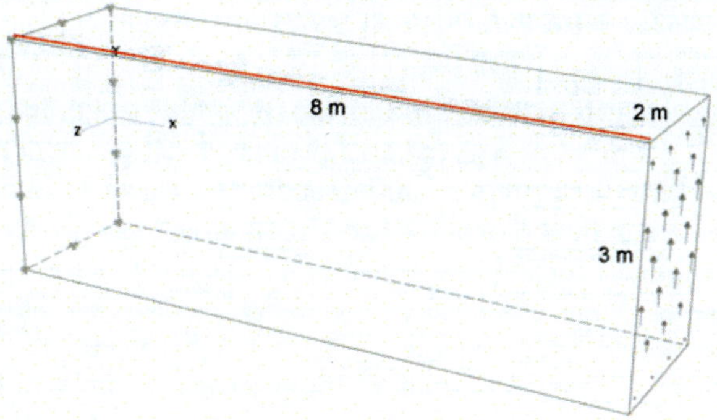

Fig. 7.2 Setup of the thick beam

Table 7.1 Convergence study for different discretizations, where $e(E) = |E - E^{exact}|/|E^{exact}|$, N is the number of particles. The exact strain energy is $E^{exact}_{strain} = 0.00277284$

Case	N	h_{min}	h_{max}	$\|u\|_{error}$	E_{strain}	$\mathcal{F}^{hg}$	$e(E_{strain})$
1	286	0.844	2.235	0.0859	3.011e–3	2.422e–6	0.0859
2	695	0.608	1.691	0.0729	2.948e–3	1.487e–6	0.0631
3	2609	0.316	1.102	0.0273	2.891e–3	7.785e–7	0.0426
4	14250	0.164	0.675	0.0208	2.819e–3	3.625e–7	0.0165

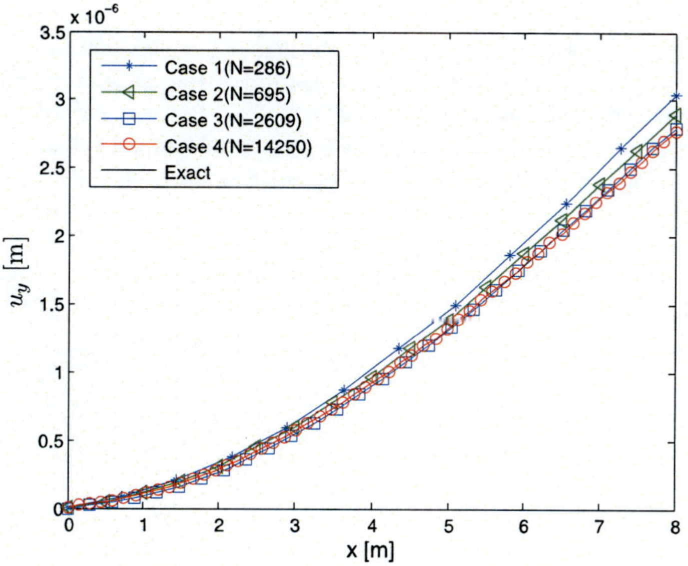

Fig. 7.3 Displacement curve in y direction for different discretizations with hourglass control

7.6.2 Plate Under Compression

In order to study the influence of variable smoothing length and hourglass penalty, we model a 1×1 m^2 plate with material parameters elastic modulus $E = 2$ Pa, Poisson ratio $\nu = 0.3$. Plane strain condition and linear elasticity in Eq. 7.44 are assumed. The particles in the bottom are fixed in all directions and the top boundary of the plate is applied with pressure $p = 2$ N/m. The plate is discretized with an irregular quadrilateral element and then the element is converted into a particle positioned at the barycenter of the element. The area of the particle is determined by that of the element. The distribution of the particles is shown in Fig. 7.4. The support domain for each particle comprises 10 nearest particles and the smoothing length is selected as the maximal distance of neighbors with respect to the particle. We test different hourglass penalties $\alpha \in \{0, 0.2, 2, 10, 100, 10^3, 10^5\}$. The contour plots of the displacement field are shown in Figs. 7.5 and 7.6. The maximal displacement in $x, y-$directions are given in Table 7.2. It can be seen that the hourglass control has a very positive effect on the formulation. Without hourglass control, the displacement field is quite poor. On the other hand, too large an hourglass penalty would make the mechanical system over-stiff, as shown in Fig. 7.6e, f.

7.6.3 3D Cantilever Tension Test

A 3D cantilever beam loaded at the end with pure tension or compression of $P_x = 1.0 \times 10^6$ Pa is considered to test the performance of hourglass control. The dimensions and material parameters of the beam are the same as that in Sect. 7.6.1, as shown in Fig. 7.2. The theoretical maximal displacement in x-direction is $(u_x)_{max} = 2.6667 \times 10^{-4}$ m. The total strain energy is $E_{strain} = 800$ J. The

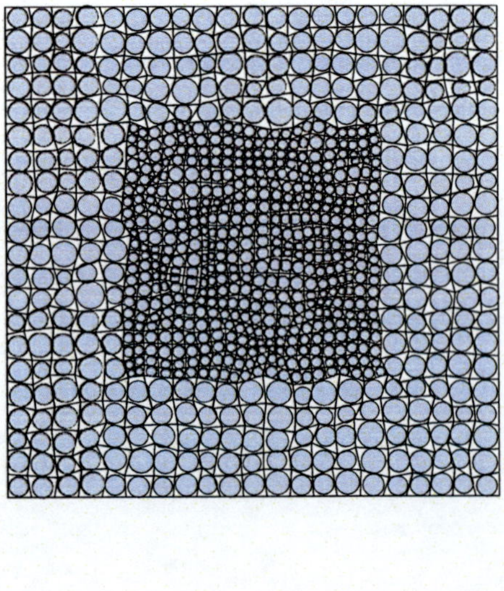

Fig. 7.4 Distribution of the particles, the radius represents the size of the particle

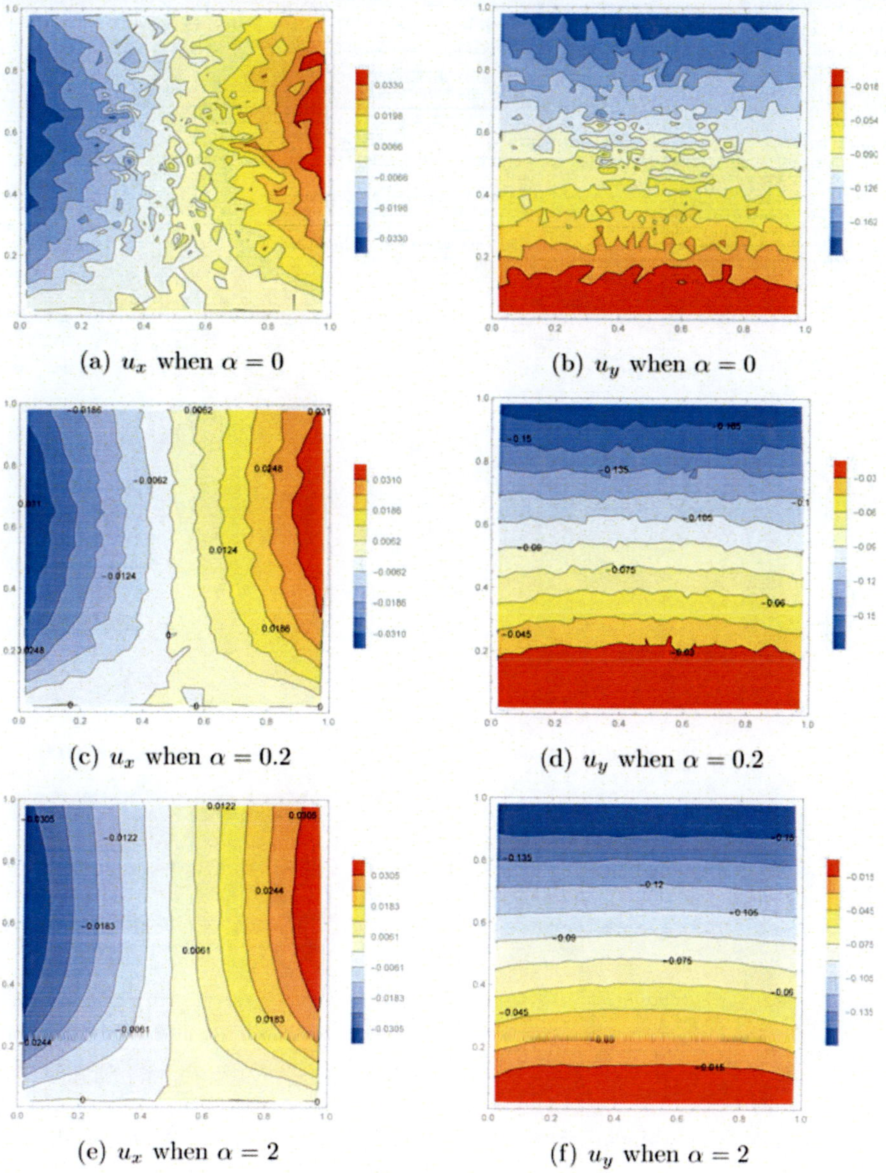

(a) u_x when $\alpha = 0$ (b) u_y when $\alpha = 0$

(c) u_x when $\alpha = 0.2$ (d) u_y when $\alpha = 0.2$

(e) u_x when $\alpha = 2$ (f) u_y when $\alpha = 2$

Fig. 7.5 Contour of displacement field for different hourglass penalties

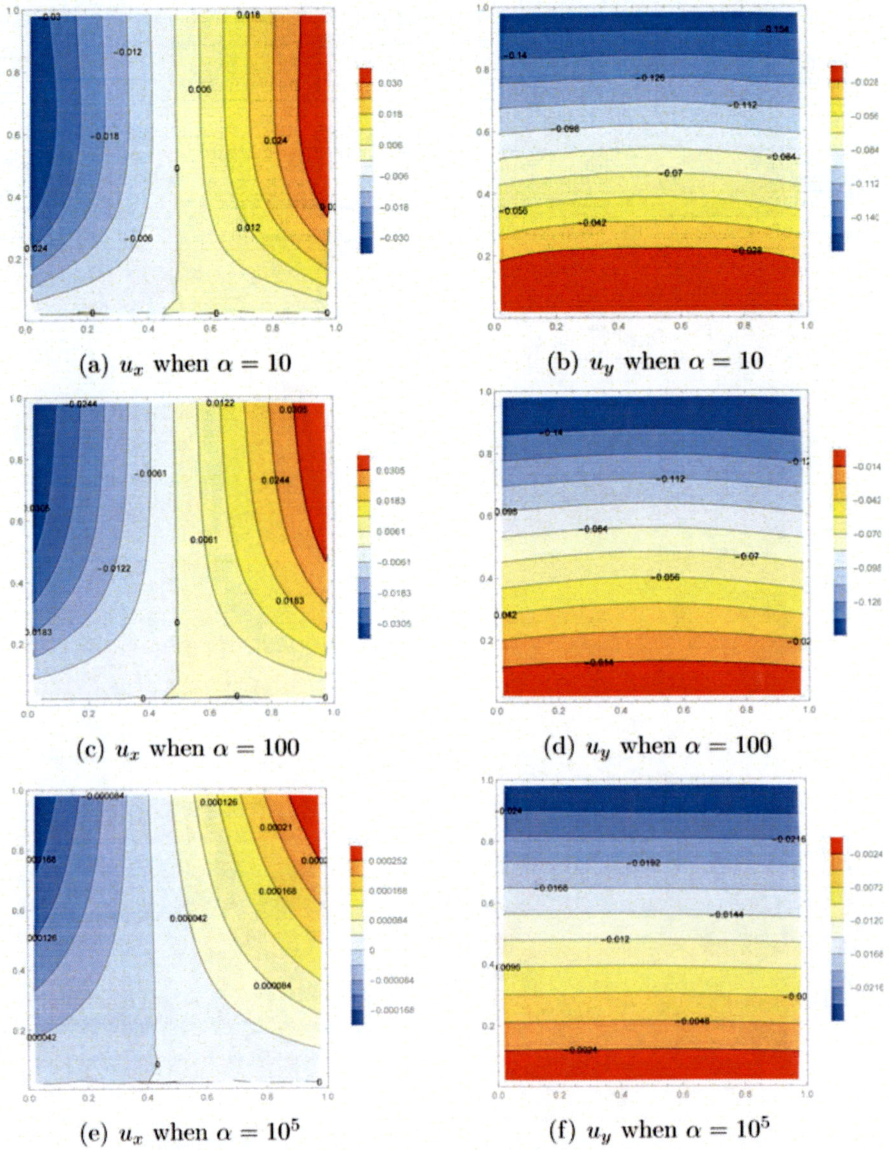

(a) u_x when $\alpha = 10$

(b) u_y when $\alpha = 10$

(c) u_x when $\alpha = 100$

(d) u_y when $\alpha = 100$

(e) u_x when $\alpha = 10^5$

(f) u_y when $\alpha = 10^5$

Fig. 7.6 Contour of displacement field for different hourglass penalties

Table 7.2 Maximal displacement for different hourglass penalties

	$\alpha = 0$	$\alpha = 0.2$	$\alpha = 2$	$\alpha = 10$	$\alpha = 100$	$\alpha = 10^3$	$\alpha = 10^5$	Abaqus
$\max(u_x)$	0.0393	0.0382	0.0372	0.037	0.0381	0.0288	0.000303	0.04
$\max(u_y)$	−0.202	−0.175	−0.1664	−0.16367	−0.1604	−0.1465	−0.0267	−0.178

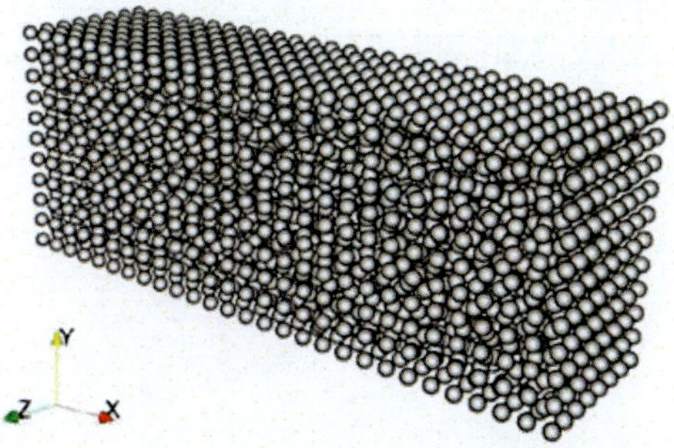

Fig. 7.7 Irregular distribution of particles

particles on the left yz−plane are fixed in x direction except one particle in (0, 0, 0)is fixed in all directions. Two discretizations the same as (a) Case 3 and (b) Case 4 in Table 7.1 with/without hourglass control are tested. As shown in Fig. 7.7, the distribution of particles is irregular; this also applies to the volume of each particle. The x displacement and hourglass energy density on the clip of $z = 1$ m are shown in Figs. 7.8 and 7.9, respectively. The total strain energy and hourglass energy are given in Table 7.3. It can be seen that the hourglass control has a significant influence on the accuracy of the solution.

Figure 7.8a, b show that the hourglass control can effectively improve the result. For the pure tension test, the strain energy density and strain component in x are evenly distributed for hourglass control, as shown in Figs. 7.10 and 7.11.

The hourglass mode is obvious on the boundaries where the Dirichlet boundary and Neumann boundary are applied. The reason is that the boundary conditions are applied on only one layer of particles and the delta property is not well satisfied.

The comparison in Fig. 7.10 shows that hourglass control can effectively eliminate the hourglass mode, and make the strain field more smooth.

7.6.4 *Influence of Smoothing Length*

One disadvantage of the implicit formula is the high cost in assembling the global stiffness matrix due to the large matrix sizes for each node. In this section, we test the effect of smoothing length for quintic kernel function on the numerical accuracy in 2D plane stress solid. A thick beam in 2D with the same material parameters and dimensions in Sect. 7.6.1 is considered. The hourglass energy control is used in all numerical examples of this section. The particles are constructed from the element

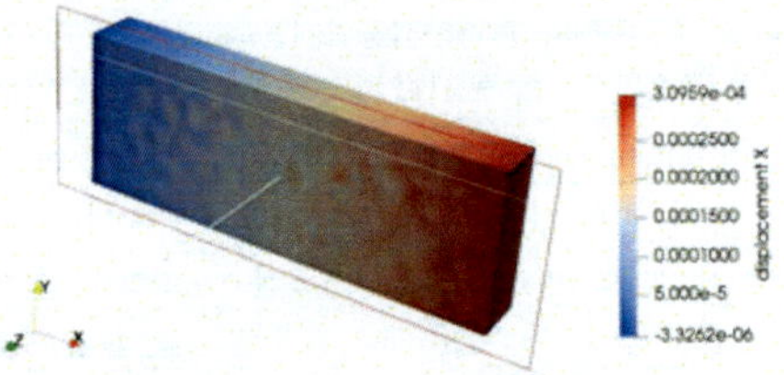

(a) Fine mesh without hourglass control

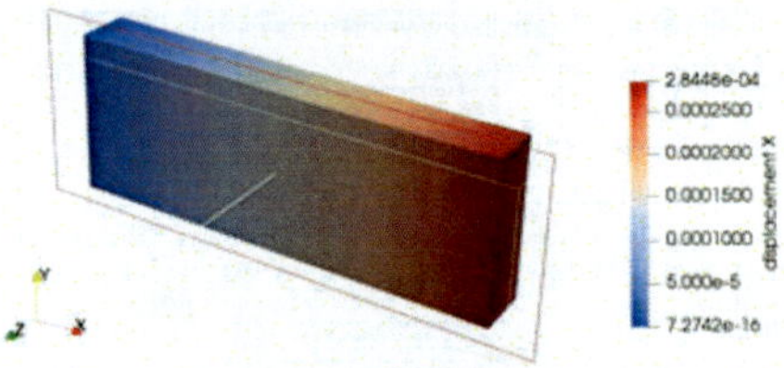

(b) Fine mesh with hourglass control

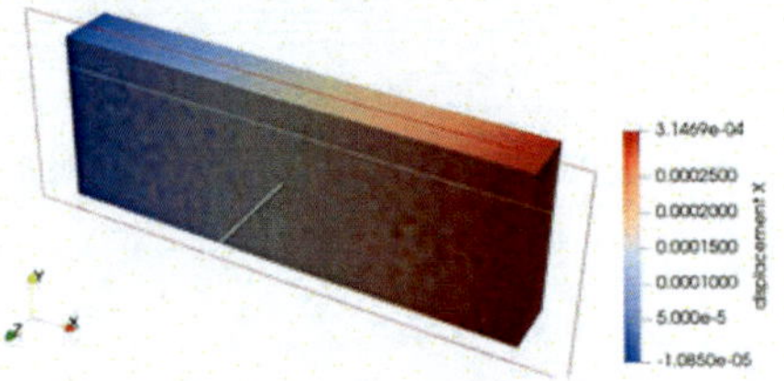

(c) Finer mesh without hourglass control

(d) Finer mesh with hourglass control

Fig. 7.8 x displacement on the clip of $z = 1$ m

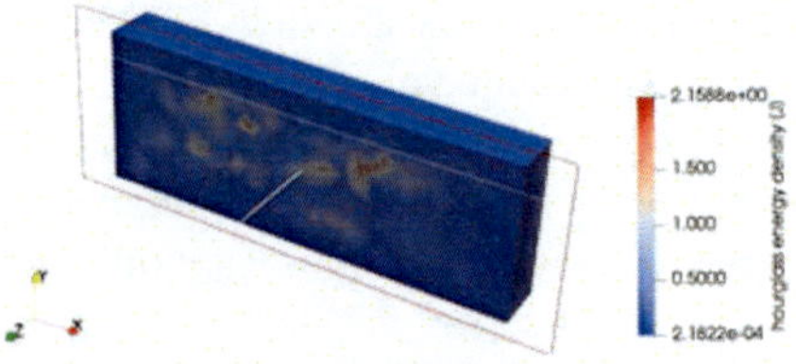

(a) Fine mesh without hourglass control

(b) Fine mesh with hourglass control

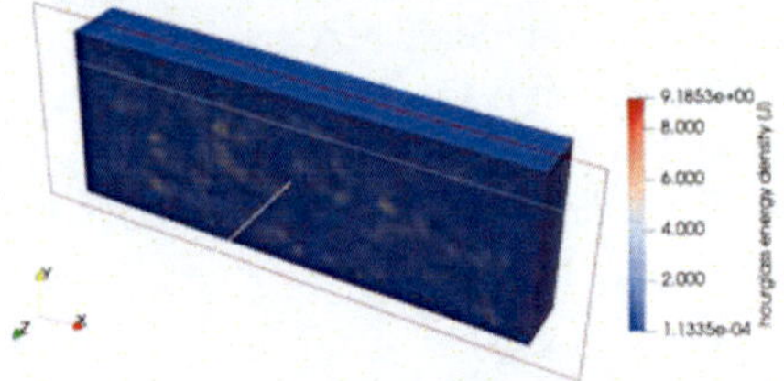

(c) Finer mesh without hourglass control

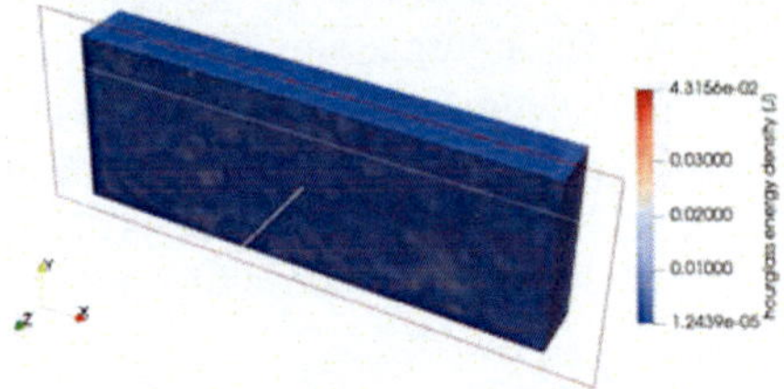

(d) Finer mesh with hourglass control

Fig. 7.9 Hourglass energy density on the clip of $z = 1$ m

Table 7.3 Energy and maximal displacement in x direction for 4 cases. HG denotes hourglass control

	(a) without HG	(a) with HG	(b) without HG	(b) with HG
$E_{strain}^{numerical}$	874.92	804.7	901.46	800.1
$E_{hourglass}^{numerical}$	6.442	0.0977	16.454	0.1051
$u_{x\,max}$	3.096E–04	2.849E–04	3.147E–04	2.752E–04

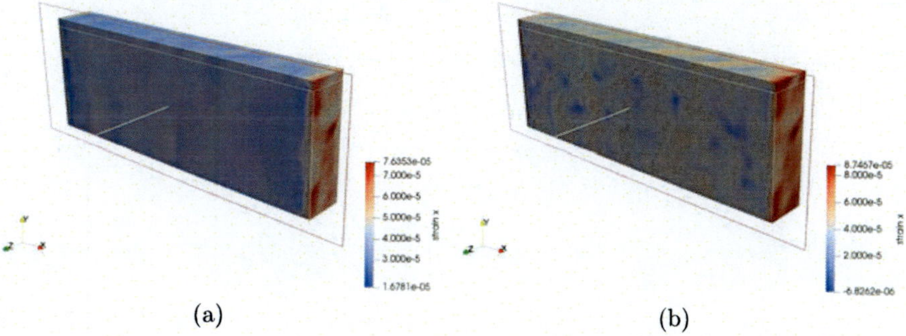

Fig. 7.10 Distribution of x component of strain tensor, **a** with hourglass control; **b** without hourglass control

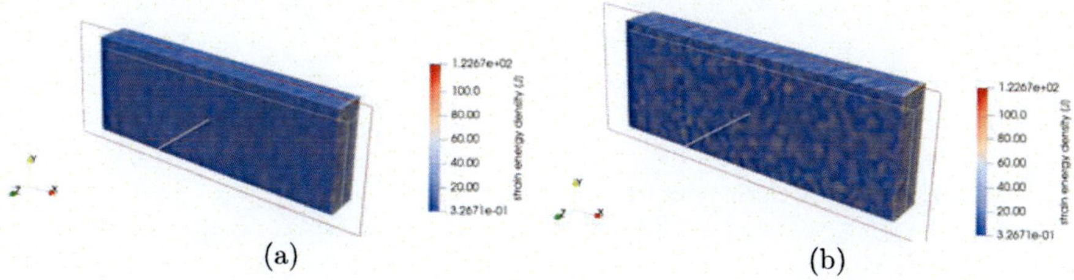

Fig. 7.11 Distribution of strain energy density, **a** with hourglass control; **b** without hourglass control

given in Fig. 7.12a by method shown in Fig. 7.12b. The particle radius is estimated by the shape of the disk. The smoothing length is selected as

$$h_i = n\Delta x_i, \forall \mathbf{X}_i \in \Omega. \tag{7.58}$$

The u_{error} and strain energy error are given in Figs. 7.13 and 7.14. The number of neighbors for different smoothing lengths is given in Table 7.4. For the case of $n = 0.9$, the minimal dimensions of the nodal stiffness matrix are 6x6, while the maximal dimensions of the nodal stiffness matrix for case $n = 3.8$ are 482x482. However, the larger smoothing length doesn't indicate a better numerical result. The "optimal" smoothing length scale for the corresponding kernel function is 2.2. When $n > 2.2$, the numerical error increases with the smoothing length. On the other hand,

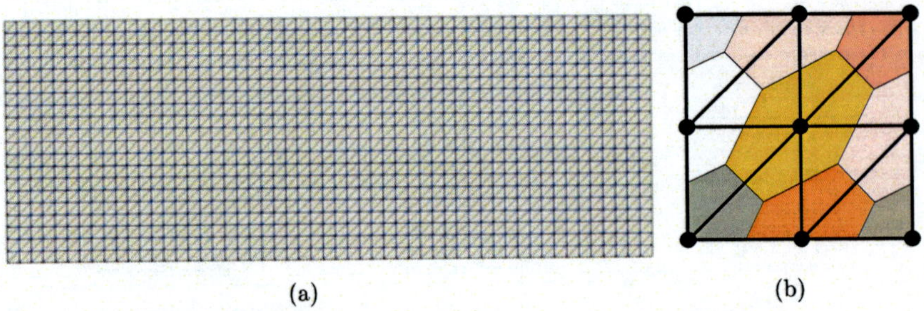

Fig. 7.12 a Mesh of 2D beam; b attach the element volume to nodes by average

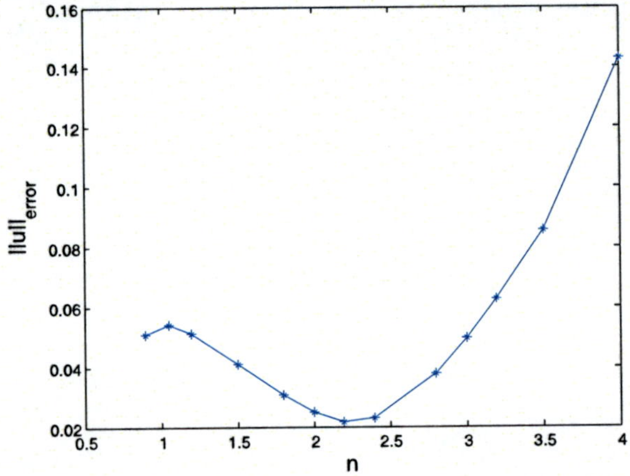

Fig. 7.13 Displacement error for smoothing length

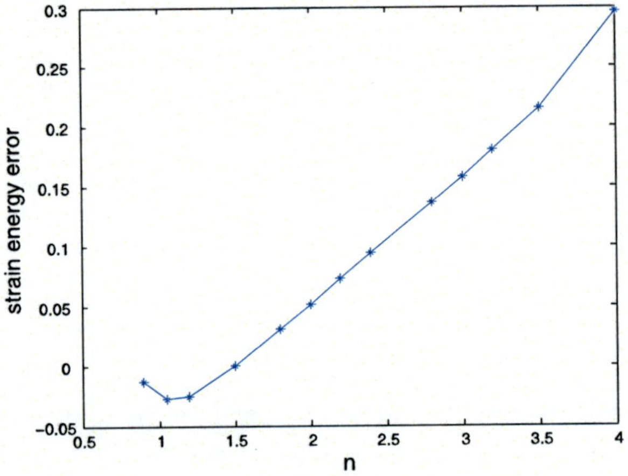

Fig. 7.14 Strain energy error for smoothing length

Table 7.4 Number of neighbors

Smoothing length	0.9	1.2	1.5	2	2.4	3	3.8
Min	2	3	3	7	10	14	21
Max	12	20	36	68	96	146	240

Fig. 7.15 Displacement field for $n = 0.9$ in Eq. 7.58

Fig. 7.16 Displacement field for $n = 3.8$ in Eq. 7.58

the smoothing length scale $n = 0.9$ offers good accuracy at the lowest computational cost. The displacement fields for $n = 0.9$ and $n = 3.8$ are given in Figs. 7.15 and 7.16, respectively.

7.6.5 *Rubber Pull Test*

In this section, we test a rubber with 500% elongation based on hyperelastic material in Eq. 7.50. The initial dimensions of the plate are $[0, 2] \times [0, 2]$ mm^2. The material parameters are elastic modulus $E = 0.2$ MPa and Poisson ratio $\nu = 0.45$. Two discretizations with regular particle distribution and irregular particle distribution are tested, as shown in Fig. 7.17. The area of each particle is selected as that of the element. Hourglass penalties with $\alpha = 0$ and $\alpha = 5\mu$ are tested, where $\mu = \frac{E}{2(1+\nu)}$. The final deformation of four cases are shown in Fig. 7.18. The smoothing length for each particle is selected as the maximal distance with respect to its 12 nearest neighbors. In order to reach the specified elongation, the up layer of the particles are displaced gradually by 10 mm in several increments. The Newton–Raphson iteration algorithm is adopted to solve the equations. For case 1, the simulation without hourglass control does not converge when the displacement on the boundary is larger than 8 mm, as shown in Fig. 7.18a. In the numerical simulation, the hourglass control stabilizes the scheme and makes the convergence easier. For Case 2, the model with

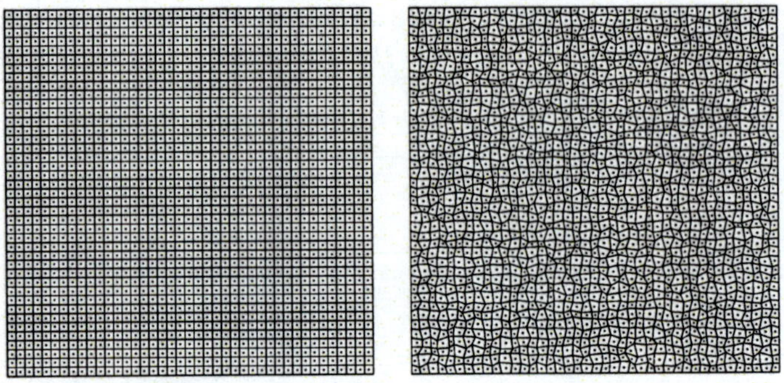

Fig. 7.17 Discretizations of the rubber

irregular particle distribution and without hourglass control becomes unstable and diverges when the displacement on the boundary is larger than 8.25 mm, as shown in Fig. 7.18c. When hourglass control is applied, the modeling based on irregular particle distribution is very stable throughout the simulation.

7.6.6 Large Deformation Problem

In this section, a cube of length $a = 1$ m with hyperelastic material given by Eq. 7.46 is modeled. This example demonstrates the capability of current formulation in solving problems involving geometric nonlinearity. The material parameters for the cube are elastic modulus $E = 10N/m^2$, Poisson ratio $\nu = 0.3$. The nodes on the plane $z = 0$ are fixed in all directions and the surface $\{x, y, z\} \in [0, a] \times [0, a/2] \times a$ are applied with pressure load $p_z = -2N/m^2$, as shown in Fig. 7.19. The cube is discretized into 21^3 particles. The load is applied on 4 sub-steps and in each sub-step the Newton–Raphson iteration is used to find the equilibrium, where the convergence is reached when the residual norm is less than 10^{-6}. The quintic kernel function is used, and smoothing length is $h_i = 2.05\Delta x_i$. The numbers of iteration for four sub-steps are (4,4,5,7) sequentially. The nonlinear effect increases with the load levels and more iteration is required to achieve the convergence. The final deformed configuration for implicit SPH and implicit FEM are given in Fig. 7.20, where the deformation is quite similar. The FEM result is implemented in the AceFEM environment (Korelc and Wriggers 2016). The largest displacement in z-direction are (−0.313 m, −0.342 m) for FEM and implicit SPH, respectively. The displacement in y, z-direction for different load level is depicted in Fig. 7.21. The difference for the maximal deformation in z-direction is approximately 9.2%, which is due to that SPH being a meshless method does not possess the Kronecker-delta property.

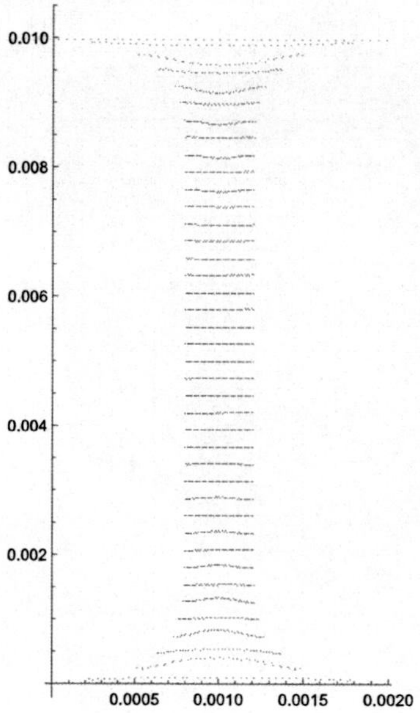

(a) $\alpha = 0$ for Case 1

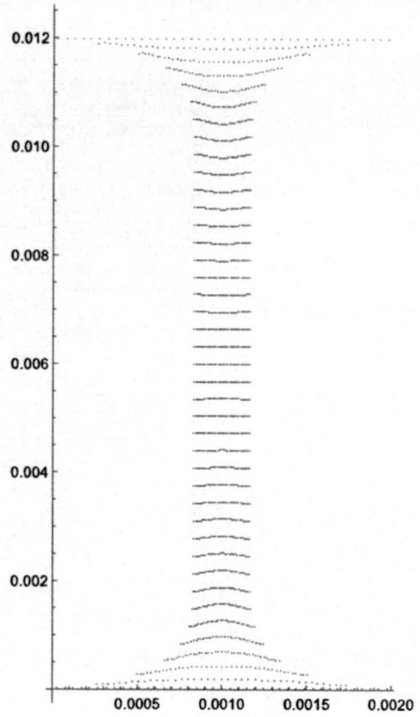

(b) $\alpha = 5\mu$ for Case 1

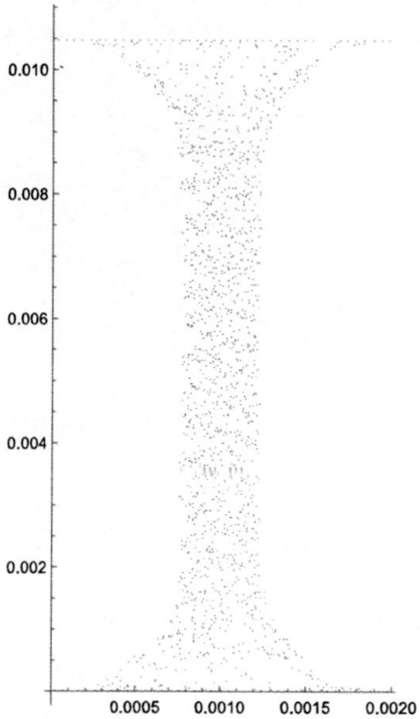

(c) $\alpha = 0$ for Case 2

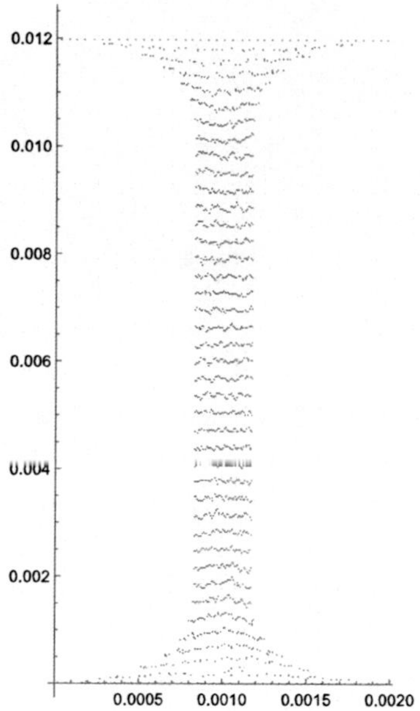

(d) $\alpha = 5\mu$ for Case 2

Fig. 7.18 Final configurations of the rubber

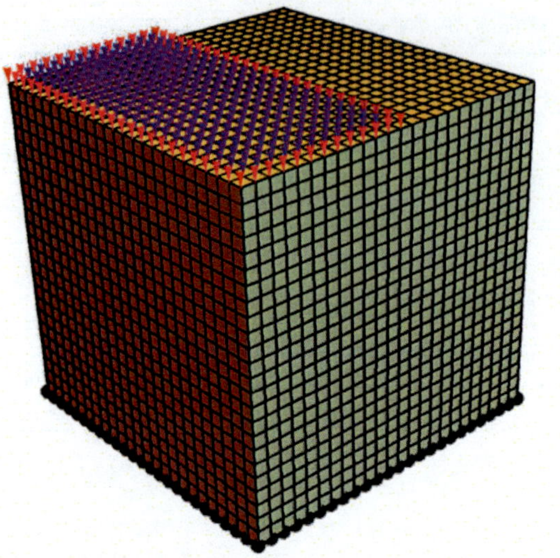

Fig. 7.19 Boundary condition of the cube, where the upper surface is applied with $p_z = -2N/m^2$ and the nodes on the bottom are fixed in all directions

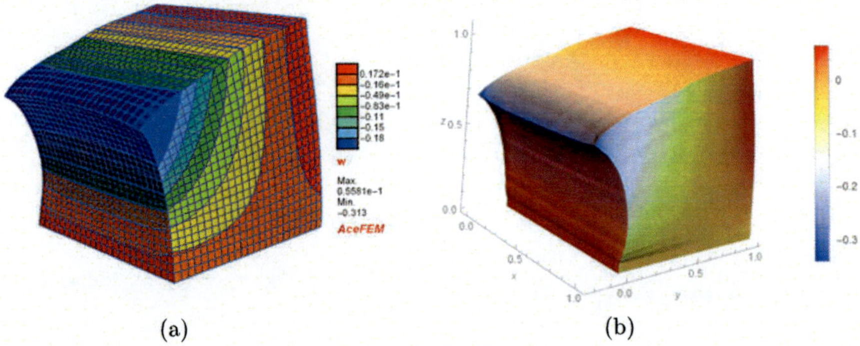

Fig. 7.20 **a** Final deformation by finite element method. **b** Final deformation by implicit SPH

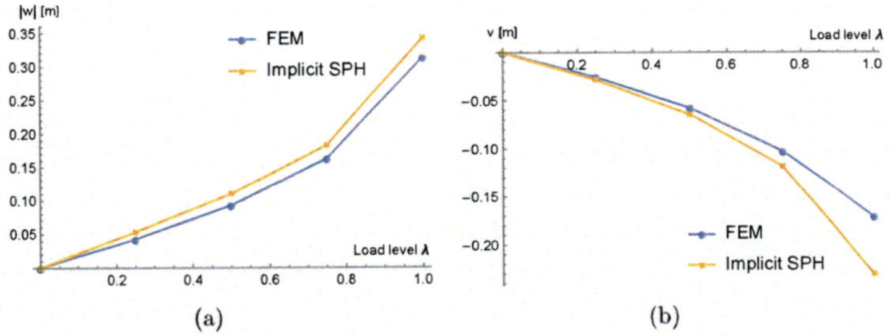

Fig. 7.21 Displacement of point $(0, 0, L)$ vs load levels. **a** Displacement in z-direction. **b** Displacement in y-direction

7.7 Conclusions

In this chapter, we derived the dual-support SPH by means of variational principle and demonstrated that the implicit form of SPH can be obtained with ease. During the evaluation of the nodal stiffness matrix, only the variation of the deformation gradient is required. We also show that the hourglass control is necessary in the SPH solid. We presented a general framework for the implicit SPH analysis which allows for material nonlinearity and geometrical nonlinearity. The fluid version of dual-support SPH is presented in the other paper. The proposed implicit SPH formulation obtains the residual and tangent stiffness matrix in a way quite similar to the finite element method. Many problems solved by FEM can be solved by the current scheme with some adaptation. For example, implicit SPH can replace the finite element formulation in the phase field fracture method (Miehe et al. 2010b; Zhou et al. 2018a, b) of solving the fracture problems.

References

Batra R, Zhang G (2004) Analysis of adiabatic shear bands in elasto-thermo-viscoplastic materials by modified smoothed-particle hydrodynamics (MSPH) method. J Comput Phys 201(1):172–190

Bonet J, Wood RD (1997) Nonlinear continuum mechanics for finite element analysis. Cambridge University Press

Caleyron F, Combescure A, Faucher V, Potapov S (2012) Dynamic simulation of damage-fracture transition in smoothed particles hydrodynamics shells. Int J Numer Methods Eng 90(6):707–738

Chen J, Beraun J, Jih C (1999) Completeness of corrective smoothed particle method for linear elastodynamics. Comput Mech 24(4):273–285

Dehnen W, Aly H (2012) Improving convergence in smoothed particle hydrodynamics simulations without pairing instability. Mon Not R Astron Soc 425(2):1068–1082

Dyka C, Ingel R (1995) An approach for tension instability in smoothed particle hydrodynamics (sph). Comput Struct 57(4):573–580

Fu Z, Chen W, Wen P, Zhang C (2018) Singular boundary method for wave propagation analysis in periodic structures. J Sound Vib 425:170–188

Maurel B, Combescure A (2008) An sph shell formulation for plasticity and fracture analysis in explicit dynamics. Int J Numer Methods Eng 76(7):949–971

Ganzenmüller GC (2015) An hourglass control algorithm for lagrangian smooth particle hydrodynamics. Comput Methods Appl Mech Eng 286:87–106

Gingold R, Monaghan J (1977) Smoothed particle hydrodynamics: theory and application to non-spherical stars. Mon Not R Astron Soc 181(3):375–389

Liu M, Liu G (2010) Smoothed particle hydrodynamics (sph): an overview and recent developments. Arch Comput Methods Eng 17(1):25–76

Korelc J, Wriggers P (2016) Automation of finite element methods. Springer

Li B, Habbal F, Ortiz M (2010) Optimal transportation meshfree approximation schemes for fluid and plastic flows. Int J Numer Methods Eng 83(12):1541–1579

Limido J, Espinosa C, Salaün M, Lacome J-L (2007) Sph method applied to high speed cutting modelling. Int J Mech Sci 49(7):898–908

Liu W, Jun S, Zhang Y (1995) Reproducing kernel particle methods. Int J Numer Methods Fluids 20(8–9):1081–1106

Liu M, Liu G (2010) Smoothed particle hydrodynamics (sph): an overview and recent developments. Arch Comput Methods Eng 17(1):25–76
Liu M, Liu G, Lam K (2006) Adaptive smoothed particle hydrodynamics for high strain hydrodynamics with material strength. Shock Waves 15(1):21–29
Lucy L (1977) A numerical approach to the testing of the fission hypothesis. Astron J 82:1013–1024
Maurel B, Combescure A (2008) An sph shell formulation for plasticity and fracture analysis in explicit dynamics. Int J Numer Methods Eng 76(7):949–971
Miehe C, Welschinger F, Hofacker M (2010) Thermodynamically consistent phase-field models of fracture: Variational principles and multi-field fe implementations. Int J Numer Methods Eng 83(10):1273–1311
Monaghan J (1989) On the problem of penetration in particle methods. J Comput Phys 82(1):1–15
Monaghan J (1992) Smoothed particle hydrodynamics. Ann Rev Astron Astrophys 30(1):543–574
Monaghan J, Huppert H, Worster M (2005) Solidification using smoothed particle hydrodynamics. J Comput Phys 206(2):684–705
Zhang G, Batra R (2009) Symmetric smoothed particle hydrodynamics (ssph) method and its application to elastic problems. Comput Mech 43(3):321–340
Rabczuk T, Belytschko T, Xiao S (2004) Stable particle methods based on lagrangian kernels. Comput Methods Appl Mech Eng 193(12):1035–1063
Randles P, Libersky L (1996) Smoothed particle hydrodynamics: some recent improvements and applications. Comput Methods Appl Mech Eng 139(1–4):375–408
Simo J, Taylor RL, Pister K (1985) Variational and projection methods for the volume constraint in finite deformation elasto-plasticity. Comput Methods Appl Mech Eng 51(1–3):177–208
Timoshenko S, Goodier J (1970) Theory of elasticity, vol 412, 3rd edn. McGraw-Hill, New York
Violeau D, Rogers BD (2016) Smoothed particle hydrodynamics (sph) for free-surface flows: past, present and future. J Hydraul Res 54(1):1–26
Zhang G, Batra R (2009) Symmetric smoothed particle hydrodynamics (ssph) method and its application to elastic problems. Comput Mech 43(3):321–340
Zhou C, Liu G, Lou K (2007) Three-dimensional penetration simulation using smoothed particle hydrodynamics. Int J Comput Methods 4(04):671–691
Zhou S, Rabczuk T, Zhuang X (2018) Phase field modeling of quasi-static and dynamic crack propagation: Comsol implementation and case studies. Adv Eng Softw 122:31–49
Zhou S, Zhuang X, Rabczuk T (2018) A phase-field modeling approach of fracture propagation in poroelastic media. Eng Geol 240:189–203
Zhuang X, Augarde C (2010) Aspects of the use of orthogonal basis functions in the element-free galerkin method. Int J Numer Methods Eng 81(3):366–380
Zienkiewicz OC, Taylor RL, Nithiarasu P, Zhu J (1977) The finite element method, vol 3. McGraw-hill, London

Chapter 8
Nonlocal Strong Forms of Thin Plate, Gradient Elasticity, Magneto–Electro-Elasticity and Phase Field Fracture by Nonlocal Operator Method

Although much progress in nonlocal methods has been achieved in the literature, the derivations of nonlocal models for many physical problems remain cumbersome and complicated; see, for example, (Chen and Chan 2020; Chowdhury et al. 2016; Javili et al. 2020; Wang et al. 2018). In this chapter, we make use of the power of NOM in deriving nonlocal models and present the explicit algorithm for solving the nonlocal models. The remainder of the chapter is outlined as follows. In Sect. 8.2, the second-order NOM in 2D/3D is formulated in detail. In Sect. 8.3, we apply the NOM scheme combined with variational principle/weighted residual method to derive the nonlocal governing equations for elasticity, thin plate, gradient elasticity, electro–magneto-elasticity and phase field fracture model. The correspondence between local form and nonlocal form for higher order problems is discussed. In Sect. 8.4, an instability criterion of nonlocal gradient is presented in the fracture modeling of linear elastic solid. The implementation of nonlocal solid and nonlocal thin plates is discussed in Sect. 8.5. Several numerical examples for solid and thin plates are used to demonstrate the accuracy and efficiency of the current method in Sect. 8.6.

8.1 Second Order Nonlocal Operator Method

NOM uses the integral form to replace the partial differential derivatives of different orders. Although NOM can solve higher order linear/nonlinear problems in 2D/3D, we restrict our discussion to second-order NOM, which is sufficient for the nonlocal derivation of the physical problems to be studied in Sect. 8.3.

T. Rabczuk et al., *Computational Methods Based on Peridynamics and Nonlocal Operators*, Computational Methods in Engineering & the Sciences,
https://doi.org/10.1007/978-3-031-20906-2_8

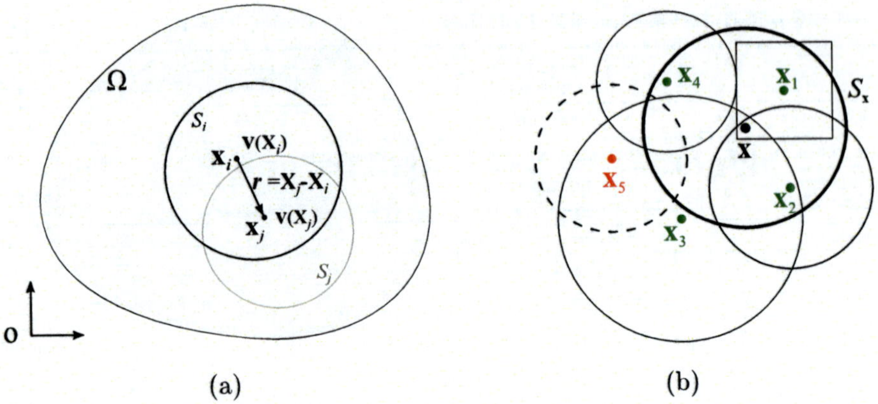

Fig. 8.1 **a** Domain and notation. **b** Schematic diagram for support and dual-support; all shapes above are supports, $\mathcal{S}_x = \{x_1, x_2, x_4\}$, $\mathcal{S}'_x = \{x_1, x_2, x_3\}$

8.1.1 Support and Dual-Support

Consider a domain as shown in Fig. 8.1a; let $\boldsymbol{x}_i$ be spatial coordinates in the domain $\boldsymbol{\Omega}$; $\mathbf{r}_{ij} := \boldsymbol{x}_j - \boldsymbol{x}_i$ is a spatial vector starting from $\boldsymbol{x}_i$ to $\boldsymbol{x}_j$; $\mathbf{v}_i := \mathbf{v}(\boldsymbol{x}_i, t)$ and $\mathbf{v}_j := \mathbf{v}(\boldsymbol{x}_j, t)$ are the field values for $\boldsymbol{x}_i$ and $\boldsymbol{x}_j$, respectively; $\mathbf{v}_{ij} := \mathbf{v}_j - \mathbf{v}_i$ is the relative field vector for spatial vector $\mathbf{r}_{ij}$.

Support $\mathcal{S}_i$ is the neighborhood of point $\boldsymbol{x}_i$. A point $\boldsymbol{x}_j$ in support $\mathcal{S}_i$ forms the spatial vector $\mathbf{r}_{ij}(= \boldsymbol{x}_j - \boldsymbol{x}_i)$. The support in NOM can be a spherical domain, a cube, semi-spherical domain and so on.

Dual-support is defined as a union of points whose supports include $\boldsymbol{x}_i$, denoted by

$$\mathcal{S}'_i = \{\boldsymbol{x}_j | \boldsymbol{x}_i \in \mathcal{S}_j\}. \tag{8.1}$$

Point $\boldsymbol{x}_j$ forms the dual-vector $\mathbf{r}_{ji}(= \boldsymbol{x}_i - \boldsymbol{x}_j = -\mathbf{r}_{ij})$ in $\mathcal{S}'_i$. On the other hand, $\mathbf{r}_{ji}$ is the spatial vector formed in $\mathcal{S}_j$. It is worth mentioning that the size of the support of each point can be different. When the support sizes for all material points are the same, the dual-support is equal to the support. On the other hand, if the size of the support varies for each point, the shape of dual-support can be quite irregular, even discontinuous for two adjacent points. One example to illustrate the support and dual-support is shown in Fig. 8.1b.

8.1.2 Dual Property of Dual-Support

Let f_{ij} be work conjugated to field difference $(u_j - u_i)$, the dual property of dual-support is

$$\int_{\Omega}\int_{\mathcal{S}_i} f_{ij}(u_j - u_i)\mathrm{d}V_j\mathrm{d}V_i = \int_{\Omega}\Big(\int_{\mathcal{S}_i'} f_{ji}\mathrm{d}V_j - \int_{\mathcal{S}_i} f_{ij}\mathrm{d}V_j\Big)u_i\mathrm{d}V_i. \tag{8.2}$$

Proof Let the domain Ω be divided into N non-overlapping particles, so that $\Omega = \sum_{i=1}^{N} \Delta V_i$, where ΔV_i is the volume assigned to particle i. Herein, N can be arbitrarily large so that the ΔV_i is infinitesimal and the double summations of discrete form converge to the double integrals in continuous form:

$$\begin{aligned}
&\int_{\Omega}\int_{\mathcal{S}_i} f_{ij}(u_j - u_i)\mathrm{d}V_j\mathrm{d}V_i\\
&\approx \sum_{\Delta V_i\in\Omega}\sum_{\Delta V_j\in\mathcal{S}_i} f_{ij}(u_j - u_i)\Delta V_j\Delta V_i\\
&= \sum_{\Delta V_i\in\Omega}\sum_{\Delta V_j\in\mathcal{S}_i} f_{ij}u_j\Delta V_j\Delta V_i - \sum_{\Delta V_i\in\Omega}\sum_{\Delta V_j\in\mathcal{S}_i} f_{ij}u_i\Delta V_j\Delta V_i\\
&= \sum_{\Delta V_i\in\Omega}\sum_{\Delta V_j\in\mathcal{S}_i'} f_{ji}u_i\Delta V_j\Delta V_i - \sum_{\Delta V_i\in\Omega}\sum_{\Delta V_j\in\mathcal{S}_i} f_{ij}u_i\Delta V_j\Delta V_i\\
&\approx \int_{\Omega}\Big(\int_{\mathcal{S}_i'} f_{ji}\mathrm{d}V_j - \int_{\mathcal{S}_i} f_{ij}\mathrm{d}V_j\Big)u_i\mathrm{d}V_i.
\end{aligned} \tag{8.3}$$

In the third step, the dual-support is considered as follows. The term f_{ij} with u_j is the physical quantity from i's support, but is added to particle j; since $j \in \mathcal{S}_i$, i belongs to the dual-support $\mathcal{S}_j'$ of j; all terms f_{ji} with u_i are collected from any material point j whose support contains i and hence form the dual-support of i. Therefore, the dual property of the dual-support is proved.

When all points have the same size of support domains, i.e. $j \in \mathcal{S}_i \leftrightarrow i \in \mathcal{S}_j$, we have $\mathcal{S}_i = \mathcal{S}_i'$ for any point i and then the dual property of dual-support by Eq. 8.2 becomes

$$\int_{\Omega}\int_{\mathcal{S}_i} f_{ij}(u_j - u_i)\mathrm{d}V_j\mathrm{d}V_i = \int_{\Omega}\int_{\mathcal{S}_i}(f_{ji} - f_{ij})u_i\mathrm{d}V_j\mathrm{d}V_i. \tag{8.4}$$

The above equation is widely used in the derivation of nonlocal strong form from weak form. Such expression is valid in the continuum form as well as in the discrete form. The dual property of dual-support is also proved in the dual-horizon peridynamics (Ren et al. 2017). A simple example with $N = 4$ to illustrate this property is given in Sect. 8.1.3.

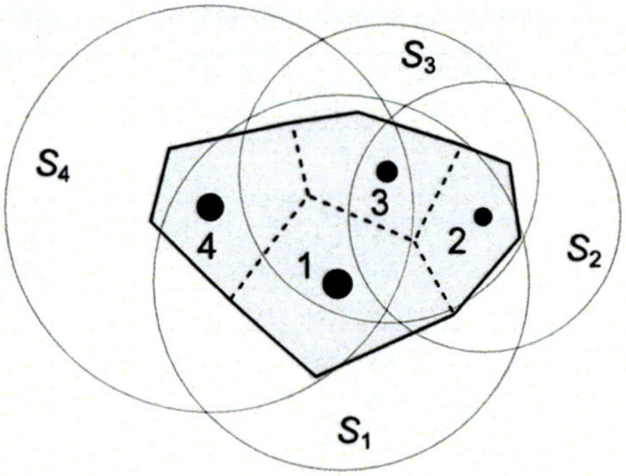

Fig. 8.2 Particles 1–4 and their supports $\mathcal{S}_i, i = \{1, 2, 3, 4\}$

8.1.3 A Simple Example to Illustrate Dual-Support

In order to facilitate the comprehension of dual-support, let us consider 4 particles in Fig. 8.2, each with particle volume $\Delta V_i, i = \{1, 2, 3, 4\}$ and $\Omega = \sum_{i=1}^{4} \Delta V_i$. Obviously, the support and dual-support can be listed as follows:

$$
\begin{aligned}
\mathcal{S}_1 &= \{2, 3, 4\}, \mathcal{S}_1' = \{3, 4\} \\
\mathcal{S}_2 &= \{3\}, \mathcal{S}_2' = \{1, 3\} \\
\mathcal{S}_3 &= \{1, 2\}, \mathcal{S}_3' = \{1, 2, 4\} \\
\mathcal{S}_4 &= \{1, 3\}, \mathcal{S}_4' = \{1\}.
\end{aligned}
$$

Here we neglect whether the shape tensor is invertible or not.

The most common formula in the derivation based on NOM and variational principle is the double integrations in support and the whole domain. Consider the double integrations

$$
\begin{aligned}
&\int_{\Omega} \int_{\mathcal{S}_i} f_{ij}(u_j - u_i) \mathrm{d}V_j \mathrm{d}V_i \\
&\approx \sum_{i=1}^{4} \Big(\sum_{j \in \mathcal{S}_i} f_{ij}(u_j - u_i) \Delta V_j \Big) \Delta V_i \\
&= \sum_{i=1}^{4} \Big(\sum_{j \in \mathcal{S}_i} f_{ij} u_j \Delta V_j \Big) \Delta V_i - \sum_{i=1}^{4} \Big(\sum_{j \in \mathcal{S}_i} f_{ij} u_i \Delta V_j \Big) \Delta V_i .
\end{aligned}
$$

Expand the double summations

$$
\begin{aligned}
&\sum_{i=1}^{4}\Big(\sum_{j\in\mathcal{S}_i} f_{ij}u_j\Delta V_j\Big)\Delta V_i \\
&=\Big(f_{12}\Delta V_2\Delta V_1 u_2+f_{13}\Delta V_3\Delta V_1 u_3+f_{14}\Delta V_4\Delta V_1 u_4\Big)+\Big(f_{23}\Delta V_2\Delta V_3 u_3\Big)\\
&\quad+\Big(f_{31}\Delta V_1\Delta V_3 u_1+f_{32}\Delta V_2\Delta V_3 u_2\Big)+\Big(f_{41}\Delta V_1\Delta V_4 u_1+f_{43}\Delta V_3\Delta V_4 u_3\Big)\\
&=\Big(f_{31}\Delta V_3+f_{41}\Delta V_4\Big)u_1\Delta V_1+\Big(f_{12}\Delta V_1+f_{32}\Delta V_3\Big)u_2\Delta V_2\\
&\quad+\Big(f_{13}\Delta V_1+f_{23}\Delta V_2+f_{43}\Delta V_4\Big)u_3\Delta V_3+\Big(f_{14}\Delta V_1\Big)u_4\Delta V_4\\
&=\sum_{j\in\mathcal{S}'_1} f_{j1}\Delta V_j\Delta V_1 u_1+\sum_{j\in\mathcal{S}'_2} f_{j2}\Delta V_j\Delta V_2 u_2+\sum_{j\in\mathcal{S}'_3} f_{j3}\Delta V_j\Delta V_3 u_3+\sum_{j\in\mathcal{S}'_4} f_{j4}\Delta V_j\Delta V_4 u_4\\
&=\sum_{i=1}^{4}\Big(\sum_{j\in\mathcal{S}'_i} f_{ji}\Delta V_j\Big)u_i\Delta V_i.
\end{aligned}
\tag{8.5}
$$

Therefore

$$
\begin{aligned}
&\sum_{i=1}^{4}\Big(\sum_{j\in\mathcal{S}_i} f_{ij}u_j\Delta V_j\Big)\Delta V_i-\sum_{i=1}^{4}\Big(\sum_{j\in\mathcal{S}_i} f_{ij}u_i\Delta V_j\Big)\Delta V_i\\
&=\sum_{i=1}^{4}\Big(\sum_{j\in\mathcal{S}'_i} f_{ji}\Delta V_j\Big)u_i\Delta V_i-\sum_{i=1}^{4}\Big(\sum_{j\in\mathcal{S}_i} f_{ij}u_i\Delta V_j\Big)\Delta V_i\\
&=\sum_{i=1}^{4}\Big(\sum_{j\in\mathcal{S}'_i} f_{ji}\Delta V_j-\sum_{j\in\mathcal{S}_i} f_{ij}\Delta V_j\Big)u_i\Delta V_i\\
&\approx\int_{\Omega}\Big(\int_{\mathcal{S}'_i} f_{ji}\,\mathrm{d}V_j-\int_{\mathcal{S}_i} f_{ij}\,\mathrm{d}V_j\Big)u_i\,\mathrm{d}V_i.
\end{aligned}
\tag{8.6}
$$

At last, we obtain

$$
\int_{\Omega}\int_{\mathcal{S}_i} f_{ij}(u_j-u_i)\,\mathrm{d}V_j\,\mathrm{d}V_i=\int_{\Omega}\Big(\int_{\mathcal{S}'_i} f_{ji}\,\mathrm{d}V_j-\int_{\mathcal{S}_i} f_{ij}\,\mathrm{d}V_j\Big)u_i\,\mathrm{d}V_i.
\tag{8.7}
$$

The above equation is widely used in the derivation of nonlocal strong form from weak form. Such expression is valid in the continuum form as well as in the discrete form.

8.1.4 Nonlocal Gradient and Hessian Operator

The local gradient operator and Hessian operator for a scalar-valued function u have the forms in 2D

$$\nabla u = \left(u_{,x}, u_{,y}\right)^T, \quad \nabla^2 u = \begin{pmatrix} u_{,xx} & u_{,xy} \\ u_{,xy} & u_{,yy} \end{pmatrix} \tag{8.8}$$

and in 3D

$$\nabla u = \left(u_{,x}, u_{,y}, u_{,z}\right)^T, \quad \nabla^2 u = \begin{pmatrix} u_{,xx} & u_{,xy} & u_{,xz} \\ u_{,xy} & u_{,yy} & u_{,yz} \\ u_{,xz} & u_{,yz} & u_{,zz} \end{pmatrix} \tag{8.9}$$

where $u_{,xx}$ denotes the partial derivative of u with respect to x twice.

In the framework of NOM, the partial derivatives can be constructed as follows. The Taylor series expansion of scalar-valued field u_j in 2D can be written as

$$u_j = u_i + (u_{i,x}, u_{i,y}, u_{i,xx}, u_{i,xy}, u_{i,yy}) \cdot (x_{ij}, y_{ij}, x_{ij}^2/2, x_{ij}y_{ij}, y_{ij}^2/2) + O(|\mathbf{r}_{ij}|^3) \tag{8.10}$$

where $\boldsymbol{r}_{ij} = (x_{ij}, y_{ij})^T = \boldsymbol{x}_j - \boldsymbol{x}_i$ and $O(|\mathbf{r}_{ij}|^3)$ denotes the higher order terms.

Let

$$u_{ij} = u_j - u_i \tag{8.11}$$

$$\boldsymbol{p}_{ij} = (x_{ij}, y_{ij}, x_{ij}^2/2, x_{ij}y_{ij}, y_{ij}^2/2)^T \tag{8.12}$$

$$\partial u_i = (u_{i,x}, u_{i,y}, u_{i,xx}, u_{i,xy}, u_{i,yy})^T. \tag{8.13}$$

The Taylor series expansion of Eq. 8.10 can be rewritten as

$$u_{ij} = \partial u_i^T \boldsymbol{p}_{ij}. \tag{8.14}$$

Tensor product with $\boldsymbol{p}_{ij}^T$ on both sides of Eq. 8.14

$$u_{ij}\boldsymbol{p}_{ij}^T = \partial u_i^T \boldsymbol{p}_{ij}\boldsymbol{p}_{ij}^T. \tag{8.15}$$

Considering the weighted integration in the support $\mathcal{S}_i$, we obtain

$$\int_{\mathcal{S}_i} \omega(\mathbf{r}_{ij}) u_{ij} \boldsymbol{p}_{ij}^T \mathrm{d}V_j = \partial u_i^T \int_{\mathcal{S}_i} \omega(\mathbf{r}_{ij}) \boldsymbol{p}_{ij} \boldsymbol{p}_{ij}^T \mathrm{d}V_j \tag{8.16}$$

where $\omega(\mathbf{r}_{ij})$ is the weight function.

Then the nonlocal operators can be obtained as

$$\tilde{\partial} u_i := \int_{\mathcal{S}_i} \omega(\mathbf{r}_{ij}) \boldsymbol{K}_i \cdot \boldsymbol{p}_{ij} u_{ij} \mathrm{d}V_j \tag{8.17}$$

where

$$K_i = \left(\int_{S_i} \omega(\mathbf{r}_{ij}) \boldsymbol{p}_{ij} \otimes \boldsymbol{p}_{ij}^T \mathrm{d}V_j \right)^{-1}. \tag{8.18}$$

Here, we use $\tilde{\Box}$ to denote the nonlocal form of the local operator $\Box$ since the definitions of the local operator and the nonlocal operator are distinct.

The Taylor series expansion of a vector field $\mathbf{u}$ can be obtained in a similar manner as

$$\boldsymbol{u}_{ij}^T = \boldsymbol{p}_{ij}^T \cdot \partial \boldsymbol{u}_i \tag{8.19}$$

$$\omega(\mathbf{r}_{ij}) \boldsymbol{p}_{ij} \otimes \boldsymbol{u}_{ij}^T = \omega(\mathbf{r}_{ij}) \boldsymbol{p}_{ij} \otimes \boldsymbol{p}_{ij}^T \cdot \partial \boldsymbol{u}_i \tag{8.20}$$

$$\int_{S_i} \omega(\mathbf{r}_{ij}) \boldsymbol{p}_{ij} \otimes \boldsymbol{u}_{ij}^T \mathrm{d}V_j = \int_{S_i} \omega(\mathbf{r}_{ij}) \boldsymbol{p}_{ij} \otimes \boldsymbol{p}_{ij}^T \cdot \partial \boldsymbol{u}_i \mathrm{d}V_j. \tag{8.21}$$

That is

$$\tilde{\partial} \boldsymbol{u}_i := \int_{S_i} \omega(\mathbf{r}_{ij}) \boldsymbol{K}_i \cdot \boldsymbol{p}_{ij} \otimes \boldsymbol{u}_{ij}^T \mathrm{d}V_j. \tag{8.22}$$

For example, consider the displacement field $\boldsymbol{u} = (u, v)^T$ in 2D space; the relative displacement vector and the nonlocal partial derivatives have the explicit forms

$$\boldsymbol{u}_{ij} = \begin{pmatrix} u_j - u_i \\ v_j - v_i \end{pmatrix}, \quad \tilde{\partial} \boldsymbol{u}_i = (\tilde{\partial} u_i, \tilde{\partial} v_i) = \begin{pmatrix} u_{i,x} & v_{i,x} \\ u_{i,y} & v_{i,y} \\ u_{i,xx} & v_{i,xx} \\ u_{i,xy} & v_{i,xy} \\ u_{i,yy} & v_{i,yy} \end{pmatrix}, \tag{8.23}$$

Let $\boldsymbol{K}_i \cdot \boldsymbol{p}_{ij}$ be denoted by

$$(g_{1j}, g_{2j}, h_{1j}, h_{2j}, h_{3j})^T = \boldsymbol{K}_i \cdot \boldsymbol{p}_{ij}. \tag{8.24}$$

The gradient vector $\boldsymbol{g}_{ij}$ and Hessian matrix $\boldsymbol{h}_{ij}$ between points i and j in 2D are, respectively,

$$\boldsymbol{g}_{ij} = (g_{1j}, g_{2j})^T, \quad \boldsymbol{h}_{ij} = \begin{pmatrix} h_{1j} & h_{2j} \\ h_{2j} & h_{3j} \end{pmatrix}. \tag{8.25}$$

In 3D case, the polynomial vector based on relative coordinates $\boldsymbol{r}_{ij} = (x_{ij}, y_{ij}, z_{ij})^T = \boldsymbol{x}_j - \boldsymbol{x}_i$ is given as

$$\boldsymbol{p}_{ij} = (x_{ij}, y_{ij}, z_{ij}, x_{ij}^2/2, x_{ij}y_{ij}, x_{ij}z_{ij}, y_{ij}^2/2, y_{ij}z_{ij}, z_{ij}^2)^T. \tag{8.26}$$

The shape tensor in 3D is constructed by Eq. 8.18 with $\boldsymbol{p}_{ij}$ in Eq. 8.26.

Let $\boldsymbol{K}_i \cdot \boldsymbol{p}_{ij}$ in 3D be denoted by

$$(g_{1j}, g_{2j}, g_{3j}, h_{1j}, h_{2j}, h_{3j}, h_{4j}, h_{5j}, h_{6j})^T = \boldsymbol{K}_i \cdot \boldsymbol{p}_{ij}. \tag{8.27}$$

The gradient vector $\boldsymbol{g}_{ij}$ and Hessian matrix $\boldsymbol{h}_{ij}$ for two points i, j in support in 3D are, respectively,

$$\boldsymbol{g}_{ij} = (g_{1j}, g_{2j}, g_{3j})^T, \quad \boldsymbol{h}_{ij} = \begin{pmatrix} h_{1j} & h_{2j} & h_{3j} \\ h_{2j} & h_{4j} & h_{5j} \\ h_{3j} & h_{5j} & h_{6j} \end{pmatrix}. \tag{8.28}$$

It is worth mentioning that for first-order NOM or peridynamics, the gradient vector can be calculated as well by

$$\boldsymbol{g}_{ij} = \left(\int_{S_i} \omega(\mathbf{r}_{ik}) \mathbf{r}_{ik} \otimes \mathbf{r}_{ik} \mathrm{d}V_k \right)^{-1} \cdot \mathbf{r}_{ij}. \tag{8.29}$$

Then the nonlocal gradient operator and Hessian operator for the vector field can be defined as

$$\tilde{\nabla} \otimes \boldsymbol{u}_i := \int_{S_i} \omega(\mathbf{r}_{ij}) \boldsymbol{u}_{ij} \otimes \boldsymbol{g}_{ij} \mathrm{d}V_j \tag{8.30}$$

$$\tilde{\nabla} \otimes \tilde{\nabla} \otimes \boldsymbol{u}_i := \int_{S_i} \omega(\mathbf{r}_{ij}) \boldsymbol{u}_{ij} \otimes \boldsymbol{h}_{ij} \mathrm{d}V_j. \tag{8.31}$$

In the case of 2-vector in 2D space, the explicit forms of $\tilde{\nabla} \otimes \boldsymbol{u}_i$ and $\tilde{\nabla} \otimes \tilde{\nabla} \otimes \boldsymbol{u}_i$ are

$$\tilde{\nabla} \otimes \boldsymbol{u}_i = \begin{pmatrix} u_{i,x} & u_{i,y} \\ v_{i,x} & v_{i,y} \end{pmatrix} \tag{8.32}$$

$$\tilde{\nabla} \otimes \tilde{\nabla} \otimes \boldsymbol{u}_i = \begin{pmatrix} \frac{\partial(\tilde{\nabla} \otimes \boldsymbol{u}_i)}{\partial x} & \frac{\partial(\tilde{\nabla} \otimes \boldsymbol{u}_i)}{\partial y} \end{pmatrix} = \left(\begin{pmatrix} u_{i,xx} & u_{i,yx} \\ v_{i,xx} & v_{i,yx} \end{pmatrix} \begin{pmatrix} u_{i,xy} & u_{i,yy} \\ v_{i,xy} & v_{i,yy} \end{pmatrix} \right). \tag{8.33}$$

For scalar-valued field, the nonlocal Laplace operator is the tensor contraction of $\tilde{\nabla} \otimes \tilde{\nabla} u_i$, e.g. $\tilde{\Delta} = \tilde{\nabla} \cdot \tilde{\nabla} = \mathrm{tr}(\tilde{\nabla} \otimes \tilde{\nabla})$, where $tr(\cdot)$ denotes the trace of a matrix. More specifically, in 2D

$$\tilde{\Delta} u_i := \int_{S_i} \omega(\mathbf{r}_{ij})(h_{1j} + 2h_{2j} + h_{3j}) u_{ij} \mathrm{d}V_j \tag{8.34}$$

and in 3D

$$\tilde{\Delta} u_i := \int_{S_i} \omega(\mathbf{r}_{ij})(h_{1j} + 2h_{2j} + 2h_{3j} + h_{4j} + 2h_{5j} + h_{6j}) u_{ij} \mathrm{d}V_j. \tag{8.35}$$

And their local counterparts for the scalar-valued field are

$$\Delta w = w_{,yy} + 2w_{,xy} + w_{,xx} \qquad \text{in 2D} \tag{8.36}$$

$$\Delta w = w_{,xx} + w_{,yy} + w_{,zz} + 2w_{,xy} + 2w_{,xz} + 2w_{,yz} \qquad \text{in 3D.} \tag{8.37}$$

8.1.5 Stability of the Second-Order Nonlocal Operators

According to (Ren et al. 2020b), the energy functional for the second-order nonlocal operator in discrete form can be written as

$$\mathcal{F}_i(\mathbf{u}) = \frac{1}{2}\frac{p^{hg}}{m_i} \int_{S_i} \omega(\mathbf{r}_{ij}) \big(u_{ij} - \boldsymbol{p}_j^T \tilde{\partial} u_i\big)^2 \mathrm{d}V_j \tag{8.38}$$

where p^{hg} is the penalty and $m_i = \int_{S_i} \omega(\mathbf{r}_{ij}) \mathrm{d}V_j$. The operator in Eq. 8.17 corresponds to the minimum of Eq. 8.38. The first variation of $\mathcal{F}_i$ is

$$\begin{aligned}
\delta \mathcal{F}_i(\mathbf{u}) &= \frac{p^{hg}}{m_i} \int_{S_i} \omega(\mathbf{r}_{ij}) \big(u_{ij} - \boldsymbol{p}_j^T \tilde{\partial} u_i\big)(\delta u_j - \delta u_i - \boldsymbol{p}_j^T \tilde{\partial} \delta u_i) \mathrm{d}V_j \\
&= \frac{p^{hg}}{m_i} \int_{S_i} \omega(\mathbf{r}_{ij}) \big(u_{ij} - \boldsymbol{p}_j^T \tilde{\partial} u_i\big)(\delta u_j - \delta u_i) \mathrm{d}V_j \\
&- \frac{p^{hg}}{m_i} \int_{S_i} \omega(\mathbf{r}_{ij}) \big(u_{ij} - \boldsymbol{p}_j^T \tilde{\partial} u_i\big)(\boldsymbol{p}_j^T \tilde{\partial} \delta u_i) \mathrm{d}V_j.
\end{aligned} \tag{8.39}$$

We can prove that

$$\begin{aligned}
& - \frac{p^{hg}}{m_i} \int_{S_i} \omega(\mathbf{r}_{ij}) \big(u_{ij} - \boldsymbol{p}_j^T \tilde{\partial} u_i\big)(\boldsymbol{p}_j^T \tilde{\partial} \delta u_i) \mathrm{d}V_j \\
&= - \frac{p^{hg}}{m_i} \int_{j \in S_i} \omega(\mathbf{r}_{ij}) \big(\boldsymbol{p}_j u_{ij} - \boldsymbol{p}_j \boldsymbol{p}_j^T \tilde{\partial} u_i\big) \mathrm{d}V_j \cdot \tilde{\partial} \delta u_i \\
&= - \frac{p^{hg}}{m_i} \underbrace{\Big(\int_{S_i} \omega(\mathbf{r}_{ij}) \boldsymbol{p}_j u_{ij} \mathrm{d}V_j - \int_{S_i} \omega(\mathbf{r}) \boldsymbol{p}_j \boldsymbol{p}_j^T \mathrm{d}V_j \cdot \tilde{\partial} u_i \Big)}_{=0 \text{ since Eq. 8.16}} \cdot \tilde{\partial} \delta u_i \\
&= 0
\end{aligned}$$

Therefore,

$$\delta\mathcal{F}_i(\mathbf{u}) = \frac{p^{hg}}{m_i}\int_{\mathcal{S}_i} \omega(\mathbf{r}_{ij})\big(u_{ij} - \boldsymbol{p}_j^T\tilde{\partial}u_i\big)(\delta u_j - \delta u_i)\mathrm{d}V_j.$$

Consider integration of $\delta\mathcal{F}_i(\mathbf{u})$ in domain

$$\int_{\Omega}\delta\mathcal{F}_i\mathrm{d}V_i = p^{hg}\underbrace{\int_{\Omega}\int_{\mathcal{S}_i}\frac{\omega(\mathbf{r}_{ij})}{m_i}(u_{ij} - \boldsymbol{p}_j^T\tilde{\partial}u_i)(\delta u_j - \delta u_i)\mathrm{d}V_j\mathrm{d}V_i}_{\text{by Eq. 8.2}}$$

$$= \int_{\Omega}\Big(\Big(\int_{\mathcal{S}_i'}\omega(\mathbf{r}_{ij})\frac{p^{hg}}{m_j}(u_{ji} - \boldsymbol{p}_i^T\tilde{\partial}u_j)\mathrm{d}V_j - \int_{\mathcal{S}_i}\omega(\mathbf{r})\frac{p^{hg}}{m_i}(u_{ij} - \boldsymbol{p}_j^T\tilde{\partial}u_i)\mathrm{d}V_j\Big)\delta u_i\mathrm{d}V_i \tag{8.40}$$

For any δu_i, $\int_{\Omega}\delta\mathcal{F}_i\mathrm{d}V_i = 0$ leads to the internal force due to the stability of the nonlocal operator

$$\int_{\mathcal{S}_i'}\omega(\mathbf{r}_{ij})\frac{p^{hg}}{m_j}(u_{ji} - \boldsymbol{p}_i^T\tilde{\partial}u_j)\mathrm{d}V_j - \int_{\mathcal{S}_i}\omega(\mathbf{r})\frac{p^{hg}}{m_i}(u_{ij} - \boldsymbol{p}_j^T\tilde{\partial}u_i)\mathrm{d}V_j. \tag{8.41}$$

Equation 8.41 is the expression for a scalar-valued field. For vector-valued field, the internal force due to the stability of the nonlocal operator is

$$\int_{\mathcal{S}_i'}\omega(\mathbf{r}_{ji})\frac{p^{hg}}{m_j}(\boldsymbol{u}_{ji} - \boldsymbol{p}_i^T\tilde{\partial}\boldsymbol{u}_j)\mathrm{d}V_j - \int_{\mathcal{S}_i}\omega(\mathbf{r}_{ij})\frac{p^{hg}}{m_i}(\boldsymbol{u}_{ij} - \boldsymbol{p}_j^T\tilde{\partial}\boldsymbol{u}_i)\mathrm{d}V_j. \tag{8.42}$$

8.2 Nonlocal Governing Equations Based on NOM

This section is devoted to the variational derivation of nonlocal strong forms of solid mechanics, including hyperelasticity, thin plate, gradient elasticity, electro–magneto-elasticity theory and phase field fracture method. The strong form is suitable for theoretical analysis as well as explicit time integration. For the fully implicit simulation of various PDEs, the reader is referred to NOM for PDEs (Rabczuk et al. 2019; Ren et al. 2020a, b, c, d, 2021).

8.2.1 *Nonlocal Form for Hyperelasticity*

Consider the energy density of a hyperelasticity as $\psi := \psi(\boldsymbol{F})$, where $\boldsymbol{F} = \nabla\boldsymbol{u} + \boldsymbol{I}$. The balance equation for the hyperelastic solid is

$$\nabla\cdot\boldsymbol{P} + \boldsymbol{b} = 0 \text{ on } \Omega \tag{8.43}$$

with boundary conditions $\boldsymbol{u} = \boldsymbol{u}_0$ on Γ_D and $\boldsymbol{P} \cdot \boldsymbol{n} = \boldsymbol{t}_0$ on Γ_N, where $\boldsymbol{u}_0$ is the specified displacement and $\boldsymbol{t}_0$ is the prescribed traction load, $\boldsymbol{P} = \frac{\partial \psi}{\partial \boldsymbol{F}}$, the first Piola–Kirchhoff stress, and $\boldsymbol{b}$ is the body force density.

Derivation based on variational principle

The variation of strain energy over the domain is

$$
\begin{aligned}
\delta \mathcal{F} &= \int_{\Omega} \delta\psi(\boldsymbol{F})\mathrm{d}V = \int_{\Omega} \frac{\partial \psi}{\partial \boldsymbol{F}} : \delta \boldsymbol{F}\mathrm{d}V \\
&= \int_{\Omega} \boldsymbol{P} : \nabla(\delta \boldsymbol{u})\mathrm{d}V \\
&= \int_{\Omega} \boldsymbol{P}_i : \int_{S_i} \omega(\mathbf{r}_{ij})\delta \boldsymbol{u}_{ij} \otimes \boldsymbol{g}_{ij}\mathrm{d}V_j\mathrm{d}V_i \\
&= \int_{\Omega}\int_{S_i} \omega(\mathbf{r}_{ij})\boldsymbol{P}_i : \delta \boldsymbol{u}_{ij} \otimes \boldsymbol{g}_{ij}\mathrm{d}V_j\mathrm{d}V_i \\
&= \int_{\Omega}\int_{S_i} \omega(\mathbf{r}_{ij})(\boldsymbol{P}_i \cdot \boldsymbol{g}_{ij}) \cdot \delta \boldsymbol{u}_{ij}\mathrm{d}V_j\mathrm{d}V_i \\
&= \underbrace{\int_{\Omega}\int_{S_i} \omega(\mathbf{r}_{ij})(\boldsymbol{P}_i \cdot \boldsymbol{g}_{ij}) \cdot (\delta \boldsymbol{u}_j - \delta \boldsymbol{u}_i)\mathrm{d}V_j\mathrm{d}V_i}_{\text{by Eq. 8.2}} \\
&= \int_{\Omega}\Big(\int_{S_i'} \omega(\mathbf{r}_{ji})\boldsymbol{P}_j \cdot \boldsymbol{g}_{ji}\mathrm{d}V_j - \int_{S_i} \omega(\mathbf{r}_{ij})\boldsymbol{P}_i \cdot \boldsymbol{g}_{ij}\mathrm{d}V_j\Big) \cdot \delta \mathbf{u}_i\mathrm{d}V_i
\end{aligned}
\tag{8.44}
$$

In above derivation, we replace the gradient operator with nonlocal gradient, e.g. $\tilde{\nabla} \otimes \boldsymbol{u}_i \to \int_{S_i} \omega(\mathbf{r}_{ij})\boldsymbol{u}_{ij} \otimes \boldsymbol{g}_{ij}\mathrm{d}V_j$ in Eq. 8.30, and the relation $\boldsymbol{A} : \boldsymbol{a} \otimes \boldsymbol{b} = (\boldsymbol{A} \cdot \boldsymbol{b}) \cdot \boldsymbol{a}$ for second-order tensor $\boldsymbol{A}$ and vectors $\boldsymbol{a}, \boldsymbol{b}$ is employed.

The variation of external body force energy

$$\delta \mathcal{F}_{ext} = \int_{\Omega} \delta \boldsymbol{u} \cdot \boldsymbol{b}\mathrm{d}V. \tag{8.45}$$

For any $\delta \boldsymbol{u}_i$, $\delta \mathcal{F} - \delta \mathcal{F}_{ext} = 0$ leads to the nonlocal governing equations for elasticity

$$\int_{S_i} \omega(\mathbf{r}_{ij})\boldsymbol{P}_i \cdot \boldsymbol{g}_{ij}\mathrm{d}V_j - \int_{S_i'} \omega(\mathbf{r}_{ji})\boldsymbol{P}_j \cdot \boldsymbol{g}_{ji}\mathrm{d}V_j + \boldsymbol{b} = 0. \tag{8.46}$$

Considering the effect of inertial force $\rho\ddot{\boldsymbol{u}}_i$ per unit volume, and replacing the dual-support with dual-horizon, we obtain the equations of motion for dual-horizon peridynamics

$$\int_{\mathcal{H}_i} \omega(\mathbf{r}_{ij}) \boldsymbol{P}_i \cdot \boldsymbol{g}_{ij} \mathrm{d}V_j - \int_{\mathcal{H}_i'} \omega(\mathbf{r}_{ji}) \boldsymbol{P}_j \cdot \boldsymbol{g}_{ji} \mathrm{d}V_j + \boldsymbol{b}_i = \rho \ddot{\boldsymbol{u}}_i. \tag{8.47}$$

If the sizes of horizons for all material points are the same, the dual-horizon peridynamics degenerates to the conventional constant horizon peridynamics.

For any specific strain energy density (for example, isotropic/anisotropic linear/nonlinear elasticity), the explicit form of $\boldsymbol{P}$ can be derived straightforwardly. In the section of numerical examples, we consider linear isotropic elasticity, which can be viewed as a special case of hyperelasticity.

Derivation based on weighted residual method

Besides the derivation based on strain energy density, the nonlocal strong form can be derived by the weighted residual method. Considering the governing equations for hyperelasticity , the weak form of Eq. 8.43 for any trial vector becomes

$$\begin{aligned}
0 &= \int_{\Omega} \boldsymbol{v} \cdot \nabla \cdot \boldsymbol{P} + \boldsymbol{v} \cdot \boldsymbol{b} \mathrm{d}V \\
&= \int_{\Omega} -\nabla \boldsymbol{v} : \boldsymbol{P} + \boldsymbol{v} \cdot \boldsymbol{b} \mathrm{d}V + \int_{\Gamma} \boldsymbol{P} \cdot \boldsymbol{n} \cdot \boldsymbol{v} \mathrm{d}S \\
&= \int_{\Omega} -\Big(\int_{\mathcal{S}_i} \omega(\mathbf{r}_{ij}) \boldsymbol{v}_{ij} \otimes \boldsymbol{g}_{ij} \mathrm{d}V_j\Big) : \boldsymbol{P}_i + \boldsymbol{v}_i \cdot \boldsymbol{b} \mathrm{d}V_i + \int_{\Gamma} \boldsymbol{P} \cdot \boldsymbol{n} \cdot \boldsymbol{v} \mathrm{d}S.
\end{aligned} \tag{8.48}$$

Let us focus on the integral in Ω; the first term in the above equation can be written as

$$\begin{aligned}
&\int_{\Omega} -\Big(\int_{\mathcal{S}_i} \omega(\mathbf{r}_{ij}) \boldsymbol{v}_{ij} \otimes \boldsymbol{g}_{ij} \mathrm{d}V_j\Big) : \boldsymbol{P}_i \mathrm{d}V_i \\
&= \underbrace{\int_{\Omega} -\Big(\int_{\mathcal{S}_i} \omega(\mathbf{r}_{ij}) \boldsymbol{P}_i \cdot \boldsymbol{g}_{ij} \cdot (\boldsymbol{v}_j - \boldsymbol{v}_i) \mathrm{d}V_j\Big) \mathrm{d}V_i}_{\text{by Eq. 8.2}} \\
&= \int_{\Omega} \Big(\int_{\mathcal{S}_i} \omega(\mathbf{r}_{ij}) \boldsymbol{P}_i \cdot \boldsymbol{g}_{ij} \mathrm{d}V_j - \int_{\mathcal{S}_i'} \omega(\mathbf{r}_{ji}) \boldsymbol{P}_j \cdot \boldsymbol{g}_{ji} \mathrm{d}V_j\Big) \cdot \boldsymbol{v}_i \mathrm{d}V_i
\end{aligned} \tag{8.49}$$

For any $\boldsymbol{v}_i$, the weak form being zero leads to

$$\int_{\mathcal{S}_i} \omega(\mathbf{r}_{ij}) \boldsymbol{P}_i \cdot \boldsymbol{g}_{ij} \mathrm{d}V_j - \int_{\mathcal{S}_i'} \omega(\mathbf{r}_{ji}) \boldsymbol{P}_j \cdot \boldsymbol{g}_{ji} \mathrm{d}V_j + \boldsymbol{b} = 0$$

which is identical to Eq. 8.46. As being more general than the energy method, the weighted residual method can be used to convert PDEs that have no energy functional to nonlocal integral forms.

8.2.2 Nonlocal Thin Plate Theory

The thin plate theory is widely used in engineering applications (Timoshenko and Woinowsky-Krieger 1959). The basic assumptions of a thin plate are as follows: (1) the thickness of the plate is much smaller than the length inside the mid-plane; (2) the deflection is much smaller than the thickness of the plate so that higher order effect is neglectable; (3) the stress along the thickness direction is assumed as zero, e.g. $\sigma_z \approx 0$ and the points in the mid-plane have no displacement parallel to the mid-plane, e.g. $u(x, y, 0) = v(x, y, 0) \approx 0$; 4) the normal of the mid-plane remains perpendicular to the mid-plane after deformation. Then the plate bending can be simplified into a 2D problem and the displacements, strain and stress can be described by the deflection on the mid-plane

$$u(x, y, z) = -z\frac{\partial w}{\partial x} \tag{8.50}$$

$$v(x, y, z) = -z\frac{\partial w}{\partial y} \tag{8.51}$$

$$w(x, y, z) \simeq w(x, y, 0) \cong w(x, y). \tag{8.52}$$

The generalized strain is the Hessian operator on the deflection

$$\boldsymbol{\kappa} = \nabla^2 w = \begin{pmatrix} w_{,xx} & w_{,xy} \\ w_{,xy} & w_{,yy} \end{pmatrix} \tag{8.53}$$

with nonlocal correspondence and its variation

$$\boldsymbol{\kappa} = \tilde{\nabla}^2 w := \int_{S_i} \omega(\mathbf{r}_{ij})\boldsymbol{h}_{ij} w_{ij} \mathrm{d}V_j \tag{8.54}$$

$$\delta\boldsymbol{\kappa} = \int_{S_i} \omega(\mathbf{r}_{ij})\boldsymbol{h}_{ij} \delta w_{ij} \mathrm{d}V_j \tag{8.55}$$

The bending moment tensor $\boldsymbol{M}$, the general stress for isotropic thin plate, is given by

$$\boldsymbol{M} = \begin{pmatrix} M_{xx} & M_{xy} \\ M_{xy} & M_{yy} \end{pmatrix} = D_0\big(\nu \,\mathrm{tr}(\boldsymbol{\kappa})\boldsymbol{I}_{2\times 2} + (1-\nu)\boldsymbol{\kappa}\big) \tag{8.56}$$

where $D_0 = \frac{Et^3}{12(1-\nu^2)}$ and t is the thickness of the plate.

Based on the principle of minimum potential energy, the energy functional for the governing equation is

$$\mathcal{F}_{int} = \int_{\Omega} \frac{1}{2} \boldsymbol{M} : \boldsymbol{\kappa} - qw\mathrm{d}S \tag{8.57}$$

and for the boundary condition can be expressed as

$$\mathcal{F}_{ext} = \int_{S_3} \bar{V}_n w\mathrm{d}\Gamma - \int_{S_2+S_3} \bar{M}_n \frac{\partial w}{\partial n} \mathrm{d}\Gamma \tag{8.58}$$

where q is the external transverse load on the mid-plane, $\bar{V}_n$ is the shear force load on boundary S_3 and $\bar{M}_n$ is the prescribed moment on boundary $S_2 + S_3$. For simplicity, we leave the integral on the boundary for later consideration. The variation of the internal energy functional is

$$\begin{aligned}
\delta\mathcal{F}_{int} &= \int_{\Omega} \boldsymbol{M} : \delta\boldsymbol{\kappa} - q\delta w\mathrm{d}S \\
&= \int_{\Omega} \boldsymbol{M}_i : \int_{S_i} \omega(\mathbf{r}_{ij})\boldsymbol{h}_{ij}\delta w_{ij}\mathrm{d}S_j - q_i\delta w_i\mathrm{d}S_i \\
&= \underbrace{\int_{\Omega}\int_{S_i} \omega(\mathbf{r}_{ij})\boldsymbol{M}_i : \boldsymbol{h}_{ij}(\delta w_j - \delta w_i)\mathrm{d}S_j}_{\text{by Eq. 8.2}} - \int_{\Omega} q_i\delta w_i\mathrm{d}S_i \\
&= \int_{\Omega}\Big(\int_{S_i'} \omega(\mathbf{r}_{ij})\boldsymbol{M}_j : \boldsymbol{h}_{ji}\mathrm{d}S_j - \int_{S_i} \omega(\mathbf{r}_{ij})\boldsymbol{M}_i : \boldsymbol{h}_{ij}\mathrm{d}S_j - q_i\Big)\delta w_i\mathrm{d}S_i
\end{aligned} \tag{8.59}$$

The variation of the external energy function is

$$\begin{aligned}
\delta\mathcal{F}_{ext} &= \int_{S_3} \bar{V}_n\delta w\mathrm{d}\Gamma - \int_{S_2+S_3} \bar{M}_n \frac{\partial\delta w}{\partial n}\mathrm{d}\Gamma \\
&= \int_{S_3} \bar{V}_n\delta w\mathrm{d}\Gamma - \int_{S_2+S_3} \bar{M}_n \nabla\delta w \cdot \boldsymbol{n}\mathrm{d}\Gamma \\
&= \int_{S_3} \bar{V}_n\delta w\mathrm{d}\Gamma - \int_{S_2+S_3} \bar{M}_{ni} \int_{S_i} \omega(\mathbf{r}_{ij})\delta w_{ij}\boldsymbol{g}_{ij}\mathrm{d}V_j \cdot \boldsymbol{n}_i\mathrm{d}\Gamma_i \\
&= \int_{S_3} \bar{V}_n\delta w\mathrm{d}\Gamma - \int_{S_2+S_3}\int_{S_i} \omega(\mathbf{r}_{ij})\bar{M}_{ni}\boldsymbol{g}_{ij} \cdot \boldsymbol{n}_i\delta w_{ij}\mathrm{d}V_j\mathrm{d}\Gamma_i \\
&= \int_{S_3} \bar{V}_n\delta w\mathrm{d}\Gamma - \int_{S_2+S_3}\Big(\int_{S_i'} \omega(\mathbf{r}_{ji})\bar{M}_{nj}\boldsymbol{g}_{ji} \cdot \boldsymbol{n}_j\mathrm{d}V_j - \int_{S_i} \omega(\mathbf{r}_{ij})\bar{M}_{ni}\boldsymbol{g}_{ij} \cdot \boldsymbol{n}_i\mathrm{d}V_j\Big)\delta w_i\mathrm{d}\Gamma_i.
\end{aligned} \tag{8.60}$$

For any δw_i, $\delta\mathcal{F}_{int} - \delta\mathcal{F}_{ext} = 0$ leads to the nonlocal thin plate equation for material point in domain Ω

$$\int_{S_i} \omega(\mathbf{r}_{ij}) \boldsymbol{M}_i : \boldsymbol{h}_{ij} \mathrm{d}V_j - \int_{S_i'} \omega(\mathbf{r}_{ij}) \boldsymbol{M}_j : \boldsymbol{h}_{ji} \mathrm{d}V_j + q_i = 0. \tag{8.61}$$

The additional nonlocal form for material point applied with the moment boundary condition is

$$\int_{S_i} \omega(\mathbf{r}_{ij}) \bar{M}_{ni} \boldsymbol{g}_{ij} \cdot \boldsymbol{n}_i \mathrm{d}V_j - \int_{S_i'} \omega(\mathbf{r}_{ji}) \bar{M}_{nj} \boldsymbol{g}_{ji} \cdot \boldsymbol{n}_j \mathrm{d}V_j = 0. \tag{8.62}$$

Based on D'Alembert's principle, the equation of motion considering the effect of inertial force $\rho t \ddot{w}_i$ per unit area is

$$\int_{S_i'} \omega(\mathbf{r}_{ij}) \boldsymbol{M}_j : \boldsymbol{h}_{ji} \mathrm{d}V_j - \int_{S_i} \omega(\mathbf{r}_{ij}) \boldsymbol{M}_i : \boldsymbol{h}_{ij} \mathrm{d}V_j + q_i = t\rho \ddot{w}_i. \tag{8.63}$$

For clamped boundary condition $w_{,n} = \nabla w \cdot \boldsymbol{n} = 0$, the nonlocal form is

$$\int_{S_i} \omega(\mathbf{r}_{ij}) w_{ij} \boldsymbol{g}_{ij} \cdot \boldsymbol{n}_i \mathrm{d}V_j = 0. \tag{8.64}$$

Compared with the local governing equation for thin plate $\nabla^2 : \boldsymbol{M} + q = t\rho\ddot{w}$, we can find the correspondence between local and nonlocal formulation

$$\nabla^2 : \boldsymbol{M} \rightarrow \tilde{\nabla}^2 : \boldsymbol{M}_i := \int_{S_i'} \omega(\mathbf{r}_{ji}) \boldsymbol{M}_j : \boldsymbol{h}_{ji} \mathrm{d}V_j - \int_{S_i} \omega(\mathbf{r}_{ij}) \boldsymbol{M}_i : \boldsymbol{h}_{ij} \mathrm{d}V_j. \tag{8.65}$$

The nonlocal derivation for thin plates can be extended to composite plate and functional gradient plate theories.

8.2.3 Nonlocal Gradient Elasticity

Gradient theories emerge from considerations of the micro-structure in the material at micro-scale, where a mass point after homogenization is not the center of a micro-volume and the rotation of the micro-volume depends on the moment stress/couple stress as well as the Cauchy stress. Gradient elasticity generalizes the elasticity theory by employing higher order terms of the deformation gradient or the gradient of the strain tensor. Generally, the energy density functional can be assumed as $\psi := \psi(\boldsymbol{F}, \nabla \boldsymbol{F}) = \psi(\nabla \boldsymbol{u}, \nabla^2 \boldsymbol{u})$, where $\boldsymbol{F} = \nabla \boldsymbol{u} + \boldsymbol{I}$. The total potential energy in the domain is

$$\mathcal{F} = \int_{\Omega} \psi - \boldsymbol{b} \cdot \boldsymbol{u} \mathrm{d}V. \tag{8.66}$$

The stress tensor and generalized stress tensor of the first Piola–Kirchhoff type are defined as

$$\boldsymbol{P} = \frac{\partial \psi}{\partial \boldsymbol{F}} \tag{8.67}$$

$$\boldsymbol{\Sigma} = \frac{\partial \psi}{\partial \nabla \boldsymbol{F}}. \tag{8.68}$$

The variation of the total internal energy is

$$\begin{aligned}\delta \mathcal{F} &= \int_{\Omega} \Big(\frac{\partial \psi}{\partial \boldsymbol{F}} : \nabla \delta \boldsymbol{u} + \frac{\partial \psi}{\partial \nabla \boldsymbol{F}} \vdots \nabla^2 \delta \boldsymbol{u} - \boldsymbol{b} \cdot \delta \boldsymbol{u}\Big) \mathrm{d}V \\ &= \int_{\Omega} \Big(\boldsymbol{P} : \nabla \delta \boldsymbol{u} + \boldsymbol{\Sigma} \vdots \nabla^2 \delta \boldsymbol{u} - \boldsymbol{b} \cdot \delta \boldsymbol{u}\Big) \mathrm{d}V. \end{aligned} \tag{8.69}$$

Based on the integration by parts, the local form can be derived by

$$\begin{aligned}\delta \mathcal{F} &= \int_{\partial \Omega} \Big(\boldsymbol{n} \cdot \boldsymbol{P} \cdot \delta \boldsymbol{u} + \boldsymbol{n} \cdot \boldsymbol{\Sigma} \vdots \nabla \delta \boldsymbol{u}\Big) \mathrm{d}S - \int_{\Omega} \Big(\nabla \cdot \boldsymbol{P} \cdot \delta \boldsymbol{u} + \nabla \cdot \boldsymbol{\Sigma} \vdots \nabla \delta \boldsymbol{u} + \boldsymbol{b} \cdot \boldsymbol{u}\Big) \mathrm{d}V \\ &= \int_{\partial \Omega} \Big(\boldsymbol{n} \cdot \boldsymbol{P} \cdot \delta \boldsymbol{u} + \boldsymbol{n} \cdot \boldsymbol{\Sigma} : \nabla \delta \boldsymbol{u} - \boldsymbol{n} \cdot \nabla \cdot \boldsymbol{\Sigma} \cdot \delta \boldsymbol{u}\Big) \mathrm{d}S - \int_{\Omega} (\nabla \cdot \boldsymbol{P} - \nabla^2 : \boldsymbol{\Sigma} + \boldsymbol{b}) \cdot \delta \boldsymbol{u} \mathrm{d}V. \end{aligned} \tag{8.70}$$

On one hand, based on D'Alembert's principle, the governing equations for dynamic gradient elasticity in local form can be written as

$$\nabla \cdot \boldsymbol{P} - \nabla^2 : \boldsymbol{\Sigma} + \boldsymbol{b} = \rho \ddot{\boldsymbol{u}} \text{ in } \Omega. \tag{8.71}$$

On the other hand, do the substitutions $\nabla \delta \boldsymbol{u} \to \int_{S_i} \omega(\mathbf{r}_{ij}) \boldsymbol{g}_{ij} \otimes \delta \boldsymbol{u}_{ij} \mathrm{d}V_j$, and $\nabla^2 \delta \boldsymbol{u} \to \int_{S_i} \omega(\mathbf{r}_{ij}) \boldsymbol{h}_{ij} \otimes \delta \boldsymbol{u}_{ij} \mathrm{d}V_j$, we get

$$\begin{aligned}\delta \mathcal{F} &= \int_{\Omega} \boldsymbol{P} : \nabla \delta \boldsymbol{u} + \boldsymbol{\Sigma} \vdots \nabla^2 \delta \boldsymbol{u} - \boldsymbol{b} \cdot \delta \boldsymbol{u} \mathrm{d}V \\ &= \int_{\Omega} \Big(\boldsymbol{P}_i : \int_{S_i} \omega(\mathbf{r}_{ij}) \boldsymbol{g}_{ij} \otimes \delta \boldsymbol{u}_{ij} \mathrm{d}V_j + \boldsymbol{\Sigma}_i \vdots \int_{S_i} \omega(\mathbf{r}_{ij}) \boldsymbol{h}_{ij} \otimes \delta \boldsymbol{u}_{ij} \mathrm{d}V_j - \boldsymbol{b} \cdot \delta \boldsymbol{u}\Big) \mathrm{d}V_i \\ &= \underbrace{\int_{\Omega} \int_{S_i} \omega(\mathbf{r}_{ij}) \boldsymbol{P}_i \vdots (\delta \boldsymbol{u}_j - \delta \boldsymbol{u}_i) \otimes \boldsymbol{g}_{ij} \mathrm{d}V_j \mathrm{d}V_i}_{\text{by Eq. 8.2}} \\ &\quad + \underbrace{\int_{\Omega} \int_{S_i} \omega(\mathbf{r}_{ij}) \boldsymbol{\Sigma}_i \vdots (\delta \boldsymbol{u}_j - \delta \boldsymbol{u}_j) \otimes \boldsymbol{h}_{ij} \mathrm{d}V_j \mathrm{d}V_i}_{\text{by Eq. 8.2}} - \int_{\Omega} \boldsymbol{b} \cdot \delta \boldsymbol{u}_i \mathrm{d}V_i \end{aligned}$$

$$
\begin{aligned}
&= \int_{\Omega} \left(\int_{S_i'} \omega(\mathbf{r}_{ji}) \boldsymbol{P}_j \cdot \boldsymbol{g}_{ji} \mathrm{d}V_j - \int_{S_i} \omega(\mathbf{r}_{ij}) \boldsymbol{P}_i \cdot \boldsymbol{g}_{ij} \mathrm{d}V_j \right) \cdot \delta \boldsymbol{u}_i \mathrm{d}V_i \\
&+ \int_{\Omega} \left(\int_{S_i'} \omega(\mathbf{r}_{ji}) \boldsymbol{\Sigma}_j : \boldsymbol{h}_{ji} \mathrm{d}V_j - \int_{S_i} \omega(\mathbf{r}_{ij}) \boldsymbol{\Sigma}_i : \boldsymbol{h}_{ij} \mathrm{d}V_j \right) \cdot \delta \boldsymbol{u}_i \mathrm{d}V_i - \int_{\Omega} \boldsymbol{b} \cdot \delta \boldsymbol{u}_i \mathrm{d}V_i
\end{aligned}
\tag{8.72}
$$

In the above derivation, we used $\boldsymbol{\Sigma} \vdots \boldsymbol{u} \otimes \boldsymbol{h} = (\boldsymbol{\Sigma} : \boldsymbol{h}) \cdot \boldsymbol{u}$. For any $\delta \boldsymbol{u}_i$, $\delta \mathcal{F} = 0$ leads to the nonlocal form of gradient elasticity

$$
\int_{S_i} \omega(\mathbf{r}_{ij})(\boldsymbol{P}_i \cdot \boldsymbol{g}_{ij} + \boldsymbol{\Sigma}_i : \boldsymbol{h}_{ij}) \mathrm{d}V_j - \int_{S_i'} \omega(\mathbf{r}_{ji})(\boldsymbol{P}_j \cdot \boldsymbol{g}_{ji} + \boldsymbol{\Sigma}_j : \boldsymbol{h}_{ji}) \mathrm{d}V_j + \boldsymbol{b} = \rho \ddot{\boldsymbol{u}}_i .
\tag{8.73}
$$

The inertial force term is added based on D'Alembert's principle.

Comparing Eqs. 8.70 and 8.72, the correspondence from local form to nonlocal form is

$$
\nabla^2 : \boldsymbol{\Sigma}_i \rightarrow \int_{S_i'} \omega(\mathbf{r}_{ji}) \boldsymbol{\Sigma}_j : \boldsymbol{h}_{ji} \mathrm{d}V_j - \int_{S_i} \omega(\mathbf{r}_{ij}) \boldsymbol{\Sigma}_i : \boldsymbol{h}_{ij} \mathrm{d}V_j .
\tag{8.74}
$$

8.2.4 *Nonlocal Form of Magneto–Electro-Elasticity*

In accordance with reference (Liu 2014), let us postulate the following form of internal energy for the energy function $\psi := \psi(\boldsymbol{F}, \nabla \boldsymbol{F}, \boldsymbol{p}, \nabla \boldsymbol{p}, \boldsymbol{m}, \nabla \boldsymbol{m})$; a function depends on the displacement gradient $\boldsymbol{F} = \nabla \boldsymbol{u} + \boldsymbol{I}$ and its second gradient $\nabla \boldsymbol{F} = \nabla^2 \boldsymbol{u}$, polarization vector $\boldsymbol{p}$ and its gradient $\nabla \boldsymbol{p}$, magnetic field $\boldsymbol{m}$ and its gradient $\nabla \boldsymbol{m}$. The total potential energy in the domain can be written as

$$
\mathcal{F} = \int_{\Omega} \psi(\boldsymbol{F}, \nabla \boldsymbol{F}, \boldsymbol{p}, \nabla \boldsymbol{p}, \boldsymbol{m}, \nabla \boldsymbol{m}) \mathrm{d}V .
\tag{8.75}
$$

This model has a strong physical background, for example, the nonlinear electro-gradient elasticity for semiconductors (Nguyen et al. 2019) and flexoelectricity (Roy et al. 2019).

The first variation of $\mathcal{F}$ is

$$
\begin{aligned}
\delta \mathcal{F} &= \int_{\Omega} \delta \psi \mathrm{d}V \\
&= \int_{\Omega} \frac{\partial \psi}{\partial \boldsymbol{F}} : \nabla \delta \boldsymbol{u} + \frac{\partial \psi}{\partial \nabla \boldsymbol{F}} \vdots \nabla^2 \delta \boldsymbol{u} + \frac{\partial \psi}{\partial \boldsymbol{p}} \cdot \delta \boldsymbol{p} + \\
&\quad \frac{\partial \psi}{\partial \nabla \boldsymbol{p}} : \nabla \delta \boldsymbol{p} + \frac{\partial \psi}{\partial \boldsymbol{m}} \cdot \delta \boldsymbol{m} + \frac{\partial \psi}{\partial \nabla \boldsymbol{m}} : \nabla \delta \boldsymbol{m} \mathrm{d}V
\end{aligned}
$$

$$= \int_{\Omega} \boldsymbol{P} : \nabla \delta \boldsymbol{u} + \boldsymbol{\Sigma} \vdots \nabla^2 \delta \boldsymbol{u} + \boldsymbol{e} \cdot \delta \boldsymbol{p}$$
$$+ \boldsymbol{E} : \nabla \delta \boldsymbol{p} + \boldsymbol{s} \cdot \delta \boldsymbol{m} + \boldsymbol{S} : \nabla \delta \boldsymbol{m} \mathrm{d}V \tag{8.76}$$

where

$$\boldsymbol{P} = \frac{\partial \psi}{\partial \boldsymbol{F}}, \boldsymbol{\Sigma} = \frac{\partial \psi}{\partial \nabla \boldsymbol{F}}, \boldsymbol{e} = \frac{\partial \psi}{\partial \boldsymbol{p}} \tag{8.77}$$
$$\boldsymbol{E} = \frac{\partial \psi}{\partial \nabla \boldsymbol{p}}, \boldsymbol{s} = \frac{\partial \psi}{\partial \boldsymbol{m}}, \boldsymbol{S} = \frac{\partial \psi}{\partial \nabla \boldsymbol{m}}. \tag{8.78}$$

Doing substitutions $\nabla \delta \boldsymbol{u}_i \rightarrow \int_{S_i} \omega(\mathbf{r}_{ij}) \delta \boldsymbol{u}_{ij} \otimes \boldsymbol{g}_{ij} \mathrm{d}V_j$, $\nabla^2 \delta \boldsymbol{u}_i \rightarrow \int_{S_i} \omega(\mathbf{r}_{ij}) \delta \boldsymbol{u}_{ij} \otimes \boldsymbol{h}_{ij} \mathrm{d}V_j$, $\nabla \delta \boldsymbol{p}_i \rightarrow \int_{S_i} \omega(\mathbf{r}_{ij}) \delta \boldsymbol{p}_{ij} \otimes \boldsymbol{g}_{ij} \mathrm{d}V_j$, $\nabla \delta \boldsymbol{m}_i \rightarrow \int_{S_i} \omega(\mathbf{r}_{ij}) \delta \boldsymbol{m}_{ij} \otimes \boldsymbol{g}_{ij} \mathrm{d}V_j$ and following the same operations in prior sections, the functional becomes

$$\delta \mathcal{F} = \int_{\Omega} \Big(\int_{S_i'} \omega(\mathbf{r}_{ji})(\boldsymbol{P}_j \cdot \boldsymbol{g}_{ji} + \boldsymbol{\Sigma}_j : \boldsymbol{h}_{ji}) \mathrm{d}V_j - \int_{S_i} \omega(\mathbf{r}_{ij})(\boldsymbol{P}_i \cdot \boldsymbol{g}_{ij} + \boldsymbol{\Sigma}_i : \boldsymbol{h}_{ij}) \mathrm{d}V_j \Big) \cdot \delta \boldsymbol{u}_i \mathrm{d}V_i$$
$$+ \int_{\Omega} \Big(\int_{S_i'} \omega(\mathbf{r}_{ji})(\boldsymbol{E}_j \cdot \boldsymbol{g}_{ji}) \mathrm{d}V_j - \int_{S_i} \omega(\mathbf{r}_{ij}) \boldsymbol{E}_i \cdot \boldsymbol{g}_{ij} \mathrm{d}V_j + \boldsymbol{e}_i \Big) \cdot \delta \boldsymbol{p}_i \mathrm{d}V_i +$$
$$\int_{\Omega} \Big(\int_{S_i'} \omega(\mathbf{r}_{ji})(\boldsymbol{S}_j \cdot \boldsymbol{g}_{ji}) \mathrm{d}V_j - \int_{S_i} \omega(\mathbf{r}_{ij}) \boldsymbol{S}_i \cdot \boldsymbol{g}_{ij} \mathrm{d}V_j + \boldsymbol{s}_i \Big) \cdot \delta \boldsymbol{m}_i \mathrm{d}V_i. \tag{8.79}$$

For any $\delta \boldsymbol{u}_i, \delta \boldsymbol{p}_i, \delta \boldsymbol{m}_i$, $\delta \mathcal{F} = 0$ leads to a general nonlocal governing equation for the mechanical field, electrical field and magnetic field, respectively,

$$\int_{S_i} \omega(\mathbf{r}_{ij})(\boldsymbol{P}_i \cdot \boldsymbol{g}_{ij} + \boldsymbol{\Sigma}_i : \boldsymbol{h}_{ij}) \mathrm{d}V_j -$$
$$\int_{S_i'} \omega(\mathbf{r}_{ji})(\boldsymbol{P}_j \cdot \boldsymbol{g}_{ji} + \boldsymbol{\Sigma}_j : \boldsymbol{h}_{ji}) \mathrm{d}V_j + \boldsymbol{b}_i = 0 \tag{8.80}$$
$$\int_{S_i} \omega(\mathbf{r}_{ij}) \boldsymbol{E}_i \cdot \boldsymbol{g}_{ij} \mathrm{d}V_j - \int_{S_i'} \omega(\mathbf{r}_{ji}) \boldsymbol{E}_j \cdot \boldsymbol{g}_{ji} \mathrm{d}V_j - \boldsymbol{e}_i = 0 \tag{8.81}$$
$$\int_{S_i} \omega(\mathbf{r}_{ij}) \boldsymbol{S}_i \cdot \boldsymbol{g}_{ij} \mathrm{d}V_j - \int_{S_i'} \omega(\mathbf{r}_{ji}) \boldsymbol{S}_j \cdot \boldsymbol{g}_{ji} \mathrm{d}V_j - \boldsymbol{s}_i = 0. \tag{8.82}$$

In the derivation, we did not specify the exact form of the energy density, whether it is of small deformation or of finite deformation. For the specified energy form, one only needs to derive the expression for $\boldsymbol{P}$, $\boldsymbol{\Sigma}$, $\boldsymbol{e}$, $\boldsymbol{E}$, $\boldsymbol{s}$, $\boldsymbol{S}$ based on the material constitutions. It can be seen that the nonlocal governing equations for the continuum magneto–electro-elasticity can be obtained with ease by using the nonlocal operator method and a variational principle. The same rule applies to many other physical problems.

8.2.5 Nonlocal Form of Phase Field Fracture Method

Phase field fracture method is powerful in fracture modeling (Miehe et al. 2010b). The difference in tensile and compressive strengths of the material can be considered by dividing the strain energy density into a tensile part affected by the phase field and a compressive part, which is independent of the phase field,

$$\psi_e(\boldsymbol{\varepsilon}(\nabla \boldsymbol{u}), s) = (1-s)^2\psi_e^+(\boldsymbol{\varepsilon}(\nabla \boldsymbol{u})) + \psi_e^-(\boldsymbol{\varepsilon}(\nabla \boldsymbol{u})) \tag{8.83}$$

where ψ_e^+ (ψ_e^-) denotes the strain energy density for a tensile (compressive) part, $\boldsymbol{u}$ is the displacement, $s \in [0, 1]$ is the phase field, $\boldsymbol{\varepsilon}$ denotes the strain and ℓ is the phase field intrinsic length scale.

The full potential functional of the phase field fracture model reads

$$\begin{aligned}\mathcal{F}_\ell(\boldsymbol{u}, s) = &\int_\Omega \Big((1-s)^2\psi_e^+(\boldsymbol{\varepsilon}(\nabla \boldsymbol{u})) + \psi_e^-(\boldsymbol{\varepsilon}(\nabla \boldsymbol{u}))\Big)\mathrm{d}V - \int_{\partial\Omega} \boldsymbol{t}^* \cdot \boldsymbol{u}\mathrm{d}A \\ &- \int_\Omega \boldsymbol{b}\cdot\boldsymbol{u}\mathrm{d}V + \int_\Omega g_c\Big(\frac{s^2}{2\ell} + \frac{\ell}{2}\nabla s \cdot \nabla s\Big)\mathrm{d}V,\end{aligned} \tag{8.84}$$

where $\boldsymbol{t}^*$ denotes the surface traction at the boundary, $\boldsymbol{b}$ is the body force density and g_c is the critical energy release rate.

For the sake of simplicity, we neglect the surface traction force and consider the first variation of $\mathcal{F}_\ell$

$$\begin{aligned}\delta\mathcal{F}_\ell &= \int_\Omega \delta\Big((1-s)^2\psi_e^+ + \psi_e^-\Big)\mathrm{d}V - \int_\Omega \boldsymbol{b}\cdot\delta\boldsymbol{u}\mathrm{d}V + \int_\Omega g_c\delta\Big(\frac{s^2}{2\ell} + \frac{\ell}{2}\nabla s\cdot\nabla s\Big)\mathrm{d}V \\
&= \int_\Omega \Big((1-s)^2\frac{\partial\psi_e^+}{\partial\boldsymbol{\varepsilon}} : \nabla\delta\boldsymbol{u} - 2\psi_e^+(1-s)\delta s + \frac{\partial\psi_e^-}{\partial\boldsymbol{\varepsilon}} : \nabla\delta\boldsymbol{u}\Big)\mathrm{d}V - \int_\Omega \boldsymbol{b}\cdot\delta\boldsymbol{u}\mathrm{d}V \\
&\quad + \int_\Omega g_c\Big(\frac{s}{\ell}\delta s + \ell\nabla s\cdot\nabla\delta s\Big)\mathrm{d}V \\
&= \int_\Omega \Big(((1-s)^2\boldsymbol{\sigma}^+ + \boldsymbol{\sigma}^-) : \nabla\delta\boldsymbol{u} - \boldsymbol{b}\cdot\delta\boldsymbol{u}\Big)\mathrm{d}V + \int_\Omega g_c\Big(\frac{s}{\ell}\delta s - 2\frac{\psi_e^+}{g_c}(1-s)\delta s + \ell\nabla s\cdot\nabla\delta s\Big)\mathrm{d}V \\
&= \int_\Omega \Big(\boldsymbol{\sigma}_i : \nabla\delta\boldsymbol{u}_i - \boldsymbol{b}_i\cdot\delta\boldsymbol{u}_i\Big)\mathrm{d}V_i + \int_\Omega g_c\Big(\frac{s_i}{\ell}\delta s_i - 2\frac{\psi_{ei}^+}{g_c}(1-s_i)\delta s_i + \ell\nabla s_i\cdot\nabla\delta s_i\Big)\mathrm{d}V_i \\
&= \int_\Omega \Big(\boldsymbol{\sigma}_i : \Big(\int_{S_i} \omega(\mathbf{r}_{ij})\delta\boldsymbol{u}_{ij}\otimes\boldsymbol{g}_{ij}\mathrm{d}V_j\Big) - \boldsymbol{b}_i\cdot\delta\boldsymbol{u}_i\Big)\mathrm{d}V_i \\
&\quad + \int_\Omega g_c\Big(\frac{s_i}{\ell}\delta s_i - 2\frac{\psi_{ei}^+}{g_c}(1-s_i)\delta s_i + \ell\nabla s_i\cdot\int_{S_i}\omega(\mathbf{r}_{ij})\delta s_{ij}\boldsymbol{g}_{ij}\mathrm{d}V_j\Big)\mathrm{d}V_i \\
&= \int_\Omega \Big(\Big(\int_{S_i'} \omega(\mathbf{r}_{ji})\boldsymbol{\sigma}_j\cdot\boldsymbol{g}_{ji}\mathrm{d}V_j - \int_{S_i}\omega(\mathbf{r}_{ij})\boldsymbol{\sigma}_i\cdot\boldsymbol{g}_{ij}\mathrm{d}V_j\Big)\cdot\delta\boldsymbol{u}_i - \boldsymbol{b}_i\cdot\delta\boldsymbol{u}_i\Big)\mathrm{d}V_i \\
&\quad + \int_\Omega g_c\Big(\frac{s_i}{\ell} - 2\frac{\psi_{ei}^+}{g_c}(1-s_i) + \int_{S_i'}\omega(\mathbf{r}_{ji})\ell\nabla s_j\cdot\boldsymbol{g}_{ji}\mathrm{d}V_j - \int_{S_i}\omega(\mathbf{r}_{ij})\ell\nabla s_i\cdot\boldsymbol{g}_{ij}\mathrm{d}V_j\Big)\delta s_i\mathrm{d}V_i
\end{aligned} \tag{8.85}$$

where

$$\boldsymbol{\sigma}^+ = \frac{\partial \psi_e^+}{\partial \boldsymbol{\varepsilon}}, \boldsymbol{\sigma}^- = \frac{\partial \psi_e^-}{\partial \boldsymbol{\varepsilon}} \tag{8.86}$$

$$\boldsymbol{\sigma} = (1-s)^2 \boldsymbol{\sigma}^+ + \boldsymbol{\sigma}^-. \tag{8.87}$$

For any $\delta \boldsymbol{u}_i, \delta s_i, \delta \mathcal{F}_\ell = 0$ leads to the nonlocal governing equations for the mechanical field and phase field

$$\int_{\mathcal{S}_i} \omega(\mathbf{r}_{ij}) \boldsymbol{\sigma}_i \cdot \boldsymbol{g}_{ij} \mathrm{d}V_j - \int_{\mathcal{S}_i'} \omega(\mathbf{r}_{ji}) \boldsymbol{\sigma}_j \cdot \boldsymbol{g}_{ji} \mathrm{d}V_j + \boldsymbol{b}_i = 0 \tag{8.88}$$

$$\frac{s_i}{\ell} - 2\frac{\psi_{ei}^+}{g_c}(1-s_i) + \int_{\mathcal{S}_i'} \omega(\mathbf{r}_{ji}) \ell \nabla s_j \cdot \boldsymbol{g}_{ji} \mathrm{d}V_j - \int_{\mathcal{S}_i} \omega(\mathbf{r}_{ij}) \ell \nabla s_i \cdot \boldsymbol{g}_{ij} \mathrm{d}V_j = 0. \tag{8.89}$$

The above examples aim at illustrating the power of the nonlocal operator method combined with the weighted residual method or variational principle in the derivation of nonlocal strong forms based on their local strong or energy forms. The derived nonlocal strong forms are variationally consistent and allow variable support sizes for each point in the model.

8.3 Instability Criterion for Fracture Modeling

Typical methods for fracture modeling are either based on diffusive crack domain in phase field methods or on direct topological modification on meshes in XFEM or bonds in PD. Direct topological modification on meshes often leads to instability issues. For example, in NOSBPD, the breakage of a bond based on the quantities derived from the stress state or strain state often introduces too much perturbation to the scheme, which may abort the calculation because of the singularity in shape tensors. These criteria include critical stretch (Silling 2000), energy-based (Foster et al. 2011) or stress-based criterion (Zhou et al. 2016a, b). Another issue in NOSBPD is that the strain energy carried by a bond is closely related to other bonds. It also depends on the direction, the length of the bond and the choice of influence functions. Removing one neighbor often gives rise to catastrophic results in the calculation. A criterion on how to remove the neighbors safely from the neighbor list remains unclear.

Damage is a process deviating from the robust mathematical expression, where the transition happens in a very narrow zone, such as the crack tip front. It is observed that around the crack tip, the gradient or strain undergoes a sharp transition within a very small zone. Most conventional numerical methods for fracture modeling focus on an accurate description of the singularity occurring around the crack tip; such a description is very hard to tackle and its evolution is inconvenient to update. This dilemma can be handled when something different from the continuous function is introduced.

In NOM, the gradient operator is defined in a "redundant" way. Around the crack tip, the deformation is irregular, and the part due to hourglass energy is comparable to the strain energy carried by a particle. More specifically, the operator energy in the nonlocal operator method describes the irregularity of a function around the crack tip. The irregularity is the part that cannot be described by the continuous function. For a continuous domain, the strain energy density is much larger than the operator energy density. However, for particles around the crack tip, the operator energy density is far from zero, and the irregularity due to the singularity around the crack tip increases comparably to the strain energy density. In this sense, the operator energy density can be viewed as an indicator for the crack tip.

Unlike the strain energy density, the hourglass energy density describes the irregular deformation around the crack tip. It depends on the penalty for the strain energy. A larger penalty improves the continuity of deformation, but the extent of hourglass energy compared with the strain energy density is hard to estimate. In this paper, we propose a special manner to estimate the critical hourglass strain. Let the critical bond strain be denoted by s_{max}, which may depend on the characteristic length scale of the support, critical energy release rate and the elastic modulus. When the maximal strain reached s_{max}, the damage process is activated and the critical hourglass strain s_{max}^{hg} is set as the maximal hourglass strain s_{ij}^{hg} for all bonds in the computational model. In the sequential calculation, when the hourglass strain of a bond is larger than s_{max}^{hg}, the damage on that bond occurs, which is mathematically described as

$$d_{ij} = \begin{cases} 0 \text{ if } s_{ij}^{hg}(t) > s_{max}^{hg}, t \in [0, T] \\ 1 \text{ otherwise} \end{cases} \tag{8.90}$$

where d_{ij} denotes the damage status between particle i and particle j.

The damage of a particle is calculated as

$$d_i = \frac{\int_{S_i} d_{ij} \mathrm{d}V_j}{\int_{S_i} \mathrm{d}V_j}. \tag{8.91}$$

Every time one particle is removed from the neighbor list, the nonlocal gradient for the central particle should be recalculated based on the remaining "healthy" neighbor. We will apply this rule to model fractures in 2D and 3D linear elastic material.

8.4 Numerical Implementation

We have applied NOM to derive the nonlocal strong forms for the traditional continuum model in Sect. 8.2. Two representative nonlocal theories, the dual-horizon peridynamics by Eq. 8.47 for fracture modeling and the nonlocal thin plate by Eq. 8.63, are selected for numerical tests. For the DH-PD, the focus is on the test of instability criterion for quasi-static fracture modeling by explicit time integration algorithm. The nonlocal thin plate is compared with the finite element method. The nonlocal derivatives can be viewed as a generalization of the local derivatives, and the nonlocal derivatives recover the local derivatives when the size of the support degenerates to zero. The range of nonlocality depends on the choice of the weighting functions and the size of the supports. One obstacle of the nonlocal models is the verification since the exact solutions of the nonlocal model are rare. For simplicity of verification, we aim at solving the local problems with nonlocal forms where the nonlocal effect is reduced by selecting certain weighting functions.

The primary step in the implementation is the calculation of internal force based on the governing equations. In the first step, the computational domain is discretized into particles:

$$\Omega = \sum_{i=1}^{N} \Delta V_i \tag{8.92}$$

where N is the number of particles in the domain. Then the support of each particle is represented by a list of particle indices,

$$\mathcal{S}_i = \{j_1, j_2, ..., j_{n_i}\} \tag{8.93}$$

where j is the global index of the particle and n_i is the number of particles in $\mathcal{S}_i$.

The gradient $\boldsymbol{g}_{ij}$ and Hessian $\boldsymbol{h}_{ij}$ for two particles i, j can be assembled by collecting terms in $\boldsymbol{K}_i \cdot \boldsymbol{p}_{ij}$ according to Eqs. 8.24 or 8.26, where

$$\boldsymbol{K}_i = \Big(\sum_{\mathcal{S}_i} \omega(\mathbf{r}_{ij}) \boldsymbol{p}_{ij} \otimes \boldsymbol{p}_{ij}^T \Delta V_j\Big)^{-1} \tag{8.94}$$

with weight function $\omega(\mathbf{r}_{ij}) = 1/|\mathbf{r}_{ij}|^2$.

The nonlocal differential derivatives at point i can be calculated as

$$\tilde{\partial} u_i = \sum_{j \in \mathcal{S}_i} \omega(\mathbf{r}_{ij}) \boldsymbol{K}_i \cdot \boldsymbol{p}_{ij} u_{ij} \Delta V_j. \tag{8.95}$$

The nonlocal operators in $\tilde{\partial} u_i$ can be used to define the strain tensor, stress tensor, bending moment and others.

In discrete form, Eqs. 8.47 and 8.61 become

$$\sum_{\mathcal{H}_i} \omega(\mathbf{r}_{ij}) \boldsymbol{P}_i \cdot \boldsymbol{g}_{ij} \Delta V_j \Delta V_i - \sum_{\mathcal{H}'_i} \omega(\mathbf{r}_{ji}) \boldsymbol{P}_j \cdot \boldsymbol{g}_{ji} \Delta V_j \Delta V_i + \boldsymbol{b}_i \Delta V_i = \rho \Delta V_i \ddot{\boldsymbol{u}}_i \tag{8.96}$$

$$\sum_{\mathcal{S}_i} \omega(\mathbf{r}_{ij}) \boldsymbol{M}_i : \boldsymbol{h}_{ij} \Delta V_j \Delta V_i - \sum_{\mathcal{S}'_i} \omega(\mathbf{r}_{ij}) \boldsymbol{M}_j : \boldsymbol{h}_{ji} \Delta V_j \Delta V_i + q_i \Delta V_i = t\rho \Delta V_i \ddot{w}_i. \tag{8.97}$$

In Eqs. 8.96 and 8.97, the volume of particle i is multiplied on both sides of the equations. It is not required to calculate the internal forces from the dual-support. Let $\boldsymbol{f}_i = \mathbf{0}, 1 \leq i \leq N$ denote the initial internal force on particle i. For each particle, one only needs to focus on the support, calculating the forces and adding the force to the particle's internal force

$$\begin{aligned}
\sum_{j \in \mathcal{S}_i} \omega(\mathbf{r}_{ij}) \boldsymbol{P}_i \cdot \boldsymbol{g}_{ij} \Delta V_j \Delta V_i &\to \boldsymbol{f}_i \\
-\omega(\mathbf{r}_{ij_1}) \boldsymbol{P}_i \cdot \boldsymbol{g}_{ij_1} \Delta V_{j_1} \Delta V_i &\to \boldsymbol{f}_{j_1} \\
-\omega(\mathbf{r}_{ij_2}) \boldsymbol{P}_i \cdot \boldsymbol{g}_{ij_2} \Delta V_{j_2} \Delta V_i &\to \boldsymbol{f}_{j_2} \\
&\dots \\
-\omega(\mathbf{r}_{ij_{n_i}}) \boldsymbol{P}_i \cdot \boldsymbol{g}_{ij_{n_i}} \Delta V_{j_{n_i}} \Delta V_i &\to \boldsymbol{f}_{j_{n_i}}
\end{aligned} \tag{8.98}$$

where $a \to b$ denotes the addition of a to b. The process of adding force $-\omega(\mathbf{r}_{ij_1}) \boldsymbol{P}_i \cdot \boldsymbol{g}_{ij} \Delta V_j \Delta V_i$ to $\boldsymbol{f}_j$ is equivalent to accumulating the internal forces from particle j's dual-support.

For the calculating of internal force of a thin plate, the same applies

$$\begin{aligned}
\sum_{j \in \mathcal{S}_i} \omega(\mathbf{r}_{ij}) \boldsymbol{M}_i : \boldsymbol{h}_{ij} \Delta V_j \Delta V_i &\to \boldsymbol{f}_i \\
-\omega(\mathbf{r}_{ij_1}) \boldsymbol{M}_i : \boldsymbol{h}_{ij_1} \Delta V_{j_1} \Delta V_i &\to \boldsymbol{f}_{j_1} \\
-\omega(\mathbf{r}_{ij_2}) \boldsymbol{M}_i : \boldsymbol{h}_{ij_2} \Delta V_{j_2} \Delta V_i &\to \boldsymbol{f}_{j_2} \\
&\dots \\
-\omega(\mathbf{r}_{ij_{n_i}}) \boldsymbol{M}_i : \boldsymbol{h}_{ij_{n_i}} \Delta V_{j_{n_i}} \Delta V_i &\to \boldsymbol{f}_{j_{n_i}}.
\end{aligned} \tag{8.99}$$

In order to maintain the stability of the nonlocal operator, the discrete form of Eq. 8.42 is

$$\sum_{\mathcal{S}'_i} \omega(\mathbf{r}) \frac{p^{hg}}{m_j} (\mathbf{u}_{ji} - \boldsymbol{p}_i^T \tilde{\partial} \mathbf{u}_j) \Delta V_j \Delta V_i - \sum_{\mathcal{S}_i} \omega(\mathbf{r}) \frac{p^{hg}}{m_i} (\mathbf{u}_{ij} - \boldsymbol{p}_j^T \tilde{\partial} \mathbf{u}_i) \Delta V_j \Delta V_i. \tag{8.100}$$

For particle i with support $\mathcal{S}_i$, the hourglass force is calculated as follows:

$$
\begin{aligned}
&\sum_{j\in S_i} \omega(\mathbf{r}_{ij})\frac{p^{hg}}{m_i}(\mathbf{u}_{ij} - \boldsymbol{p}_j^T\tilde{\partial}\boldsymbol{u}_i)\Delta V_j \Delta V_i \rightarrow \boldsymbol{f}_i \\
&-\omega(\mathbf{r}_{ij_1})\frac{p^{hg}}{m_i}(\mathbf{u}_{ij_1} - \boldsymbol{p}_{j_1}^T\tilde{\partial}\boldsymbol{u}_i)\Delta V_{j_1} \Delta V_i \rightarrow \boldsymbol{f}_{j_1} \\
&-\omega(\mathbf{r}_{ij_2})\frac{p^{hg}}{m_i}(\mathbf{u}_{ij_2} - \boldsymbol{p}_{j_2}^T\tilde{\partial}\boldsymbol{u}_i)\Delta V_{j_2} \Delta V_i \rightarrow \boldsymbol{f}_{j_2} \\
&\qquad \cdots \\
&-\omega(\mathbf{r}_{ij_{n_i}})\frac{p^{hg}}{m_i}(\mathbf{u}_{ij_{n_i}} - \boldsymbol{p}_{j_{n_i}}^T\tilde{\partial}\boldsymbol{u}_i)\Delta V_{j_{n_i}} \Delta V_i \rightarrow \boldsymbol{f}_{j_{n_i}}.
\end{aligned} \tag{8.101}
$$

When the internal force is attained and the contribution of the external force boundary condition or body force is accumulated, the basic Verlet algorithm (Verlet 1967) outlined as follows is used to update the displacement:

$$
\boldsymbol{u}_i(t+\Delta t) = \boldsymbol{u}_i(t) + \boldsymbol{v}_i(t)\Delta t + \frac{1}{2}\boldsymbol{a}_i(t)\Delta t^2 \tag{8.102}
$$

$$
\boldsymbol{v}_i(t+\Delta t) = \boldsymbol{v}_i(t) + \frac{1}{2}\Big(\boldsymbol{a}_i(t) + \boldsymbol{a}_i(t+\Delta t)\Big)\Delta t \tag{8.103}
$$

where $\boldsymbol{u}_i$ denotes the displacement or deflection, $\boldsymbol{v}_i$ the velocity and $\boldsymbol{a}_i = \frac{\boldsymbol{f}_i}{m_i}$ the acceleration for particle i with mass m_i subject to net force $\boldsymbol{f}_i$. For the detailed implementation and the numerical examples, the reader can find the open-source code on Github https://github.com/hl-ren/Nonlocal_elasticity, and https://github.com/hl-ren/Nonlocal_thin_plate.

8.5 Numerical Examples

8.5.1 *Accuracy of Nonlocal Hessian Operator*

We first test the accuracy of the nonlocal Hessian operator. Thus, consider the analytical derivatives of the field

$$
w(x, y) = e^{xy}\big(\sin 3(x-y) - \cos 2(x+y)\big), \text{ with } x \in [-1, 1], y \in [-1, 1]. \tag{8.104}
$$

The domain $[-1, 1]^2$ is discretized with different number of particles, $N \in \{20^2, 40^2, 60^2, 80^2, 100^2, 160^2, 180^2, 200^2\}$. The number of neighbors in support is selected as $n = 14$. The L_2 norm of the nonlocal Hessian operator is calculated as

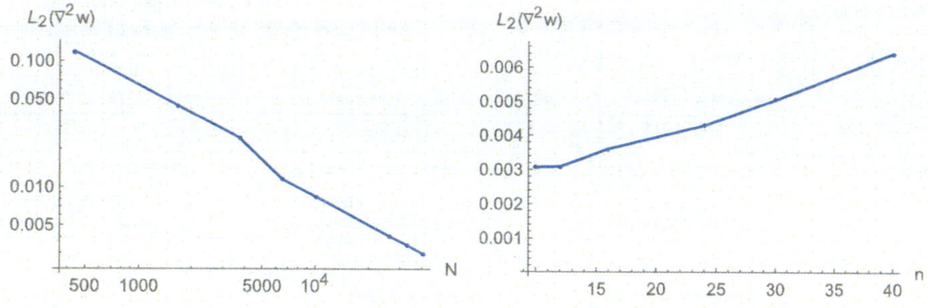

(a) (b)

Fig. 8.3 L_2 norm of the nonlocal Hessian operator **a** for N, the number of particles with $n = 14$ and **b** for n, the number of particles in support with $N = 180^2$

$$L_2(\nabla^2 w) = \sqrt{\frac{\sum_{i=1}^{N}(\nabla^2 w_i - \tilde{\nabla}^2 w_i) : (\nabla^2 w_i - \tilde{\nabla}^2 w_i)\Delta V_i}{\sum_{i=1}^{N}(\nabla^2 w_i) : (\nabla^2 w_i)\Delta V_i}}. \tag{8.105}$$

For different discretizations, the L_2 norm is plotted in Fig. 8.3a with a convergence rate of 0.835. We also tested the influence of the support size. For fixed discretization $N = 180^2$, the nonlocal effect increases with the number of neighbors in the support, as shown in Fig. 8.3b.

8.5.2 Square Thin Plate Subject to Pressure

The dimensions of the plate are $0.5 \times 0.5\ \text{m}^2$ with a thickness of 0.01 m. The material parameters are elastic modulus $E = 210$ GPa and Poisson ratio $\nu = 0.3$. The plate is applied with a static pressure load of $p = 10^3$ Pa. Two different boundary conditions are taken into account: (a) four sides are all simply supported and (b) four sides are all clamped. The case of clamped boundaries constrains the rotation as well as the deflection. The reference result is calculated by 64×64 S4R elements in ABAQUS without considering the geometrical nonlinearity. For the simply supported boundary conditions, the particles on the boundaries of the plate are fixed. The enforcement of clamped boundary conditions requires some special treatment. As shown in Fig. 8.4, the actual physical model of the plate is denoted by the black rectangular particles, and a fictitious domain of two layers of particles outside the physical domain is generated where the particle's deflections are set to zero. the particles outside of the blue rectangle are applied with penalty $p^{hg} = 400E$ while the particles inside the blue rectangle with penalty $p^{hg} = 0$. The deflection for a simply supported plate at different times is plotted in Fig. 8.5. The deflections for a clamped plate at different times are depicted in Fig. 8.6. The deflection of the central point of the plate is monitored and compared with the result by ABAQUS, as shown in Fig. 8.7a, b, where good agreement with the FEM model is observed.

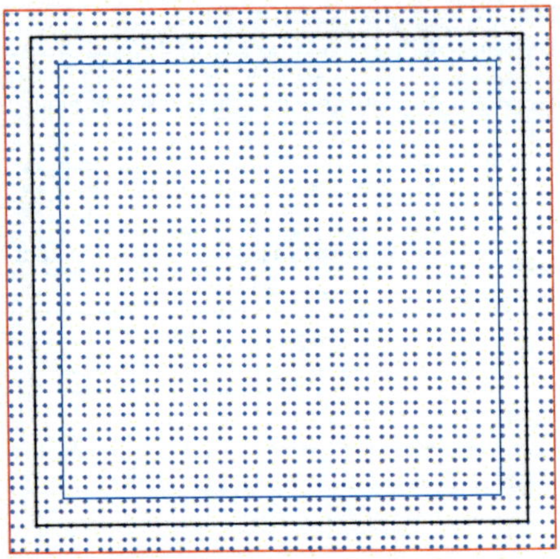

Fig. 8.4 The implementation of clamped boundary conditions. The particles in black rectangle represent the physical model and particles outside of the blue rectangle are applied with penalty $p^{hg} = 400E$

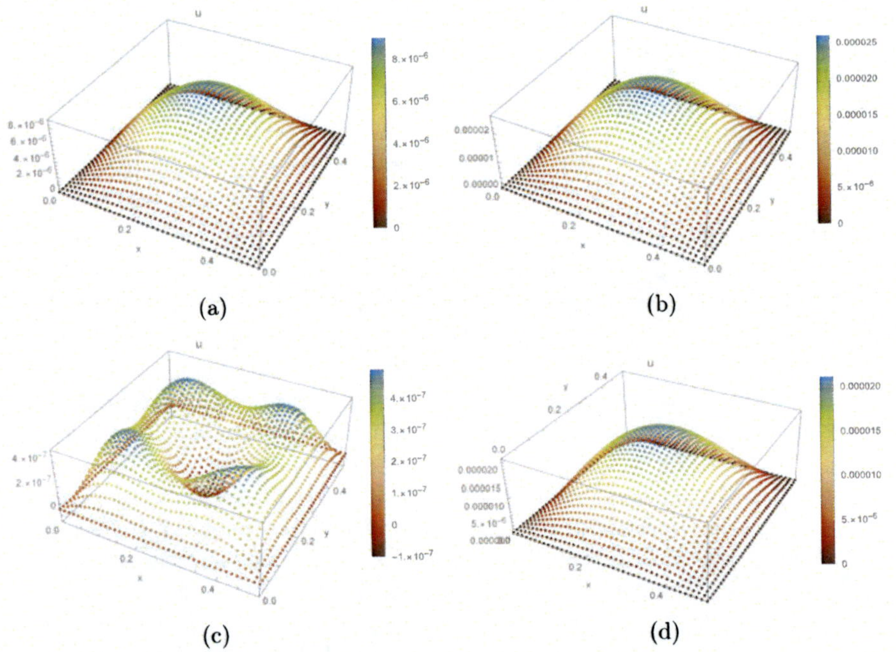

Fig. 8.5 Deflection of simply supported plate at **a** $t = 0.97$ ms, **b** $t = 2.9$ ms, **c** $t = 4.87$ ms and **d** $t = 6.77$ ms

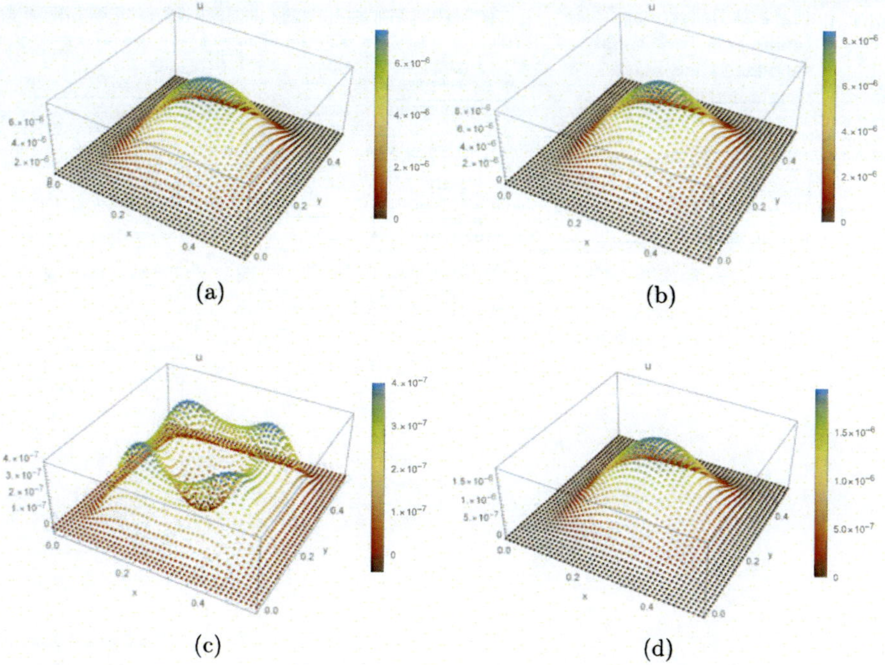

Fig. 8.6 Deflection of clamped plate at **a** $t = 0.966$ ms, **b** $t = 1.44$ ms, **c** $t = 2.42$ ms and **d** $t = 2.90$ ms

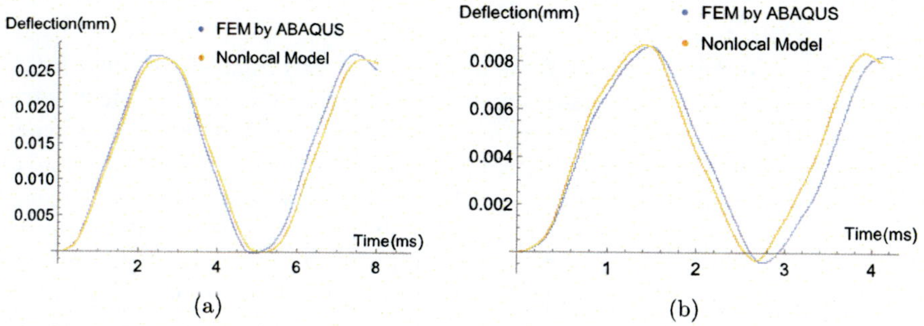

Fig. 8.7 Deflection of central point for **a** simply support plate and **b** clamped plate

For the simply supported plate, the deflection of the central point for four different weight functions is shown in Fig. 8.8. It can be seen that the weight functions barely influence the results.

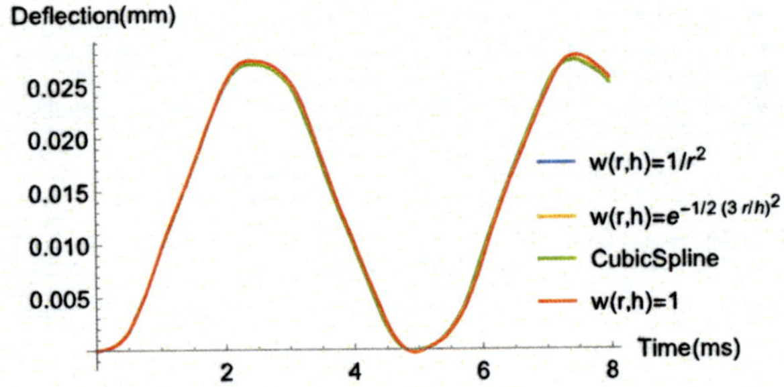

Fig. 8.8 Deflection of the central point for a simply supported plate under 4 weight functions

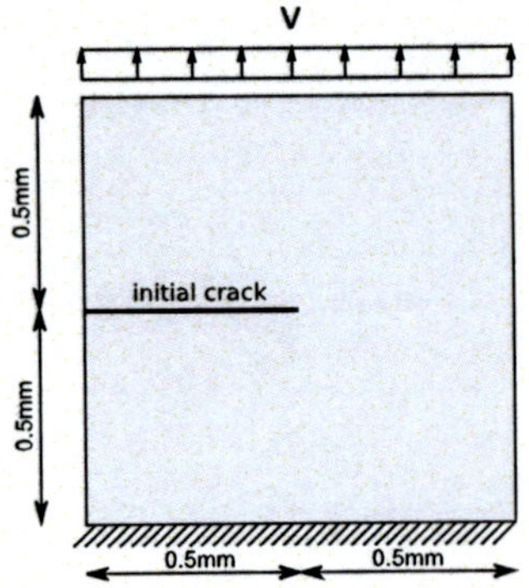

Fig. 8.9 Setup of the plate

8.5.3 *Single-Edge Notched Tension Test*

In this example, we tested the nonlocal elasticity by Eq. 8.46 for single-edge notched tension in 2D under plane stress conditions. For the case of linear elasticity, the first Piola–Kirchhoff stress is the same as the Cauchy stress. The geometry setup is given in Fig. 8.9. The bottom is fixed while the top of the plate is applied with velocity boundary condition $v = 1$ m/s, which can achieve the quasi-static condition. The material parameters are $E = 210$ GPa, $\nu = 0.3$ and critical strain is set as $s_{max} = 0.02$. The plate is discretized into 100×100 particles. Each particle's support consists of 33 nearest neighbors. The initial crack is created by modifying the neighbor list when searching for the nearest neighbors. The support for each particle is constructed by finding the k-nearest neighbors, and the size of the support is determined by the farthest particle in the support. Obviously, the size of the support can be different from each other. The fixed number of neighbors in support results in particles near the boundary with relatively large support sizes and particles at the center of the plate with small support sizes. A duration of $T = 6.5 \times 10^{-6}$ s is integrated by approximately 4500 steps at a time increment of $\Delta t = 1.5418 \times 10^{-9}$ s. Fixed velocity and displacement boundary conditions are applied to one layer of particles.

Figure 8.10a is the displacement field u_y at full damage, where the interaction of internal force between the two half-planes is cut and rigid body displacement dominates. Figure 8.10b is the distribution of hourglass energy. We can observe that

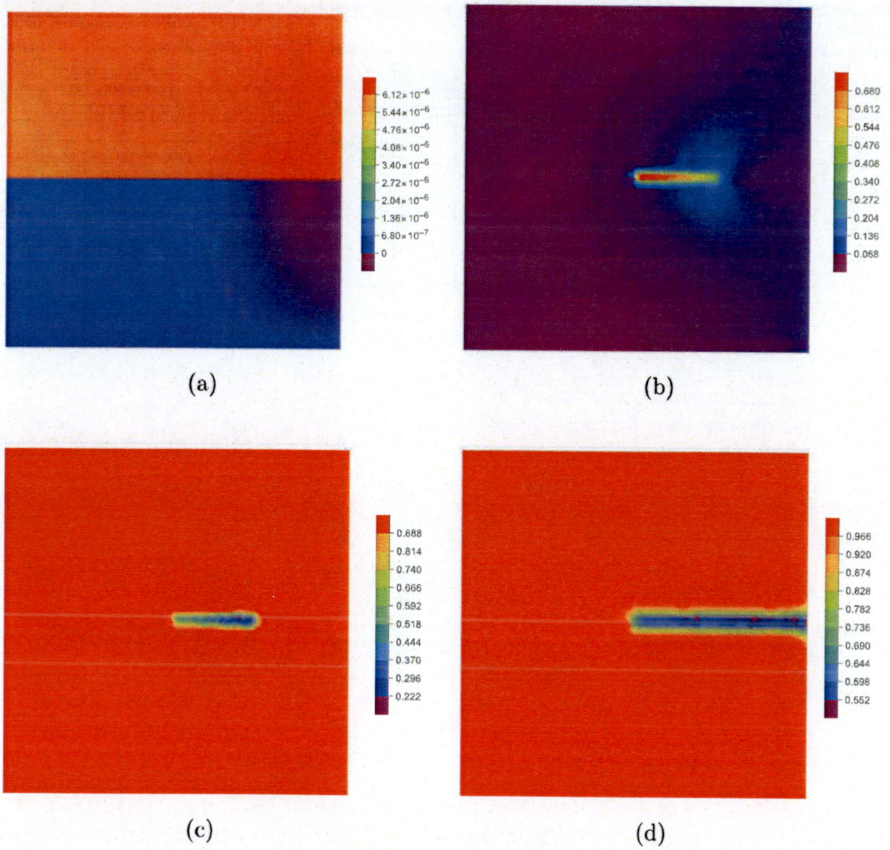

Fig. 8.10 **a** Displacement u_y at full damage, **b** operator energy at $u_y = 5.5 \times 10^{-3}$ mm, **c** damage field at $u_y = 5.5 \times 10^{-3}$ mm and **d** damage field at $u_y = 6.2 \times 10^{-3}$ mm

the hourglass energy is concentrated on the crack surface and crack tip. Figure 8.10c, d are the snapshots of the damage field, which confirms that the instability criterion in Sect. 8.3 is stable for fracture modeling.

Although the plate is solved by an explicit dynamic method, the kinetic energy is much lower than the strain energy as shown in Fig. 8.11a. The dynamic load curve agrees well with that of the finite element method in (Miehe et al. 2010b), as shown in Fig. 8.11b. One possible reason for the difference in reaction force increment is due to the explicit algorithm and nonlocal effect of the current formulation.

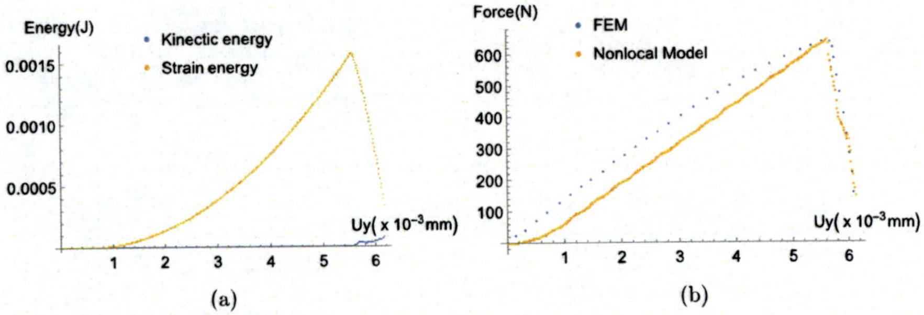

Fig. 8.11 **a** Energy curve on displacement; **b** load curve on displacement

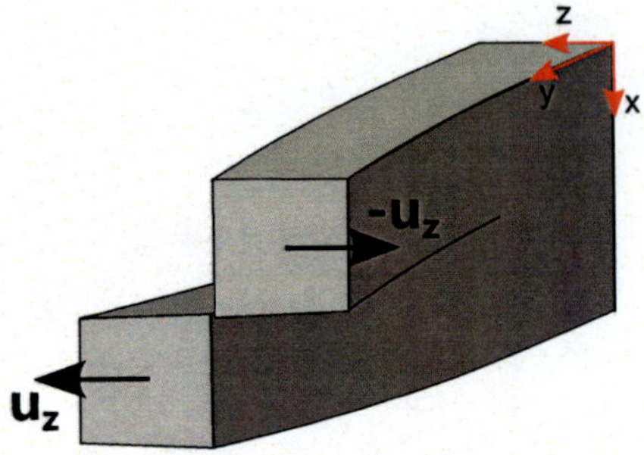

Fig. 8.12 Illustration of out-of-plane shear fracture

8.5.4 *Out-of-Plane Shear Fracture in 3D*

For brittle fracture, the basic modes of fracture are tensile fracture, in-plane shear fracture and out-of-plane shear fracture. In this section, we apply the instability damage criterion to the out-of-plane shear fracture, as shown in Fig. 8.12. The dimensions of the specimen are $5 \times 2.5 \times 1$ mm^3, as shown in Fig. 8.13. The size of the initial crack surface is 2.5×1 mm^2. The velocity boundary conditions $u_z = 1$ m/s are applied. The model is discretized into 86,961 particles with particle size $\Delta x = 0.05$ mm. Each particle has 102 neighbors in its support. Material parameters include elastic modulus $E = 210 \times 10^9$ Pa, Poisson's ratio $\nu = 0.3$ and density $\rho = 7800$ kg/m^3. The time step is selected as $\Delta t = 7.7 \times 10^{-9}$ s. A total of 3000 steps are calculated. The crack surface starts to propagate at step 1550. The crack surface at different steps is depicted in Fig. 8.14.

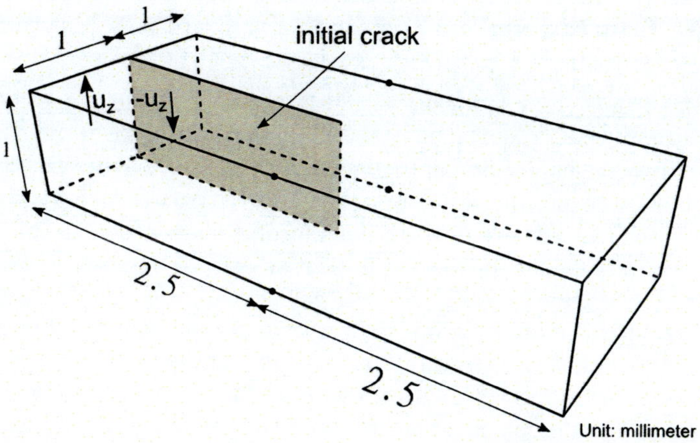

Fig. 8.13 Setup of the specimen

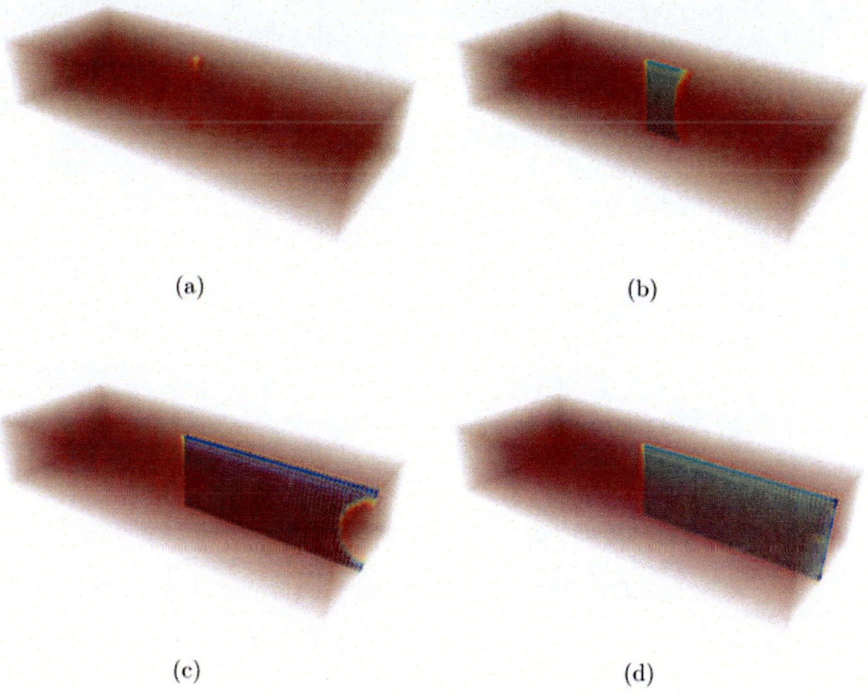

Fig. 8.14 Crack surfaces at **a** step 1550, **b** step 2050, **c** step 2950 and **d** step 3000

8.6 Conclusion

In this chapter, we employ the recently proposed NOM to derive the nonlocal strong forms for various physical models, including elasticity, thin plate, gradient elasticity, electro–magneto-elastic coupled model and phase field fracture model. These models require a second-order partial derivative at most, and we make use of the second-order NOM scheme, which contains the nonlocal gradient and nonlocal Hessian operator. Considering the fact that most physical models are compatible with the variational principle/weighted residual method, we start from the energy form/weak form of the problem, by inserting the nonlocal expression of the gradient/Hessian operator into the weak form; based on the dual property of the dual-support in NOM, the nonlocal strong form is obtained with ease. Such a process can be extended to many other physical problems in other fields. The derived strong forms are variationally consistent and allow an elegant description for inhomogeneous nonlocality in both theoretical derivation and numerical implementation.

We also propose an instability criterion in nonlocal elasticity or dual-horizon state-based peridynamics for fracture modeling. The criterion is formulated as the functional of nonlocal gradient in support, which minimizes the zero-energy deformation that cannot be described by the nonlocal gradient. Such an operator functional approaches zero for continuous fields but has a comparable value to the strain energy density for the deformation around the crack tip. During the fracture modeling by removing particles from the neighbor list, it is safer to delete the particle with larger zero-energy deformation. The numerical examples for 2D/3D fracture modeling confirm the feasibility and robustness of this criterion. The instability criterion is applicable to anisotropic elastic materials and hyperelastic materials.

References

Chen H, Chan W (2020) Higher-order peridynamic material correspondence models for elasticity. J Elast 142(1):135–161

Chowdhury S, Roy P, Roy D, Reddy J (2016) A peridynamic theory for linear elastic shells. Int J Solids Struct 84:110–132

Foster J, Silling S, Chen W (2011) An energy based failure criterion for use with peridynamic states. Int J Multiscale Comput Eng 9(6)

Javili A, Firooz S, McBride A, Steinmann P (2020) The computational framework for continuum-kinematics-inspired peridynamics. Comput Mech 66(4):795–824

Liu L (2014) An energy formulation of continuum magneto-electro-elasticity with applications. J Mech Phys Solids 63:451–480

Miehe C, Welschinger F, Hofacker M (2010) Thermodynamically consistent phase-field models of fracture: Variational principles and multi-field fe implementations. Int J Numer Methods Eng 83(10):1273–1311

Nguyen B, Zhuang X, Rabczuk T (2019) Nurbs-based formulation for nonlinear electro-gradient elasticity in semiconductors. Comput Methods Appl Mech Eng 346:1074–1095

Rabczuk T, Ren H, Zhuang X (2019) A nonlocal operator method for partial differential equations with application to electromagnetic waveguide problem. Comput Mater Continua 59(1)

Ren H, Zhuang X, Rabczuk T (2020a) A nonlocal operator method for solving partial differential equations. Comput Methods Appl Mech Eng 358:112621. ISSN 0045-7825. https://doi.org/10.1016/j.cma.2019.112621

Ren H, Zhuang X, Rabczuk T (2017) Dual-horizon peridynamics: A stable solution to varying horizons. Comput Methods Appl Mech Eng 318:762–782

Ren H, Zhuang X, Rabczuk T (2020) A higher order nonlocal operator method for solving partial differential equations. Comput Methods Appl Mech Eng 367:113132

Ren H, Zhuang X, Rabczuk T (2020) Nonlocal operator method with numerical integration for gradient solid. Comput Struct 233:106235

Ren H, Zhuang X, Trung N, Rabczuk T (2020d) Nonlocal operator method for the cahn-hilliard phase field model. Commun Nonlinear Sci Numer Simul 105687

Ren H, Zhuang X, Trung N-T, Rabczuk T (2021) A nonlocal operator method for finite deformation higher-order gradient elasticity. Comput Methods Appl Mech Eng 384:113963

Roy P, Roy D, Reddy J (2019) A conformal gauge theory of solids: Insights into a class of electromechanical and magnetomechanical phenomena. J Mech Phys Solids 130:35–55

Silling S (2000) Reformulation of elasticity theory for discontinuities and long-range forces. J Mech Phys Solids 48(1):175–209

Timoshenko S, Woinowsky-Krieger S (1959) Theory of plates and shells. McGraw-hill

Verlet L (1967) Computer "experiments" on classical fluids. i. thermodynamical properties of Lennard-Jones molecules. Phys Rev 159(1):98

Wang Y, Zhou X, Wang Y, Shou Y (2018) A 3-D conjugated bond-pair-based peridynamic formulation for initiation and propagation of cracks in brittle solids. Int J Solids Struct 134:89–115. ISSN 0020-7683. https://doi.org/10.1016/j.ijsolstr.2017.10.022

Zhou X, Wang Y, Qian Q (2016) Numerical simulation of crack curving and branching in brittle materials under dynamic loads using the extended non-ordinary state-based peridynamics. Eur J Mech-A/Solids 60:277–299

Zhou X, Wang Y, Xu X (2016b) Numerical simulation of initiation, propagation and coalescence of cracks using the non-ordinary state-based peridynamics. Int J Fract 201(2):213–234. ISSN 1573-2673. https://doi.org/10.1007/s10704-016-0126-6

Chapter 9
Nonlocal Operator Method for Dynamic Brittle Fracture Based on an Explicit Phase Field Model

Similar to non-ordinary state-based peridynamics, NOM can model fractures by modifying the neighbors in the support. However, there is no general or robust criterion for deleting the neighbors from the neighbor list for generic problems. Deleting neighbors often results in a singular shape tensor that finally terminates the calculation. Phase field fracture models (Hakim and Karma 2009; Karma et al. 2001) show good robustness due to their elegant mathematical form and variational (and thermodynamic) consistency. Phase field models for fracture enjoy Γ-convergence of the approximations of free discontinuity problems (Ambrosio and Tortorelli 1990) indicating that the integral on the volume domain with a length scale approaching zero is equal to the integral on the sharp discontinuity interface. In this chapter, we adapt the explicit phase field model in the framework of NOM to compensate for the difficulties of NOM in modeling dynamic brittle fracture. This chapter is organized as follows: In Sect. 9.1, the nonlocal operator method is briefly summarized and the nonlocal strong form of linear elasticity is derived by a variational principle. In Sect. 9.2, we briefly review the theoretical background of the phase field model and discuss the explicit phase field model with sub-stepping. We derive the nonlocal strong form for the phase field model for fracture in Sect. 9.3. Section 9.4 presents the particle-based numerical implementation of the nonlocal explicit phase field model while Sect. 9.5 presents several numerical examples.

9.1 Nonlocal Operator Method

9.1.1 Basic Principle

Consider a solid in the initial and current configuration as shown in Fig. 9.1a. Let $\boldsymbol{X}_i$ be the material coordinates of point i in the initial configuration Ω_0 and $\mathbf{x}_i = \boldsymbol{X}_i + \mathbf{u}_i$ the current coordinates of point i due to displacement $\mathbf{u}_i$. Let $\boldsymbol{X}_{ij} := \boldsymbol{X}_j - \boldsymbol{X}_i$ and $\mathbf{x}_{ij} := \mathbf{x}_j - \mathbf{x}_i$ be the relative position vector in the initial

T. Rabczuk et al., *Computational Methods Based on Peridynamics and Nonlocal Operators*, Computational Methods in Engineering & the Sciences,
https://doi.org/10.1007/978-3-031-20906-2_9

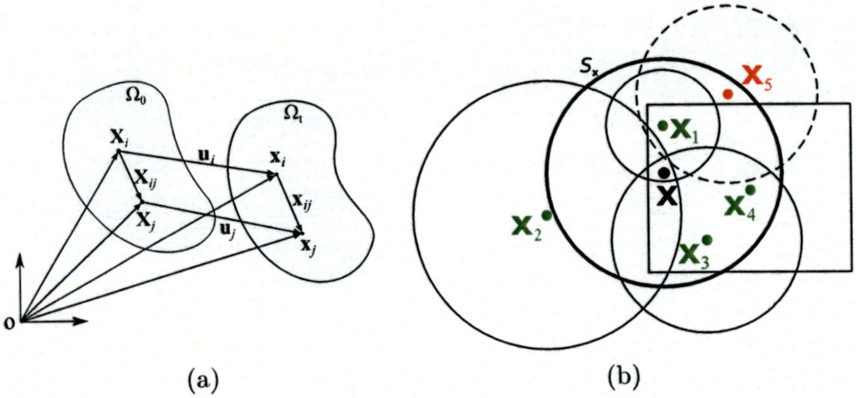

Fig. 9.1 **a** Deformation and notations; **b** schematic diagram for support and dual-support of X; the support and dual-support of X can be written as $\mathcal{S} = \{X_1, X_3, X_4, X_5\}$ *and* $\mathcal{S}' = \{X_1, X_2, X_3, X_4\}$, respectively

configuration and current configuration, respectively. The relative displacement vector for point i and j is defined as $\mathbf{u}_{ij} := \mathbf{u}_j - \mathbf{u}_i$. Obviously, $\mathbf{x}_{ij} = \boldsymbol{X}_{ij} + \mathbf{u}_{ij}$. Let $X_{ij} := \|\boldsymbol{X}_{ij}\| = \sqrt{\boldsymbol{X}_{ij} \cdot \boldsymbol{X}_{ij}}$, $x_{ij} := \|\mathbf{x}_{ij}\|$ be the length of the vector $\boldsymbol{X}_{ij}$, $\mathbf{x}_{ij}$, respectively. Two important concepts in the nonlocal operator method are the **support** and the **dual-support**.

Support of $\boldsymbol{X}_i$ is a neighborhood with a finite length scale of $\boldsymbol{X}_i$, denoted by $\mathcal{S}_i$. For example, in Fig. 9.1b, $\boldsymbol{X}_1, \boldsymbol{X}_3, \boldsymbol{X}_4, \boldsymbol{X}_5$ are included in the support domain $\mathcal{S}$. Therefore, we can write $\mathcal{S} = \{\boldsymbol{X}_1, \boldsymbol{X}_3, \boldsymbol{X}_4, \boldsymbol{X}_5\}$. The support is the basis to define the nonlocal operators. The size or the shape of the support for each particle can be different. In this paper, the support is defined based on the initial configuration, although the support could be defined in the current configuration as well.

The **dual-support** of $\boldsymbol{X}_i$ is defined as a set of points whose supports include $\boldsymbol{X}_i$,

$$\mathcal{S}_i' = \{\boldsymbol{X}_j | \boldsymbol{X}_i \in \mathcal{S}_j\}. \tag{9.1}$$

In Fig. 9.1b, the dual-support of $\boldsymbol{X}$ can be written as $\mathcal{S}' = \{\boldsymbol{X}_1, \boldsymbol{X}_2, \boldsymbol{X}_3, \boldsymbol{X}_4\}$. The dual-support is a natural consequence of the variational derivation of the energy functional, which will be demonstrated in the following sections.

The nonlocal gradient for vector field $\mathbf{u}$ at point i is defined in the support $\mathcal{S}_i$ as (Ren et al. 2020a)

$$\tilde{\nabla}\mathbf{u}_i := \int_{\mathcal{S}_i} w(\boldsymbol{X}_{ij})\mathbf{u}_{ij} \otimes \boldsymbol{X}_{ij} \mathrm{d}V_j \cdot \boldsymbol{K}_i^{-1} \tag{9.2}$$

with shape tensor $\boldsymbol{K}_i := \int_{S_i} w(\boldsymbol{X}_{ij})\boldsymbol{X}_{ij} \otimes \boldsymbol{X}_{ij}\mathrm{d}V_j$, where $w(\boldsymbol{X}) = 1/|\boldsymbol{X}|^2$ is the weight function. The nonlocal gradient for a scalar field s at point $\boldsymbol{X}_i$ is defined as

$$\tilde{\nabla} s_i := \int_{S_i} w(\boldsymbol{X}_{ij}) s_{ij}\, \boldsymbol{X}_{ij}\mathrm{d}V_j \cdot \boldsymbol{K}_i^{-1}. \tag{9.3}$$

The variation of the nonlocal gradient is

$$\tilde{\nabla}\delta\mathbf{u}_i = \int_{S_i} w(\boldsymbol{X}_{ij})(\delta\mathbf{u}_j - \delta\mathbf{u}_i) \otimes \boldsymbol{X}_{ij}\mathrm{d}V_j \cdot \boldsymbol{K}_i^{-1}. \tag{9.4}$$

The variation of the nonlocal gradient for scalar field s is

$$\tilde{\nabla}\delta s_i = \int_{S_i} w(\boldsymbol{X}_{ij})(\delta s_j - \delta s_i)\boldsymbol{K}_i^{-1}\boldsymbol{X}_{ij}\mathrm{d}V_j. \tag{9.5}$$

The NOM can make use of the energy functional or equivalent weak form in conventional local methods to obtain the nonlocal strong form.

9.1.2 Nonlocal Form of Linear Elasticity

For a linear elastic solid with small deformation, the equivalent energy functional is

$$\Pi(\boldsymbol{u}) = \int_\Omega \psi_e(\varepsilon(\nabla\boldsymbol{u}))\mathrm{d}V - \int_{\partial\Omega} \boldsymbol{t}^* \cdot \boldsymbol{u}\mathrm{d}A - \int_\Omega \mathbf{b} \cdot \boldsymbol{u}\mathrm{d}V \tag{9.6}$$

where ψ_e is strain energy density, $\boldsymbol{u}$ is the displacement field, $\boldsymbol{t}^*$ the traction force vector and $\mathbf{b}$ the body force density. Replacing the local gradient $\nabla\boldsymbol{u}$ with nonlocal gradient $\tilde{\nabla}\boldsymbol{u}$, we have

$$\Pi(\boldsymbol{u}) = \int_\Omega \psi_e(\varepsilon(\tilde{\nabla}\boldsymbol{u}))\mathrm{d}V - \int_{\partial\Omega} \boldsymbol{t}^* \cdot \boldsymbol{u}\mathrm{d}A - \int_\Omega \mathbf{b} \cdot \boldsymbol{u}\mathrm{d}V. \tag{9.7}$$

The first variation of $\Pi(\boldsymbol{u})$ leads to

$$\begin{aligned}\delta\Pi(\boldsymbol{u}) &= \int_\Omega \frac{\partial\psi_e}{\partial\varepsilon} : \tilde{\nabla}\delta\boldsymbol{u}\mathrm{d}V - \int_{\partial\Omega} \boldsymbol{t}^* \cdot \delta\boldsymbol{u}\mathrm{d}A - \int_\Omega \mathbf{b} \cdot \delta\boldsymbol{u}\mathrm{d}V \\ &= \int_\Omega \boldsymbol{\sigma} : \tilde{\nabla}\delta\boldsymbol{u}\mathrm{d}V - \int_{\partial\Omega} \boldsymbol{t}^* \cdot \delta\boldsymbol{u}\mathrm{d}A - \int_\Omega \mathbf{b} \cdot \delta\boldsymbol{u}\mathrm{d}V\end{aligned}$$

$$
\begin{aligned}
&= \int_{\Omega} \boldsymbol{\sigma}_i : \tilde{\nabla} \delta \boldsymbol{u}_i \mathrm{d}V_i - \int_{\partial\Omega} \boldsymbol{t}^* \cdot \delta \boldsymbol{u}_i \mathrm{d}A_i - \int_{\Omega} \mathbf{b} \cdot \delta \boldsymbol{u}_i \mathrm{d}V_i \\
&= \int_{\Omega} \boldsymbol{\sigma}_i : \Big(\int_{\mathcal{S}_i} w(\boldsymbol{X}_{ij}) \delta \boldsymbol{u}_{ij} \otimes \boldsymbol{X}_{ij} \mathrm{d}V_j \cdot \boldsymbol{K}_i^{-1} \Big) \mathrm{d}V_i - \int_{\partial\Omega} \boldsymbol{t}^* \cdot \delta \boldsymbol{u}_i \mathrm{d}A_i - \int_{\Omega} \mathbf{b} \cdot \delta \boldsymbol{u}_i \mathrm{d}V_i \\
&= \int_{\Omega} \Big(\int_{\mathcal{S}_i} w(\boldsymbol{X}_{ij}) \boldsymbol{\sigma}_i \boldsymbol{K}_i^{-1} \boldsymbol{X}_{ij} \cdot \delta \boldsymbol{u}_{ij} \mathrm{d}V_j \Big) \mathrm{d}V_i - \int_{\partial\Omega} \boldsymbol{t}^* \cdot \delta \boldsymbol{u}_i \mathrm{d}A_i - \int_{\Omega} \mathbf{b} \cdot \delta \boldsymbol{u}_i \mathrm{d}V_i \\
&= \int_{\Omega} \Big(\int_{\mathcal{S}_i} w(\boldsymbol{X}_{ij}) \boldsymbol{\sigma}_i \boldsymbol{K}_i^{-1} \boldsymbol{X}_{ij} \cdot (\delta \boldsymbol{u}_j - \delta \boldsymbol{u}_i) \mathrm{d}V_j \Big) \mathrm{d}V_i - \int_{\partial\Omega} \boldsymbol{t}^* \cdot \delta \boldsymbol{u}_i \mathrm{d}A_i - \int_{\Omega} \mathbf{b} \cdot \delta \boldsymbol{u}_i \mathrm{d}V_i \\
&= \int_{\Omega} \Big(\int_{\mathcal{S}'_i} w(\boldsymbol{X}_{ji}) \boldsymbol{\sigma}_j \boldsymbol{K}_j^{-1} \boldsymbol{X}_{ji} \cdot \delta \boldsymbol{u}_i \mathrm{d}V_j - \int_{\mathcal{S}_i} w(\boldsymbol{X}_{ij}) \boldsymbol{\sigma}_i \boldsymbol{K}_i^{-1} \boldsymbol{X}_{ij} \cdot \delta \boldsymbol{u}_i \mathrm{d}V_j \Big) \mathrm{d}V_i - \\
&\quad \int_{\partial\Omega} \boldsymbol{t}^* \cdot \delta \boldsymbol{u}_i \mathrm{d}A_i - \int_{\Omega} \mathbf{b} \cdot \delta \boldsymbol{u}_i \mathrm{d}V_i
\end{aligned}
$$

where $\boldsymbol{\sigma} := \frac{\partial \psi_e}{\partial \boldsymbol{\epsilon}}$ is the Cauchy stress. For any $\delta \boldsymbol{u}_i$, $\delta \Pi = 0$ leads to

$$
\int_{\mathcal{S}_i} w(\boldsymbol{X}_{ij}) \boldsymbol{\sigma}_i \boldsymbol{K}_i^{-1} \boldsymbol{X}_{ij} \mathrm{d}V_j - \int_{\mathcal{S}'_i} w(\boldsymbol{X}_{ji}) \boldsymbol{\sigma}_j \boldsymbol{K}_j^{-1} \boldsymbol{X}_{ji} \mathrm{d}V_j + \mathbf{b}_i = 0 \quad \forall \boldsymbol{x}_i \in \Omega. \tag{9.8}
$$

For a point on $\partial\Omega$, the surface traction needs to be added to Eq. 9.8. For dynamic problems, the inertial force should be added to the right side of the equation

$$
\int_{\mathcal{S}_i} w(\boldsymbol{X}_{ij}) \boldsymbol{\sigma}_i \boldsymbol{K}_i^{-1} \boldsymbol{X}_{ij} \mathrm{d}V_j - \int_{\mathcal{S}'_i} w(\boldsymbol{X}_{ji}) \boldsymbol{\sigma}_j \boldsymbol{K}_j^{-1} \boldsymbol{X}_{ji} \mathrm{d}V_j + \mathbf{b}_i = \rho_i \ddot{\boldsymbol{u}}_i \quad \forall \boldsymbol{x}_i \in \Omega. \tag{9.9}
$$

It can be seen that for a linear elastic solid, the nonlocal form and local form are correlated with

$$
\int_{\mathcal{S}_i} w(\boldsymbol{X}_{ij}) \boldsymbol{\sigma}_i \boldsymbol{K}_i^{-1} \boldsymbol{X}_{ij} \mathrm{d}V_j - \int_{\mathcal{S}'_i} w(\boldsymbol{X}_{ji}) \boldsymbol{\sigma}_j \boldsymbol{K}_j^{-1} \boldsymbol{X}_{ji} \mathrm{d}V_j \rightleftharpoons \nabla \cdot \boldsymbol{\sigma}_i. \tag{9.10}
$$

Equation 9.9 is the nonlocal governing equation for a linear elastic solid and is identical to dual-horizon peridynamics. It can be seen that the nonlocal governing equation is variationally consistent, and the dual-support is a natural consequence of the variational principle. However, Eq. 9.9 suffers from the zero-energy mode and the penalty force from the nonlocal operator functional should be added.

9.1.3 *Operator Energy Functional for Vector Field and Scalar Field*

A smooth vector field should be consistent with its nonlocal gradient. For a vector field $\mathbf{u}$ with nonlocal gradients defined by Eq. 9.2, the inconsistency between $\mathbf{u}_{ij}$ and

$\tilde{\nabla}\mathbf{u}_i\boldsymbol{X}_{ij}$, the prediction by the nonlocal gradient, can be formulated as a quadratic functional, the operator energy functional of vector field $\mathbf{u}$, which can be defined as

$$\mathcal{F}_i = \frac{\alpha_v}{2m_i}\int_{\mathcal{S}_i} w(\boldsymbol{X}_{ij})(\tilde{\nabla}\mathbf{u}_i\boldsymbol{X}_{ij} - \mathbf{u}_{ij})^T(\tilde{\nabla}\mathbf{u}_i\boldsymbol{X}_{ij} - \mathbf{u}_{ij})\mathrm{d}V_j \tag{9.11}$$

with $m_i = \mathrm{tr}[\boldsymbol{K}_i]$ and α_v is a penalty factor. The variation of $\mathcal{F}_i$ can be derived as follows.

$$\begin{aligned}\delta\mathcal{F}_i &= \frac{\alpha_v}{m_i}\int_{\mathcal{S}_i} w(\boldsymbol{X}_{ij})(\tilde{\nabla}\mathbf{u}_i\boldsymbol{X}_{ij} - \mathbf{u}_{ij})^T(\tilde{\nabla}\delta\mathbf{u}_i\boldsymbol{X}_{ij} - \delta\mathbf{u}_{ij})\mathrm{d}V_j\\ &= \frac{\alpha_v}{m_i}\int_{\mathcal{S}_i} w(\boldsymbol{X}_{ij})(\tilde{\nabla}\mathbf{u}_i\boldsymbol{X}_{ij} - \mathbf{u}_{ij})^T\tilde{\nabla}\delta\mathbf{u}_i\boldsymbol{X}_{ij}\mathrm{d}V_j - \frac{\alpha_v}{m_i}\int_{\mathcal{S}_i} w(\boldsymbol{X}_{ij})(\tilde{\nabla}\mathbf{u}_i\boldsymbol{X}_{ij} - \mathbf{u}_{ij})^T\delta\mathbf{u}_{ij}\mathrm{d}V_j.\end{aligned} \tag{9.12}$$

It can be shown that the first term in Eq. 9.12 vanishes.

$$\begin{aligned}&\int_{\mathcal{S}_i} w(\boldsymbol{X}_{ij})(\tilde{\nabla}\mathbf{u}_i\boldsymbol{X}_{ij} - \mathbf{u}_{ij})^T\tilde{\nabla}\delta\mathbf{u}_i\boldsymbol{X}_{ij}\mathrm{d}V_j\\ =&\int_{\mathcal{S}_i} w(\boldsymbol{X}_{ij})(\tilde{\nabla}\mathbf{u}_i\boldsymbol{X}_{ij} - \mathbf{u}_{ij})\otimes\boldsymbol{X}_{ij}\mathrm{d}V_j : \tilde{\nabla}\delta\mathbf{u}_i\\ =&\Big(\tilde{\nabla}\mathbf{u}_i\int_{\mathcal{S}_i} w(\boldsymbol{X}_{ij})\boldsymbol{X}_{ij}\otimes\boldsymbol{X}_{ij}\mathrm{d}V_j - \int_{\mathcal{S}_i} w(\boldsymbol{X}_{ij})\mathbf{u}_{ij}\otimes\boldsymbol{X}_{ij}\mathrm{d}V_j\Big) : \delta\tilde{\nabla}\mathbf{u}_i\\ =&\Big(\tilde{\nabla}\mathbf{u}_i\boldsymbol{K}_i - \tilde{\nabla}\mathbf{u}_i\boldsymbol{K}_i\Big) : \delta\tilde{\nabla}\mathbf{u}_i = 0.\end{aligned} \tag{9.13}$$

Hence,

$$\begin{aligned}\delta\mathcal{F}_i &= -\frac{\alpha_v}{m_i}\int_{\mathcal{S}_i} w(\boldsymbol{X}_{ij})(\tilde{\nabla}\mathbf{u}_i\boldsymbol{X}_{ij} - \mathbf{u}_{ij})^T\delta\mathbf{u}_{ij}\mathrm{d}V_j\\ &= \frac{\alpha_v}{m_i}\int_{\mathcal{S}_i} w(\boldsymbol{X}_{ij})(\mathbf{u}_{ij} - \tilde{\nabla}\mathbf{u}_i\boldsymbol{X}_{ij})^T(\delta\mathbf{u}_j - \delta\mathbf{u}_i)\mathrm{d}V_j.\end{aligned}$$

The variation of $\int_\Omega \mathcal{F}_i\mathrm{d}V_i$ is

$$\begin{aligned}&\int_\Omega \delta\mathcal{F}_i\mathrm{d}V_i = \int_\Omega\frac{\alpha_v}{m_i}\int_{\mathcal{S}_i} w(\boldsymbol{X}_{ij})(\mathbf{u}_{ij} - \tilde{\nabla}\mathbf{u}_i\boldsymbol{X}_{ij})^T(\delta\mathbf{u}_j - \delta\mathbf{u}_i)\mathrm{d}V_j\mathrm{d}V_i\\ =&\int_\Omega\Big(\int_{\mathcal{S}'_i}\frac{\alpha_v}{m_j}w(\boldsymbol{X}_{ji})(\mathbf{u}_{ji} - \tilde{\nabla}\mathbf{u}_j\boldsymbol{X}_{ji})\mathrm{d}V_j - \int_{\mathcal{S}_i}\frac{\alpha_v}{m_i}w(\boldsymbol{X}_{ij})(\mathbf{u}_{ij} - \tilde{\nabla}\mathbf{u}_i\boldsymbol{X}_{ij})\Big)\delta\mathbf{u}_i\mathrm{d}V_i.\end{aligned} \tag{9.14}$$

Therefore, the internal force due to operator energy functional is

$$\int_{\mathcal{S}'_i}\boldsymbol{T}_{ji}\mathrm{d}V_j - \int_{\mathcal{S}_i}\boldsymbol{T}_{ij}\mathrm{d}V_j \tag{9.15}$$

where $T_{ij} := \frac{\alpha_v}{m_i} w(X_{ij})(\mathbf{u}_{ij} - \tilde{\nabla}\mathbf{u}_i X_{ij})$ is the zero-energy internal force for vector X_{ij}. Similarly, for a scalar field, the correction flux can be written as

$$\int_{S_i'} t_{ji} \mathrm{d}V_j - \int_{S_i} t_{ij} \mathrm{d}V_j \tag{9.16}$$

with $t_{ij} = \frac{\alpha_s}{m_i} w(X_{ij})(s_{ij} - \tilde{\nabla} s_i X_{ij})$ with penalty coefficient α_s. The operator functional-enhanced governing equation for a linear elastic solid can be written as

$$\nabla \cdot \boldsymbol{\sigma} \underset{\text{Nonlocal to Local}}{\overset{\text{Local to Nonlocal}}{\rightleftharpoons}} \int_{S_i} (w(X_{ij})\boldsymbol{\sigma}_i \cdot K_i^{-1} X_{ij} + T_{ij}) \mathrm{d}V_j - \int_{S_i'} (w(X_{ji})\boldsymbol{\sigma}_j \cdot K_j^{-1} X_{ji} + T_{ji}) \mathrm{d}V_j. \tag{9.17}$$

9.2 Outline of Phase Field Fracture Model

9.2.1 Phase Field Model

In this section, we briefly review the thermodynamically consistent phase field model presented by Miehe *et al.* (Miehe et al. 2010b). Let the phase field $s = 1$ denote the completely damaged material while $s = 0$ indicates the total intact material. The potential functional of the phase field and mechanical field can be written as

$$\begin{aligned} \Pi_l(\boldsymbol{u}, s) = & \int_\Omega ((1-s)^2 \psi_e^+ + \psi_e^-) \mathrm{d}V - \int_{\partial\Omega} \boldsymbol{t}^* \cdot \boldsymbol{u} \, \mathrm{d}A \\ & - \int_\Omega \mathbf{b} \cdot \boldsymbol{u} \mathrm{d}V + \int_\Omega g_c \left(\frac{s^2}{2l} + \frac{l}{2} \nabla s \cdot \nabla s\right) \mathrm{d}V, \end{aligned} \tag{9.18}$$

where $\psi_e^\pm$ denotes the tensile/compressive strain energy density, g_c is the critical energy release rate, $\boldsymbol{u}$ the displacement, $\boldsymbol{\varepsilon}$ the strain, $\boldsymbol{t}^*$ the surface traction and $\mathbf{b}$ the body force density; $\psi_e^\pm$ is defined as

$$\psi_e^\pm(\boldsymbol{\varepsilon}) := \lambda \langle \varepsilon_1 + \varepsilon_2 + \varepsilon_3 \rangle_\pm^2 / 2 + \mu (\langle \varepsilon_1 \rangle_\pm^2 + \langle \varepsilon_2 \rangle_\pm^2 + \langle \varepsilon_3 \rangle_\pm^2), \tag{9.19}$$

where λ, μ are the Lamé constants, $\langle x \rangle_\pm := (x \pm |x|)/2$.

It can be shown that the strong form can be expressed as

$$\nabla \cdot \boldsymbol{\sigma}^s + \mathbf{b} = \rho \ddot{\boldsymbol{u}}, \tag{9.20}$$

$$g_c(s - l^2\nabla^2 s) = 2l(1-s)\mathcal{H}(x, T), \tag{9.21}$$

where $\mathcal{H}(x, T) := \max_{t\in[0,T]} \psi^+(\boldsymbol{\varepsilon}(\boldsymbol{x}, t))$ (Miehe et al. 2010b). $\boldsymbol{\sigma}^s$ is the Cauchy stress defined as

$$\boldsymbol{\sigma}^s = (1-s)^2 \frac{\partial \psi_e^+}{\partial \boldsymbol{\varepsilon}} + \frac{\partial \psi_e^-}{\partial \boldsymbol{\varepsilon}} = (1-s)^2 \boldsymbol{\sigma}^+ + \boldsymbol{\sigma}^-, \tag{9.22}$$

$$\begin{aligned} \boldsymbol{\sigma}^\pm = & \frac{\partial \psi_e^\pm}{\partial \boldsymbol{\varepsilon}} = \lambda \langle \varepsilon_1 + \varepsilon_2 + \varepsilon_3 \rangle_\pm I_{3\times 3} \\ & + 2\mu(\langle \varepsilon_1 \rangle_\pm \boldsymbol{n}_1 \otimes \boldsymbol{n}_1 + \langle \varepsilon_2 \rangle_\pm \boldsymbol{n}_2 \otimes \boldsymbol{n}_2 + \langle \varepsilon_3 \rangle_\pm \boldsymbol{n}_3 \otimes \boldsymbol{n}_3), \end{aligned} \tag{9.23}$$

where $I_{3\times 3}$ is the identity matrix and $\boldsymbol{n}_1, \boldsymbol{n}_2, \boldsymbol{n}_3$ are the eigenvectors of the associated principal strains $\varepsilon_1, \varepsilon_2, \varepsilon_3$ of $\boldsymbol{\varepsilon}(= \sum_{i=1}^3 \varepsilon_i \boldsymbol{n}_i \otimes \boldsymbol{n}_i)$. For more details, the reader is referred to (Miehe et al. 2010b; Ren et al. 2019).

9.2.2 Evolution Equations in Gradient Damage Mechanics

Let us consider a "total" pseudo-energy density W per unit volume (Miehe 2011)

$$W(\boldsymbol{\varepsilon}, s, \nabla s) = W_{bulk}(\boldsymbol{\varepsilon}; s) + W_{frac}(s, \nabla s) \tag{9.24}$$

where $W_{bulk}(\boldsymbol{\varepsilon}; s)$ is a degrading elastic bulk energy and $W_{frac}(s, \nabla s)$ is the accumulated dissipative energy. Equation 9.24 "converts" the bulk energy to the surface energy density (Frémond and Nedjar 1996; Fremond and Shitikova 2002; Pham et al. 2011). Based on Griffith's criterion and Γ-convergence, the dissipative energy due to fracture is written as

$$W_{frac}(s, \nabla s) = g_c \gamma_l(s, \nabla s) = \frac{g_c}{2l}(s^2 + l^2 \nabla s \cdot \nabla s). \tag{9.25}$$

The evolution of the damage variable s is determined by the variational derivative of W (Miehe 2011). The irreversibility of the rate-independent phase field evolution is guaranteed by the Kuhn–Tucker-type equations

$$\dot{s} \geq 0, \quad -\delta_s W \leq 0, \quad \dot{s}(-\delta_s W) = 0. \tag{9.26}$$

9.2.3 Phase Field Evolution with and without Threshold

Miehe*et al.* (Miehe et al. 2015) developed a generalized formulation for the evolution of the phase field $s(X, t)$ for regularized crack surfaces. The formulation is applicable to different constitutive models of energetic and non-energetic driving forces. The evolution of crack phase field reads

$$\underbrace{\eta \dot{s}}_{evolution} = \underbrace{(1-s)\mathcal{H}}_{driving\,force} - \underbrace{(s - l^2 \nabla \cdot \nabla s)}_{geometric\,resistance}, \ \mathcal{H}(X, T) := \max_{t \in [0,T]} \mathcal{D}(X, t) \tag{9.27}$$

where $\eta \geq 0$ is a material parameter that characterizes the viscosity of the crack propagation; $\eta = 0$ governs the rate-independent propagation. The terms of the driving force and geometric resistance form a generalized Ginzburg–Landau-type structure (Miehe et al. 2015). The way to derive the driving force leads to different damage criteria in phase field models. For the variationally consistent phase field formulation (Miehe et al. 2010b), the crack driving state function is

$$\mathcal{D} = \frac{2\phi^+(\varepsilon)}{g_c/l} \tag{9.28}$$

where ϕ^+ is the tensile part of the strain energy in Eq. 9.19. However, Eq. 9.28 allows a damage-type degradation at low stress level. In order to prevent the undesired damage at low loads, (Miehe et al. 2015) introduced a strain-based threshold

$$\mathcal{D} = \langle \frac{\phi^+(\varepsilon)}{\phi_c} - 1 \rangle_+. \tag{9.29}$$

This criterion is independent of the length scale l (Miehe et al. 2015).

In analogy with the strain criteria, the stress criteria (Miehe et al. 2015) with and without threshold are

$$\mathcal{D} = \frac{2\phi^{*+}(\sigma)}{g_c/l} \text{ and } \mathcal{D} = \langle \frac{\phi^{*+}(\sigma)}{\phi_c^*} - 1 \rangle_+. \tag{9.30}$$

For anisotropic elastic materials, a principal tensile stress with threshold (Miehe et al. 2015) can be defined as

$$\phi^{*+}(\sigma) = \frac{1}{2E}\|\sigma^+\|^2 = \frac{1}{2E}\sum_{i=1}^{3}\langle \sigma_i \rangle_+^2 \tag{9.31}$$

with

$$\boldsymbol{\sigma} = \boldsymbol{\sigma}^+ + \boldsymbol{\sigma}^-, \text{ with } \boldsymbol{\sigma}^+ := \sum_{i=1}^{3} \langle \sigma_a \rangle_+ \boldsymbol{n}_i \otimes \boldsymbol{n}_i. \tag{9.32}$$

Threshold energies for stresses and strains can be written as

$$\phi_c = \frac{E}{2}\varepsilon_c^2, \phi_c^* = \frac{\sigma_c^2}{2E} \tag{9.33}$$

where $\sigma_c = \frac{9}{16}E\varepsilon_c$ with $\varepsilon_c = \sqrt{\frac{g_c/l}{3E}}$.

9.2.4 Explicit Phase Field Model with Sub-Step

Equation 9.27 depends on the phase field viscosity η, which is difficult to use in explicit algorithms (Ren et al. 2019). Based on Eq. 9.27, Ziaei and Shen (Vahid and Shen 2016) studied the explicit phase field using the phase field viscosity η and observed that the phase field evolution is different from the static model. Ren et al. (2019) proposed a phase field increment scheme with sub-steps. The method formulates the energy functional on the nodes rather than on the elements. Since the NOM used in this paper can be viewed as a particle-based method, the sub-step scheme in (Ren et al. 2019) can be adapted into NOM with ease. Replacing the phase field rate in Eq. 9.27 with the phase field increment, the sub-step scheme can be written as

$$\Delta s = \frac{1}{n_s}\langle \beta \rangle_+, \text{ with } \beta = (1-s)\mathcal{H} - (s - l^2\nabla^2 s) \tag{9.34}$$

where β is the "phase field residual", n_s refers to the phase field mobility (Kim et al. 1999) that controls the increment of the phase field and Δs denotes the phase field increment for each sub-step. Similar to the phase field mobility in the Cahn–Hilliard equation or solidification of binary alloys (Kim et al. 1999), n_s is a proportional constant between the driving force of the phase field and the interface moving velocity. Equation 9.34 results in an explicit, irreversible and rate-independent phase field scheme, where the phase field is fully developed at each step. The sub-step scheme adaptively reduces the residual β. Ren et al. (2019) shows that $n_s = 4, 5$ can achieve fast convergence in terms of L^∞ norm or L^2-norm of the phase field residual in several sub-steps. The L^∞ norm of the phase field residual is defined as

$$L^\infty(\beta) = \max_i \beta_i. \tag{9.35}$$

The use of L^∞ is based on the fact that the crack tip is quite small compared with the whole domain. It would be more practical to track the local maximal phase field residual than to deal with other norms of the phase field residual.

9.3 Nonlocal Form of the Phase Field Model

According to (Miehe et al. 2010b), the regularized crack functional is given as

$$\Gamma(s) \approx \int_{\Omega} (\frac{s^2}{2l} + \frac{l}{2}\nabla s \cdot \nabla s) \mathrm{d}V, \tag{9.36}$$

where $\Gamma(s)$ is the crack surface. The variation derivation of crack functional results in

$$\begin{aligned} s - l^2\nabla^2 s = 0 \text{ in } \Omega \\ \nabla s \cdot \boldsymbol{n} = 0 \text{ on } \partial\Omega, \end{aligned} \tag{9.37}$$

where $\boldsymbol{n}$ is the outward normal on the domain boundary $\partial\Omega$. Based on the local form of the regularized crack functional by Eq. 9.36 and replacing the local gradient ∇s with nonlocal gradient $\tilde{\nabla} s$, we obtain the nonlocal form of the regularized crack functional

$$\Gamma(s) \approx \int_{\Omega} (\frac{s^2}{2l} + \frac{l}{2}\tilde{\nabla} s \cdot \tilde{\nabla} s) \mathrm{d}V. \tag{9.38}$$

The variation of the regularized crack functional is

$$\begin{aligned} \delta\Gamma(s) &\approx \int_{\Omega} (\frac{s_i}{l}\delta s_i + l\tilde{\nabla}\delta s_i \cdot \tilde{\nabla} s_i) \mathrm{d}V_i \\ &= \int_{\Omega} (\frac{s_i}{l}\delta s_i + l\int_{S_i} (w(\boldsymbol{X}_{ij})(\delta s_j - \delta s_i)\boldsymbol{K}_i^{-1}\boldsymbol{X}_{ij}\mathrm{d}V_j) \cdot \tilde{\nabla} s_i) \mathrm{d}V_i \\ &= \int_{\Omega} (\frac{s_i}{l}\delta s_i + l\int_{S_i} (w(\boldsymbol{X}_{ij})(\delta s_j - \delta s_i)\tilde{\nabla} s_i \boldsymbol{K}_i^{-1}\boldsymbol{X}_{ij}\mathrm{d}V_j)) \mathrm{d}V_i \\ &= \int_{\Omega} (\frac{s_i}{l} - l\int_{S_i} w(\boldsymbol{X}_{ij})\tilde{\nabla} s_i \boldsymbol{K}_i^{-1}\boldsymbol{X}_{ij}\mathrm{d}V_j + l\int_{S_i'} w(\boldsymbol{X}_{ji})\tilde{\nabla} s_j \boldsymbol{K}_j^{-1}\boldsymbol{X}_{ji}\mathrm{d}V_j)\delta s_i \mathrm{d}V_i. \end{aligned} \tag{9.39}$$

For any δs_i, $\delta\Gamma(s) = 0$ leads to

$$s_i - l^2\int_{S_i} w(\boldsymbol{X}_{ij})\tilde{\nabla} s_i \boldsymbol{K}_i^{-1}\boldsymbol{X}_{ij}\mathrm{d}V_j + l^2\int_{S_i'} w(\boldsymbol{X}_{ji})\tilde{\nabla} s_j \boldsymbol{K}_j^{-1}\boldsymbol{X}_{ji}\mathrm{d}V_j = 0 \forall \boldsymbol{x}_i \in \Omega \tag{9.40}$$

which is the nonlocal form of phase field. Comparing Eqs. 9.40 with 9.37, we find the correspondence between the local and nonlocal form

$$\nabla^2 s \underset{\text{Nonlocal to Local}}{\overset{\text{Local to Nonlocal}}{\rightleftharpoons}} \int_{\mathcal{S}_i} (w(\boldsymbol{X}_{ij})\tilde{\nabla} s_i \boldsymbol{K}_i^{-1}\boldsymbol{X}_{ij} + t_{ij})\mathrm{d}V_j - \int_{\mathcal{S}_i'} (w(\boldsymbol{X}_{ji})\tilde{\nabla} s_j \boldsymbol{K}_j^{-1}\boldsymbol{X}_{ji} + t_{ji})\mathrm{d}V_j \tag{9.41}$$

where $t_{ij} = \frac{\alpha_s}{m_i} w(\boldsymbol{X}_{ij})(s_{ij} - \tilde{\nabla} s_i \boldsymbol{X}_{ij})$ is the stabilized term for phase field; a penalty factor of $\alpha_s = 1$ is used in this work. In order to obtain the nonlocal strong form of the phase field model, we only need to replace the local differential form with the nonlocal integral form. Based on Eq. 9.41, the nonlocal form of Eq. 9.21 is

$$\beta_i = (1 - s_i)\mathcal{H}_i - \left(s_i - l^2 \int_{\mathcal{S}_i} (w(\boldsymbol{X}_{ij})\tilde{\nabla} s_i \boldsymbol{K}_i^{-1}\boldsymbol{X}_{ij} + t_{ij})\mathrm{d}V_j + l^2 \int_{\mathcal{S}_i'} (w(\boldsymbol{X}_{ji})\tilde{\nabla} s_j \boldsymbol{K}_j^{-1}\boldsymbol{X}_{ji} + t_{ji})\mathrm{d}V_j\right). \tag{9.42}$$

Based on Eq. 9.17, the nonlocal form of Eq. 9.20 is

$$\int_{\mathcal{S}_i} (w(\boldsymbol{X}_{ij})\boldsymbol{\sigma}_i \boldsymbol{K}_i^{-1}\boldsymbol{X}_{ij} + \boldsymbol{T}_{ij})\mathrm{d}V_j - \int_{\mathcal{S}_i'} (w(\boldsymbol{X}_{ji})\boldsymbol{\sigma}_j \boldsymbol{K}_j^{-1}\boldsymbol{X}_{ji} + \boldsymbol{T}_{ij})\mathrm{d}V_j + \boldsymbol{b}_i = \rho_i \ddot{\boldsymbol{u}}_i \tag{9.43}$$

with $\boldsymbol{\sigma}_i = (1 - s_i)^2 \boldsymbol{\sigma}_i^+ + \boldsymbol{\sigma}_i^-$, $\boldsymbol{T}_{ij} = (1 - s_i)\frac{\alpha_v}{m_i}(\mathbf{u}_{ij} - \tilde{\nabla} u_i \boldsymbol{X}_{ij})$, where $\boldsymbol{T}_{ij}$ is used to suppress zero-energy modes. The penalty factor α_v is selected as the shear modulus of the solid. The phase field length scale is chosen as in the conventional phase field model. We employ $l = (1 \sim 3)h$ where h is the diameter of the particle. The initial phase field boundary condition is enforced by setting the phase field to 1.

9.4 Numerical Implementation

We use particles to discretize the computational domain

$$\Omega = \sum_{i=1}^{N} \Delta V_i, \tag{9.44}$$

where N is the number of particles and ΔV_i is the volume associated to particle i. For each particle, the support is denoted by

$$\mathcal{S}_i = \{i, j_1, j_2, ..., j_{n_i}\}, \tag{9.45}$$

where n_i is the number of neighbors in the i's support domain. The discrete form of the shape tensor can be expressed as

$$\boldsymbol{K}_i = \sum_{\mathcal{S}_i} w(\boldsymbol{X}_{ij})\boldsymbol{X}_{ij} \otimes \boldsymbol{X}_{ij} \Delta V_j.$$

The equation of motion at the particles is given by

$$\boldsymbol{M}_i \boldsymbol{a}_i = \boldsymbol{F}_i - \boldsymbol{f}_i, \tag{9.46}$$

where M_i is the mass of the particle i, $\boldsymbol{a}_i$ represents the acceleration, $\boldsymbol{F}_i$ is the external force vector and $\boldsymbol{f}_i$ denotes the internal force vector. The velocity and current displacement can be updated via the Verlet-Velocity algorithm (Verlet 1967):

$$\begin{aligned}
\boldsymbol{v}^{n+1/2} &= \boldsymbol{v}^n + \frac{\Delta t}{2M}(\boldsymbol{F}^n - \boldsymbol{f}^n) \\
\boldsymbol{u}^{n+1} &= \boldsymbol{u}^n + \Delta t \boldsymbol{v}^{n+1/2} \\
\boldsymbol{v}^{n+1} &= \boldsymbol{v}^{n+1/2} + \frac{\Delta t}{2M}(\boldsymbol{F}^{n+1} - \boldsymbol{f}^{n+1}).
\end{aligned} \tag{9.47}$$

The superscript denotes the time step; $\boldsymbol{v}$, $\boldsymbol{u}$ denote velocity and displacement, respectively; and $\boldsymbol{F}^n$, $\boldsymbol{f}^n$ are the resultant external and internal force vectors, respectively. For numerical stability of the explicit algorithm, a maximal time increment of $\Delta t_{max} = \Delta x_{min}/C_{sound}$ should not be exceeded, where Δx_{min} is the minimal particle spacing and C_{sound} the sound speed of the material. In all our simulations, we used $\Delta t = 0.8\Delta t_{max}$ where $\alpha = 0.8$ is the Courant number. The discrete form of Eq. 9.48 reads

$$\sum_{j\in\mathcal{S}_i}(w(\boldsymbol{X}_{ij})\boldsymbol{\sigma}_i\boldsymbol{K}_i^{-1}\boldsymbol{X}_{ij} + \boldsymbol{T}_{ij})\Delta V_j - \sum_{\mathcal{S}'_i}(w(\boldsymbol{X}_{ji})\boldsymbol{\sigma}_j\boldsymbol{K}_j^{-1}\boldsymbol{X}_{ji} + \boldsymbol{T}_{ij})\Delta V_j + \boldsymbol{b}_i = \rho_i\ddot{\boldsymbol{u}}_i \tag{9.48}$$

where $\varepsilon_i = \frac{1}{2}(\tilde{\nabla}\mathbf{u}_i + \tilde{\nabla}\mathbf{u}_i^T)$, $\tilde{\nabla}\mathbf{u}_i := \sum_{\mathcal{S}_i} w(\boldsymbol{X}_{ij})\mathbf{u}_{ij} \otimes \boldsymbol{X}_{ij}\Delta V_j \cdot \boldsymbol{K}_i^{-1}$ and $\boldsymbol{\sigma}_a^{\pm}$ are given in Eq. 9.23. It can be shown that the discrete form of Eq. 9.42 is given by

$$\begin{aligned}
\Delta s_i = \frac{1}{n_s}\langle\beta_i\rangle_+, \text{ with } \beta_i &= l^2\sum_{j\in\mathcal{S}_i}(w(\boldsymbol{X}_{ij})\tilde{\nabla}s_i\boldsymbol{K}_i^{-1}\boldsymbol{X}_{ij} + t_{ij})\Delta V_j \\
&- l^2\sum_{j\in\mathcal{S}'_i}(w(\boldsymbol{X}_{ji})\tilde{\nabla}s_j\boldsymbol{K}_j^{-1}\boldsymbol{X}_{ji} + t_{ji})\Delta V_j - s_i + (1 - s_i)\mathcal{H}_i
\end{aligned} \tag{9.49}$$

with $\tilde{\nabla}s_i = \sum_{\mathcal{S}_i} w(\boldsymbol{X}_{ij})s_{ij}\boldsymbol{X}_{ij}\Delta V_j \cdot \boldsymbol{K}_i^{-1}$. It is worth mentioning that β_i is proportional to l^2. When l is too large, the phase field at a point may exceed 1 quickly, which leads to unrealistic or divergent results. Based on the numerical tests, $1.5\Delta x \le l \le 3\Delta x$ is recommended.

Equations 9.48 and 9.49 contain two important parameters: the intrinsic length scale of the phase field l and the number of neighbors in the support. The particle size is determined by the discretization. Based on the K-nearest searching algorithm, the support of each particle can be constructed. The support size is assumed as the distance of the furthest particle in support. To some extent, the number of neighbors in support is similar to the number of nodes in one element in FEM and the size of support can be viewed as the mesh size in FEM.

The implementation of NOM requires calculating the terms from support and dual-support. Based on the concept of dual-support, the terms from dual-support for one particle can be accumulated when calculating terms from another particle's support.

The key steps for the explicit NOM and phase field with sub-steps are outlined in Box 1 and Box 2. One can find that in Box 1 and Box 2, the terms from dual-support are obtained by adding a minus sign to the terms from other particle's support.

Box 1. Explicit NOM with explicit phase field

1. For each particle, evaluate $\tilde{\nabla}\mathbf{u}_i = \sum_{\mathcal{S}_i} w(\boldsymbol{X}_{ij})\mathbf{u}_{ij} \otimes \boldsymbol{X}_{ij}\Delta V_j \cdot \boldsymbol{K}_i^{-1}$.
2. For each particle $\varepsilon_i = \frac{1}{2}(\tilde{\nabla}\boldsymbol{u}_i + \tilde{\nabla}\boldsymbol{u}_i^T)$, eigenvalue decomposition $\boldsymbol{\varepsilon} = \sum_{i=1}^{3} \varepsilon_i \boldsymbol{n}_i \otimes \boldsymbol{n}_i$, evaluate $\psi(\boldsymbol{\varepsilon})^+, \boldsymbol{\sigma}_i^{\pm}$.
3. Evaluate $\mathcal{H}_i$ by Eqs. 9.28 or 9.29 and update phase field s_i with sub-steps in Box 2.
4. For each particle, evaluate $\boldsymbol{\sigma}_i = (1 - s_i)^2\boldsymbol{\sigma}_i^+ + \boldsymbol{\sigma}_i^-$.
5. For each particle, initiate $\boldsymbol{f}_i = \mathbf{0}$; for each neighbor $j \in \mathcal{S}_i$, evaluate

$$f_{ij} = \Delta V_i \Delta V_j \Big(w(\boldsymbol{X}_{ij})\sigma_i \boldsymbol{K}_i^{-1}\boldsymbol{X}_{ij} + (1 - s_i)\frac{\alpha_v}{m_i} w(\boldsymbol{X}_{ij})(\mathbf{u}_{ij} - \tilde{\nabla}\mathbf{u}_i \boldsymbol{X}_{ij}) \Big),$$

add $\boldsymbol{f}_{ij}$ to $\boldsymbol{f}_i$ and add $-\boldsymbol{f}_{ij}$ to $\boldsymbol{f}_j$.

Box 2. Update phase field with sub-steps

1. For each particle, calculate $\tilde{\nabla}s_i = \sum_{\mathcal{S}_i} w(\boldsymbol{X}_{ij}) s_{ij} \boldsymbol{X}_{ij} \Delta V_j \cdot \boldsymbol{K}_i^{-1}$.
2. For each particle, initiate $f_i = 0$; for each neighbor $j \in \mathcal{S}_i$, evaluate

$$f_{ij} = \Delta V_i \Delta V_j \Big(w(\boldsymbol{X}_{ij})\tilde{\nabla}s_i \cdot \boldsymbol{K}_i^{-1}\boldsymbol{X}_{ij} + \frac{\alpha_s}{m_i} w(\boldsymbol{X}_{ij})(s_{ij} - \tilde{\nabla}s_i \boldsymbol{X}_{ij}) \Big)$$

   add f_{ij} to f_i and $-f_{ij}$ to f_j.
3. For each particle, evaluate phase field residual $\beta_i = (1 - s_i)\mathcal{H}_i + l_i^2 \frac{f_i}{\Delta V_i} - s_i$; evaluate Δs_i; update $s_i = s_i + \Delta s_i$.
4. If $L^{\infty}(\beta_i) < 1.0 \times 10^{-3}$, update finish; else go to Step 1 in Box 2.

9.5 Numerical Examples

We implemented the current method in the environment of Mathematica (Wolfram 1999). The results are processed by the Open Visualization Tool (OVITO) (Stukowski 2009). Five numerical examples are presented to demonstrate the performance of the method. In the plate tensile test, the parameters in the explicit phase field model are analyzed and the convergence of the phase field residual is discussed for different intrinsic length scales. The third numerical example tests the method for complex fracture patterns involving crack branching. The fourth example is the Kalthoff–Winkler experiment and the last one shows the capability of our method in modeling complex 3D fracture patterns.

9.5.1 Convergence of Sub-Step Scheme

In order to test the performance of the sub-step scheme, a 2D plate of 1 x 1 mm^2 discretized by 100 x 100 particles with initial phase field boundary conditions is considered. There are two methods to apply the initial crack. One can modify the neighbors in the support domain as shown in Fig. 9.2a, or assign 1 to the phase field degrees of freedom for particles in the crack domain as shown in Fig. 9.2b. For the first method, removing one neighbor from the support domain is equivalent to cutting the interaction between two particles, which is in line with the fracture model in peridynamics. The NOM phase field approach can adopt any number of neighbors in the support, independent of the radius of the support domain. Variable support radii and variable number of neighbors are allowed thanks to the dual-support. However, a large number of neighbors is computationally expensive. Based on our numerical experience, 8 (16) neighbors in 2D (3D) seems suitable for defining the nonlocal operators.

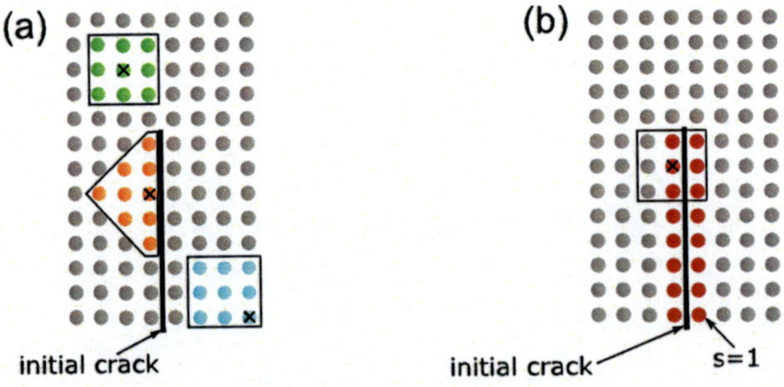

Fig. 9.2 Representation of initial crack by **a** modifying the neighbor list based on the crack and by **b** assigning the phase field $s = 1$ to the particle in the crack domain

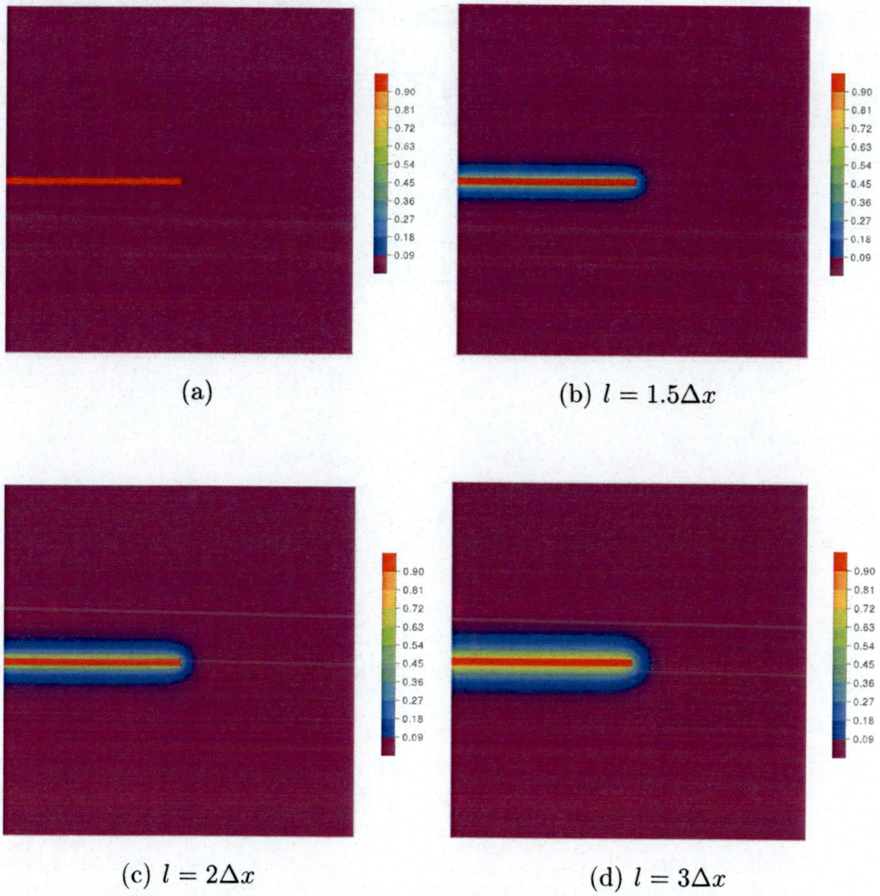

Fig. 9.3 **a** Initial phase field boundary conditions, **b–d** phase field distributions for different phase field lengths after sub-step scheme of phase field

In this section, we use the method in Fig. 9.2b to represent the initial crack. Therefore, the phase field value of two layers of particles in the left middle is set to 1 in Fig. 9.3a. The sub-step scheme will diffuse the phase field in order to guarantee the local equilibrium of the phase field. We test the influence of the phase field length scale on the reduction of the phase field residual. We select $l = \{1, 1.5, 2, 3\} \times \Delta x$, where Δx is the grid space of the particles. Contour plots of the phase field are depicted in Fig. 9.3. When $l = 2\Delta x$, the thickness of the phase field domain is approximately $6\Delta x$. Note that the phase field modulus at the first few steps should be large so that the phase field increment is significantly smaller than 1. The statistics of the maximal phase field residual are given in Table 9.1. The sub-step scheme achieves a smaller residual with adequate l as long as enough sub-steps are carried out.

Table 9.1 Statistics of the phase field residual

$l/\Delta x$	1	1.5	2	3
$L^\infty(\beta)$	7.45e–10	8.19e–10	7.30e–10	0

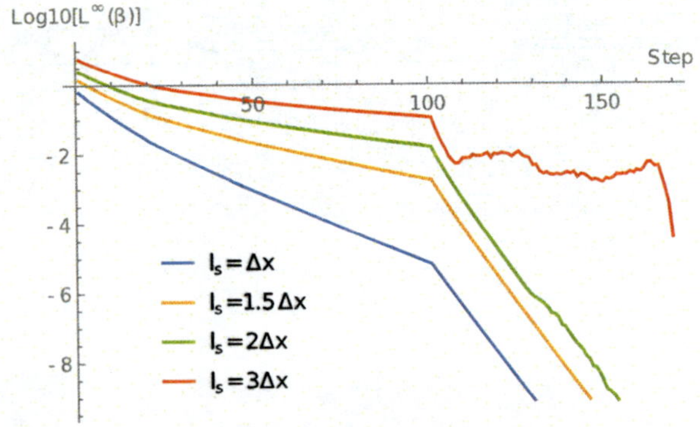

Fig. 9.4 The evolution of maximal phase field residual

The evolution of the maximal phase field residual is depicted in Fig. 9.4. When $l \leq 3\Delta x$, the sub-step scheme can reduce the phase field residual to 10^{-8} with around 100 sub-steps. It also shows that the smaller l leads to the faster convergence of the phase field residual.

The numerical tests show that $l < 4\Delta x$ is numerically stable and it can achieve a relatively sharp phase field distribution for the crack. Large phase field length scale can achieve wider phase field distribution, but is not suitable for the explicit phase scheme due to numerical instability. Therefore, $l = (1.5 \sim 3)\Delta x$ is recommended for the sub-step scheme.

We tested the influence of the number of neighbors on the phase field representation of the crack surface and the convergence of maximal phase field residual. Based on the same discretization, Fig. 9.5 depicts the evolution of the maximal phase field residual for different numbers of neighbors. More particles in the support slightly contribute to the faster reduction of the phase field residual. Large number of neighbors in the support is beneficial for a more accurate estimation of the nonlocal gradient but it increases the computational cost.

9.5.2 *Single-Edge Notched Tension Test*

In this example, we tested our formulation for a single-edge notched 2D specimen in tension. We assumed plane stress conditions and the geometric setup can be found in Fig. 9.6. The plate is discretized with 100 x 100 particles and 200 x 200 particles, respectively. The bottom is fixed while a velocity boundary condition of $v = 1$ m/s is applied at the top of the plate. The material parameters are $E = 210$ GPa, $\nu = 0.3$ and

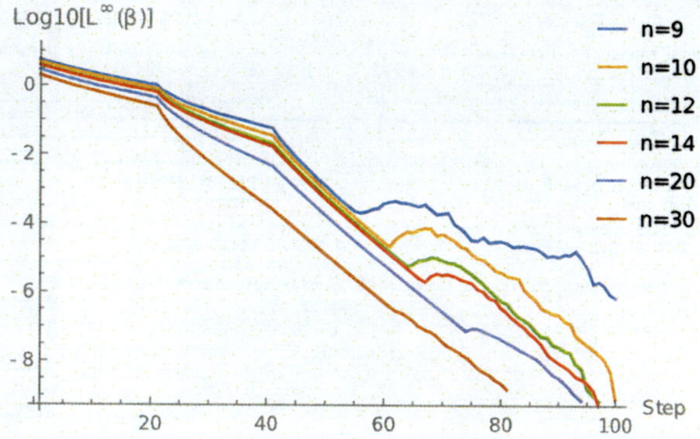

Fig. 9.5 The evolution of maximal phase field residual for different numbers of neighbors in support. The phase field length scale is selected as $l = 3\Delta x$

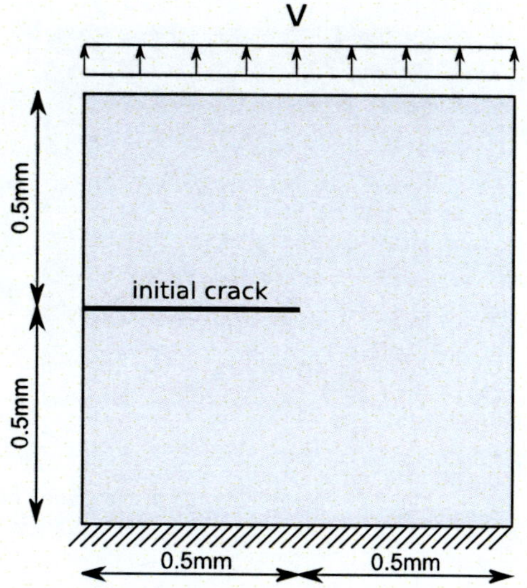

Fig. 9.6 Setup of plate

critical strain energy release rate $g_c = 2700$ J/m^2. The phase field based on damage rule Eqs. 9.28 and 9.29 are shown in Fig. 9.7. The phase field internal length is selected as $l = 1.5\Delta x$. The damage rule according to Eq. (9.29) limits the evolution of the phase field to the crack tip region, while (Eq. 9.28) yields to damage under low stress conditions.

The load–deflection curve is shown in Fig. 9.8a and compared to FEM results with identical length scale ($l = 0.0375$ mm); the FE results are extracted from (Miehe et al. 2010b). Although the plate is solved by an explicit algorithm, the kinetic energy is much lower than the strain energy as shown by Fig. 9.8c, d. Figure 9.8 also shows that this formulation is not sensitive to the discretizations.

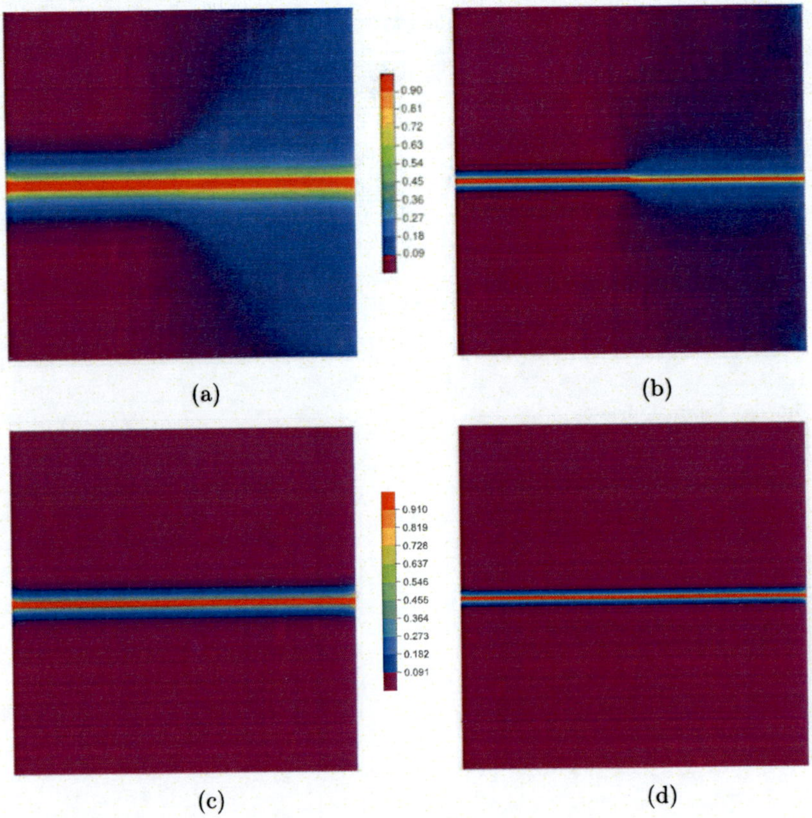

Fig. 9.7 Phase field with discretization of **a** 100 × 100 by Eq. 9.28, **b** 200 × 200 by Eq. 9.28, **c** 100 × 100 by Eq. 9.29, **e** 200 × 200 by Eq. 9.29

9.5.3 Dynamic Crack Branching

In this example, a pre-notched rectangular plate loaded dynamically in tension is modeled. This benchmark example is widely studied in the literature (Belytschko et al. 2003; Song et al. 2006; Xu and Needleman 1994). The geometry and the traction boundary conditions are shown in Fig. 9.9. The constant traction of $\sigma = 1$ MPa is applied on the top and bottom of the plate. The material parameters include density $\rho = 2450$ kg/m^3, elastic modulus $E = 32$ GPa, Poisson's ratio $\nu = 0.2$ and critical energy release rate $g_c = 3$ J/m^2. The plate is discretized with 64000 particles and a grid spacing of $\Delta x = 2.5 \times 10^{-4}$ m. The phase field length scale is $l = 2.5\Delta x = 6.25 \times 10^{-4}$ m. Phase field damage criteria by Eqs. 9.28 and 9.29 are tested. The displacement field u_y for phase field damage (a) without threshold and (b) with threshold is depicted in Fig. 9.10. The final phase field fractures for two criteria are shown in Fig. 9.11a, b. For criterion Eq. 9.28, the damage occurred not only on the

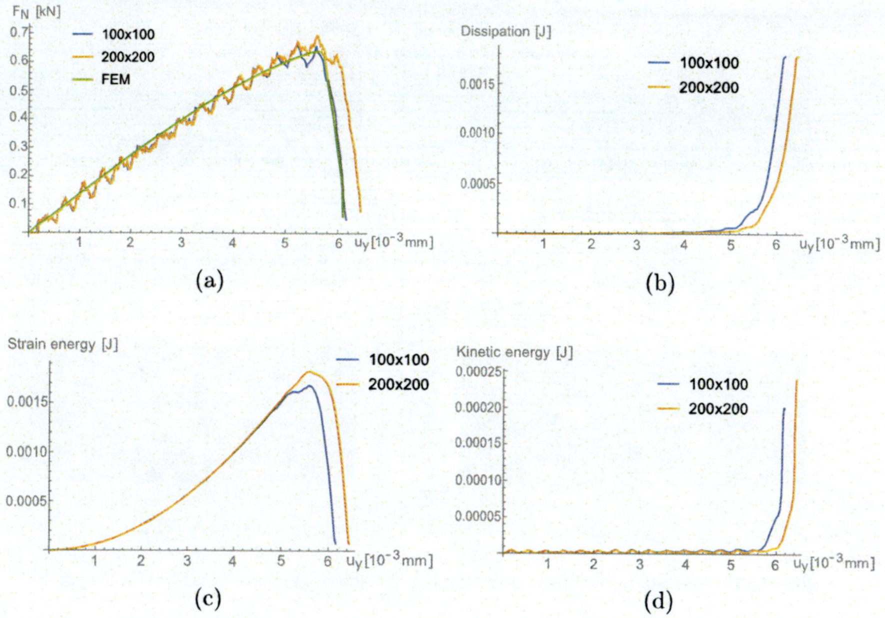

Fig. 9.8 **a** Load curves and **b–d** energy statistics; result by FEM with length scale $l = 0.0375$ mm comes from Fig. 12a of (Miehe et al. 2010b)

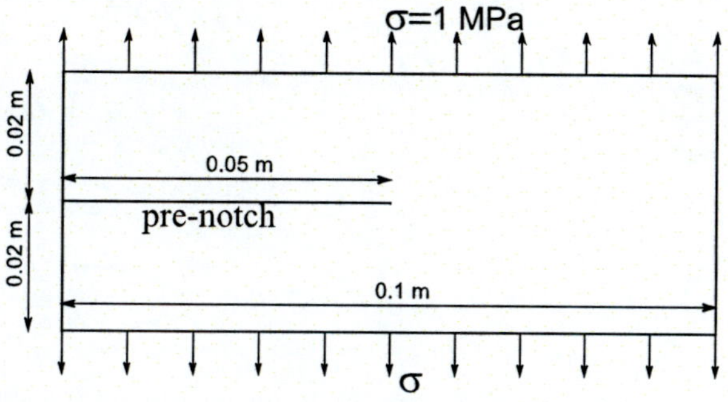

Fig. 9.9 Setup of the pre-cracked plate under tension load

crack path but also on the rest with low stress condition. Criterion without threshold dissipated more energy compared with that with threshold, as shown in Fig. 9.13. The branching angle with respect to the horizontal line for two criteria is in line with (Borden et al. 2012; Nguyen and Wu 2018; Schlüter et al. 2014). Figure 9.12 depicts the crack propagation speed, which shows that it is comparable to the one obtained by the FEM-based explicit phase field approach presented in (Ren et al. 2019). The FEM phase field is based on an averaged deformation gradient scheme and Delaunay mesh in 3D. The maximal crack propagation speed of $v_{max} = 752$ m/s is approximately 35.5% of the Rayleigh wave speed $c_R = 2119$ m/s; the latter one is given by (Freund 1998)

Fig. 9.10 Displacement u_y for phase field damage **a** without threshold and **b** with threshold

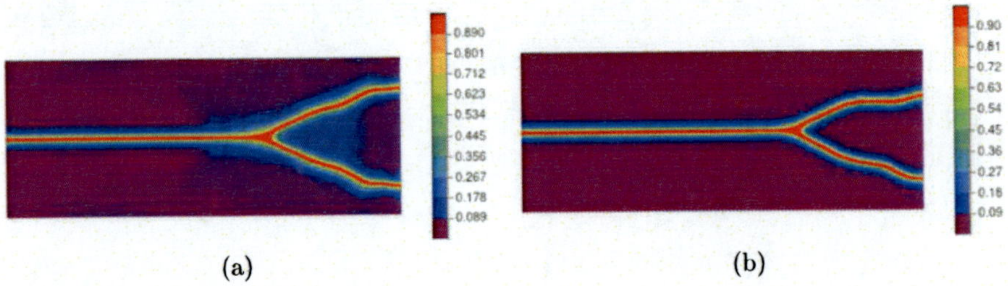

Fig. 9.11 Phase field for phase field damage **a** without threshold and **b** with threshold

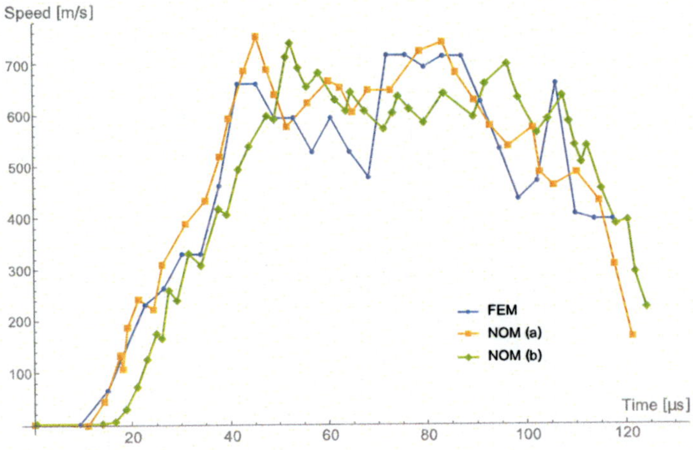

Fig. 9.12 The propagation speed of crack tip; FEM: Fig. 8 in (Ren et al. 2019); NOM **a** without threshold; NOM **b** with threshold

$$c_R = c_S \frac{0.862 + 1.14\nu}{1+\nu} \text{ with } c_S = \sqrt{G/\rho}, \tag{9.50}$$

where c_S denotes the shear waves (transverse waves) speed and G the shear modulus.

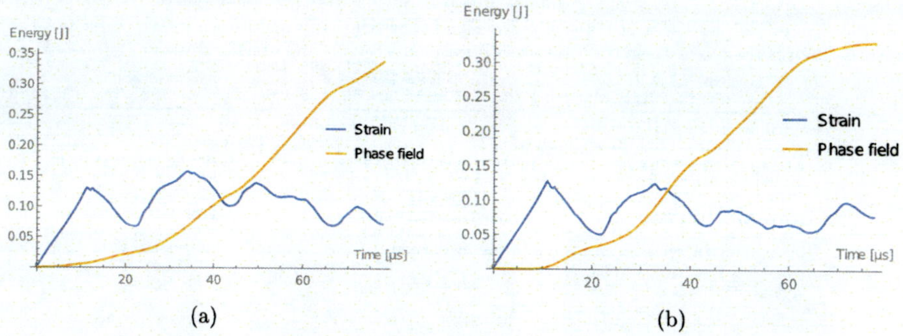

Fig. 9.13 Statistic energy for phase field damage **a** without threshold and **b** with threshold

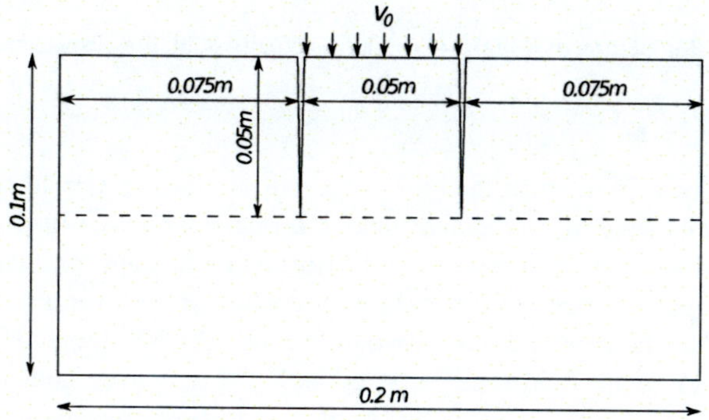

Fig. 9.14 Setup for the Kalthoff–Winkler experiment

9.5.4 Kalthoff–Winkler Experiment in 2D

The Kalthoff-Winkler experiment (Kalthoff and Winkler 1988) is a classical benchmark problem for dynamic fracture modeling (Belytschko et al. 2003; Li et al. 2002; Nguyen and Wu 2018; Song et al. 2006). For different impact velocities, the fracture can be brittle or ductile. For low impact velocities, the dynamic brittle fracture propagates from the crack tip at an angle of around 70° vs. the initially horizontal crack. When the impact velocity is increased further, a ductile failure phenomenon occurs and a shear band is observed. In this section, the brittle failure is studied. The velocity applied on the top of the plate starts from 0 to $v_y = 16.5$m/s in a period of 10^{-6} s and remains constant thereafter (Hesch and Weinberg 2014). The dimension of the plate is 0.2×0.1 m^2 as shown in Fig. 9.14. The material parameters are $E = 190$ GPa, $\nu = 0.3$, $g_c = 2.213 \times 10^4$ J/m^2. The plate is discretized with 800x400 particles and a grid spacing of $\Delta x = 2.5 \times 10^{-4}$ m. The initial crack is represented by modifying the neighbors in support. The number of neighbors for each particle is restricted to 8 for the purpose of numerical efficiency. The length scale is selected as $l = 2\Delta x = 5 \times 10^{-4}$ m. The phase field criterion of Eq. 9.29 is used.

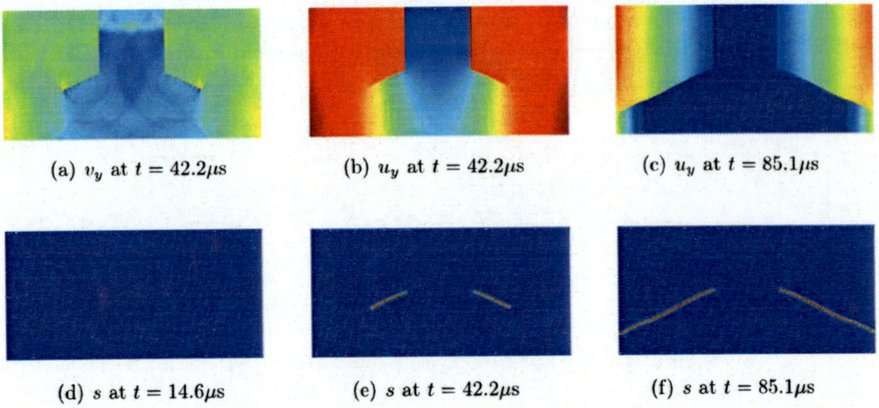

Fig. 9.15 Simulation of Kalthoff–Winkler **a** velocity field; **b–c** displacement field; **e–f** phase field

The crack initiates at the end of the initial crack at $t = 14.6\mu s$. Contour plots of the velocity field, displacement field and phase field are illustrated in Fig. 9.15. The angle of the crack with respect to the initial crack is approximately 65.6°. The crack tip propagation speed of several methods can be found in Fig. 9.16b, where "FEM implicit" is the conventional phase field model by FEM and "FEM explicit" denotes the explicit phase field scheme based on FEM with averaged deformation on nodes. The maximal crack speed in the simulation is 2018 m/s, about 72% of the Rayleigh speed. The crack speed of our method agrees with the result by XFEM (Belytschko et al. 2003).

Time histories of the kinetic, strain and dissipated energy are plotted in Fig. 9.16a. The slope of the dissipated energy at time $t = (33 - 55)\mu s$ is larger than that in $t = (60 - 80)\mu s$, which is consistent with the crack propagation speed. In (Wang et al. 2019), the physical time for complete crack propagation for an impact velocity of $v_y = 20$m/s is reported as $100\mu s$. The crack propagation of our method ended at $85.1\mu s$, which is very close to the time of $87\mu s$ in (Borden et al. 2012) by an implicit phase field method. It was reported in (Wang et al. 2019) that the phase field viscosity affects the crack propagation speed, while our method reduces the phase field residual adaptively by sub-steps and thus obtains a rate-independent phase field model.

9.5.5 Cylinder Under Impact

This example shows the capability of NOM to model complex 3D fracture patterns. Therefore, consider a cylinder shown in Fig. 9.17. It is discretized with 10.36 million linear tetrahedron elements and 1.788 million nodes by Abaqus. The mesh created by Abaqus is then imported into Mathematica. The volume of each node is determined

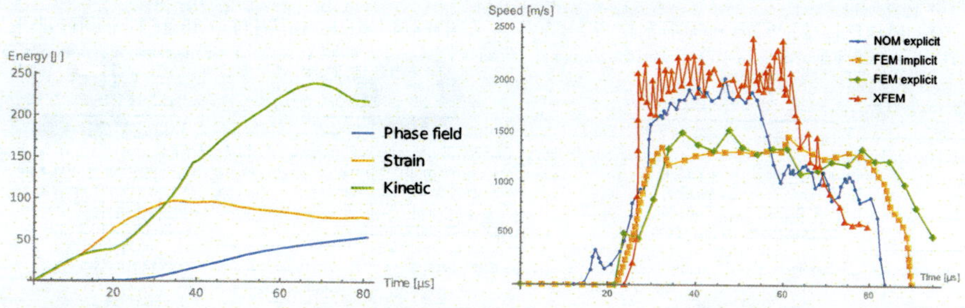

(a) (b)

Fig. 9.16 **a** The statistics of energies; **b** the crack propagation speed; XFEM: Fig. 16 of (Belytschko et al. 2003); FEM explicit: Fig. 13 of (Ren et al. 2019); FEM implicit: Fig. 9 of (Hesch and Weinberg 2014)

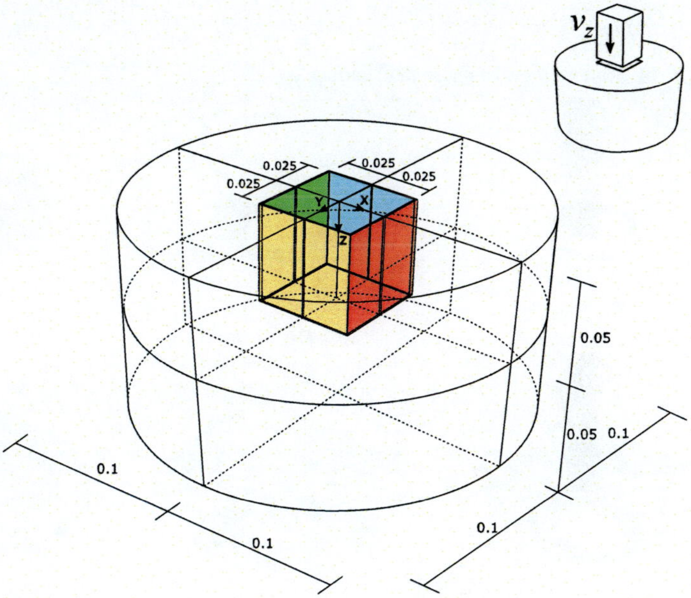

Fig. 9.17 Geometry of the cylinder under impact with length unit in meters. The areas with colors are the initial cracks

by accumulating volumes from the elements. Note that the mesh is unstructured and some particles have larger volumes than others. The maximal time increment in each step is determined by the size of the smallest particle, and in this example, the constant time increment is selected as $\Delta t = 5.978\text{e-}8$ s. The number of particles in the support for each particle is selected as 16. Four initial crack surfaces denoted with four different colors between the cube domain and the cylinder are formed by the method in Fig. 9.2a. The bottom of the cube is connected to the cylinder. The particle on the surface of the cylinder has a larger support radius due to all neighbors being on one side while the particles inside the domain have a smaller support radius. The

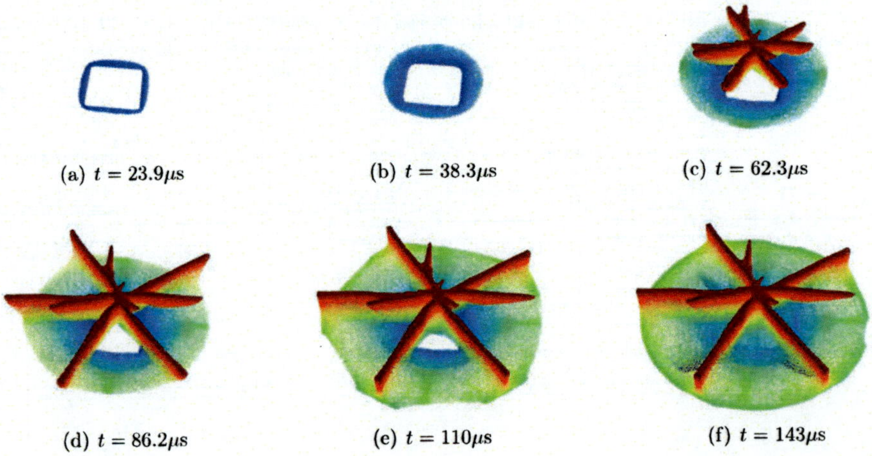

Fig. 9.18 Bottom view of phase field surface

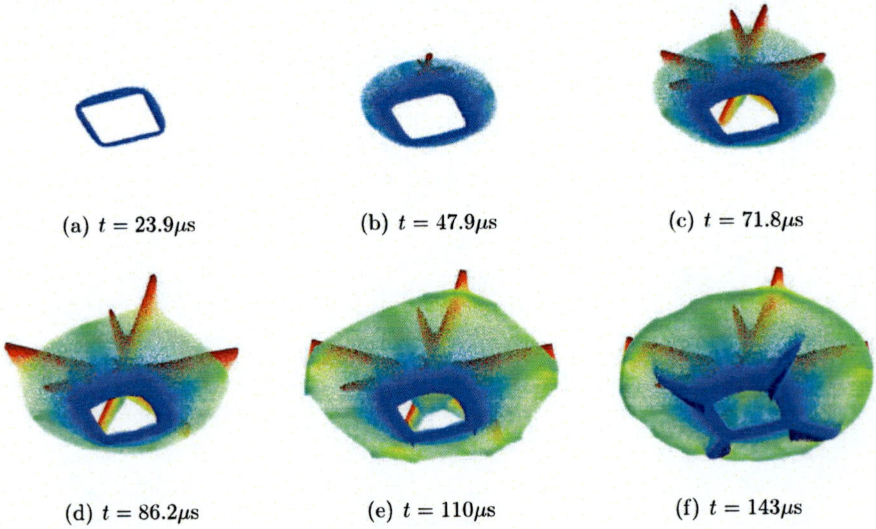

Fig. 9.19 Top view of phase field distribution

material parameters are the same as that in Sect. 9.5.4. The particles in the square of the xy-plane are applied with velocity condition v_z arising from 0 to 16.5m/s in 1 μs and remaining constant thereafter. Hence, the contact modeling between the impactor and the cylinder is avoided.

The crack surfaces at different points in time are depicted in Figs. 9.18 and 9.19, where the particles have similar phase field values (e.g. $s > 0.75$) and different colors indicate different z-coordinates showing the depth of the surface. The crack initiated at 17.9 μs at the middle edge of the initial square crack tip. The 3D crack surface expanded in the shape of a cone. The phase field at the center of the bottom plane

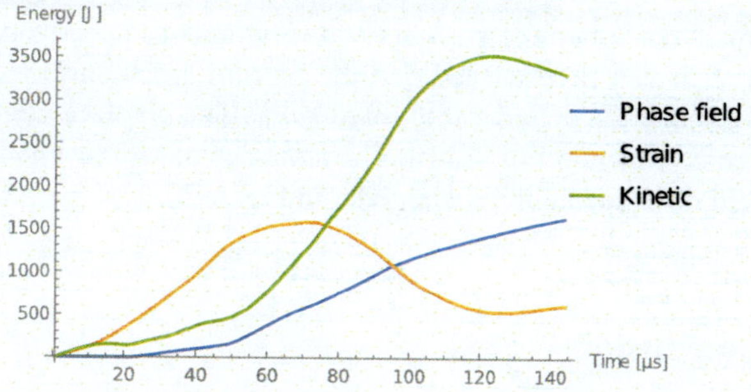

Fig. 9.20 Statistics of kinetic energy, strain energy and dissipated fracture energy

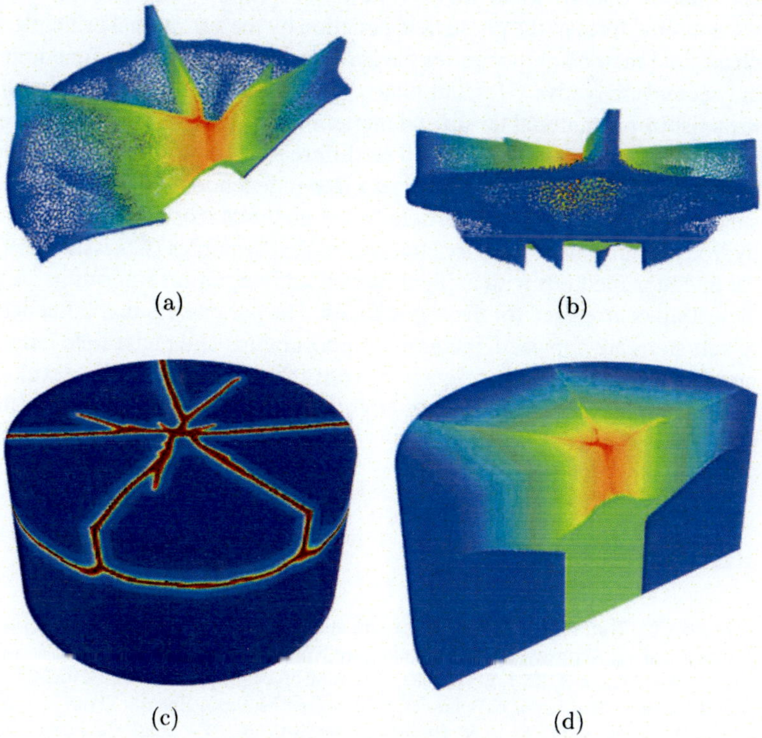

Fig. 9.21 Crack surface **a–b**, phase field **c** and displacement **d** u_z

began at 47.9 μs. Due to the complex stress state of the bottom center particles, several crack surfaces with multiple branches formed and propagated toward the crack cone surface. The bottom crack surface intersected with the cone surface at 91.9 μs. The bottom crack propagated faster than the cone crack surface and reached the cylindrical surface at 88.7 μs. The bottom vertical crack surfaces intersected and formed a bowl, as shown in Fig. 9.21a. At the corners of the initial square crack tip, four vertical crack surfaces formed.

The final crack surface with bottom view and perspective view is given in Fig. 9.21a, b, c. It can be observed that the bottom crack surfaces and the crack cone surfaces divided the cylinder into seven parts. The displacement in z-direction at the final stage is shown in Fig. 9.21d. The statistics of kinetic energy, strain energy and phase field dissipated energy are given in Fig. 9.20.

9.6 Conclusion

In this chapter, we have proposed an explicit phase field model with sub-steps for brittle dynamic fracture in the framework of nonlocal operator method. We also derived the nonlocal form of the phase field equation by the variational principle. The explicit phase field method only requires the first gradient of the energy functional rather than the second derivative (Hessian matrix) in the implicit algorithm and hence avoids the anisotropic material tensor and numerical difficulty in terms of convergence. The phase field sub-step scheme based on different damage criteria can reduce the phase field residual adaptively and achieve a rate-independent phase field model. The sub-step scheme is suitable for particle-based methods (for example, smoothed particle hydrodynamics, non-ordinary state-based peridynamics (Gu et al. 2019) and nonlocal operator method) with explicit numerical integration algorithms. A very concise algorithm is proposed for these two fields. The parameters that influence the number of sub-steps are analyzed, including support radius and phase field length scale. Five numerical examples are presented to verify the method. Further development of the current method may incorporate large deformation of hyper-elastic materials and elastic–plastic materials.

References

Ambrosio L, Tortorelli V (1990) Approximation of functional depending on jumps by elliptic functional via t-convergence. Commun Pure Appl Math 43(8):999–1036

Belytschko T, Chen H, Xu J, Zi G (2003) Dynamic crack propagation based on loss of hyperbolicity and a new discontinuous enrichment. Int J Numer Methods Eng 58(12):1873–1905

Borden M, Verhoosel C, Scott M, Hughes T, Landis C (2012) A phase-field description of dynamic brittle fracture. Comput Methods Appl Mech Eng 217:77–95
Borden M, Verhoosel C, Scott M, Hughes T, Landis C (2012) A phase-field description of dynamic brittle fracture. Comput Methods Appl Mech Eng 217:77–95
Frémond M, Nedjar B (1996) Damage, gradient of damage and principle of virtual power. Int J Solids Struct 33(8):1083–1103

Hakim V, Karma A (2009) Laws of crack motion and phase-field models of fracture. J Mech Phys Solids 57(2):342–368

Hesch C, Weinberg K (2014) Thermodynamically consistent algorithms for a finite-deformation phase-field approach to fracture. Int J Numer Methods Eng 99(12):906–924
Hesch C, Weinberg K (2014) Thermodynamically consistent algorithms for a finite-deformation phase-field approach to fracture. Int J Numer Methods Eng 99(12):906–924

Kalthoff J, Winkler S (1988) Failure mode transition at high rates of shear loading. DGM Informat Impact Load Dyn Behav Mater 1:185–195

Karma A, Kessler D, Levine H (2001) Phase-field model of mode III dynamic fracture. Phys Rev Lett 87(4):045501

Kim S, Kim W, Suzuki T (1999) Phase-field model for binary alloys. Phys Rev E 60(6):7186

Li S, Liu W, Rosakis A, Belytschko T, Hao W (2002) Mesh-free Galerkin simulations of dynamic shear band propagation and failure mode transition. Int J Solids Struct 39(5):1213–1240

Li S, Liu W, Rosakis A, Belytschko T, Hao W (2002) Mesh-free Galerkin simulations of dynamic shear band propagation and failure mode transition. Int J Solids Struct 39(5):1213–1240

Miehe C (2011) A multi-field incremental variational framework for gradient-extended standard dissipative solids. J Mech Phys Solids 59(4):898–923

Miehe C (2011) A multi-field incremental variational framework for gradient-extended standard dissipative solids. J Mech Phys Solids 59(4):898–923

Miehe C, Welschinger F, Hofacker M (2010) Thermodynamically consistent phase-field models of fracture: Variational principles and multi-field fe implementations. Int J Numer Methods Eng 83(10):1273–1311

Miehe C, Welschinger F, Hofacker M (2010) Thermodynamically consistent phase-field models of fracture: Variational principles and multi-field fe implementations. Int J Numer Methods Eng 83(10):1273–1311

Nguyen V, Wu J (2018) Modeling dynamic fracture of solids with a phase-field regularized cohesive zone model. Comput Methods Appl Mech Eng 340:1000–1022

Pham K, Amor H, Marigo J, Maurini C (2011) Gradient damage models and their use to approximate brittle fracture. Int J Damage Mech 20(4):618–652

Ren H, Zhuang X, Anitescu C, Rabczuk T (2019) An explicit phase field method for brittle dynamic fracture. Comput Struct 217:45–56

Ren H, Zhuang X, Anitescu C, Rabczuk T (2019) An explicit phase field method for brittle dynamic fracture. Comput Struct 217:45–56

Schlüter A, Willenbücher A, Kuhn C, Müller R (2014) Phase field approximation of dynamic brittle fracture. Comput Mech 54(5):1141–1161

Song JH, Areias PM, Belytschko T (2006) A method for dynamic crack and shear band propagation with phantom nodes. Int J Numer Methods Eng 67(6):868–893

Stukowski A (2009) Visualization and analysis of atomistic simulation data with ovito-the open visualization tool. Model Simul Mater Sci Eng 18(1):015012

Vahid Z, Shen Y (2016) Massive parallelization of the phase field formulation for crack propagation with time adaptivity. Comput Methods Appl Mech Eng 312:224–253

Wang T, Ye X, Liu Z, Chu D, Zhuang Z (2019) Modeling the dynamic and quasi-static compression-shear failure of brittle materials by explicit phase field method. Comput Mech 1–20

Wolfram S (1999) The mathematica book. Assembly Automation

Xu X, Needleman A (1994) Numerical simulations of fast crack growth in brittle solids. J Mech Phys Solids 42(9):1397–1434

Chapter 10
A Nonlocal Operator Method for Finite Deformation Higher-Order Gradient Elasticity

Gradient elasticity as a generalization of classical elasticity includes the contribution of strain gradients in the strain energy. Different from classical elasticity theory, such consideration enables gradient elasticity to model some interesting phenomena (such as size effect, the stress and strain effects on surface physics, nonlocal effect at micrometer/nanometer scale). In this chapter, we propose a strain energy density with objectivity. The energy form is based on the second Piola-Kirchhoff stress and is invariant under rigid body transformation. The number of gradient order is extended to 5 in 2D and 3 in 3D. For the first time, the geometrical nonlinear fifth-order gradient elasticity in 2D and third-order gradient elasticity in 3D are studied by numerical experiments based on nonlocal operator method.

The content of this chapter is outlined as follows. The general strain energy density for large deformations is proposed for the fourth-order gradient elasticity in Sect. 10.1. In Sect. 10.2, we derive the governing equations and the associated boundary conditions for the third-order gradient elasticity by using variational principles and exploiting integration by parts on surfaces. In Sect. 10.3, the framework of the particle-based nonlocal operator method is briefly summarized and its implementation for solving higher order gradient solids presented. In Sect. 10.4, several representative numerical tests, including a point displacement load, point force load and the influence of the length scale in linear/nonlinear gradient elasticity, are presented to study the physical response of higher order gradient elasticity.

10.1 Higher Order Gradient Solid with Finite Deformation

Let us denote the material coordinates (in the initial configuration Ω) by $\boldsymbol{X}$, the spatial coordinates (in the current configuration Ω_t) by $\boldsymbol{x}$ and the displacement field by $\boldsymbol{u} := \boldsymbol{x} - \boldsymbol{X}$. The deformation gradient $\boldsymbol{F}$, right Cauchy Green tensor $\boldsymbol{C}$ and Green-Lagrange strain tensor $\boldsymbol{E}$ are written as

T. Rabczuk et al., *Computational Methods Based on Peridynamics and Nonlocal Operators*, Computational Methods in Engineering & the Sciences,
https://doi.org/10.1007/978-3-031-20906-2_10

$$F = \frac{\partial x}{\partial X} = \nabla x = \nabla u + I \tag{10.1}$$

$$C = F^T \cdot F, E = \frac{1}{2}(C - I) \tag{10.2}$$

where I is the identity matrix and $\nabla\Box = \frac{\partial\Box}{\partial X}$. The principle of frame indifference requires the quantity remain invariant under rigid body transformation $x' = Q(t)x + c(t)$, where $c(t)$ is the rigid translation vector and $Q(t)$ is the (orthogonal) rotation matrix satisfying $QQ^T = Q^TQ = I$. Vectors and second-order tensors are objective if they are related by the rotation tensor as

$$u' = Q \cdot u \tag{10.3}$$

$$T' = Q \cdot T \cdot Q^T \tag{10.4}$$

The deformation gradient under rigid body transformation is related by

$$F' = \frac{\partial x'}{\partial X} = \frac{\partial x'}{\partial x} \cdot \frac{\partial x}{\partial X} = Q \cdot F \tag{10.5}$$

Quantities are invariant if they remain unchanged by the rigid body transformation. Apparently, the right Cauchy tensor is invariant as $C' = F'^T F' = F^T Q^T Q F = F^T F = C$. Let $F_{,i} := \frac{\partial F}{\partial X_i}$ denote the partial derivative of F with respect to X_i. The derivative of F' and Q can be written as

$$F'_{,i} = Q_{,i}F + QF_{,i} \tag{10.6}$$

$$Q^T_{,i}Q + Q^TQ_{,i} = 0 \tag{10.7}$$

The gradient of C' can be derived as

$$\begin{aligned}
C'_{,i} &= (F'^T F')_{,i} = F'^T_{,i}F' + F'^T F'_{,i} \\
&= (F^T Q^T_{,i} + F^T_{,i} Q^T)QF + F^T Q^T (Q_{,i}F + QF_{,i}) \\
&= F^T Q^T_{,i} QF + F^T_{,i} Q^T QF + F^T Q^T Q_{,i} F + F^T Q^T Q F_{,i} \\
&= -F^T Q^T Q_{,i} F + F^T Q^T Q_{,i} F + F^T F_{,i} + F^T_{,i} F \\
&= (F^T F)_{,i} = C_{,i}.
\end{aligned} \tag{10.8}$$

Therefore ∇C is invariant. Note that ∇F is neither objective nor invariant since

$$F'_{,i} = (QF)_{,i} = Q_{,i}F + QF_{,i} \neq F_{,i} \tag{10.9}$$

$$\nabla F{:}\nabla F = \sum_i F_{,i} : F_{,i} \neq \sum_i F'_{,i} : F'_{,i}. \tag{10.10}$$

Hence $\nabla \boldsymbol{F}$ cannot be used directly to define the energy density. However, the invariant property makes $\nabla \boldsymbol{C}$ a good choice. Different orders of strain gradient can be written as

$$\boldsymbol{H} = \frac{\partial \boldsymbol{E}}{\partial \boldsymbol{X}} = \nabla \boldsymbol{E} = \frac{1}{2}(\boldsymbol{F}^T \nabla \boldsymbol{F} + \nabla \boldsymbol{F}^T \boldsymbol{F}) \tag{10.11}$$

for the first gradient strain tensor,

$$\boldsymbol{G} = \frac{\partial \boldsymbol{H}}{\partial \boldsymbol{X}} = \frac{\partial^2 \boldsymbol{E}}{\partial \boldsymbol{X}^2} = \nabla^2 \boldsymbol{E} \tag{10.12}$$

for the second-gradient strain tensor, and

$$\boldsymbol{L} = \frac{\partial \boldsymbol{G}}{\partial \boldsymbol{X}} = \frac{\partial^3 \boldsymbol{E}}{\partial \boldsymbol{X}^3} = \nabla^3 \boldsymbol{E} \tag{10.13}$$

for the third-gradient strain tensor and with $\nabla^n = \underbrace{\nabla \otimes ... \otimes \nabla}_{n\ times}$. Most 'computational' contributions focus on applications with small strain gradients. In this case, strain gradients of different orders are decoupled. Note that the expression in $\boldsymbol{L}$ contains derivatives ranging from $\nabla \boldsymbol{u}$ to $\nabla^4 \boldsymbol{u}$, which is due to geometric nonlinearities arising from finite deformation. The exact forms of $\boldsymbol{L}$, $\boldsymbol{G}$ can be derived with Mathematica. It can be verified that $\boldsymbol{C}$, $\boldsymbol{E}$, $\boldsymbol{H}$, $\boldsymbol{G}$, $\boldsymbol{L}$ are invariants under rigid body translations and rotations. Thus, these quantities can be used to define the stresses and generalized stresses

$$S_{ij} = \mathbb{D}_{ijkl} E_{kl} \text{ with } \mathbb{D}_{ijkl} = \frac{\partial S_{ij}}{\partial E_{kl}} \tag{10.14a}$$

$$R_{ijk} = \mathbb{E}_{ijklmn} H_{lmn} \text{ with } \mathbb{E}_{ijklmn} = \frac{\partial R_{ijk}}{\partial H_{lmn}} \tag{10.14b}$$

$$Q_{ijkl} = \mathbb{F}_{ijklmnst} G_{mnst} \text{ with } \mathbb{F}_{ijklmnst} = \frac{\partial Q_{ijkl}}{\partial G_{mnst}} \tag{10.14c}$$

$$P_{ijklm} = \mathbb{G}_{ijklm\alpha\beta\gamma\eta\delta} L_{\alpha\beta\gamma\eta\delta} \text{ with } \mathbb{G}_{ijklm\alpha\beta\gamma\eta\delta} = \frac{\partial P_{ijklm}}{\partial L_{\alpha\beta\gamma\eta\delta}} \tag{10.14d}$$

Equation 10.14a has the exact form of the Saint Venant–Kirchhoff model. R_{ijk}, Q_{ijkl} and P_{ijklm} are defined as the generalization of the Saint Venant-Kirchhoff model.

The strain energy density in the initial configuration can be assumed as

$$\phi = \frac{1}{2}(S_{ij} E_{ij} + R_{ijk} H_{ijk} + Q_{ijkl} G_{ijkl} + P_{ijklm} L_{ijklm}) \tag{10.15}$$

$$= \frac{1}{2}(E_{ij} \mathbb{D}_{ijkl} E_{kl} + H_{ijk} \mathbb{E}_{ijklmn} H_{lmn} + G_{ijkl} \mathbb{F}_{ijklmnst} G_{mnst} + L_{ijklm} \mathbb{G}_{ijklm\alpha\beta\gamma\eta\delta} L_{\alpha\beta\gamma\eta\delta}) \tag{10.16}$$

where $\mathbb{D}$ is a 4th-order tensor, $\mathbb{E}$ is a 6th-order tensor, $\mathbb{F}$ is an 8th-order tensor and $\mathbb{G}$ is a 10th-order tensor. The second Piola-Kirchhoff stress is work conjugate to the Green-Lagrange strain (Bonet and Wood 1997; Korelc and Wriggers 2016). Therefore, the generalized second Piola-Kirchhoff stresses define the strain energy density in the initial configuration and are objective under any rigid body transformations. The strain energy density, given by Eq. 10.16, is among the most simplified quadratic energy functionals. This energy functional is required to be positive, which can be satisfied provided that the material tensors $\mathbb{D}$, $\mathbb{E}$, $\mathbb{F}$ and $\mathbb{G}$ are positive definite. For an n dimensional space, a kth-order tensor has n^k entries, for example $\mathbb{G}$ has $2^{10} = 1\,024$ elements in 2D and $3^{10} = 59\,049$ elements in 3D. However, when symmetry conditions are exploited, the number of elements can be greatly reduced. We discussed the symmetry of a 6th-order tensor in B.1 and in Voigt notation in B.2. For a third-order gradient solid, there are thousands of material parameters to be determined, which are complicated to resolve experimentally. For simplicity, we introduce only three material length scales.

$$\phi(\boldsymbol{E}, \nabla\boldsymbol{E}, \nabla^2\boldsymbol{E}, \nabla^3\boldsymbol{E}) = \frac{1}{2}\Big(\boldsymbol{S} : \boldsymbol{E} + l_1^2 \nabla\boldsymbol{S} \vdots \nabla\boldsymbol{E} + l_2^4 \nabla^2\boldsymbol{S} \cdot^{(4)} \nabla^2\boldsymbol{E} + l_3^6 \nabla^3\boldsymbol{S} \cdot^{(5)} \nabla^3\boldsymbol{E}\Big) \tag{10.17}$$

where $\cdot^{(k)}$ is the generalization of the inner product, for example, $\cdot^{(1)} = \cdot$, $\cdot^{(2)} =:$, $\cdot^{(3)} = \vdots$, and the stresses can be written as

$$\boldsymbol{S} = \mathbb{D} : \boldsymbol{E} \tag{10.18a}$$

$$\boldsymbol{R} = \nabla\boldsymbol{S} \tag{10.18b}$$

$$\boldsymbol{Q} = \nabla^2\boldsymbol{S} \tag{10.18c}$$

$$\boldsymbol{P} = \nabla^3\boldsymbol{S}. \tag{10.18d}$$

The total internal strain energy in the domain can be expressed as

$$\mathcal{F}_{int} = \int_{\Omega} \phi(\boldsymbol{E}, \nabla\boldsymbol{E}, \nabla^2\boldsymbol{E}, \nabla^3\boldsymbol{E}) \tag{10.19}$$

We recall that Ω is the initial configuration. Here we used $\int_{\Omega}\{\cdot\} := \int_{\Omega}\{\cdot\}\mathbf{d}V$. When small deformations are assumed, the Green-Lagrange strain and second Piola-Kirchhoff stress degenerate to the linear strain and Cauchy stress tensor:

$$\boldsymbol{E} \to \varepsilon = \frac{1}{2}\Big((\nabla\boldsymbol{u})^T + \nabla\boldsymbol{u}\Big) \tag{10.20}$$

$$\boldsymbol{S} \to \sigma = \mathbb{D} : \varepsilon \tag{10.21}$$

Then Eq. 10.17 can be written as

$$\phi(\varepsilon, \nabla\varepsilon, \nabla^2\varepsilon, \nabla^3\varepsilon) = \frac{1}{2}(\sigma : \varepsilon + l_1^2 \nabla\sigma \vdots \nabla\varepsilon + l_2^4 \nabla^2\sigma \cdot^{(4)} \nabla^2\varepsilon + l_3^6 \nabla^3\sigma \cdot^{(5)} \nabla^3\varepsilon) \tag{10.22}$$

Based on the simplified higher gradient elasticity in Eq. 10.17, the general nth-gradient elasticity can be written as

$$\phi(\boldsymbol{E}, \nabla\boldsymbol{E}, \nabla^2\boldsymbol{E}, ..., \nabla^n\boldsymbol{E}) = \frac{1}{2}\Big(\boldsymbol{S} : \boldsymbol{E} + \sum_{k=1}^{n} (l_k)^{2k} \nabla^k \boldsymbol{S} \cdot^{(k+2)} \nabla^k \boldsymbol{E}\Big) \tag{10.23}$$

where l_k is the internal length scale of kth-order. We employ E^n elasticity ($n = 0, 1, 2, 3, 4, 5$) to abbreviate the nth-gradient hyperelasticity theory wherein $\nabla^n \boldsymbol{E}$ is used to define the energy density functional. Accordingly, the conventional hyperelasticity is denoted by E^0 elasticity, gradient elasticity is abbreviated by E^1 elasticity and the second-gradient elasticity by E^2 elasticity. The highest order of the energy form of E^n-elasticity is $(n+1)$, while that of the strong form is $2(n+1)$. The highest order of gradient elasticity we implemented is the E^5 elasticity in 2D. For E^5 elasticity, the governing equations are a set of 12th-order nonlinear partial differential equations (PDEs).

10.2 Governing Equations of Second-Gradient Solid

10.2.1 Integration by Parts on Close Surface

Before delving into the variational derivation of second-gradient solids, we briefly review the integration by parts in domains and on surfaces using the following abbreviations $\int_\Omega\{\cdot\} := \int_\Omega\{\cdot\}\mathbf{d}V$ and $\int_{\partial\Omega}\{\cdot\} := \int_{\partial\Omega}\{\cdot\}\mathbf{d}\Gamma$. The integration by parts for tensor fields in Ω is

$$\int_\Omega \boldsymbol{S} . \nabla\boldsymbol{u} - \int_{\partial\Omega} \boldsymbol{n} \cdot \boldsymbol{S} \cdot \boldsymbol{u} - \int_\Omega \nabla \cdot \boldsymbol{S} \cdot \boldsymbol{u} \tag{10.24}$$

where $\boldsymbol{S}$ is the second-order tensor field and $\boldsymbol{u}$ is the vector field.

According to Refs (Gurtin and Murdoch 1975; Shen and Hu 2010), the integration by parts under the assumption of smooth surfaces can be expressed as

$$\int_{\partial\Omega} \boldsymbol{S} : \nabla\boldsymbol{u} = \int_{\partial\Omega} \boldsymbol{S} : \nabla_n\boldsymbol{u} + \int_{\partial\Omega} \boldsymbol{S} : \nabla_t\boldsymbol{u} = \int_{\partial\Omega} \Big(\boldsymbol{S} : \nabla_n\boldsymbol{u} + (g\boldsymbol{n} \cdot \boldsymbol{S} - \nabla_t \cdot \boldsymbol{S}) \cdot \boldsymbol{u}\Big) \tag{10.25}$$

where $g = \nabla_t \cdot \boldsymbol{n}$ is the Gauss curvature, ∇_t and ∇_n are the tangential and normal gradient

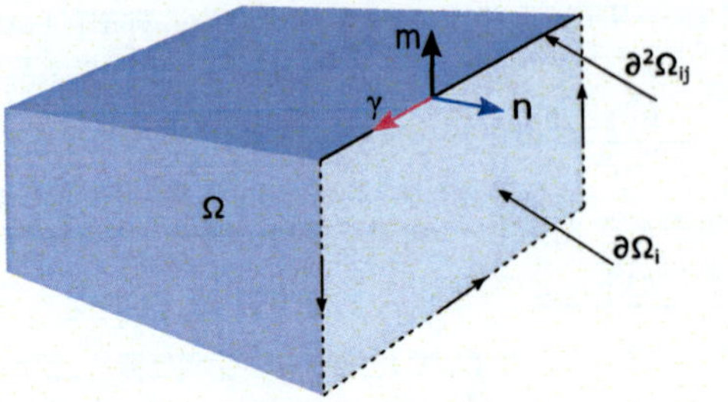

Fig. 10.1 Domain with piecewise smooth surfaces; $\boldsymbol{n}$ is the outward unit normal direction of surface boundary $\partial\Omega$, γ is the tangent direction of line boundary $\partial^2\Omega_{ij}$, $\boldsymbol{m} = \gamma \times \boldsymbol{n}$ is the outward normal unit vector of line boundary

$$\nabla_t = (1 - \boldsymbol{n}\boldsymbol{n}^T) \cdot \nabla, \quad \nabla_n = \boldsymbol{n}\boldsymbol{n}^T \cdot \nabla \tag{10.26}$$

In the above derivation, $\nabla \boldsymbol{u}$ needs to be divided into the tangential and normal parts with respect to the surface, i.e.$\nabla \boldsymbol{u} = \nabla_t \boldsymbol{u} + \nabla_n \boldsymbol{u}$. For the case of piecewise smooth surfaces shown in Fig. 10.1, the boundary term should be considered explicitly,

$$\int_{\partial\Omega} \boldsymbol{S} : \nabla \boldsymbol{u} = \int_{\partial\Omega} \Big(\boldsymbol{S} : \nabla_n \boldsymbol{u} + (g\boldsymbol{n} \cdot \boldsymbol{S} - \nabla_t \cdot \boldsymbol{S}) \cdot \boldsymbol{u}\Big) + \int_{\partial^2\Omega} \boldsymbol{m} \cdot \boldsymbol{S} \cdot \boldsymbol{u} \tag{10.27}$$

where $\boldsymbol{m}$ is the outward normal direction of $\partial\Omega$ in the tangent plane defined by $\boldsymbol{n}$. If both $\boldsymbol{S}$ and $\nabla \boldsymbol{u}$ are defined based on the tangent space of the surface, i.e. $\boldsymbol{n} \cdot \boldsymbol{S} = \boldsymbol{0}$ and $\nabla_n \boldsymbol{u} = \boldsymbol{0}$, the integration by parts on surface is the same as Eq. 10.24.

10.2.2 Variational Derivation of Second-Gradient Solid

In this chapter, we only consider the higher order bulk energy. For the boundary (surface and curve) energies, the reader is referred to (Javili et al. 2013) for more details. The second-gradient solid for linear elasticity with second velocity gradient inertia can be found in (Polizzotto 2013a, b). Let $\phi := \phi(\boldsymbol{u}, \nabla \boldsymbol{u}, \nabla^2 \boldsymbol{u}, \nabla^3 \boldsymbol{u})$ denote the internal energy density of a second-gradient solid (E^2 elasticity). The variation of the internal energy in Ω is then given by

$$\begin{aligned}\delta \mathcal{F}_{int} &= \int_{\Omega} \underbrace{\frac{\partial \phi}{\partial \boldsymbol{u}}}_{\boldsymbol{b}} \cdot \delta \boldsymbol{u} + \underbrace{\frac{\partial \phi}{\partial \nabla \boldsymbol{u}}}_{\boldsymbol{S}_1} : \nabla \delta \boldsymbol{u} + \underbrace{\frac{\partial \phi}{\partial \nabla^2 \boldsymbol{u}}}_{\boldsymbol{S}_2} \vdots \nabla^2 \delta \boldsymbol{u} + \underbrace{\frac{\partial \phi}{\partial \nabla^3 \boldsymbol{u}}}_{\boldsymbol{S}_3} :: \nabla^3 \delta \boldsymbol{u} \\ &= \int_{\Omega} \underbrace{\boldsymbol{b} \cdot \delta \boldsymbol{u}}_{p1} + \underbrace{\boldsymbol{S}_1 : \nabla \delta \boldsymbol{u}}_{p2} + \underbrace{\boldsymbol{S}_2 \vdots \nabla^2 \delta \boldsymbol{u}}_{p3} + \underbrace{\boldsymbol{S}_3 :: \nabla^3 \delta \boldsymbol{u}}_{p4}\end{aligned} \tag{10.28}$$

where $\boldsymbol{b}$ can be viewed as the body force density, $\boldsymbol{S}_i$ denotes the work conjugate to $\nabla^i \boldsymbol{u}$, $(i = 1, 2, 3)$ and $\boldsymbol{S}_i$ is the general stress defined in the initial configuration:

$$S_3 = \frac{\partial \phi}{\partial \nabla^3 u} = \frac{\partial \phi}{\partial \nabla^2 E} :: \frac{\partial \nabla^2 E}{\partial \nabla^3 u} \tag{10.29a}$$

$$S_2 = \frac{\partial \phi}{\partial \nabla^2 u} = \frac{\partial \phi}{\partial \nabla E} \vdots \frac{\partial \nabla E}{\partial \nabla^2 u} + \frac{\partial \phi}{\partial \nabla^2 E} :: \frac{\partial \nabla^2 E}{\partial \nabla^2 u} \tag{10.29b}$$

$$S_1 = \frac{\partial \phi}{\partial \nabla u} = \frac{\partial \phi}{\partial E} : \frac{\partial E}{\partial \nabla u} + \frac{\partial \phi}{\partial \nabla E} \vdots \frac{\partial \nabla E}{\partial \nabla u} + \frac{\partial \phi}{\partial \nabla^2 E} :: \frac{\partial \nabla^2 E}{\partial \nabla u} \tag{10.29c}$$

where E, ∇E, $\nabla^2 E$ have the forms

$$E = \frac{1}{2}\left(\nabla u^T + \nabla u + \nabla u^T \nabla u\right) \tag{10.30a}$$

$$\nabla E = \frac{1}{2}\left(\nabla(\nabla u)^T + \nabla^2 u + \nabla(\nabla u)^T \nabla u + \nabla u^T \nabla^2 u\right) \tag{10.30b}$$

$$\begin{aligned} \nabla^2 E = &\frac{1}{2}\Big(\nabla^2(\nabla u)^T + \nabla^3 u + \nabla^2(\nabla u)^T \nabla u + \\ &\nabla(\nabla u)^T \nabla^2 u + \nabla(\nabla u)^T \nabla^2 u + (\nabla u)^T \nabla^3 u\Big) \end{aligned} \tag{10.30c}$$

Obviously, higher order strain gradients contain a low order gradient due to the existence of a geometric nonlinearity. The explicit form of S_1, S_2, S_3 can be derived by mathematical software such as Mathematica. Since the order of the derivatives is reduced by one each time when applying integration by parts, we integrate the terms from high order to low order so that the derived low order term with the existing low order term can be handled together. In other words, the term with the third-order gradient is firstly integrated by parts, then the second-order gradient term and the accumulated gradient terms at last.

The $p4$ part in Eqs. 10.28 via 10.24 can be written as

$$\begin{aligned} \int_{\Omega} S_3 :: \nabla^3 \delta u &= \int_{\partial\Omega} n \cdot S_3 \vdots \nabla^2 \delta u + \int_{\Omega} -\nabla \cdot S_3 \vdots \nabla^2 \delta u \\ &= \int_{\partial\Omega} \underbrace{n \cdot S_3 \vdots \nabla_t(\nabla \delta u)}_{p7} + \int_{\partial\Omega} \underbrace{n \cdot S_3 \vdots \nabla_n(\nabla \delta u)}_{p8} + \int_{\Omega} \underbrace{-\nabla \cdot S_3 \vdots \nabla^2 \delta u}_{p6} \end{aligned} \tag{10.31}$$

The $p7$ part in Eq. 10.31 is obtained via integration by parts on surface

$$\begin{aligned} \int_{\partial\Omega} n \cdot S_3 \vdots \nabla_t(\nabla \delta u) &= \int_{\partial\Omega} \underbrace{\Big(gn \cdot (n \cdot S_3) - \nabla_t \cdot (n \cdot S_3)\Big)}_{V_1} : \nabla \delta u \\ &= \int_{\partial\Omega} \underbrace{V_1 : \nabla \delta u}_{p9} \end{aligned} \tag{10.32}$$

In above derivation, we abbreviate the long expression as $\boldsymbol{V}_1$. Based on Eq. 10.25, the $p9$ part reads

$$\int_{\partial\Omega} \boldsymbol{V}_1 : \nabla\delta\boldsymbol{u} = \int_{\partial\Omega} \boldsymbol{V}_1 : \nabla_n\delta\boldsymbol{u} + (g\boldsymbol{n}\cdot\boldsymbol{V}_1 - \nabla_t\cdot\boldsymbol{V}_1)\cdot\delta\boldsymbol{u} \tag{10.33}$$

Hence, $p4$ has the form

$$\int_{\Omega} \boldsymbol{S}_3 :: \nabla^3\delta\boldsymbol{u} = \int_{\partial\Omega} \boldsymbol{V}_1 : \nabla\delta\boldsymbol{u} + \int_{\partial\Omega} \underbrace{\boldsymbol{n}\cdot\boldsymbol{S}_3 \vdots \nabla_n(\nabla\delta\boldsymbol{u})}_{p8} + \int_{\Omega} \underbrace{-\nabla\cdot\boldsymbol{S}_3 \vdots \nabla^2\delta\boldsymbol{u}}_{p6} \tag{10.34}$$

The $p6$ term has the same form as part $p3$ in Eq. 10.28, and it can be estimated that a new surface term similar to the $p9$ part will arise. The summation of p_3, p_6, p_9 can be written as

$$\begin{aligned}
&\int_{\Omega} \boldsymbol{S}_2 \vdots \nabla^2\delta\boldsymbol{u} - \nabla\cdot\boldsymbol{S}_3 \vdots \nabla^2\delta\boldsymbol{u} + \int_{\partial\Omega} \boldsymbol{V}_1 : \nabla\delta\boldsymbol{u} = \underbrace{\int_{\Omega} (\boldsymbol{S}_2 - \nabla\cdot\boldsymbol{S}_3) \vdots \nabla^2\delta\boldsymbol{u}}_{by\ Eq.10.24} + \int_{\partial\Omega} \boldsymbol{V}_1 : \nabla\delta\boldsymbol{u} \\
&= \underbrace{\int_{\partial\Omega} \boldsymbol{n}\cdot(\boldsymbol{S}_2 - \nabla\cdot\boldsymbol{S}_3) : \nabla\delta\boldsymbol{u} + \int_{\partial\Omega} \boldsymbol{V}_1 : \nabla\delta\boldsymbol{u}}_{by\ Eq.10.25} - \int_{\Omega} \nabla\cdot(\boldsymbol{S}_2 - \nabla\cdot\boldsymbol{S}_3) : \nabla\delta\boldsymbol{u} \\
&= \int_{\partial\Omega} \underbrace{(g\boldsymbol{n}\cdot\boldsymbol{V}_1 - \nabla_t\cdot\boldsymbol{V}_1 + g\boldsymbol{n}\cdot(\boldsymbol{n}\cdot(\boldsymbol{S}_2 - \nabla\cdot\boldsymbol{S}_3)) - \nabla_t\cdot(\boldsymbol{n}\cdot(\boldsymbol{S}_2 - \nabla\cdot\boldsymbol{S}_3)))}_{\boldsymbol{V}_2} \cdot\delta\boldsymbol{u} \\
&+ \int_{\partial\Omega} (\boldsymbol{V}_1 + \boldsymbol{n}\cdot(\boldsymbol{S}_2 - \nabla\cdot\boldsymbol{S}_3)) : \nabla_n\delta\boldsymbol{u} + \int_{\Omega} \underbrace{-\nabla\cdot(\boldsymbol{S}_2 - \nabla\cdot\boldsymbol{S}_3) : \nabla\delta\boldsymbol{u}}_{p10} \\
&= \int_{\partial\Omega} \boldsymbol{V}_2\cdot\delta\boldsymbol{u} + \int_{\partial\Omega} (\boldsymbol{V}_1 + \boldsymbol{n}\cdot(\boldsymbol{S}_2 - \nabla\cdot\boldsymbol{S}_3)) : \nabla_n\delta\boldsymbol{u} + \int_{\Omega} -\nabla\cdot(\boldsymbol{S}_2 - \nabla\cdot\boldsymbol{S}_3) : \nabla\delta\boldsymbol{u}
\end{aligned} \tag{10.35a}$$

In the above derivation, the long expression in the second line is abbreviated as $\boldsymbol{V}_2$. Both integration by parts in the domain and integration by parts on the surface are used in the derivation.

The gradient order of $p10$ is identical to that in the $p2$ part. For simplicity, the integration by parts of $p10 + p2$ is

$$\int_{\Omega} \underbrace{(\boldsymbol{S}_1 - \nabla\cdot(\boldsymbol{S}_2 - \nabla\cdot\boldsymbol{S}_3))}_{\boldsymbol{S}_4} : \nabla\delta\boldsymbol{u} = \int_{\partial\Omega} \boldsymbol{n}\cdot\boldsymbol{S}_4\cdot\delta\boldsymbol{u} - \int_{\Omega} \nabla\cdot\boldsymbol{S}_4\cdot\delta\boldsymbol{u} \tag{10.36}$$

By using integration by parts in Eqs. 10.24 and 10.25 several times, the variation of the internal energy can be finally written as

$$\delta\mathcal{F}_{int} = \int_{\Omega}(\boldsymbol{b} - \nabla\cdot\boldsymbol{S}_4)\cdot\delta\boldsymbol{u} + \int_{\partial\Omega}(\boldsymbol{V}_1 + \boldsymbol{n}\cdot(\boldsymbol{S}_2 - \nabla\cdot\boldsymbol{S}_3)) : \nabla_n\delta\boldsymbol{u}$$
$$+ \int_{\partial\Omega}\boldsymbol{V}_2\cdot\delta\boldsymbol{u} + \int_{\partial\Omega}\boldsymbol{n}\cdot\boldsymbol{S}_4\cdot\delta\boldsymbol{u} + \int_{\partial\Omega}\boldsymbol{n}\cdot\boldsymbol{S}_3\vdots\nabla_n(\nabla\delta\boldsymbol{u}) \tag{10.37}$$
$$= \int_{\Omega}(\boldsymbol{b} - \nabla\cdot\boldsymbol{S}_4)\cdot\delta\boldsymbol{u} + \int_{\partial\Omega}(\boldsymbol{V}_2 + \boldsymbol{n}\cdot\boldsymbol{S}_4)\cdot\delta\boldsymbol{u}$$
$$+ \int_{\partial\Omega}(\boldsymbol{V}_1 + \boldsymbol{n}\cdot(\boldsymbol{S}_2 - \nabla\cdot\boldsymbol{S}_3)) : \nabla_n\delta\boldsymbol{u} + \int_{\partial\Omega}\boldsymbol{n}\cdot\boldsymbol{S}_3\vdots\nabla_n(\nabla\delta\boldsymbol{u}) \tag{10.38}$$

where

$$\boldsymbol{S}_4 = \boldsymbol{S}_1 - \nabla\cdot(\boldsymbol{S}_2 - \nabla\cdot\boldsymbol{S}_3) \tag{10.39a}$$
$$\boldsymbol{V}_1 = g\boldsymbol{n}\cdot(\boldsymbol{n}\cdot\boldsymbol{S}_3) - \nabla_t\cdot(\boldsymbol{n}\cdot\boldsymbol{S}_3) \tag{10.39b}$$
$$\boldsymbol{V}_2 = g\boldsymbol{n}\cdot\boldsymbol{V}_1 - \nabla_t\cdot\boldsymbol{V}_1 + g\boldsymbol{n}\cdot(\boldsymbol{n}\cdot(\boldsymbol{S}_2 - \nabla\cdot\boldsymbol{S}_3)) - \nabla_t\cdot(\boldsymbol{n}\cdot(\boldsymbol{S}_2 - \nabla\cdot\boldsymbol{S}_3)) \tag{10.39c}$$

In Eq. 10.38, we used $\partial\Omega$ to denote the boundaries, which should be tailored based on the actual boundary conditions. The variation of the internal energy yields the work-conjugate pairs on the boundaries. The expression for $\boldsymbol{V}_2$ describes the contribution from the curvature related terms (e.g. curvature and curvature gradient on surface) as well as the generalized stresses of different orders. The contribution from the surface curvature indicates that the gradient effect or nonlocal effect of the solid may be significant at the 'micro'-scale, where the surface-to-volume ratio is much larger compared to the macroscale and the surface curvature for small-scale objects is huge. This relation is consistent with the fact that the strength of a material at micro-scale is much larger than that at macro-scale.

For the gradient elasticity, the boundary conditions may contain essential boundary conditions such as translation, the gradient of the translation and force boundary conditions like stress and couple stress. The gradient of the translation is similar to the prescribed rotation on the boundary in plate/shell theory, while the couple stress is the work conjugate to the gradient of the translation.

Based on the boundary work-conjugate pairs in the variation of the internal energy, the external energy can be constructed as

$$\mathcal{F}_{ext} = \int_{\partial\Omega_D^0}\boldsymbol{P}\cdot(\boldsymbol{u} - \bar{\boldsymbol{u}}) + \int_{\partial\Omega_N^0}\bar{\boldsymbol{P}}\cdot\boldsymbol{u}$$
$$+ \int_{\partial\Omega_D^1}\boldsymbol{Q} : (\nabla_n\boldsymbol{u} - \overline{\nabla_n\boldsymbol{u}}) + \int_{\partial\Omega_N^1}\bar{\boldsymbol{Q}} : \nabla_n\boldsymbol{u}$$
$$+ \int_{\partial\Omega_D^2}\boldsymbol{R}\vdots(\nabla_n\nabla\boldsymbol{u} - \overline{\nabla_n\nabla\boldsymbol{u}}) + \int_{\partial\Omega_N^2}\bar{\boldsymbol{R}}\vdots\nabla_n\nabla\boldsymbol{u} \tag{10.40}$$

where $\boldsymbol{P} = (\boldsymbol{V}_2 + \boldsymbol{n}\cdot\boldsymbol{S}_4)$, $\boldsymbol{Q} = (\boldsymbol{V}_1 + \boldsymbol{n}\cdot(\boldsymbol{S}_2 - \nabla\cdot\boldsymbol{S}_3))$, $\boldsymbol{R} = \boldsymbol{n}\cdot\boldsymbol{S}_3$, $\partial\Omega_D^i$, $\partial\Omega_N^i$, $(i = 0, 1, 2)$ refer to the Dirichlet and Neumann boundary conditions for $\boldsymbol{u}$ of different partial derivative orders; $\partial\Omega_D^0$ designates the constraints of the displacement, $\partial\Omega_D^1$ denotes the constraints of the displacement gradient (e.g. fixed rotation state) and $\partial\Omega_D^1$ describes the displacement second-gradient; $\bar{\boldsymbol{P}}$ is the traction load, $\bar{\boldsymbol{Q}}$ refers to the couple stress load and $\bar{\boldsymbol{R}}$ is the higher order couple stress load.

Gradient elasticity deals not only with the gradient of the deformation but also the gradient of inertia terms (Askes and Aifantis 2009; Mindlin 1964; Polizzotto 2013a, b). The kinetic energy with velocity gradient can be written as (Polizzotto 2013b)

$$\mathcal{K} = \int_{t_0}^{t_1} \int_{\Omega} \frac{1}{2}\rho\dot{\boldsymbol{u}} \cdot \dot{\boldsymbol{u}} + \frac{1}{2}\rho l_d^2 \nabla\dot{\boldsymbol{u}} : \nabla\dot{\boldsymbol{u}} \tag{10.41}$$

The variation of the kinetic energy can be written as

$$\begin{aligned}
\delta\mathcal{K} &= \int_{t_0}^{t_1} \int_{\Omega} \rho\dot{\boldsymbol{u}} \cdot \delta\dot{\boldsymbol{u}} + \rho l_d^2 \nabla\dot{\boldsymbol{u}} : \nabla\delta\dot{\boldsymbol{u}} \\
&= \int_{t_0}^{t_1} \int_{\Omega} -\rho\ddot{\boldsymbol{u}} \cdot \delta\boldsymbol{u} - \rho l_d^2 \nabla\ddot{\boldsymbol{u}} : \nabla\delta\boldsymbol{u} \\
&= \int_{t_0}^{t_1} \int_{\Omega} -\rho\ddot{\boldsymbol{u}} \cdot \delta\boldsymbol{u} + \rho l_d^2 \nabla \cdot \nabla\ddot{\boldsymbol{u}} \cdot \delta\boldsymbol{u}
\end{aligned} \tag{10.42}$$

For any $\delta\boldsymbol{u}$, $\delta\nabla_n\boldsymbol{u}$, $\delta\nabla_n\nabla\boldsymbol{u}$, the Hamilton principle

$$\delta\mathcal{K} - \int_{t_0}^{t_1} \delta\mathcal{F}_{int} + \int_{t_0}^{t_1} \delta\mathcal{F}_{ext} = 0 \tag{10.43}$$

leads to

$$\rho\ddot{\boldsymbol{u}} - l_d^2\nabla \cdot \nabla\ddot{\boldsymbol{u}} = -\boldsymbol{b} + \nabla \cdot (\boldsymbol{S}_1 - \nabla \cdot (\boldsymbol{S}_2 - \nabla \cdot \boldsymbol{S}_3)) \text{ in } \Omega \tag{10.44a}$$

$$\boldsymbol{u} = \bar{\boldsymbol{u}} \text{ on } \partial\Omega_D^0 \tag{10.44b}$$

$$\boldsymbol{P} = \bar{\boldsymbol{P}} \text{ on } \partial\Omega_N^0 \tag{10.44c}$$

$$\nabla_n\boldsymbol{u} = \overline{\nabla_n\boldsymbol{u}} \text{ on } \partial\Omega_D^1 \tag{10.44d}$$

$$\boldsymbol{Q} = \bar{\boldsymbol{Q}} \text{ on } \partial\Omega_N^1 \tag{10.44e}$$

$$\nabla_n\nabla\boldsymbol{u} = \overline{\nabla_n\nabla\boldsymbol{u}} \text{ on } \partial\Omega_D^2 \tag{10.44f}$$

$$\boldsymbol{R} = \bar{\boldsymbol{R}} \text{ on } \partial\Omega_N^2 \tag{10.44g}$$

The derivation of energies based on variational principle leads very naturally to the governing equations and various boundary conditions. The maximal order of derivatives in Eq. 10.44a and $\boldsymbol{V}_2$ is 6, 5, respectively. Similarly, the variation of the strain energy density of strain gradient elasticity (E^1 elasticity) can be derived. By setting $\boldsymbol{S}_3 = 0$ in Eq. 10.44, we obtain the governing equations and boundary conditions of gradient elasticity

$$\rho\ddot{\boldsymbol{u}} - l_d^2\nabla\cdot\nabla\ddot{\boldsymbol{u}} = -\boldsymbol{b} + \nabla\cdot(\boldsymbol{S}_1 - \nabla\cdot\boldsymbol{S}_2) \text{ in } \Omega \tag{10.45a}$$

$$\boldsymbol{u} = \bar{\boldsymbol{u}} \text{ on } \partial\Omega_D^0 \tag{10.45b}$$

$$\boldsymbol{P} = \bar{\boldsymbol{P}} \text{ on } \partial\Omega_N^0 \tag{10.45c}$$

$$\nabla_n\boldsymbol{u} = \overline{\nabla_n\boldsymbol{u}} \text{ on } \partial\Omega_D^1 \tag{10.45d}$$

$$\boldsymbol{Q} = \bar{\boldsymbol{Q}} \text{ on } \partial\Omega_N^1 \tag{10.45e}$$

where $\boldsymbol{Q}=\boldsymbol{n}\cdot\boldsymbol{S}_2$, $\boldsymbol{P}=(g\boldsymbol{n}\cdot(\boldsymbol{n}\cdot\boldsymbol{S}_2)-\nabla_t\cdot(\boldsymbol{n}\cdot\boldsymbol{S}_2)+\boldsymbol{n}\cdot(\boldsymbol{S}_1-\nabla\cdot\boldsymbol{S}_2))$, $\partial\Omega_D^i$, $\partial\Omega_N^i$, $(i = 0, 1)$ indicate the Dirichlet and Neumann boundary conditions for $\boldsymbol{u}$ of different partial derivative orders.

10.3 Numerical Implementation

10.3.1 Review of Nonlocal Operator Method

NOM uses the integral form to replace the partial differential derivatives of different orders. We adopted a Total Lagrangian description of motion for the higher order gradient elasticity NOM. Consider a domain as shown in Fig. 10.2a, let $\boldsymbol{X}_i$ be spatial coordinates in the domain $\boldsymbol{\Omega}$; $\mathbf{r} := \boldsymbol{X}_j - \boldsymbol{X}_i$ is a spatial vector ranging from $\boldsymbol{X}_i$ to $\boldsymbol{X}_j$; $\mathbf{v}_i := \mathbf{v}(\boldsymbol{X}_i, t)$ and $\mathbf{v}_j := \mathbf{v}(\boldsymbol{X}_j, t)$ are the field values for $\boldsymbol{X}_i$ and $\boldsymbol{X}_j$, respectively; $\mathbf{v}_{ij} := \mathbf{v}_j - \mathbf{v}_i$ is the relative field vector for spatial vector $\mathbf{r}$.

Support $\mathcal{S}_i$ of point $\boldsymbol{X}_i$ is the neighborhood of point $\boldsymbol{X}_i$. A point $\boldsymbol{X}_j$ in support $\mathcal{S}_i$ forms the spatial vector $\mathbf{r}(= \boldsymbol{X}_j - \boldsymbol{X}_i)$. The support in the NOM can be a spherical domain, a cube, semi-spherical domain and so on.

Dual-support is defined as a union of points whose supports include $\boldsymbol{X}$, denoted by

$$\mathcal{S}_i' = \{\boldsymbol{X}_j | \boldsymbol{X}_i \in \mathcal{S}_j\}. \tag{10.46}$$

Point $\boldsymbol{X}_j$ forms the dual-vector $\mathbf{r}'(= \boldsymbol{X}_i - \boldsymbol{X}_j = -\mathbf{r})$ in $\mathcal{S}_i'$. On the other hand, $\mathbf{r}'$ is the spatial vector formed in $\mathcal{S}_j$. One example to illustrate the support and dual-support is shown in Fig. 10.2b.

The first-order nonlocal operator method uses the basic nonlocal operators to replace the local operator in calculus such as the gradient, divergence and curl operators. The functional formulated by the local differential operator can be used to construct the residual or tangent stiffness matrix by replacing the local operator with the corresponding nonlocal operator. The nonlocal gradient of a vector field $\mathbf{v}$ for point $\boldsymbol{X}_i$ in support $\mathcal{S}_i$ is defined as

$$\tilde{\nabla}\mathbf{v}_i := \int_{\mathcal{S}_i} w(\mathbf{r})\mathbf{v}_{ij} \otimes \mathbf{r} dV_j \cdot \Big(\int_{\mathcal{S}_i} w(\mathbf{r})\mathbf{r} \otimes \mathbf{r} dV_j\Big)^{-1}. \tag{10.47}$$

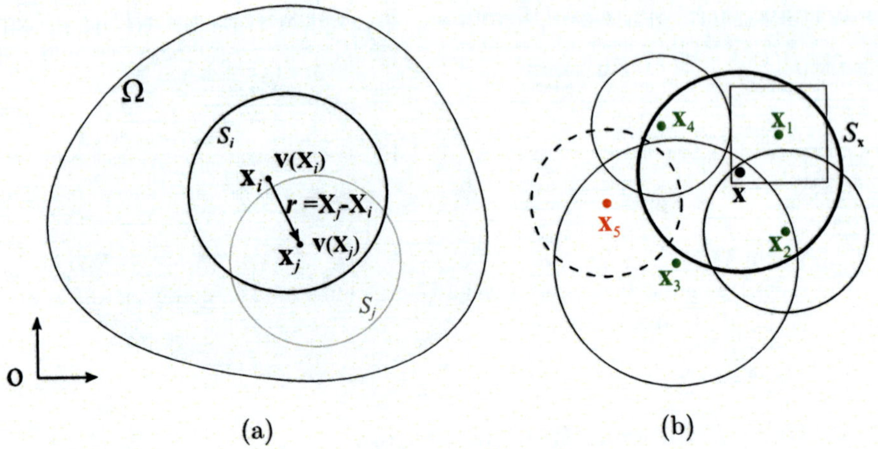

Fig. 10.2 **a** Domain and notation. **b** Schematic diagram for support and dual-support, all shapes above are supports, $\mathcal{S}_X = \{X_1, X_2, X_4\}$, $\mathcal{S}'_X = \{X_1, X_2, X_3\}$

The nonlocal gradient operator and its variation in discrete form are given by

$$\tilde{\nabla}\mathbf{v}_i = \sum_{j\in\mathcal{S}_i} w(\mathbf{r})\mathbf{v}_{ij} \otimes \mathbf{r}\Delta V_j \cdot \Big(\sum_{j\in\mathcal{S}_i} w(\mathbf{r})\mathbf{r} \otimes \mathbf{r}\Delta V_j\Big)^{-1}, \tag{10.48}$$

$$\tilde{\nabla}\delta\mathbf{v}_i = \sum_{j\in\mathcal{S}_i} w(\mathbf{r})\delta\mathbf{v}_{ij} \otimes \mathbf{r}\Delta V_j \cdot \Big(\sum_{j\in\mathcal{S}_i} w(\mathbf{r})\mathbf{r} \otimes \mathbf{r}\Delta V_j\Big)^{-1}. \tag{10.49}$$

The operator energy functional for a vector field at point $\mathbf{x}_i$ is

$$\mathcal{F}_i^{hg} = p^{hg} \int_{\mathcal{S}_i} w(\mathbf{r})(\tilde{\nabla}\mathbf{v}_i \cdot \mathbf{r} - \mathbf{v}_{ij}) \cdot (\tilde{\nabla}\mathbf{v}_i \cdot \mathbf{r} - \mathbf{v}_{ij})\mathbf{d}V_j \tag{10.50}$$

where p^{hg} is a penalty coefficient. The residual and tangent stiffness matrix of $\mathcal{F}_i^{hg}$ can be obtained with ease, see (Ren et al. 2020a) for more details. For problems that require higher order continuity, the higher order NOM is needed. According to Ref (Ren et al. 2020b), a scalar field u_j at a point $j \in \mathcal{S}_i$ can be obtained by a Taylor series expansion at u_i in d dimensions with maximal derivative order not higher than n,

$$u_j = u_i + \sum_{(n_1,\ldots,n_d)\in\alpha_d^n} \frac{r_1^{n_1}\ldots r_d^{n_d}}{n_1!\ldots n_d!} u_{i,n_1\ldots n_d} + O(r^{|\alpha|+1}) \tag{10.51}$$

with

$$\mathbf{r} = (r_1, ..., r_d) = (X_{j1} - X_{i1}, ..., X_{jd} - X_{id}) \tag{10.52a}$$

$$u_{i,n_1...n_d} = \frac{\partial^{n_1+...+n_d} u_i}{\partial X_{i1}^{n_1} ... \partial X_{id}^{n_d}} \tag{10.52b}$$

$$|\alpha| = \max\,(n_1 + ... + n_d) \tag{10.52c}$$

α_d^n being the list of multi-indexes, given by

$$\alpha_d^n = \{(n_1, ..., n_d) | 1 \le \sum_{i=1}^{d} n_i \le n,\ n_i \in \mathbb{N}^0,\ 1 \le i \le d\} \tag{10.53}$$

and $\mathbb{N}^0 = \{0, 1, 2, 3, ...\}$. The number of multi-indices in α_d^n is $(n + d)!/(n!d!) - 1$ and all elements in α_d^n of Eq. 10.53 can be obtained easily by Mathematica (Ren et al. 2020b). For any multi-index $(n_1, ..., n_d) \in \alpha_d^n$, the partial derivative and the polynomial are

$$u_{i,n_1...n_d},\ \frac{r_1^{n_1} ... r_d^{n_d}}{n_1! ... n_d!},\quad \forall (n_1, ..., n_d) \in \alpha_d^n. \tag{10.54}$$

When the length scale of support $\mathcal{S}_i$ at u_i is taken into account, the Taylor series expansion in Eq. 10.51 can be written as

$$\begin{aligned} u_j &= u_i + \sum_{(n_1,...,n_d) \in \alpha_d^n} \frac{r_1^{n_1} ... r_d^{n_d}}{h_i^{n_1+...+n_d}} \left(\frac{h_i^{n_1+...+n_d}}{n_1! ... n_d!} u_{i,n_1...n_d} \right) + O(r^{n+1}) \\ &= u_i + \sum_{(n_1,...,n_d) \in \alpha_d^n} \frac{r_1^{n_1} ... r_d^{n_d}}{h_i^{n_1+...+n_d}} u_{i,n_1...n_d}^h + O(r^{n+1}) \end{aligned} \tag{10.55}$$

where h_i is the characteristic length of $\mathcal{S}_i$, and

$$u_{i,n_1...n_d}^h = \frac{h_i^{n_1+...+n_d}}{n_1! ... n_d!} u_{i,n_1...n_d} \tag{10.56}$$

Let $\boldsymbol{p}_j^h$, $\partial_\alpha^h u_i$ and $\partial_\alpha u_i$ denote the list of the polynomials, scaled partial derivatives, partial derivatives, respectively, based on multi-index notation α_d^n in Eq. 10.53,

$$\boldsymbol{p}_j^h = (\frac{r_d}{h}, ..., \frac{r_1^{n_1} ... r_d^{n_d}}{h^{n_1+...+n_d}}, ..., \frac{r_1^n}{h^n})^T \tag{10.57a}$$

$$\partial_\alpha^h u_i = (u_{i,0...1}^h, ..., u_{i,n_1...n_d}^h, ..., u_{i,n...0}^h)^T \tag{10.57b}$$

$$\partial_\alpha u_i = (u_{i,0...1}, ..., u_{i,n_1...n_d}, ..., u_{i,n...0})^T. \tag{10.57c}$$

$\partial_\alpha^h u_i$ and $\partial_\alpha u_i$ are related by

$$\partial_\alpha u_i = \boldsymbol{H}_i^{-1}\partial_\alpha^h u_i, \text{ with } \boldsymbol{H}_i = \text{diag}\Big[h_i, ..., \frac{h_i^{n_1+...+n_d}}{n_1!...n_d!}, ..., \frac{h_i^n}{n!}\Big] \tag{10.58}$$

where diag$[a_1, ..., a_n]$ denotes a diagonal matrix whose diagonal entries starting in the upper left corner are $a_1, ..., a_n$. Therefore, the Taylor series expansion with u_i being moved to the left side of the equation can be written as

$$u_{ij} = (\partial_\alpha^h u_i)^T \boldsymbol{p}_j^h, \forall j \in \mathcal{S}_i \tag{10.59}$$

where $u_{ij} = u_j - u_i$. Integrating u_{ij} with weighted coefficient $w(\mathbf{r})(\boldsymbol{p}_j^h)^T$ in support $\mathcal{S}_i$, we obtain

$$\begin{aligned}\int_{\mathcal{S}_i} w(\mathbf{r})u_{ij}(\boldsymbol{p}_j^h)^T \mathbf{d}V_j &= (\partial_\alpha^h u_i)^T \int_{\mathcal{S}_i} w(\mathbf{r})\boldsymbol{p}_j^h \otimes (\boldsymbol{p}_j^h)^T \mathbf{d}V_j \\ &= (\partial_\alpha u_i)^T \boldsymbol{H}_i \int_{\mathcal{S}_i} w(\mathbf{r})\boldsymbol{p}_j^h \otimes (\boldsymbol{p}_j^h)^T \mathbf{d}V_j\end{aligned} \tag{10.60}$$

where $w(\mathbf{r})$ is the weight function. Thus, the nonlocal operator $\tilde{\partial}_\alpha u_i$ can be obtained as

$$\tilde{\partial}_\alpha u_i := \boldsymbol{H}_i^{-1}\Big(\int_{\mathcal{S}_i} w(\mathbf{r})\boldsymbol{p}_j^h \otimes (\boldsymbol{p}_j^h)^T \mathbf{d}V_j\Big)^{-1}\int_{\mathcal{S}_i} w(\mathbf{r})u_{ij}\boldsymbol{p}_j^h \mathbf{d}V_j = \boldsymbol{K}_i \cdot \int_{\mathcal{S}_i} w(\mathbf{r})\boldsymbol{p}_j^h u_{ij}\mathbf{d}V_j \tag{10.61}$$

with

$$\boldsymbol{K}_i := \boldsymbol{H}_i^{-1}\Big(\int_{\mathcal{S}_i} w(\mathbf{r})\boldsymbol{p}_j^h \otimes (\boldsymbol{p}_j^h)^T \mathbf{d}V_j\Big)^{-1}. \tag{10.62}$$

The variation of $\tilde{\partial}_\alpha u_i$ is

$$\tilde{\partial}_\alpha \delta u_i := \boldsymbol{K}_i \cdot \int_{\mathcal{S}_i} w(\mathbf{r})\boldsymbol{p}_j^h(\delta u_j - \delta u_i)\mathbf{d}V_j \tag{10.63}$$

In the continuous form, the number of dimensions of $\partial\delta u_i$ is infinite and a discretization is required. After discretization of the domain by particles, the whole domain is represented by

$$\Omega = \sum_{i=1}^{N} \Delta V_i \tag{10.64}$$

where i is the global index of volume ΔV_i and N is the number of particles in Ω. Particles in $\mathcal{S}_i$ are represented by

$$\mathcal{S}_i = \{j_1, ..., j_k, ..., j_{n_i}\} \tag{10.65}$$

where $j_1, ..., j_k, ..., j_{n_i}$ are the global indices of neighboring particles of i and n_i is the number of neighbors of i in $\mathcal{S}_i$. The discrete form of Eq. 10.61 and its variation are

$$\tilde{\partial}_\alpha u_i = \boldsymbol{K}_i \cdot \sum_{j\in\mathcal{S}_i} u_{ij} w(\mathbf{r}_j) \boldsymbol{p}_j^h \Delta V_j = \boldsymbol{K}_i \boldsymbol{p}_{wi}^h \Delta\mathbf{u}_i \tag{10.66a}$$

$$\tilde{\partial}_\alpha \delta u_i = \boldsymbol{K}_i \cdot \sum_{j\in\mathcal{S}_i} \delta u_{ij} w(\mathbf{r}_j) \boldsymbol{p}_j^h \Delta V_j = \boldsymbol{K}_i \boldsymbol{p}_{wi}^h \delta\Delta\mathbf{u}_i \tag{10.66b}$$

with

$$\boldsymbol{K}_i = \boldsymbol{H}_i^{-1} \Big(\sum_{j\in\mathcal{S}_i} w(\mathbf{r}) \boldsymbol{p}_j^h \otimes (\boldsymbol{p}_j^h)^T \Delta V_j \Big)^{-1}, \tag{10.67a}$$

$$\boldsymbol{p}_{wi}^h = \Big(w(\mathbf{r}_{j_1}) \boldsymbol{p}_{j_1}^h \Delta V_{j_1}, ..., w(\mathbf{r}_{j_{n_i}}) \boldsymbol{p}_{j_{n_i}}^h \Delta V_{j_{n_i}} \Big) \tag{10.67b}$$

$$\Delta\mathbf{u}_i = (u_{ij_1}, ..., u_{ij_k}, ..., u_{ij_{n_i}})^T \tag{10.67c}$$

The nonlocal operator provides all partial derivatives with maximal order for a single index up to n. The set of derivatives in PDEs of real application is a subset of the nonlocal operator. Together with the weak formulation (weighted residual method or variational principles (i.e.(Ren et al. 2020a))), Eq. 10.66a can be employed to solve many linear (nonlinear) PDEs. Equation 10.66a can be written more concisely as

$$\tilde{\partial}_\alpha u_i = \boldsymbol{K}_i \boldsymbol{p}_{wi}^h \Delta\mathbf{u}_i = \boldsymbol{B}_{\alpha i} \boldsymbol{u}_i \tag{10.68}$$

with $\boldsymbol{B}_{\alpha i}$ being the operator matrix for point i based on multi-index α_d^n

$$\boldsymbol{B}_{\alpha i} = \begin{bmatrix} -(1, \ldots, 1)_{n_p} \boldsymbol{K}_i \boldsymbol{p}_{wi}^h \\ \boldsymbol{K}_i \boldsymbol{p}_{wi}^h \end{bmatrix} \tag{10.69}$$

$$\boldsymbol{u}_i = (u_i, u_{j_1}, u_{j_2}, \ldots, u_{j_{n_i}})^T \tag{10.70}$$

where $(1, \ldots, 1)_{n_p} \boldsymbol{K}_i \boldsymbol{p}_{wi}^h$ is the column sum of $\boldsymbol{K}_i \boldsymbol{p}_{wi}^h$, n_p is the length of α_d^n. The operator matrix obtains all partial derivatives of maximal order less than $|\alpha| + 1$ by the nodal values in the support. For real applications, one can select the specific rows in the operator matrix based on the partial derivatives contained in the specific PDEs. For example, if the order of derivatives in the given PDEs are $\tilde{\partial} u_i = (u_{i,Y}, u_{i,YY}, u_{i,XY})^T \subset \tilde{\partial}_\alpha u_i$, one can select the blue lines in Eq. 10.71 to form the actually operator matrix given by $\boldsymbol{B}_i$ in Eq. 10.72.

$$\underbrace{\begin{bmatrix} u_{i,Y} \\ u_{i,X} \\ u_{i,YY} \\ u_{i,XY} \\ u_{i,XX} \\ \vdots \end{bmatrix}}_{\tilde{\partial}_\alpha u_i} = \underbrace{\begin{bmatrix} b_{11} & b_{12} & \cdots & b_{1(n+1)} \\ b_{21} & b_{22} & \cdots & b_{2(n+1)} \\ b_{31} & b_{32} & \cdots & b_{3(n+1)} \\ b_{41} & b_{42} & \cdots & b_{4(n+1)} \\ \vdots & \vdots & \ddots & \vdots \\ b_{m1} & b_{m2} & \cdots & b_{m(n+1)} \end{bmatrix}}_{B_{\alpha i}} \underbrace{\begin{bmatrix} u_i \\ u_{j_1} \\ \vdots \\ u_{j_n} \end{bmatrix}}_{u_i} \tag{10.71}$$

$$\underbrace{\begin{bmatrix} u_{i,Y} \\ u_{i,YY} \\ u_{i,XY} \end{bmatrix}}_{\tilde{\partial} u_i} = \underbrace{\begin{bmatrix} b_{11} & b_{12} & \cdots & b_{1(n+1)} \\ b_{31} & b_{32} & \cdots & b_{3(n+1)} \\ b_{41} & b_{42} & \cdots & b_{4(n+1)} \end{bmatrix}}_{B_i} \underbrace{\begin{bmatrix} u_i \\ u_{j_1} \\ \vdots \\ u_{j_n} \end{bmatrix}}_{u_i} \tag{10.72}$$

For a given maximal differential order and number of space dimensions, NOM offers the derivatives of all orders in discrete form 'automatically'. These nonlocal derivatives are similar to the derivatives of the shape functions in IGA. When the selected derivatives are inserted into the equivalent functional of the physical problem in discrete form, the residual and tangent stiffness matrix of the functional can be derived.

Besides considering the functional for the physical problem, the functional for the nonlocal operators should be considered explicitly. The energy functional for all nonlocal operators is defined as (Ren et al. 2020b)

$$\mathcal{F}_i(\mathbf{u}) = \sum_{j\in\mathcal{S}_i} w(\mathbf{r})\big(u_{ij} - (\boldsymbol{p}_j^h)^T \tilde{\partial}_\alpha^h u_i\big)^2 \Delta V_j \tag{10.73}$$

Based on Eq. 10.66a, $\mathcal{F}_i(\mathbf{u})$ can be simplified as

$$\begin{aligned}\mathcal{F}_i(\mathbf{u}) = & \sum_{j\in\mathcal{S}_i} w(\mathbf{r}) u_{ij}^2 \Delta V_j - \Delta\mathbf{u}_i^T (\boldsymbol{p}_{wi}^h)^T \Big(\sum_{j\in\mathcal{S}_i} w(\mathbf{r}) \boldsymbol{p}_j^h (\boldsymbol{p}_j^h)^T \Delta V_j\Big)^{-1} \boldsymbol{p}_{wi}^h \Delta\mathbf{u}_i \\ = & \Delta\mathbf{u}_i^T \Big(\mathrm{diag}\big[w(\mathbf{r}_{j_1})\Delta V_{j_1}, ..., w(\mathbf{r}_{j_{n_i}})\Delta V_{j_{n_i}}\big] - (\boldsymbol{p}_{wi}^h)^T \Big(\sum_{j\in\mathcal{S}_i} w(\mathbf{r}) \boldsymbol{p}_j^h (\boldsymbol{p}_j^h)^T \Delta V_j\Big)^{-1} \boldsymbol{p}_{wi}^h\Big) \Delta\mathbf{u}_i \\ = & \Delta\mathbf{u}_i^T \boldsymbol{M}_i \Delta\mathbf{u}_i \end{aligned} \tag{10.74}$$

with

$$\boldsymbol{M}_i = \mathrm{diag}\big[w(\mathbf{r}_{j_1})\Delta V_{j_1}, ..., w(\mathbf{r}_{j_{n_i}})\Delta V_{j_{n_i}}\big] - (\boldsymbol{p}_{wi}^h)^T \Big(\sum_{j\in\mathcal{S}_i} w(\mathbf{r}) \boldsymbol{p}_j^h (\boldsymbol{p}_j^h)^T \Delta V_j\Big)^{-1} \boldsymbol{p}_{wi}^h \tag{10.75}$$

Apparently, M_i is a symmetric matrix. The expression of $\mathcal{F}_i(\mathbf{u})$ is quadratic, and its Hessian matrix can be extracted as

$$K_i^{hg} = \frac{p_{hg}}{m_i}\begin{bmatrix} \sum \mathbf{v}_i & -\mathbf{v}_i^T \\ -\mathbf{v}_i & M_i \end{bmatrix} \tag{10.76}$$

where $\mathbf{v}_i(j) = \sum_{k=1}^{n_i} M_i(j,k)$ is the sum of the row of M_i; the first row (column) denotes the entries for point i, while the neighbors start from the second row (column), p_{hg} is a penalty coefficient and m_i the normalization coefficient given by $m_i = \sum_{j\in S_i} w(\mathbf{r})\mathbf{r}\cdot\mathbf{r}\Delta V_j$. The reader is referred to (Ren et al. 2020b) for more details of the NOM.

10.3.2 Newton-Raphson Method

The governing equations and boundary conditions in Eq. 10.44 are quite complicated. The highest continuity in $\boldsymbol{Q}$ is C^4 and the gradient and Hessian matrix of the functional on boundary $\partial\Omega_D^1$ are cumbersome. Note that NOM does not satisfy the Kronecker-Delta property, and the order of NOM should be at least C^5 in order to satisfy the Dirichlet boundary conditions on $\partial\Omega_D^0$, where the continuity order in $\boldsymbol{P}$ is C^5. Therefore, we employ the penalty method to enforce both Dirichlet boundary conditions and the normal Dirichlet boundary conditions. The equivalent energy functional of second-gradient elasticity then becomes

$$\begin{aligned}\mathcal{F} = &\int_\Omega \phi(\boldsymbol{u}, \nabla\boldsymbol{u}, \nabla^2\boldsymbol{u}, \nabla^3\boldsymbol{u})\mathbf{d}V + \int_{\partial\Omega_D^0} \alpha_1(\boldsymbol{u}-\bar{\boldsymbol{u}})\cdot(\boldsymbol{u}-\bar{\boldsymbol{u}})\mathbf{d}S - \int_{\partial\Omega_N^0} \bar{\boldsymbol{P}}\cdot\boldsymbol{u}\mathbf{d}S \\ &+ \int_{\partial\Omega_D^1} \alpha_2(\nabla_n\boldsymbol{u} - \overline{\nabla_n\boldsymbol{u}}) : (\nabla_n\boldsymbol{u} - \overline{\nabla_n\boldsymbol{u}})\mathbf{d}S - \int_{\partial\Omega_N^1} \bar{\boldsymbol{Q}} : \nabla_n\boldsymbol{u}\mathbf{d}S\end{aligned} \tag{10.77}$$

where α_1, α_2 are penalty parameters. One advantage of the penalty method is that the highest order of partial derivatives is 4 for third-gradient elasticity, while the formulation based on the modified variational principle requires C^7 continuity. We neglect the terms on $\partial\Omega_D^2$ and $\partial\Omega_N^2$ for simplicity.

After discretization, the discrete form of the functional in Eq. 10.77 becomes

$$\begin{aligned}\mathcal{F} = &\sum_{\Delta V_i\in\Omega} \phi(\boldsymbol{u}_i, \nabla\boldsymbol{u}_i, \nabla^2\boldsymbol{u}_i, \nabla^3\boldsymbol{u}_i)\Delta V_i \\ &+ \sum_{\Delta S_i\in\partial\Omega_D^0} \alpha_1(\boldsymbol{u}_i-\bar{\boldsymbol{u}}_i)\cdot(\boldsymbol{u}_i-\bar{\boldsymbol{u}}_i)\Delta S_i - \sum_{\Delta S_i\in\partial\Omega_N^0} \bar{\boldsymbol{P}}_i\cdot\boldsymbol{u}_i\Delta S_i \\ &+ \sum_{\Delta S_i\in\partial\Omega_D^1} \alpha_2(\nabla_n\boldsymbol{u}_i - \overline{\nabla_n\boldsymbol{u}_i}) : (\nabla_n\boldsymbol{u}_i - \overline{\nabla_n\boldsymbol{u}_i})\Delta S_i - \sum_{\Delta S_i\in\partial\Omega_N^1} \bar{\boldsymbol{Q}}_i : \nabla_n\boldsymbol{u}_i\Delta S_i\end{aligned} \tag{10.78}$$

The differential derivatives in ϕ of E^3 elasticity with unknowns $\boldsymbol{u} = (u, v)^T$ in material coordinates $\boldsymbol{X} = (X, Y)$ are

$$\begin{aligned}\partial U_{2d} =\Big(& u_{,Y}, v_{,Y}, u_{,X}, v_{,X}, u_{,YY}, v_{,YY}, u_{,XY}, v_{,XY}, u_{,XX}, v_{,XX}, u_{,YYY}, v_{,YYY},\\ & u_{,XYY}, v_{,XYY}, u_{,XXY}, v_{,XXY}, u_{,XXX}, v_{,XXX}, u_{,YYYY}, v_{,YYYY}, u_{,XYYY},\\ & v_{,XYYY}, u_{,XXYY}, v_{,XXYY}, u_{,XXXY}, v_{,XXXY}, u_{,XXXX}, v_{,XXXX}\Big)^T \end{aligned} \quad (10.79)$$

The differential derivatives in $\mathcal{F}$ of E^3 elasticity with unknowns $\boldsymbol{u} = (u, v, w)^T$ in material coordinates $\boldsymbol{X} = (X, Y, Z)$ are

$$\partial U_{3d} =\Big(\underbrace{u_{,Z}, v_{,Z}, w_{,Z}, ..., v_{,XY}, ..., u_{,XYZ}, ..., w_{,XXXX}}_{102\ terms}\Big)^T \quad (10.80)$$

In NOM, the differential derivatives can be written as $\partial \boldsymbol{U}_s = \boldsymbol{B}_s \boldsymbol{U}, s \in \{2d, 3d\}$, where $\boldsymbol{B}_s$ is the operator matrix constructed with steps similar to that in Eq. 10.71 and $\boldsymbol{U}$ is the vector form of all unknowns in one support. In other words, for a given equivalent energy functional, the independent derivatives of various orders can be extracted, which form a subset of the list of nonlocal derivatives provided by NOM. The operator matrix $\boldsymbol{B}_s$ is formed by selecting a row with the same index of derivative in $\tilde{\partial}_\alpha u_i$. The residual vector and tangent stiffness matrix for one particle can be obtained as

$$\boldsymbol{R}_i = \frac{\partial \phi_i}{\partial \boldsymbol{U}} = \frac{\partial(\partial \boldsymbol{U})}{\partial \boldsymbol{U}} \frac{\partial \phi_i}{\partial(\partial \boldsymbol{U})} = \boldsymbol{B}^T \frac{\partial \phi_i}{\partial(\partial \boldsymbol{U})} \quad (10.81)$$

$$\boldsymbol{K}_i = \frac{\partial \boldsymbol{R}_i}{\partial \boldsymbol{U}} = \boldsymbol{B}^T \frac{\partial^2 \phi_i}{\partial(\partial \boldsymbol{U})^2} \boldsymbol{B} \quad (10.82)$$

The explicit forms of $\frac{\partial \phi}{\partial(\partial \boldsymbol{U})}$ and $\frac{\partial^2 \phi}{\partial(\partial \boldsymbol{U})^2}$ can be obtained by softwares such as Mathematica (Wolfram 1999) allowing symbolic operations. For simplicity, we omit these lengthy expressions in the paper. However, the code will be made available. One can see that the construction of the residual vector and tangent stiffness matrix for each particle is a series of matrix multiplications. The global tangent stiffness matrix for the functional in domain Ω can be expressed as

$$\boldsymbol{R}_\Omega = \sum_{\Delta V_i \in \Omega} \boldsymbol{R}_i \Delta V_i, \qquad \boldsymbol{K}_\Omega = \sum_{\Delta V_i \in \Omega} \boldsymbol{K}_i \Delta V_i \quad (10.83)$$

The assembly of $\boldsymbol{R}_i, \boldsymbol{K}_i$ is based on the global indices of all unknowns in one support. The global tangent stiffness matrices (e.g. $\boldsymbol{K}_{\partial\Omega_D^0}, \boldsymbol{K}_{\partial\Omega_D^1}$) and residuals (e.g. $\boldsymbol{R}_{\partial\Omega_D^0}, \boldsymbol{R}_{\partial\Omega_D^1}$) for functionals on boundaries $\partial\Omega_D^0, \partial\Omega_D^1$ can be obtained in the same manner. The Neumann boundary condition on $\partial\Omega_N^0$ can be applied directly on the particles. The moment boundary condition on $\partial\Omega_N^1$ can be enforced by calculating the residual

$$\boldsymbol{R}_{\partial\Omega_N^1} = \sum_{\Delta S_i \in \partial\Omega_N^1} \frac{\partial \phi_N}{\partial(\partial \boldsymbol{U}_N)} \frac{\partial(\partial \boldsymbol{U}_N)}{\partial \boldsymbol{U}} \Delta S_i = \sum_{\Delta S_i \in \partial\Omega_N^1} \boldsymbol{B}_N^T \frac{\partial \phi_N}{\partial(\partial \boldsymbol{U})} \Delta S_i \tag{10.84}$$

where $\phi_N = \bar{\boldsymbol{Q}} : \nabla_n \boldsymbol{u}$ and $\partial \boldsymbol{U}_N = \nabla \otimes \boldsymbol{u} = \boldsymbol{B}_N \boldsymbol{U}$; $\boldsymbol{B}_N$ are constructed by selecting the first 2 rows or 3 rows of $\boldsymbol{B}$ in 2D or 3D, respectively; $\frac{\partial \phi_N}{\partial(\partial \boldsymbol{U}_N)}$ can be obtained by Mathematica (Wolfram 1999). Then, the global tangent stiffness matrix and residual are

$$\boldsymbol{R} = \boldsymbol{R}_{\Omega} + \boldsymbol{R}_{\partial\Omega_D^0} + \boldsymbol{R}_{\partial\Omega_D^1} + \boldsymbol{R}_{\partial\Omega_N^1} \tag{10.85}$$

$$\boldsymbol{K} = \boldsymbol{K}_{\Omega} + \boldsymbol{K}_{\partial\Omega_D^0} + \boldsymbol{K}_{\partial\Omega_D^1} \tag{10.86}$$

With the global residual and tangent stiffness available, a standard Newton Raphson method can be used to find the solution.

10.4 Numerical Examples

In this section, we present several representative numerical examples to study the property of the E^n elasticity theory. The setup of the 2D/3D examples and the associated boundary conditions are outlined in Fig. 10.3. The domain is discretized with a Cartesian grid. The particles in the void domain are removed to form holes. Each particle has the same number of neighbouring particles in the support, and the support size is selected as the distance between the furthest neighbour particle and the master particle in the support. So the central particles have a smaller support size compared to the particles close to the boundary. The number of neighbours in support is selected as $1/2(n^2 + 33n + 32)$ in 2D, where n is the order of gradient elasticity. The material parameters and length scales will be given in the subsections.

10.4.1 Convergence of Strain Energy in E^3 Elasticity

The first example tests the strain energy distribution of E^3 elasticity for different discretizations. The material parameters are elastic modulus $E = 30$ GPa and Poisson's ratio $\nu = 0.3$. Plane stress conditions are assumed. The internal length scales are set to $l_1 = l_2 = l_3 = 0.05$. The geometry and boundary conditions are depicted in Fig. 10.3a. The left side of the plate is fixed in all directions and a uniform tension load of $p = 1$ GPa·m is applied on the right side, which results in a moderate deformation. Different discretizations such as $40^2, 60^2, 80^2, 96^2, 120^2, 160^2, 200^2$ particles are used to study the distribution of the strain energy of different orders. The

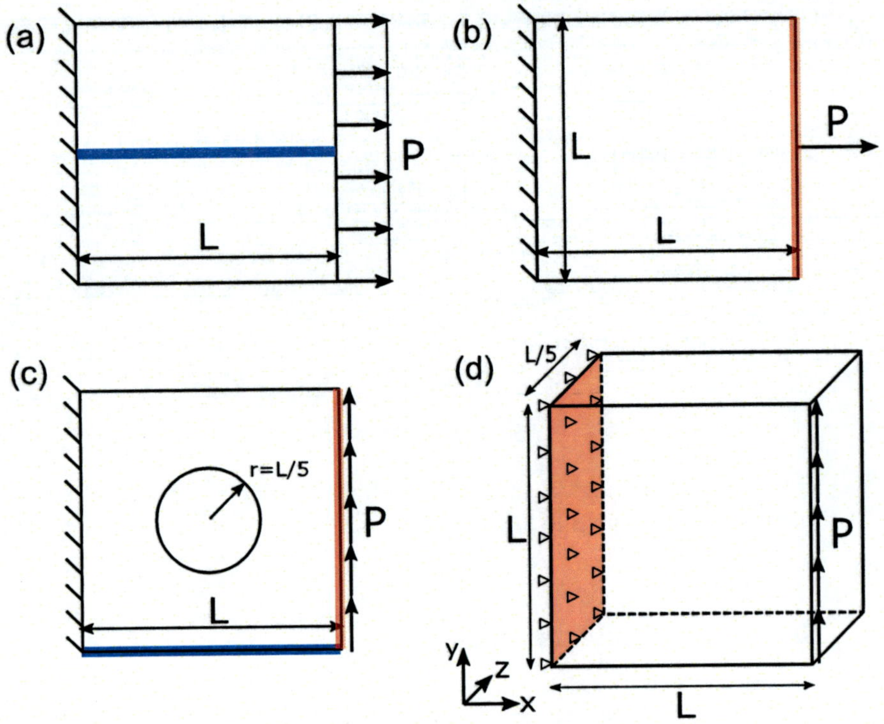

Fig. 10.3 The setup of the 2D plate and 3D plate and boundary conditions

engineering strain is approximately 0.0288 as depicted in Fig. 10.4a. The distribution of total strain energy density on each particle is given in Fig. 10.4b. The maximal strain energy density occurs around the corners. The total strain energies on different strain gradient orders can be found in Fig. 10.5. With increasing the number of particles, the energies of different levels converge. The strain energy is dominant while the higher order energies tend to decrease with increasing gradient orders. Indeed, the deformation under pure tension load is "uniform" for this numerical example.

10.4.2 2D Plate with Uniform Deformation

The second example tests the influence of E^n gradient elasticity subjected to a uniform load; E^0, E^1, E^2 elasticity theories are implemented. The geometry and boundary conditions are illustrated in Fig. 10.3a. A plate with dimensions of $1 \times 1\, m^2$ is discretized into 81^2 particles. The material parameters are elastic modulus $E = 30$ GPa and Poisson's ratio $\nu = 0.3$. The internal length scales are $l_1 = l_2 = 0.05$. The left side of the plate is fixed in all directions, and the right side is subjected to a

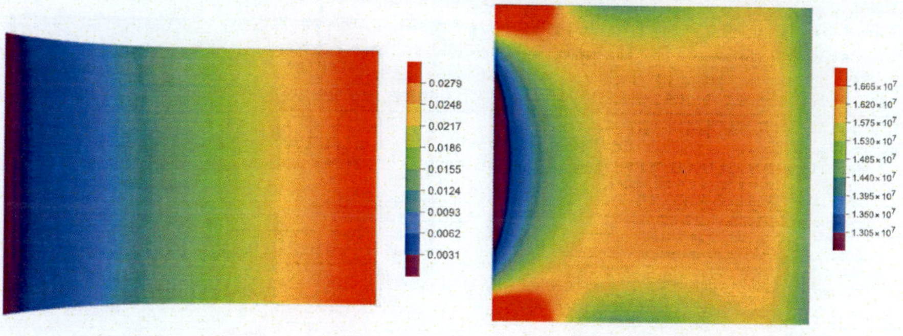

(a) (b)

Fig. 10.4 **a** Deformation with unit of meters in x-direction, scaled by 10 times and **b** the distribution of total strain energy density with units of Joule per unit volume for discretization of 120^2 particles

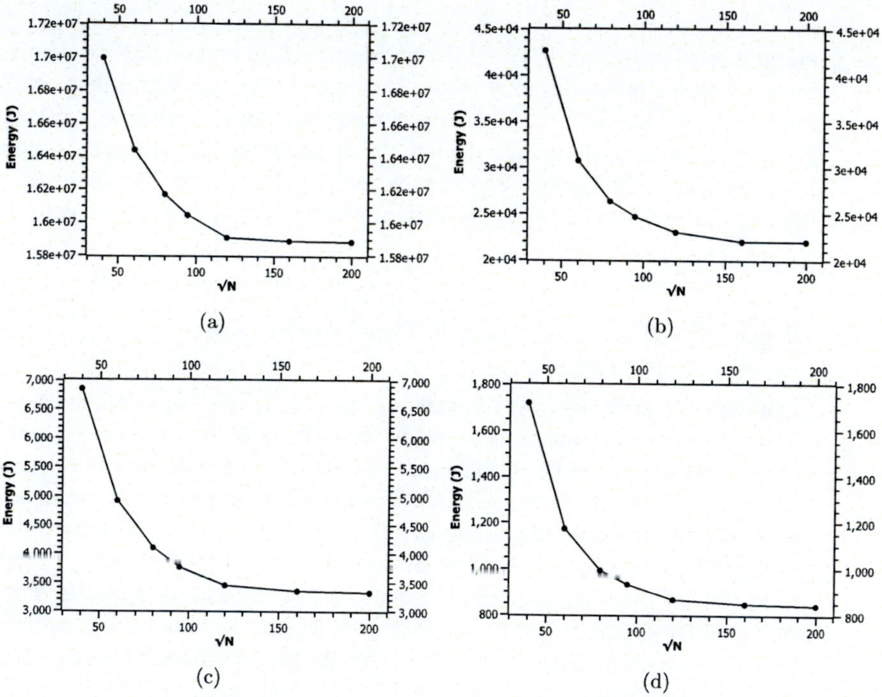

(a) (b)

(c) (d)

Fig. 10.5 Total strain energy on level **a** $\frac{1}{2}\boldsymbol{S} : \boldsymbol{E}$; **b** $\frac{1}{2}l_1^2\nabla\boldsymbol{S}\vdots\nabla\boldsymbol{E}$; **c** $\frac{1}{2}l_2^4\nabla^2\boldsymbol{S} :: \nabla^2\boldsymbol{E}$ and **d** $\frac{1}{2}l_3^6\nabla^3\boldsymbol{S} \cdot^{(5)} \nabla^3\boldsymbol{E}$; N is the total number of particles

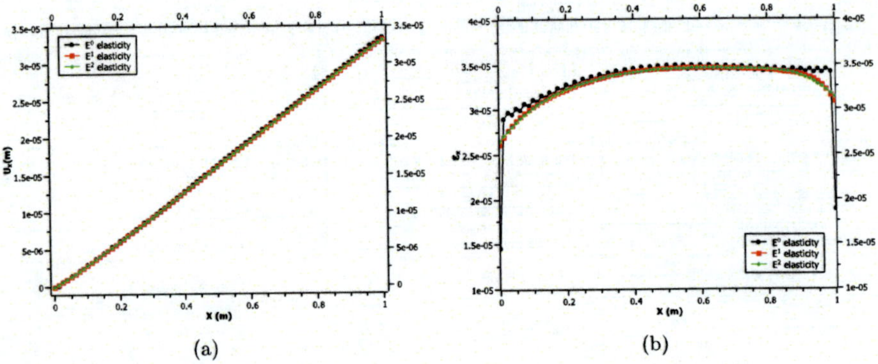

Fig. 10.6 **a** Displacement in x-direction and **b** strain in x-direction for particles on middle horizontal line (Blue line in Fig. 10.3a)

uniform tension load of $p = 1$ MPa/m. Figure 10.6a shows that the displacement based on higher order elasticity theory is identical to conventional elasticity for uniform deformations since the higher order strain components are quite small such that their contribution to the energy density can be neglected. However, the higher order terms make the deformation smoother as shown in Fig. 10.6b. This indicates that the higher order gradient elasticity should be tested with in-homogeneous deformations.

10.4.3 2D Plate Subjected to Point Force

Let us test the capability of gradient theory for point loads. We adopt the dimensions of the plate and its material parameters from the previous subsections. However, one particle in the middle of the right side boundary of the plate is subjected to a point force of 1000 N. The geometry and boundary conditions are depicted in Fig. 10.3b. The plate is discretized into 81x81 particles. $E^1 - E^4$ elasticity theories are considered. The deformations of the plate for E^1, E^2, E^3, E^4 elasticity can be found in Fig. 10.7. Obviously, gradient elasticity can 'withstand' point loads. The higher order gradient elasticity has a smoother displacement field compared with gradient elasticity. This observation is consistent with the numerical analysis by FEM (Reiher et al. 2017) and by IGA (Makvandi et al. 2018). Comparisons of the displacement in x-direction of particles on the right side boundary (i.e.the red line in Fig. 10.3b) are plotted in Fig. 10.8. The first-order and second-order derivatives of the displacement in x-direction of particles are shown in Fig. 10.9. The derivative of the displacement in E^1 elasticity changes sharply, in contrast to the smooth transition of the displacement gradient in higher order gradient elasticity.

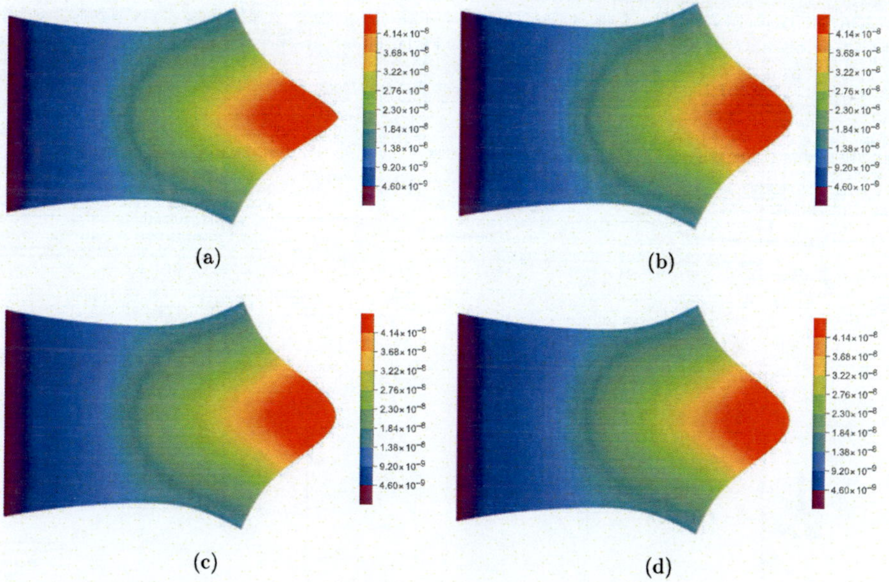

Fig. 10.7 Displacement with unit of meters in x-direction of the plate with deformation scaled by 10^7 times for **a** E^1 elasticity, **b** E^2 elasticity, **c** E^3 elasticity and **d** E^4 elasticity

Fig. 10.8 The deformation of all particles on the red line in Fig. 10.3b

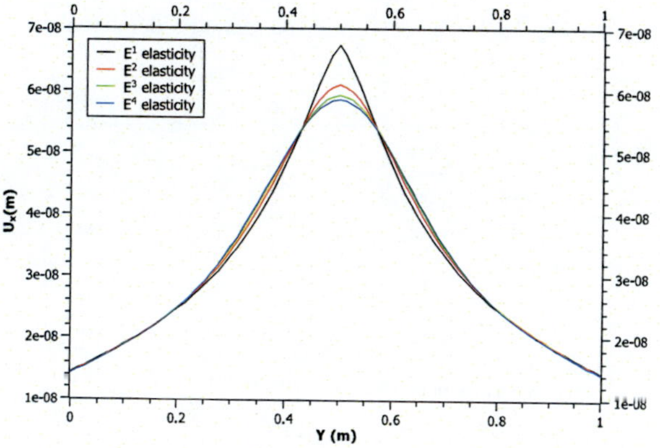

10.4.4 Plate with a Hole: Influence of Length Scales

This example deals with a plate with a hole of radius 0.2 m located at the center. The geometry and boundary conditions are depicted in Fig. 10.3c. The same material parameters as before are used. The plate is discretized into 81x81 particles and then the particles falling inside the circle are removed. With different length scales, the displacement in x-direction of all particles on the right boundary of the plate based on (a) E^1 elasticity and (b) E^2 elasticity are shown in Fig. 10.10. Higher length scale

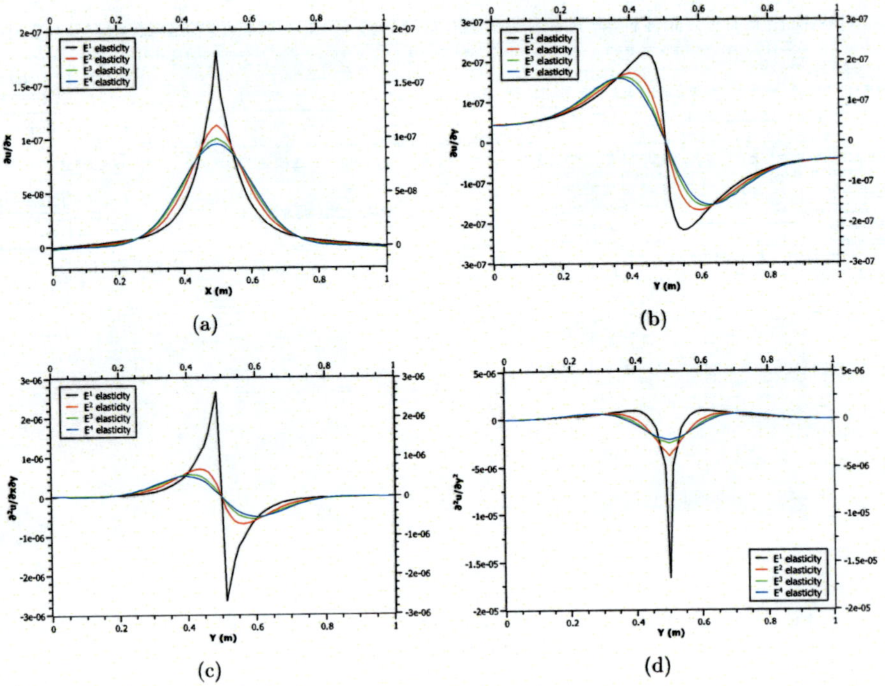

Fig. 10.9 The first-order and second-order derivative of deformation of all particles on the red line in Fig. 10.3b

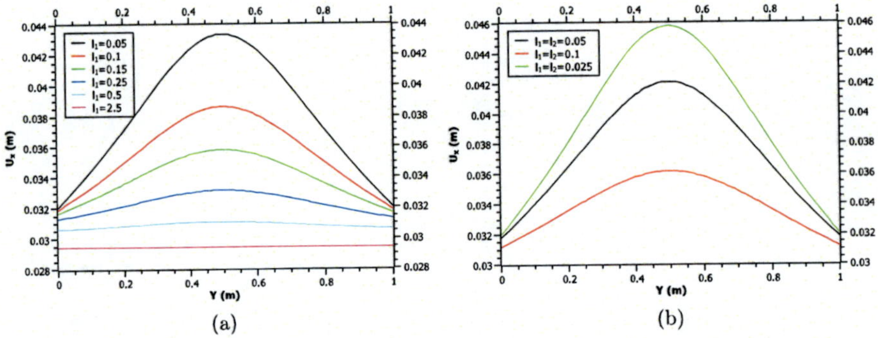

Fig. 10.10 The displacement in x-direction of particles on right side boundary of the plate based on **a** E^1 elasticity and **b** E^2 elasticity

parameters can significantly reduce the stress concentration induced by the hole. The gradient of the displacement in x-direction of E^1 elasticity is shown in Fig. 10.11, which means that a larger l_1 smoothes the strain field. The displacement gradient in x-direction of E^2 elasticity can be found in Fig. 10.12. The larger l_i, the smaller the strain field.

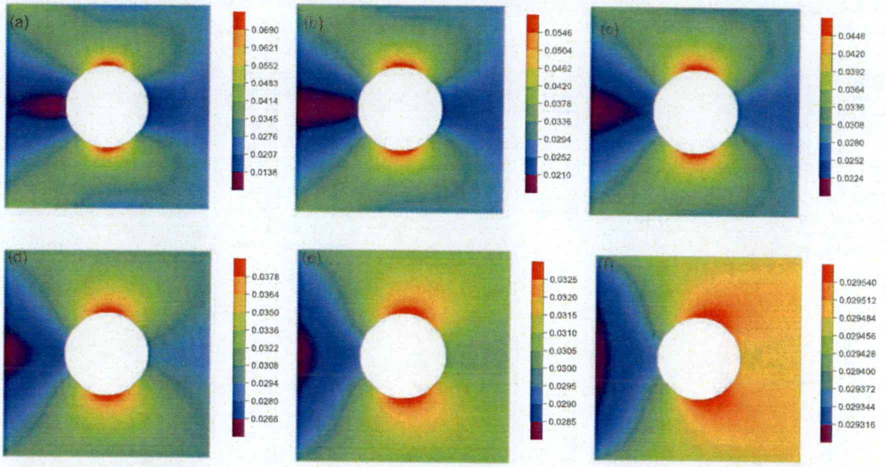

Fig. 10.11 $\frac{\partial u}{\partial X}$ distribution based on E^1 elasticity for **a** $l_1 = 0.05$, **b** $l_1 = 0.1$, **c** $l_1 = 0.15$, **d** $l_1 = 0.25$, **e** $l_1 = 0.5$, **f** $l_1 = 2.5$

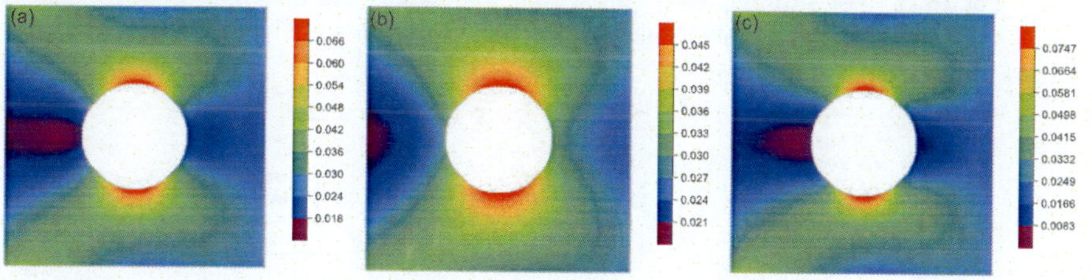

Fig. 10.12 $\frac{\partial u}{\partial X}$ distribution based on E^2 elasticity for **a** $l_1 = l_2 = 0.05$, **b** $l_1 = l_2 = 0.1$, **c** $l_1 = l_2 = 0.025$

10.4.5 Large Deformation of 2D Plate with a Hole

Again, we adopt the material parameters and plate dimensions from the previous example and study the deformation of a 2D plate with a hole based on E^n elasticity, with n=(0,1,2,3,4,5). The geometry and boundary conditions are illustrated in Fig. 10.3c. The plate is discretized into 81x81 particles, and the particles in the hole are removed. A shear load $p = T$ GPa·m, where $T \in [0, 3]$ is the time step, is applied on the right side boundary of the plate. The length scales are selected as $l_1 = l_2 = 0.05$ for E^2 elasticity, $l_1 = l_2 = l_3 = l_4 = 0.05$ for E^4 elasticity and $l_1 = l_2 = l_3 = l_4 = l_5 = 0.05$ for E^5 elasticity. The displacements at step $T = 1$ are plotted in Fig. 10.13, where E^0 elasticity has the largest deformation and the deformations by E^2, E^3, E^4, E^5 elasticity are similar.

Figure 10.14 shows that the displacement in y-direction of particle on the bottom line (e.g. the blue line in Fig. 10.3c). It can be seen that the higher order gradient

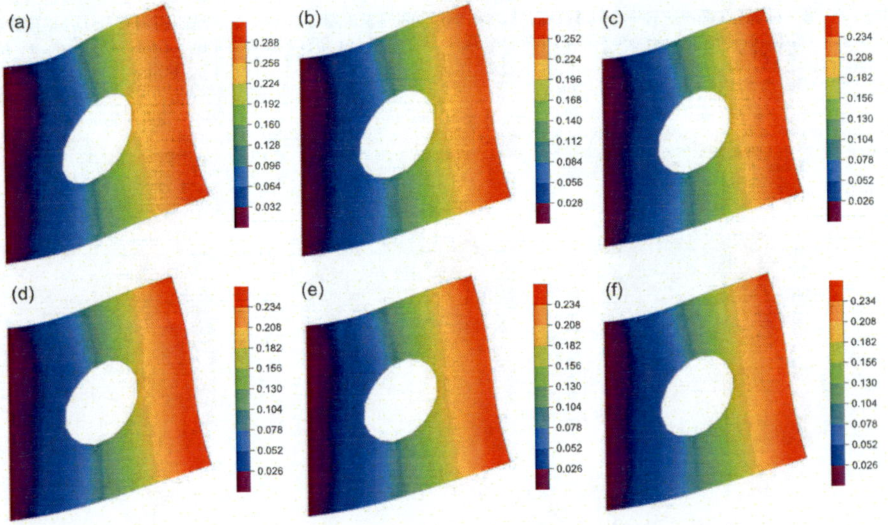

Fig. 10.13 The deformations of the plate at $T = 1$ for **a** E^0 elasticity, **b** E^1 elasticity, **c** E^2 elasticity, **d** E^3 elasticity, **e** E^4 elasticity and **f** E^5 elasticity, respectively. Unit:meters

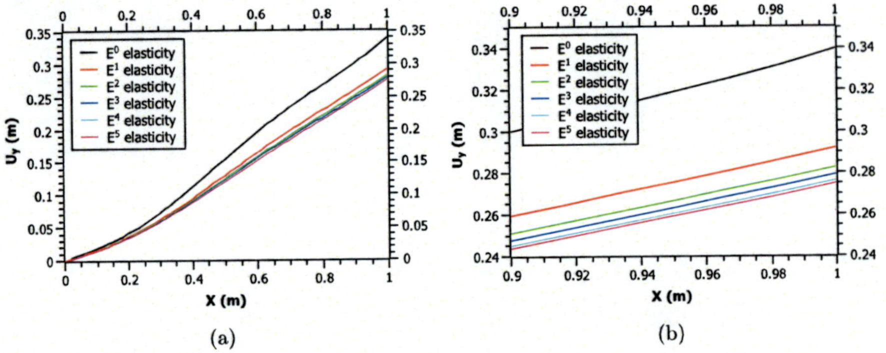

Fig. 10.14 The displacement in y-direction of all particles on the right side boundary of the plate with load level $T = 1$, where the lines in **b** are magnified from **a**

theory has smaller deformation. The difference becomes smaller when the order of gradient elasticity increases.

Contour plots of displacement gradients for E^0, ..., E^5 elasticity are plotted in Figs. 10.15 and 10.16. Higher order elasticity exhibits a very smooth displacement gradient. The gradient of the displacement field for hyperelasticity (E^0 elasticity) is not smooth around the internal line. This is due to the fact that the first order NOM is used, which is continuous in the displacement but discontinuous in its derivative.

Although the deformations for different elasticity theories at $T = 1$ are similar, the final converged deformations are different. The final load level of the plate occurs approximately at $T_0 = 0.96875$, $T_1 = 2.0$, $T_2 = 2.25$, $T_3 = 2.5$, $T_4 = 2.5$, $T_5 = 2.5$

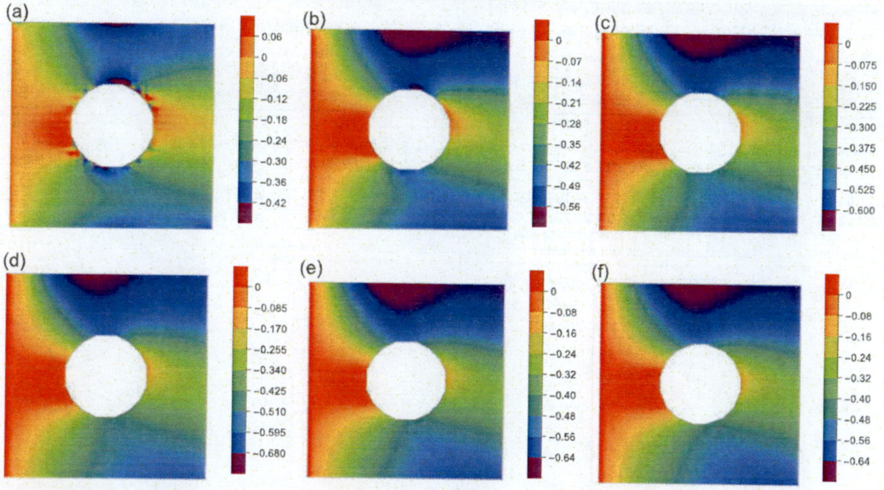

Fig. 10.15 $\frac{\partial u}{\partial Y}$ at $T = 1$ for E^0, ..., E^5 elasticity

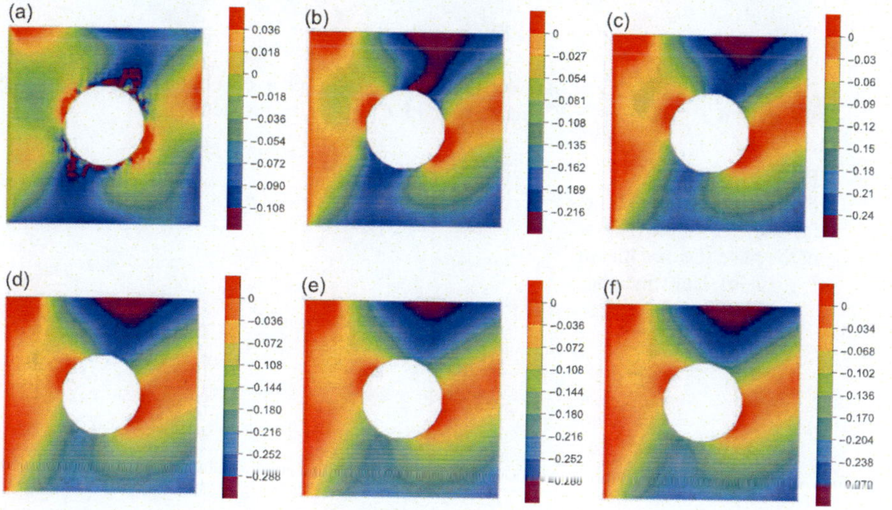

Fig. 10.16 $\frac{\partial v}{\partial Y}$ at $T = 1$ for E^0, ..., E^5 elasticity

for E^0, E^1, ..., E^5 elasticity, respectively. The final configurations of the plate can be found in Fig. 10.17. The displacement in y-direction for particles on the right side of the plate is depicted in Fig. 10.18.

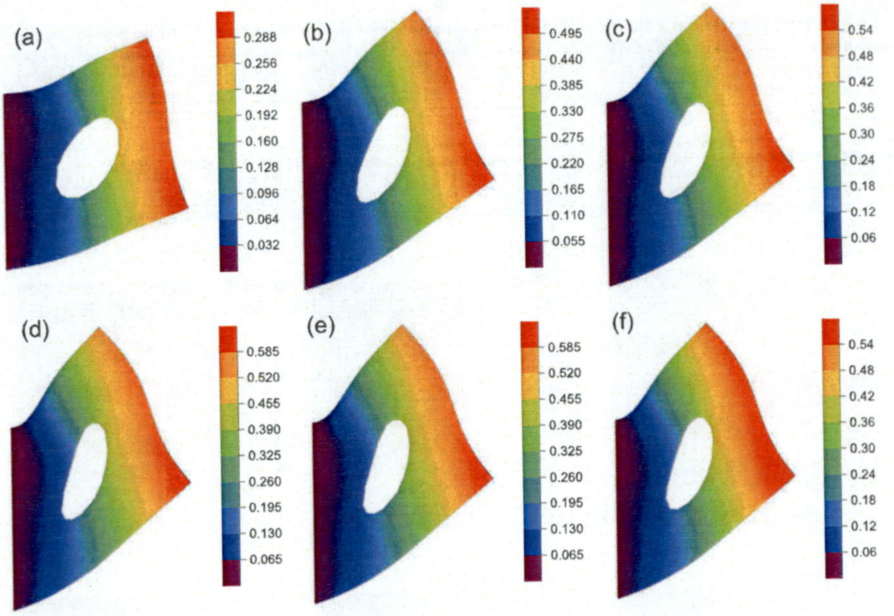

Fig. 10.17 The converged final deformations of plate for **a** E^0 elasticity, **b** E^1 elasticity, **c** E^2 elasticity, **d** E^3 elasticity, **e** E^4 elasticity and **f** E^5 elasticity, respectively

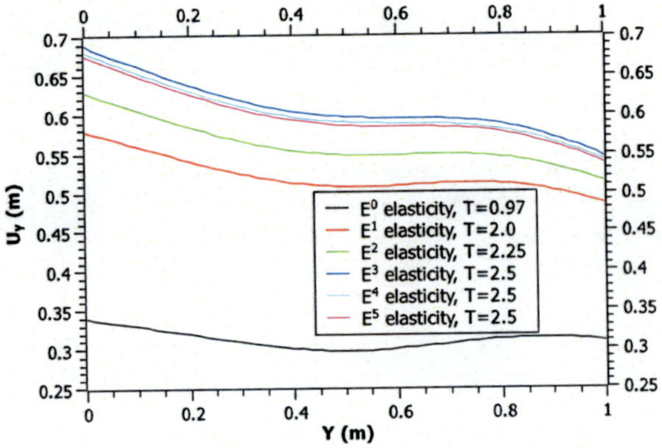

Fig. 10.18 The displacement in y-direction for particles on the right side of the plate (i.e. red line in Fig. 10.3c). Units:meters

10.4.6 Large Deformation of 3D Plate Subjected to Line Load

Finally, we present a large deformation example in 3D, i.e. a 3D thick plate based on E^3 elasticity. The geometry and boundary conditions are depicted in Fig. 10.3d. The particles in the red segment are fixed in all directions. A line force density of $p = 10^8$ N/m is applied on the particles located on the line. The load level T increases from 0 to 3. The differential operators in 3D E^3 elasticity are given in Eq. 10.80. The plate with dimensions of $1 \times 1 \times 0.2\,\mathrm{m}^3$ is discretized into $41 \times 41 \times 9 = 15129$

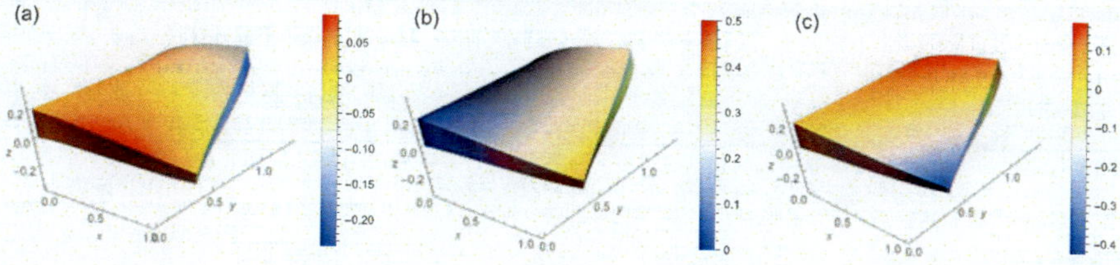

Fig. 10.19 The displacement field of (x, y, z) directions in deformed configuration. Unit:meters

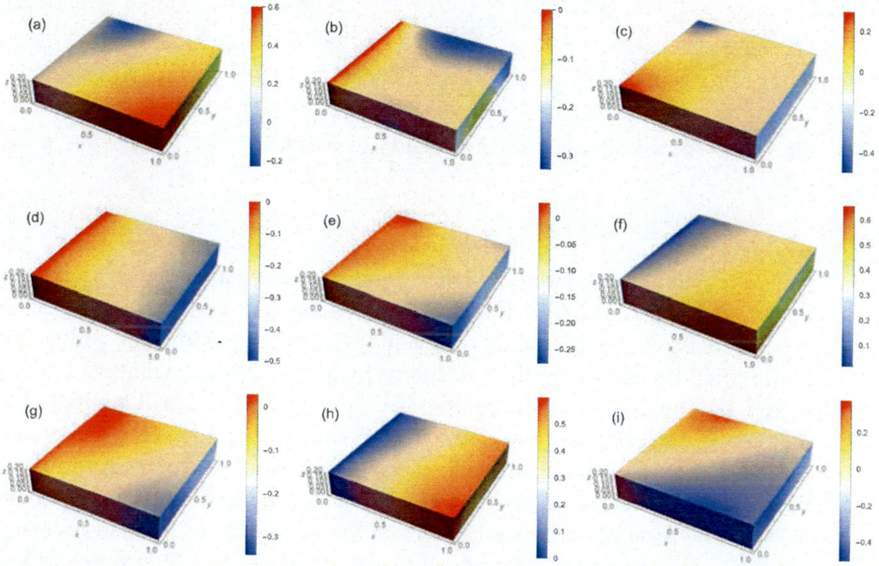

Fig. 10.20 The gradient of displacement field **a** $\frac{\partial u}{\partial Z}$, **b** $\frac{\partial u}{\partial Y}$, **c** $\frac{\partial u}{\partial X}$, **d** $\frac{\partial v}{\partial Z}$, **e** $\frac{\partial v}{\partial Y}$, **f** $\frac{\partial v}{\partial X}$, **g** $\frac{\partial w}{\partial Z}$, **h** $\frac{\partial w}{\partial Y}$, **i** $\frac{\partial w}{\partial X}$

particles. The support of each particle consists of 124 nearest neighbors. The material parameters from the previous examples are adopted. The length scale parameters are selected as $l_1 = l_2 = l_3 = 0.05$. The final deformation is plotted in Fig. 10.19 with the displacement fields shown in each sub-figure. The displacement gradient fields are plotted in Fig. 10.20. The displacement second-gradient fields are plotted in Fig. 10.21.

10.5 Conclusions

We have proposed an objective energy functional for finite deformation higher order gradient elasticity. The energy functional is based on the setting of the second Piola-Kirchhoff stress which is invariant under rigid body transformations. More specifi-

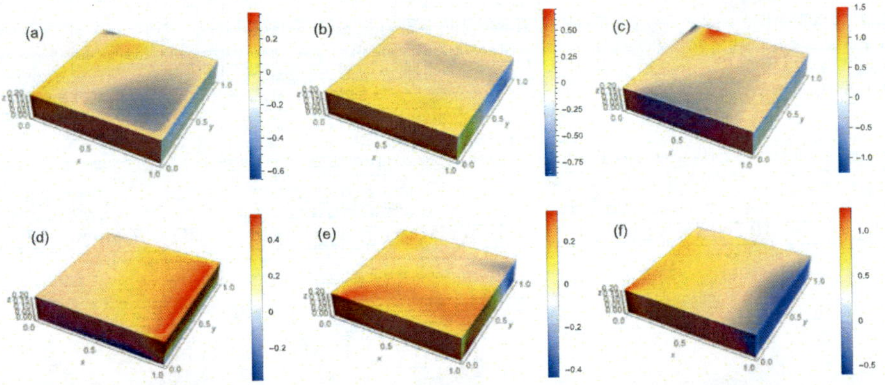

Fig. 10.21 The second-order derivatives of displacement field **a** $\frac{\partial^2 u}{\partial Z^2}$, **b** $\frac{\partial^2 u}{\partial Y^2}$, **c** $\frac{\partial^2 u}{\partial X^2}$, **d** $\frac{\partial^2 v}{\partial Z^2}$, **e** $\frac{\partial^2 v}{\partial Y^2}$, **f** $\frac{\partial^2 v}{\partial X^2}$

cally, the geometric nonlinear higher order gradient elasticity theory is formulated on the gradients of the right Cauchy Green tensor. The general form of higher order gradient elasticity may contain thousands of material parameters, and we proposed a simplified version of gradient elasticity. Such simplification reduces the number of material parameters from 10 thousands to less than 10. A small number of material parameters can greatly simplify the experiment measurement and numerical implementation. The framework of gradient elasticity also allows for other forms of simplification of material parameters. We employed the nonlocal local operator method and Newton Raphson iteration method to find the numerical solution of higher gradient elasticity. The properties of gradient elasticity are studied by a series of numerical experiments. The numerical tests show that gradient elasticity can sustain point/-line load without stress singularity. The mechanical response greatly depends on the internal length scales of gradient elasticity. Larger internal length scale induces a smaller and smoother deformation. Higher order gradient elasticity is numerically more stable and allows for larger ultimate load for the same structure. In the next stage, more physics-related research including the calibration of material parameters by experiments and numerical simulation, and the size effect, surface effect in metamaterials and gradient elasticity will be pursued. Some outlooks based on current research include, for example,

1. The higher order gradient elastoplasticity theory (Reiher and Bertram 2020) and its numerical implementation. Current research is restricted to elasticity with finite deformation, and it cannot be applied to a dissipated system involving permanent deformation or irreversible process. The extension of higher order elasticity to higher order plasticity can broaden the range of plasticity theory.
2. More clear relationship between metamaterial and gradient elasticity is expected (Khakalo et al. 2018; Yang et al. 2020). One salient feature of gradient elasticity is the micro-structure, which is essential to the theory of metamaterials as well. Direct simulation of micro-structure requires tremendous computer power.

Gradient elasticity circumvents these difficulties by introducing certain internal length scales, which however require more sophisticated measurement.

3. The wave propagation analysis of gradient elasticity. Gradient elasticity has the capability to account for interesting phenomena such as size effect, surface effect and nonlocal effect. These features may give rise to some abnormal physical phenomenon, which can be exploited to design some smart devices.

References

Askes H, Aifantis EC (2009) Gradient elasticity and flexural wave dispersion in carbon nanotubes. Phys Rev B 80(19):195412

Bonet J, Wood RD (1997) Nonlinear continuum mechanics for finite element analysis. Cambridge University Press

Gurtin ME, Murdoch AI (1975) A continuum theory of elastic material surfaces. Arch Ration Mech Anal 57(4):291–323

Javili A, dell'Isola F, Steinmann P (2013) Geometrically nonlinear higher-gradient elasticity with energetic boundaries. J Mech Phys Solids 61(12):2381–2401

Khakalo S, Balobanov V, Niiranen J (2018) Modelling size-dependent bending, buckling and vibrations of 2d triangular lattices by strain gradient elasticity models: applications to sandwich beams and auxetics. Int J Eng Sci 127:33–52

Korelc J, Wriggers P (2016) Automation of finite element methods. Springer

Makvandi R, Reiher JC, Bertram A, Juhre D (2018) Isogeometric analysis of first and second strain gradient elasticity. Comput Mech 61(3):351–363

Mindlin RD (1964) Micro-structure in linear elasticity. Arch Ration Mech Anal 16:51–78. ISSN 0003-9527. https://doi.org/10.1007/BF00248490

Polizzotto C (2013) A second strain gradient elasticity theory with second velocity gradient inertia-part i: Constitutive equations and quasi-static behavior. Int J Solids Struct 50(24):3749–3765

Polizzotto C (2013) A second strain gradient elasticity theory with second velocity gradient inertia-part ii: Dynamic behavior. Int J Solids Struct 50(24):3766–3777

Reiher JC, Bertram A (2020) Finite third-order gradient elastoplasticity and thermoplasticity. J Elast 138(2):169–193

Reiher JC, Giorgio I, Bertram A (2017) Finite-element analysis of polyhedra under point and line forces in second-strain gradient elasticity. J Eng Mech 143(2):04016112

Reiher JC, Giorgio I, Bertram A (2017) Finite-element analysis of polyhedra under point and line forces in second-strain gradient elasticity. J Eng Mech 143(2):04016112

Ren H, Zhuang X, Rabczuk T (2020) A higher order nonlocal operator method for solving partial differential equations. Comput Methods Appl Mech Eng 367:113132

Shen S, Hu S (2010) A theory of flexoelectricity with surface effect for elastic dielectrics. J Mech Phys Solids 58(5):665–677

Khakalo S, Balobanov V, Niiranen J (2018) Modelling size-dependent bending, buckling and vibrations of 2d triangular lattices by strain gradient elasticity models: applications to sandwich beams and auxetics. Int J Eng Sci 127:33–52

Yang H, Timofeev D, Giorgio I, Müller WH (2020) Effective strain gradient continuum model of metamaterials and size effects analysis. Continuum Mech Thermodyn 1–23

Appendix A
Preliminary of Mathematica

A.1 Preliminary of Mathematica

Wolfram Mathematica (abbreviation: Mathematica) is a scientific computing software, sometimes called a computer algebra system, widely used in science, engineering, mathematics, computing and other fields. It was conceived by British physicist Stephen Wolfram and developed by the Wolfram Research Company (located in Champaign, Illinois, USA) under his leadership. It has powerful numerical calculation and symbolic calculation capabilities, and is one of the most widely used mathematical software so far. In this chapter, we will very briefly describe some basic functions in Mathematica.

When the Wolfram System is first started, it displays an empty notebook with a blinking cursor. You can start typing right away. The Wolfram Language by default will interpret your text as input. You enter Wolfram Language input into the notebook, then type "Shift + Enter" to make the Wolfram Language process your input. (To type Shift + Enter, hold down the Shift key, then press Enter.) You can use the standard editing features of your graphical interface to prepare your input, which may go on for several lines. Shift + Enter tells the Wolfram System that you have finished your input. If your keyboard has a numeric keypad, you can use its Enter key instead of Shift + Enter. We can add a comment with "(*your comment*)" anywhere in the notebook. In the following, we will briefly describe some useful functions in Mathematica.

Mathematica can handle symbolic computations

```
In[1]:= Expand[(x+1)^2](*Expand a function*)
        Factor[1+ 2 x+x^2](*factor a polynomial*)
        Factor[x^10+x^5+1](*Try to factorize the expression x^10+x^5+1*)
Out[1]= 1+2 x+x^2
Out[2]= (1+x)^2
Out[3]= (1+x+x^2) (1-x+x^3-x^4+x^5-x^7+x^8)
```

We use variables to represent data. Names like x, y, x3 and myfunc are all fine.

T. Rabczuk et al., *'Computational Methods based on Peridynamics and Nonlocal Operators*, Computational Methods in Engineering & the Sciences,
https://doi.org/10.1007/978-3-031-20906-2

The symbolic language paradigm of the Wolfram Language takes the concept of variables and functions to a new level. In the Wolfram Language, a variable can not only stand for a value but can also be used purely symbolically. And building on the Wolfram Language's powerful pattern language, "functions" can be defined not just to take arguments but also to transform a pattern with any structure. Primarily, there are several equalities in Mathematica.

- x=...—set a variable.
- f[x_]:=...—define a function that takes any single argument.
- Set (=) —immediate assignment (right-hand side evaluated immediately).
- SetDelayed (:=)—delayed assignment (right-hand side evaluated only when used).
- Unset (=.) —unset a variable.
- Clear —clear a function definition.

```
In[4]:= x=Random[](*immediate assignment*)
Out[4]= 0.232356
In[5]:= x(*no change*)
Out[5]= 0.232356
In[6]:= x2:=Random[](*delayed assignment*)
In[7]:= x2(*the value of x2 is computed when running x2*)
Out[7]= 0.00472814
In[8]:= x2 (*the value of x2 is computed when running x2*)
Out[8]= 0.348469
```

List is the way Mathematica handles information. Roughly speaking, a list is a collection of objects. The objects could be of any type and pattern. Let us start with an example of a list:

```
In[9]:= p={1,-2/3,stuff,1-2 x+x^2,Cuboid[]}
Out[9]= {1,-(2/3),stuff,0.589277,Cuboid[{0,0,0}]}
```

We can access the element in a list with index

```
In[10]:= p[[1]](*get the first element in p*)
         p[[-2]](*get the last second element in p*)
         p[[2;;3]](*get the 2~3 elements*)
         p[[{1,3,5}]](*get the first, third and 5th elements in p*)
         Length[p](*the length of p*)
Out[10]= 1
Out[11]= 0.589277
Out[12]= {-(2/3),stuff}
Out[13]= {1,stuff,Cuboid[{0,0,0}]}
Out[14]= 5
```

Create a list with Range

```
In[15]:= Range[10](*range 1-10*)
         Range[12,15](*range from 12-15*)
         Range[2,11,2](*range from 2 to 11 with an increment of 2*)
Out[15]= {1,2,3,4,5,6,7,8,9,10}
Out[16]= {12,13,14,15}
Out[17]= {2,4,6,8,10}
```

Create a table or a list

```
In[18]:= Table[i,{i,10}](*1D list from 1-10*)
         Table[10,{4}] (*1D list with each element being 10*)
         Table[2,{2},{3}](*2D list with dimensions of 2x3*)
Out[18]= {1,2,3,4,5,6,7,8,9,10}
Out[19]= {10,10,10,10}
Out[20]= {{2,2,2},{2,2,2}}
```

Two ways to add an element to a list:

```
In[41]:= Append[{a,b,c},d]
Out[41]= {a,b,c,d}
In[42]:= A={a,b,c};AppendTo[A,d];A
Out[42]= {a,b,c,d}
```

Create a constant array by **ConstantArray**

```
In[21]:= ConstantArray[10,2]
         ConstantArray[1,{2,3}]
         ConstantArray[2,{2,3,4}]
Out[21]= {10,10}
Out[22]= {{1,1,1},{1,1,1}}
Out[23]= {{{2,2,2,2},{2,2,2,2},{2,2,2,2}},{{2,2,2,2},{2,2,2,2},{2,2,2,2}}}
```

Create a list of polynomial

```
In[24]:= Clear[x];
         Table[(1+x)^i,{i,5}]
Out[25]= {1+x,(1+x)^2,(1+x)^3,(1+x)^4,(1+x)^5}
```

The equivalent shorthand to apply a function to a list is "@" as follows:

```
In[26]:= Sqrt/@{a,b,c}
Out[26]= {Sqrt[a],Sqrt[b],Sqrt[c]}
```

Applying a list of functions to the same variable is "Through"

```
In[27]:= Through[{f,g,h}[x]]
Out[27]= {f[x],g[x],h[x]}
```

Applying a function to each list of 2D list

```
In[28]:= f@@@{{1,2},{3,4}}
Out[28]= {f[1,2],f[3,4]}
```

Thread[f[args]] denotes "threads" f over any lists that appear in args:

```
In[1]:= Thread[f[{a,b,c}]]
Out[1]= {f[a],f[b],f[c]}
In[2]:= Thread[{a,b,c}>= 1]
Out[2]= {a>=1,b>=1,c>=1}
In[3]:= Thread[{a,b,c}==0]
```

```
Out[3]= {a==0,b==0,c==0}
In[4]:= Thread[{{a,b},{c,d}}==1]
Out[4]= {{a,b}==1,{c,d}==1}
In[5]:= Thread[Rule[{a,b,c},2]]
Out[5]= {a->2,b->2,c->2}
In[6]:= Thread[Rule[{a,b,c},{d,e,f}]]
Out[6]= {a->d,b->e,c->f}
```

A similar function to List is Array, which has a different usage.
Array[f, n] generates a list of length n, with elements x[i]:

```
In[7]:= Array[x,8](*It creates 8 variables from x.*)
Out[7]= {x[1],x[2],x[3],x[4],x[5],x[6],x[7],x[8]}
```

It is worth mentioning that x[i] and x[j] (j≠i) are independent variables.
Generate a 3*2 array:

```
In[8]:= Array[f,{3,2}]
Out[8]= {{f[1,1],f[1,2]},{f[2,1],f[2,2]},{f[3,1],f[3,2]}}

In[9]:= Clear[x];
x=Array[x,8];(*We cannot assign Array of x to the variable x.*)
During evaluation of In[9]:= $RecursionLimit::reclim2:
Recursion depth of 1024 exceeded during evaluation of x[1].
```

In[12]:= X=Array[x,8];(*But we can use a different variable X to denote the list of x*) Then X[[1]] denotes x[1]

```
In[13]:= X[[1]]
Out[13]= x[1]
```

With the aid of "Array", we can define a list of independent variables, for example, integer programming:

```
In[22]:= Clear[x,xi,nx,Mi];nx=10;
      Mi={{-16,-10,9,0,-20,5,-17,15,-15,-11},{2,0,14,-2,0,-2,-15,-5,-15,-6},
          {-20,-7,-4,12,7,2,-2,3,0,-10},{-4,-12,9,20,-1,-6,17,9,12,8},
          {18,8,16,16,-8,17,20,-18,-4,14},{7,-10,13,18,0,13,-16,9,-12,-20},
          {7,3,3,-14,10,13,0,0,9,19},{1,2,13,4,-16,19,-9,19,20,-6},
          {11,2,1,11,-2,7,-19,10,17,5},{-11,-6,-14,-3,14,-4,3,-1,9,18}};
      xi=FindMinimum[{Total[Array[x,nx]],Join[Thread[Mi.Array[x,nx]>1],
         Thread[Array[x,nx]>0],{Element[Array[x,nx],Integers]}]},Array[x,nx]]
Out[24]= {8.,{x[1]->0,x[2]->0,x[3]->2,x[4]->0,x[5]->2,
         x[6]->1,x[7]->0,x[8]->2,x[9]->0,x[10]->1}}
```

Defining functions and anonymous functions: Sometimes, we need to "define a function as we go" and use it on the spot. Here is an anonymous function equivalent to $f(x) = x^2 + 4$:

```
In[29]:= (#^2+4)&[5]
Out[29]= 29
```

The expression $(\#^2 + 4)\&$ defines a nameless function. As usual, we can plug in data in place of #. The symbol & determines where the definition of the function is completed. We proceed by defining the function $g(x) = x + cos(x)$

```
In[30]:= g[x_]:=x+Cos[x]
         g[2]
Out[31]= 2+Cos[2]
```

One can define the functions of several variables. Here is a simple example defining $f(x, y) = (x^2 + y^2)^{(1/2)}$

```
In[32]:= f[x_,y_]:=Sqrt[x^2+y^2]
f[3,4]
Out[33]= 5
```

Mod[m, n] gives the remainder on the division of m by n.

```
In[34]:= Mod[30,11]
Out[34]= 8
```

Quotient[m, n] gives the integer quotient of m and n.

```
In[35]:= Quotient[30,11]
Out[35]= 2
```

Mathematica has a very consistent way of dealing with any expression. Any expression in Mathematica has the following presentation "head[arg1, arg2, ..., argn]" where head and arg could be expressions themselves. We can use the command FullForm to get the true face of the expression:

```
In[36]:= FullForm[a+b+c]
Out[36]//FullForm= Plus[a,b,c]
In[37]:= FullForm[a*b*c]
Out[37]//FullForm= Times[a,b,c]
In[38]:= FullForm[{a,b,c}]
Out[38]//FullForm= List[a,b,c]
In[39]:= Clear[x]
In[40]:= ReplaceAll[Sin[x],Sin->Cos](*change the head*)
Out[40]= Cos[x]
```

Sum calculates the summation of a list:

```
In[43]:= Sum[s[i],{i,1,8}]
         Sum[s[i],{i,1,k}]
Out[43]= s[1]+s[2]+s[3]+s[4]+s[5]+s[6]+s[7]+s[8]
```
Out[44]= $\sum_{i=1}^{k} s[i]$

The command **Product** performs in the same way:

```
In[45]:= Product[s[i],{i,1,8}]
         Product[s[i],{i,1,k}]
Out[45]= s[1] s[2] s[3] s[4] s[5] s[6] s[7] s[8]
```
Out[46]= $\prod_{i=1}^{k} s[i]$

Loop

- Looping is a core concept in programming. The Wolfram Language provides powerful primitives for specifying and controlling looping
- Do—evaluate an expression looping over a variable: Do[expr,i,n].
- While—evaluate an expression while a criterion is true: While[crit,expr].
- For—a "for loop": For[init,test,incr,body].
- Table—build up a table by looping over variables: Table[expr,i,n].

Do uses the standard Wolfram Language iteration specification. You can use **Return, Break, Continue, and Throw** inside **Do**. Print the first four squares

```
In[47]:= Do[Print[n^2],{n,4}]
         1
         4
         9
         16
```

n goes from −3 to 5 in steps of 2:

```
In[48]:= Do[Print[n],{n,-3,5,2}]
         -3
         -1
         1
         3
         5
```

While[test, body] evaluates test, then body, repetitively, until the test first fails to give True. If Break[] is generated in the evaluation of body, the While loop exits. Continue[] exits the evaluation of body, and continues the loop. Print and increment n while the condition $n < 4$ is satisfied:

```
In[49]:= n=1;While[n<4,Print[n];n++]
         1
         2
         3
```

For[start, test, incr, body]
executes start, then repeatedly evaluates body and incr until the test fails to give True.

```
In[50]:= For[i=0,i<4,i++,Print[i]]
         0
         1
         2
         3
```

A comma delimits the parts of **For**; a semicolon delimits the parts of procedures:

```
In[51]:= For[i=1;t=x,i^2<10,i++,t=t^2+i;Print[t]]
        1+x^2
        2+(1+x^2)^2
        3+(2+(1+x^2)^2)^2
```

Do is much faster than **For** and **While**. Let us do some experiments:

```
In[52]:= Timing[Do[,{10^6}]]
Out[52]= {0.015625,Null}
In[53]:= Timing[i=1;While[i<10^6,i++]]
Out[53]= {0.609375,Null}
In[54]:= Timing[i=1;While[i<1000000,i++]]
Out[54]= {0.4375,Null}
In[55]:= Timing[For[i=1,i<10^6,i++]]
Out[55]= {0.640625,Null}
In[56]:= Timing[Do[,{1000000}]]
Out[56]= {0.015625,Null}
```

ReplaceAll

Replace a variable with a value:

```
In[57]:= {x,x^2,y,z}/.x->1
Out[57]= {1,1,y,z}
```

Replace heads of expressions:

```
In[58]:= {a,b,c}/.List->f
         Sin[x]/.Sin->Cos
Out[58]= f[a,b,c]
Out[59]= Cos[x]
```

A.1.1 Function Compile

The **Compile function** takes Mathematica code and allows you to pre-declare the types (real, complex, etc.) and structures (value, list, matrix, etc.) of input arguments. By doing so, Mathematica can optimize the program and create a byte code to run on its own virtual machine, which can get a really big speedup. Let's see some examples:

```
In[2]:= arg = Range[-50, 50, 0.25];
In[24]:= fn[x_] := Block[{sum = 1.0, inc = 1.0},
         Do[inc=x Cos[x*i];sum += inc, {i, 10000}];sum];
In[25]:= Table[fn[i], {i, arg}]; // AbsoluteTiming
Out[25]= {4.94008, Null}
```

Using **Compile** instead of **Function** makes the execution over 22.65 times faster

```
In[26]:= cfn = Compile[{x},
   Block[{sum = 1.0, inc = 1.0},
```

```
   Do[inc = x Cos[x*i]; sum += inc, {i, 10000}]; sum]];
In[28]:= Table[cfn[i], {i, arg}]; // AbsoluteTiming
Out[28]= {0.218017, Null}
```

When the **Listable** attribute is activated, the speedup can be increased with multicores.

```
In[27]:= cfn2 = Compile[{x},
   Block[{sum =1.0, inc=1.0}, Do[inc= x Cos[x*i]; sum+= inc,{i,10000}];
     sum], RuntimeAttributes -> {Listable}, Parallelization -> True];
In[34]:= cfn2[arg]; // AbsoluteTiming
Out[34]= {0.0735768, Null}
```

A.1.2 Velocity Verlet Algorithm

Velocity Verlet algorithm as an explicit time integration is often used in peridynamics and explicit NOM. The coordinates, velocity and acceleration are updated as follows:

$$\boldsymbol{x}(t+\Delta t)=\boldsymbol{x}(t)+\boldsymbol{v}(t)\,\Delta t+\frac{1}{2}\,\boldsymbol{a}(t)\Delta t^2, \tag{A.1}$$

$$\boldsymbol{v}(t+\Delta t)=\boldsymbol{v}(t)+\frac{\boldsymbol{a}(t)+\boldsymbol{a}(t+\Delta t)}{2}\Delta t. \tag{A.2}$$

A simple test for one degree of freedom written by Mathematica language is

```
In[62]:= (*https://en.wikipedia.org/wiki/Verlet_integration
          Test Verlet integration algorithm for one dof vibration*)
         outputInterval=2;j=1;
         tInc=4 10.^-3;
         coord=0.1;
         velo=0.;acc=0.;accNew=0.;
         coord0=0.;kk=1000;pmass=1;Ul={};Vl={};Al={};ctime=0.;
         Do[coord+=tInc velo+(0.5 tInc^2)acc;
              uvw=coord-coord0;
              AppendTo[Ul,uvw];
              internalF=-kk uvw;
              accNew=internalF/pmass;
              AppendTo[Al,accNew];
              velo+=0.5 tInc (acc+accNew);
              AppendTo[Vl,velo];
              acc=accNew;
              ctime+=tInc;j++;
         ,{i,200}];
         {ListPlot[Ul],ListPlot[Vl],ListPlot[Al]}
```

This program outputs the displacement, velocity and acceleration over time as
Out[68]=

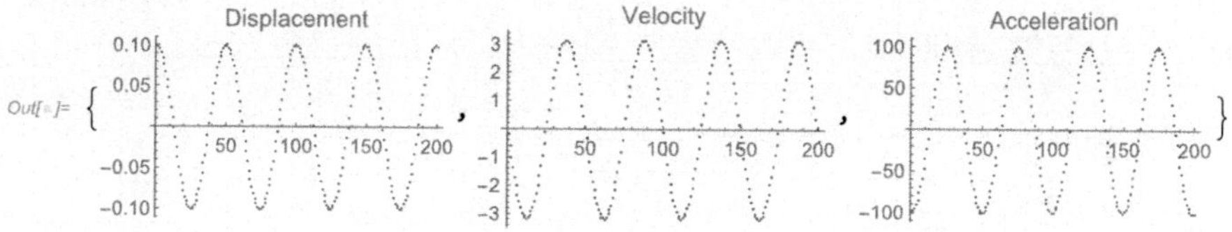

A.1.3 Fast Taylor Series Expansion

Fast Taylor series expansion for multivariable function can be obtained with Mathematica code:

```
TaylorSeries[f_, {X_, X0_, n_}] :=
  Block[{vars = Flatten[{X}], vars0 = Flatten[{X0}],
    d, alist, xlist, dflist, xx},
    d = Length[vars]; xx = vars - vars0;
    alist = MultiIndexList[d, n];
    xlist=Table[1/Apply[Times,Factorial[nn]]
     Apply[Times,xx^nn],{nn,alist}];
    dflist=Table[D[f,Sequence@@Transpose[{vars,nn}]]/.
    (Rule@@@(Transpose[{vars,vars0}])), {nn,alist}];
    {xlist, dflist}];
```

where

```
(*
d--the number of space dimension,
n--the highest order of partial differential derivative
*)
MultiIndexList[d_,n_]:=Block[{a,b,c},a=Subsets[Range[d+n],{d}];
 Do[c=a[[i]];b=c-1;b[[2;;]]-=c[[1;;-2]];a[[i]]=b,{i,Length[a]}];a];
```

Function "TaylorSeries" outputs the coefficients and derivatives separately. The key to the fast Taylor series expansion of multivariable function is the generation of the multi-index list by function "MultiIndexList".

```
(*Some Performance test*)
  MultiIndexList[5, 20] // Length // AbsoluteTiming
  MultiIndexList[5, 40] // Length // AbsoluteTiming
  MultiIndexList[10, 10] // Length // AbsoluteTiming
  MultiIndexList[200, 3] // Length // AbsoluteTiming
  MultiIndexList[200, 1] // Length // AbsoluteTiming
(*
  {0.5810 seconds,53130 terms}
  {9.777 seconds,1221759 terms}
  {1.583 seconds,184756 terms}
  {89.013 seconds,1373701 terms}
  {0.01451 seconds,201 terms}
*)
```

Test for single variable function:

```
In[84]:= TaylorSeries[f[x, y], {x, 1, 5}] // TeXForm
Out[84]//TeXForm=
```

$$\begin{pmatrix} 1 & x-1 & \frac{1}{2}(x-1)^2 & \frac{1}{6}(x-1)^3 & \frac{1}{24}(x-1)^4 & \frac{1}{120}(x-1)^5 \\ f(1,y) & f^{(1,0)}(1,y) & f^{(2,0)}(1,y) & f^{(3,0)}(1,y) & f^{(4,0)}(1,y) & f^{(5,0)}(1,y) \end{pmatrix}.$$

Test for multivariable function:

```
In[82]:= TaylorSeries[f[x, y], {{x, y}, {1, 1}, 2}] // TeXForm
Out[82]//TeXForm=
```

$$\begin{pmatrix} 1 & y-1 & \frac{1}{2}(y-1)^2 & x-1 & (x-1)(y-1) & \frac{1}{2}(x-1)^2 \\ f(1,1) & f^{(0,1)}(1,1) & f^{(0,2)}(1,1) & f^{(1,0)}(1,1) & f^{(1,1)}(1,1) & f^{(2,0)}(1,1) \end{pmatrix}.$$

Some performance tests:

```
In[73]:= tf = TaylorSeries[ f[x1, x2, x3, x4, x5, x6, x7],
         {{x1,x2,x3,x4,x5,x6,x7},Table[0,{7}],4}];//AbsoluteTiming
Out[73]= {0.027772, Null}
In[76]:= tf =TaylorSeries[f[x1, x2, x3, x4, x5, x6, x7],
         {{x1,x2,x3,x4,x5,x6,x7},Table[0,{7}],10}];//AbsoluteTiming
Out[76]= {1.4329, Null}
In[55]:= tf = TaylorSeries[f[x1, x2, x3, x4, x5, x6, x7],
         {{x1,x2,x3,x4,x5,x6,x7},Table[0,{7}],20}];//AbsoluteTiming
Out[55]= {84.3272, Null}
```

A.1.4 NOM Functions

In the implementation of NOM, we need to construct the support by finding k-nearest neighbors. Mathematica offers an inbuilt function "Nearest" to do that. Based on the coordinates of all particles, "nearestNeighbours" finds the k-nearest neighbors for each particle very efficiently:

```
(*
  coord--all particles coordinates
  numNei--the number of neighbors
*)
nearestNeighbours[coord_List, numNei_] :=
  Block[{cfun, nei},
   If[$VersionNumber > 10,
    Nearest[coord -> Automatic, coord, numNei + 1][[All, 2 ;;]],
    cfun = Nearest[coord -> Automatic];
    Table[cfun[coord[[i]], numNei + 1],
    {i, Length[coord]}][[All,2 ;;]]]];
```

Another important function in NOM is the calculation of operator matrix B and the operator tangent stiffness matrix Khg, which can be obtained with function "Khg":

```
(*
 Functionn Khgs output the operator matrix (B matrix ) and
 the operator tangent stiffness matrix (Khg matrix,
 to eliminiate hourglass mode)

 v1--list of coordinates, {{x1,y1},{x2,y2},...}
 pfun--polynomial function, 2D second order is
    pfun[{x0_,y0_},h_]:=With[{x=x0/h,y=y0/h},{x,y,x^2,x y,y^2}]
 wfun--weigh function, for example, wfun[r_,h_]:=1/r^2 or
    wfun[r_,h_]:=(1-r/h)^2
 vol--list of particle volumes in support
 penalty--is assumed as 1 in the calculation of nonlocal derivatives
    and operator tangent stiffness matrix
 h--the size of the support, the furthest distance of particle in support.
*)
Khgs[v1_List, pfun_, wfun_, vol_List, penalty_, h_] :=
  Module[{len = Length[pfun[v1[[1]], 1.]](*test the length of pfun result*),
    num = Length[v1],(*number of points in support*) k, p,
    trH = 0., r1, p1, w1, pkp, sc, ssc, khg, pw0, pw, wl},
    pkp = ConstantArray[0., {num, num}];
    k = ConstantArray[0., {len, len}];
    p = ConstantArray[0., {num, len}];
    khg = ConstantArray[0., {num + 1, num + 1}];
    wl = Norm /@ v1;
    Do[wl[[i]] = wfun[wl[[i]], 1.1 h] vol[[i]], {i, num}];
       wl *= 1.0/Total[wl];
       Do[r1 = Norm[v1[[i]]]; w1 = wl[[i]];
         trH += w1(* r1 r1*);
         pkp[[i, i]] = w1;
         p1 = pfun[v1[[i]], h]; p[[i]] = w1 p1;
         k += w1 TensorProduct[p1, p1]
    , {i, num}];
    pw0 = Inverse[k].Transpose[p];
    pkp -= p.pw0;
    pkp *= penalty/trH;

    (*Construct Khg, operator tangent stiffness matrix*)
    sc = -Total[pkp, {1}];
    ssc = -Total[sc];
    khg[[1, 1]] = ssc;
    khg[[1, 2 ;; num + 1]] = sc;
    khg[[2 ;; num + 1, 1]] = sc;
    khg[[2 ;; num + 1, 2 ;; num + 1]] = pkp;

    (*construct pw, B matrix*)
    pw = ConstantArray[0., {len, num + 1}];
    pw[[All, 1]] = -Total[pw0, {2}];
    pw[[All, 2 ;; num + 1]] = pw0;
   {pw, khg}];
```

Appendix B
Higher Order Tensors and Their Symmetry

B.1 Symmetry of Higher Order Tensors

For the fourth-order elasticity tensor, the symmetry can significantly reduce the number of material parameters. The symmetry of the Cauchy stress tensor ($\sigma_{ij} = \sigma_{ji}$ and the generalized Hooke's laws ($\sigma_{ij} = C_{ijkl}\varepsilon_{kl}$) implies that $C_{ijkl} = C_{jikl}$. Similarly, the symmetry of the infinitesimal strain tensor implies that $C_{ijkl} = C_{ijlk}$. These symmetries are called the minor symmetries. In addition, since the displacement gradient and the Cauchy stress are work conjugate, the stress–strain relation can be derived from a strain energy density functional (U); i.e.

$$\sigma_{ij} = \frac{\partial U}{\partial \varepsilon_{ij}} \quad \Longrightarrow \quad C_{ijkl} = \frac{\partial^2 U}{\partial \varepsilon_{ij} \partial \varepsilon_{kl}}. \tag{B.1}$$

The arbitrariness of the order of differentiation implies that $C_{ijkl} = C_{klij}$.

The stiffness matrix $\boldsymbol{C}$ satisfies a given symmetry condition if it does not change when subjected to the corresponding orthogonal transformation, which may represent symmetry with respect to a point, an axis or a plane.

According to Forte and Vianello (1996) and Geymonat and Weller (2002), the orthogonal transformation of a tensor of any order can be written as

$$T(\boldsymbol{M}) := (\boldsymbol{Q} \star \boldsymbol{M})_{...ijk...} = \cdots Q_{ip} Q_{jq} Q_{kr} \cdots M_{...pqr...} \tag{B.2}$$

where $\boldsymbol{Q}$ is an orthogonal matrix given by

$$\mathbb{O}(n, R) = \left\{ \boldsymbol{Q} \in \mathrm{GL}(n, R) | \boldsymbol{Q}^T \boldsymbol{Q} = \boldsymbol{Q} \boldsymbol{Q}^T = \boldsymbol{I} \right\}, \tag{B.3}$$

with $\mathrm{GL}(n, R)$ being the set of all real $n \times n$ matrices and $\boldsymbol{I}$ the identity matrix.

T. Rabczuk et al., *'Computational Methods based on Peridynamics and Nonlocal Operators*, Computational Methods in Engineering & the Sciences,
https://doi.org/10.1007/978-3-031-20906-2

The symmetry of certain orthogonal transformations of a tensor requires

$$\boldsymbol{Q} \star \boldsymbol{M} = \boldsymbol{M}, \text{ with } \boldsymbol{Q} \in \mathbb{O}(n, R). \tag{B.4}$$

The common orthogonal transformations in 3D include reflection and rotation. The transformation matrices for symmetry planes are

$$\boldsymbol{A}_1 = \begin{bmatrix} -1 & 0 & 0 \\ 0 & 1 & 0 \\ 0 & 0 & 1 \end{bmatrix}, \boldsymbol{A}_2 = \begin{bmatrix} 1 & 0 & 0 \\ 0 & -1 & 0 \\ 0 & 0 & 1 \end{bmatrix}, \boldsymbol{A}_3 = \begin{bmatrix} 1 & 0 & 0 \\ 0 & 1 & 0 \\ 0 & 0 & -1 \end{bmatrix}. \tag{B.5}$$

Rotation matrix in 3D:

$$R_x(\theta) = \begin{bmatrix} 1 & 0 & 0 \\ 0 & \cos\theta & -\sin\theta \\ 0 & \sin\theta & \cos\theta \end{bmatrix}, R_y(\theta) = \begin{bmatrix} \cos\theta & 0 & \sin\theta \\ 0 & 1 & 0 \\ -\sin\theta & 0 & \cos\theta \end{bmatrix}, R_z(\theta) = \begin{bmatrix} \cos\theta & -\sin\theta & 0 \\ \sin\theta & \cos\theta & 0 \\ 0 & 0 & 1 \end{bmatrix}. \tag{B.6}$$

The general rotation matrix $\boldsymbol{R}$ can be written as

$$R = R_z(\alpha)\, R_y(\beta)\, R_x(\gamma).$$

The coordinate transform of a vector in matrix and tensor notation is

$$\boldsymbol{v}' = \boldsymbol{Q} \cdot \boldsymbol{v} \qquad \text{and} \qquad v'_i = \lambda_{ij} v_j. \tag{B.7}$$

The coordinate transform of a tensor in matrix and tensor notation is

$$\boldsymbol{\sigma}' = \boldsymbol{Q} \cdot \boldsymbol{\sigma} \cdot \boldsymbol{Q}^T \qquad \text{and} \qquad \sigma'_{mn} = \lambda_{mi} \lambda_{nj} \sigma_{ij}. \tag{B.8}$$

The coordinate transform of a fourth-order tensor is

$$\boldsymbol{C}' = \boldsymbol{Q} \cdot \boldsymbol{Q} \cdot \boldsymbol{C} \cdot \boldsymbol{Q}^T \cdot \boldsymbol{Q}^T, C'_{ijkl} = \lambda_{im} \lambda_{jn} \lambda_{ko} \lambda_{lp} C_{mnop}. \tag{B.9}$$

The coordinate transform of a sixth-order tensor is

$$\boldsymbol{H}' = \boldsymbol{Q} \cdot \boldsymbol{Q} \cdot \boldsymbol{Q} \cdot \boldsymbol{H} \cdot \boldsymbol{Q}^T \cdot \boldsymbol{Q}^T \cdot \boldsymbol{Q}^T, H'_{ijklmn} = \lambda_{io} \lambda_{jp} \lambda_{kq} \lambda_{lr} \lambda_{ms} \lambda_{nt} H_{opqrst}. \tag{B.10}$$

Solving Eqs. B.9, B.10 by Mathematica (Wolfram 1999), we can obtain the independent variables in high-order tensor.

There are $3^6 = 729$ terms in $\boldsymbol{H}$. The Minor symmetry reduces $\boldsymbol{H}$ into 171 independent terms.

The orthotropy requires $\boldsymbol{H}' = \boldsymbol{H}, \boldsymbol{C}' = \boldsymbol{C}$ for three reflection symmetries $\boldsymbol{A}_1, \boldsymbol{A}_2, \boldsymbol{A}_3$. The case of orthotropy (the symmetry of a brick) has 51 independent elements.

The isotropy property requires $\boldsymbol{H}' = \boldsymbol{H}, \boldsymbol{C}' = \boldsymbol{C}$ for any rotation. This requirement reduces the number of independent terms in $\boldsymbol{H}'$ from 171 to 5.

B.2 Matrix Form of Strain Gradient Energy by Voigt Notations

The tensor form of higher order tensor contains many repeated terms when symmetry property is considered. In terms of numerical implementation, it is more convenient to use the matrix form than to use the tensor form. In conventional mechanics, the Voigt notation is an efficient method to formulate the matrix form. Let us take the strain gradient linear elasticity as an example. The other higher order tensor can be formulated in the same manner. The material constitutive for couple stresses can be written as

$$\overrightarrow{\nabla\sigma} = \mathcal{H}\overrightarrow{\nabla\varepsilon}, \quad \sigma_{ijk} = h_{ijklmn}\varepsilon_{lmn}. \tag{B.11}$$

The strain-gradient energy function is

$$\mathcal{F} = \frac{1}{2}\sigma_{ijk}\varepsilon_{ijk} = \frac{1}{2}\varepsilon_{ijk}h_{ijklmn}\varepsilon_{lmn} = \frac{1}{2}\overrightarrow{\nabla\boldsymbol{\varepsilon}}^T\mathcal{H}\overrightarrow{\nabla\boldsymbol{\varepsilon}} \tag{B.12}$$

where $\varepsilon_{ijk}, \sigma_{ijk}$ are defined as

$$\varepsilon_{ijk} = \frac{\partial\varepsilon_{jk}}{\partial x_i}, \quad \sigma_{ijk} = \frac{\partial\sigma_{jk}}{\partial x_i}. \tag{B.13}$$

The vectorial forms of couple stress and strain gradient can be written as

$$\begin{aligned}\overrightarrow{\nabla\sigma} = (&\sigma_{111}, \sigma_{122}, \sigma_{133}, \sigma_{123}, \sigma_{113}, \sigma_{112}, \sigma_{211}, \sigma_{222}, \sigma_{233},\\ &\sigma_{223}, \sigma_{213}, \sigma_{212}, \sigma_{311}, \sigma_{322}, \sigma_{333}, \sigma_{323}, \sigma_{313}, \sigma_{312})\end{aligned} \tag{B.14}$$

$$\begin{aligned}\overrightarrow{\nabla\boldsymbol{\varepsilon}} = (&\varepsilon_{111}, \varepsilon_{122}, \varepsilon_{133}, 2\varepsilon_{123}, 2\varepsilon_{113}, 2\varepsilon_{112}, \varepsilon_{211}, \varepsilon_{222}, \varepsilon_{233},\\ &2\varepsilon_{223}, 2\varepsilon_{213}, 2\varepsilon_{212}, \varepsilon_{311}, \varepsilon_{322}, \varepsilon_{333}, 2\varepsilon_{323}, 2\varepsilon_{313}, 2\varepsilon_{312}).\end{aligned} \tag{B.15}$$

Based on Voigt notation

$$\begin{array}{llllllll} ij = & 11 & 22 & 33 & 23, 32 & 13, 31 & 12, 21 & \\ & \Downarrow & \Downarrow & \Downarrow & \Downarrow & \Downarrow & \Downarrow & . \\ \alpha = & 1 & 2 & 3 & 4 & 5 & 6 & \end{array} \tag{B.16}$$

we write the couple stress and strain gradient as

$$\epsilon_{i\alpha} = \frac{\partial \varepsilon_\alpha}{\partial x_i}, \quad \sigma_{i\alpha} = \frac{\partial \sigma_\alpha}{\partial x_i} \tag{B.17}$$

$$h_{ijklmn} \rightarrow h_{i\alpha l\beta} \tag{B.18}$$

where α, β are the Voigt notations of jk and mn, respectively.

Then the vectorial forms of couple stress and strain gradient can be written as

$$\overrightarrow{\nabla\varepsilon} = (\epsilon_{11}, \epsilon_{12}, \epsilon_{13}, 2\epsilon_{14}, 2\epsilon_{15}, 2\epsilon_{16}, \epsilon_{21}, \epsilon_{22}, \epsilon_{23}, 2\epsilon_{24}, 2\epsilon_{25}, 2\epsilon_{26}, \epsilon_{31}, \epsilon_{32}, \epsilon_{33}, 2\epsilon_{34}, 2\epsilon_{35}, 2\epsilon_{36}) \tag{B.19}$$

$$\overrightarrow{\nabla\sigma} = (\sigma_{11}, \sigma_{12}, \sigma_{13}, \sigma_{14}, \sigma_{15}, \sigma_{16}, \sigma_{21}, \sigma_{22}, \sigma_{23}, \sigma_{24}, \sigma_{25}, \sigma_{26}, \sigma_{31}, \sigma_{32}, \sigma_{33}, \sigma_{34}, \sigma_{35}, \sigma_{36}). \tag{B.20}$$

Based on the symmetry calculation in Sect. B.1, the matrix form of isotropic gradient elasticity can be derived accordingly:

$$\begin{aligned}
\mathcal{H}^{iso} = \{&\{h_{1111}, h_{1112}, h_{1112}, 0, 0, 0, 0, 0, 0, 0, 0, h_{1126}, 0, 0, 0, 0, h_{1126}, 0\} \\
&\{h_{1112}, -h_{1111}+2h_{1112}+4h_{1616}, h_{1112}-2h_{1623}, 0,0,0,0,0,0,0,0, h_{1111}-h_{1112}-h_{1126}-2h_{1616}, 0,0,0,0, h_{1623}, 0\} \\
&\{h_{1112}, h_{1112}-2h_{1623}, -h_{1111}+2h_{1112}+4h_{1616}, 0,0,0,0,0,0,0,0, h_{1623}, 0,0,0,0, h_{1111}-h_{1112}-h_{1126}-2h_{1616}, 0\} \\
&\{0,0,0,-h_{1111}+h_{1112}+2h_{1616}+h_{1623}, 0,0,0,0,0,0, h_{1111}-h_{1112}-h_{1126}-h_{1616}-h_{1623}, 0,0,0,0,0,0, h_{1111}-h_{1112}-h_{1126}-h_{1616}-h_{1623}\} \\
&\{0, 0, 0, 0, h_{1616}, 0, 0, 0, 0, h_{1126}-h_{1623}, 0, 0, h_{1111}-h_{1112}-h_{1126}-2h_{1616}, h_{1623}, h_{1126}, 0, 0, 0\} \\
&\{0, 0, 0, 0, 0, h_{1616}, h_{1111}-h_{1112}-h_{1126}-2h_{1616}, h_{1126}, h_{1623}, 0, 0, 0, 0, 0, 0, h_{1126}-h_{1623}, 0, 0\} \\
&\{0,0,0,0,0, h_{1111}-h_{1112}-h_{1126}-2h_{1616}, -h_{1111}+2h_{1112}+4h_{1616}, h_{1112}, h_{1112}-2h_{1623}, 0,0,0,0,0,0, h_{1623}, 0,0\} \\
&\{0, 0, 0, 0, 0, h_{1126}, h_{1112}, h_{1111}, h_{1112}, 0, 0, 0, 0, 0, 0, h_{1126}, 0, 0\} \\
&\{0,0,0,0,0, h_{1623}, h_{1112}-2h_{1623}, h_{1112}, -h_{1111}+2h_{1112}+4h_{1616}, 0,0,0,0,0,0, h_{1111}-h_{1112}-h_{1126}-2h_{1616}, 0,0\} \\
&\{0, 0, 0, 0, h_{1126}-h_{1623}, 0, 0, 0, 0, h_{1616}, 0, 0, h_{1623}, h_{1111}-h_{1112}-h_{1126}-2h_{1616}, h_{1126}, 0, 0, 0\} \\
&\{0,0,0, h_{1111}-h_{1112}-h_{1126}-h_{1616}-h_{1623}, 0,0,0,0,0,0, h_{1112}+2h_{1616}+h_{1623}, 0,0,0,0,0,0, h_{1111}-h_{1112}-h_{1126}-h_{1616}-h_{1623}\} \\
&\{h_{1126}, h_{1111}-h_{1112}-h_{1126}-2h_{1616}, h_{1623}, 0,0,0,0,0,0,0,0, h_{1616}, 0,0,0,0, -h_{1623}, 0\} \\
&\{0,0,0,0, h_{1111}-h_{1112}-h_{1126}-2h_{1616}, 0,0,0,0, h_{1623}, 0,0, -h_{1111}+2h_{1112}+4h_{1616}, h_{1112}-2h_{1623}, h_{1112}, 0,0,0\} \\
&\{0,0,0,0, h_{1623}, 0,0,0,0, h_{1111}-h_{1112}-h_{1126}-2h_{1616}, 0,0, h_{1112}-2h_{1623}, -h_{1111}+2h_{1112}+4h_{1616}, h_{1112}, 0,0,0\} \\
&\{0, 0, 0, 0, h_{1126}, 0, 0, 0, 0, h_{1126}, 0, 0, h_{1112}, h_{1112}, h_{1111}, 0, 0, 0\} \\
&\{0, 0, 0, 0, 0, h_{1126}-h_{1623}, h_{1623}, h_{1126}, h_{1111}-h_{1112}-h_{1126}-2h_{1616}, 0, 0, 0, 0, 0, 0, h_{1616}, 0, 0\} \\
&\{h_{1126}, h_{1623}, h_{1111}-h_{1112}-h_{1126}-2h_{1616}, 0, 0, 0, 0, 0, 0, 0, 0, -h_{1623}, 0, 0, 0, 0, h_{1616}, 0\} \\
&\{0,0,0, h_{1111}-h_{1112}-h_{1126}-h_{1616}-h_{1623}, 0,0,0,0,0,0, h_{1111}-h_{1112}-h_{1126}-h_{1616}-h_{1623}, 0,0,0,0,0,0, h_{1112}-h_{1111}+2h_{1616}+h_{1623}\}\}.
\end{aligned} \tag{B.21}$$

The matrix form of $\boldsymbol{H}^{orth}$ is written as $\mathcal{H}^{orth}$ of 18×18

$$
\begin{aligned}
\mathcal{H}^{orth} =& \{\{h_{1111}, h_{1112}, h_{1113}, 0, 0, 0, 0, 0, 0, 0, 0, h_{1126}, 0, 0, 0, 0, h_{1135}, 0\}\\
&\{h_{1112}, h_{1212}, h_{1213}, 0, 0, 0, 0, 0, 0, 0, 0, h_{1226}, 0, 0, 0, 0, h_{1235}, 0\}\\
&\{h_{1113}, h_{1213}, h_{1313}, 0, 0, 0, 0, 0, 0, 0, 0, h_{1326}, 0, 0, 0, 0, h_{1335}, 0\}\\
&\{0, 0, 0, h_{1414}, 0, 0, 0, 0, 0, 0, h_{1425}, 0, 0, 0, 0, 0, 0, h_{1436}\}\\
&\{0, 0, 0, 0, h_{1515}, 0, 0, 0, 0, h_{1524}, 0, 0, h_{1531}, h_{1532}, h_{1533}, 0, 0, 0\}\\
&\{0, 0, 0, 0, 0, h_{1616}, h_{1621}, h_{1622}, h_{1623}, 0, 0, 0, 0, 0, 0, h_{1634}, 0, 0\}\\
&\{0, 0, 0, 0, 0, h_{1621}, h_{2121}, h_{2122}, h_{2123}, 0, 0, 0, 0, 0, 0, h_{2134}, 0, 0\}\\
&\{0, 0, 0, 0, 0, h_{1622}, h_{2122}, h_{2222}, h_{2223}, 0, 0, 0, 0, 0, 0, h_{2234}, 0, 0\}\\
&\{0, 0, 0, 0, 0, h_{1623}, h_{2123}, h_{2223}, h_{2323}, 0, 0, 0, 0, 0, 0, h_{2334}, 0, 0\}\\
&\{0, 0, 0, 0, h_{1524}, 0, 0, 0, 0, h_{2424}, 0, 0, h_{2431}, h_{2432}, h_{2433}, 0, 0, 0\}\\
&\{0, 0, 0, h_{1425}, 0, 0, 0, 0, 0, 0, h_{2525}, 0, 0, 0, 0, 0, 0, h_{2536}\}\\
&\{h_{1126}, h_{1226}, h_{1326}, 0, 0, 0, 0, 0, 0, 0, 0, h_{2626}, 0, 0, 0, 0, h_{2635}, 0\}\\
&\{0, 0, 0, 0, h_{1531}, 0, 0, 0, 0, h_{2431}, 0, 0, h_{3131}, h_{3132}, h_{3133}, 0, 0, 0\}\\
&\{0, 0, 0, 0, h_{1532}, 0, 0, 0, 0, h_{2432}, 0, 0, h_{3132}, h_{3232}, h_{3233}, 0, 0, 0\}\\
&\{0, 0, 0, 0, h_{1533}, 0, 0, 0, 0, h_{2433}, 0, 0, h_{3133}, h_{3233}, h_{3333}, 0, 0, 0\}\\
&\{0, 0, 0, 0, 0, h_{1634}, h_{2134}, h_{2234}, h_{2334}, 0, 0, 0, 0, 0, 0, h_{3434}, 0, 0\}\\
&\{h_{1135}, h_{1235}, h_{1335}, 0, 0, 0, 0, 0, 0, 0, 0, h_{2635}, 0, 0, 0, 0, h_{3535}, 0\}\\
&\{0, 0, 0, h_{1436}, 0, 0, 0, 0, 0, 0, h_{2536}, 0, 0, 0, 0, 0, 0, h_{3636}\}\}.
\end{aligned} \tag{B.22}
$$

The independent terms in $\mathcal{H}^{orth}$ are

$$
\begin{aligned}
&\{h_{1111}, h_{1112}, h_{1113}, h_{1126}, h_{1135}, h_{1212}, h_{1213}, h_{1226}, h_{1235}, h_{1313},\\
&h_{1326}, h_{1335}, h_{1414}, h_{1425}, h_{1436}, h_{1515}, h_{1524}, h_{1531}, h_{1532}, h_{1533},\\
&h_{1616}, h_{1621}, h_{1622}, h_{1623}, h_{1634}, h_{2121}, h_{2122}, h_{2123}, h_{2134}, h_{2222},\\
&h_{2223}, h_{2234}, h_{2323}, h_{2334}, h_{2424}, h_{2431}, h_{2432}, h_{2433}, h_{2525}, h_{2536},\\
&h_{2626}, h_{2635}, h_{3131}, h_{3132}, h_{3133}, h_{3232}, h_{3233}, h_{3333}, h_{3434}, h_{3535}, h_{3636}\}.
\end{aligned} \tag{B.23}
$$

The independent terms in $\mathcal{H}^{iso}$ are

$$
\{h_{1111}, h_{1112}, h_{1126}, h_{1616}, h_{1623}\}. \tag{B.24}
$$

In order to maintain a positive strain energy density for any deformation, it is required that the matrix $\mathcal{H}$ be symmetric and positive definite.

Positive definite or semidefinite matrix: a symmetric matrix A whose eigenvalues are positive ($\lambda > 0$) is called positive definite. The coefficient h_{ijkl} must be selected so that all eigenvalues of $\mathcal{H}$ are positive.

References

Forte S, Vianello M (1996) Symmetry classes for elasticity tensors. J Elast 43(2):81–108. ISSN 1573-2681. https://doi.org/10.1007/BF00042505

Geymonat G, Weller T (2002) Classes de symétrie des solides piézoélectriques. Comptes Rendus Mathematique 335(10):847–852

Wolfram S (1999) The mathematica book. Assembly Automation